OF June 12
AR Jun 14

Electricity and Basic Electronics

Seventh Edition

by
Stephen R. Matt

Publisher
The Goodheart-Willcox Company, Inc.
Tinley Park, Illinois
www.g-w.com

Manufactured in the United States of America.

Library of Congress Catalog Card Number 2007036134

ISBN 978-1-59070-877-4

1 2 3 4 5 6 7 8 9 – 09 – 13 12 11 10 09 08

The Goodheart-Willcox Company, Inc. Brand Disclaimer: Brand names, company names, and illustrations for products and services included in this text are provided for educational purposes only and do not represent or imply endorsement or recommendation by the author or the publisher.

The Goodheart-Willcox Company, Inc. Safety Notice: The reader is expressly advised to carefully read, understand, and apply all safety precautions and warnings described in this book or that might also be indicated in undertaking the activities and exercises described herein to minimize risk of personal injury or injury to others. Common sense and good judgment should also be exercised and applied to help avoid all potential hazards. The reader should always refer to the appropriate manufacturer's technical information, directions, and recommendations; then proceed with care to follow specific equipment operating instructions. The reader should understand these notices and cautions are not exhaustive.

The publisher makes no warranty or representation whatsoever, either expressed or implied, including but not limited to equipment, procedures, and applications described or referred to herein, their quality, performance, merchantability, or fitness for a particular purpose. The publisher assumes no responsibility for any changes, errors, or omissions in this book. The publisher specifically disclaims any liability whatsoever, including any direct, indirect, incidental, consequential, special, or exemplary damages resulting, in whole or in part, from the reader's use or reliance upon the information, instructions, procedures, warnings, cautions, applications, or other matter contained in this book. The publisher assumes no responsibility for the activities of the reader.

Library of Congress Cataloging-in-Publication Data

Matt, Stephen R.
Electricity and basic electronics / by Stephen R. Matt. — [7th ed.].
p. cm.
Includes index.
ISBN 978-1-59070-877-4
1. Electric engineering. 2. Electronics. I. Title.

TK146.M376 2007
621.3--dc22 2007036134

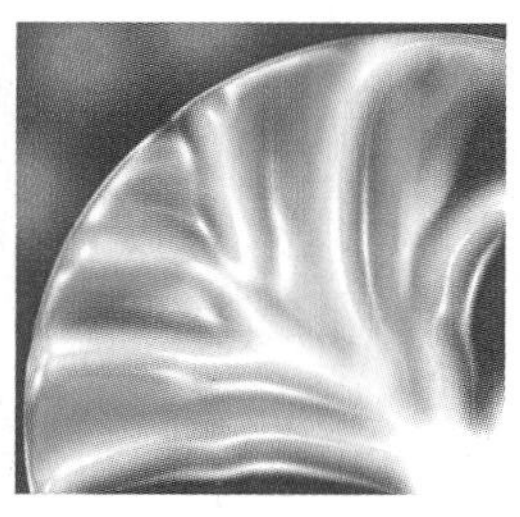

Introduction

Electricity and Basic Electronics reflects the growing and expanding interest in electricity and electronics. This first course teaches the fundamentals and is a starting point for a career in this most important area.

This text is clearly written and well-illustrated, making it easy for beginning students to understand. Chapters are organized so that content is presented in the logical order of use. They are laid out in a sequence that facilitates learning. Each new concept builds on the information covered in the previous chapter. Safe work habits are stressed throughout the text.

To avoid confusion and make learning easier, each new electrical or electronic term is explained as much as possible when it is introduced. In some cases, the explanation is quite brief. However, students are also told which chapter gives complete details on that particular subject area. This method of treating new ideas will help students build a strong understanding of the subject with only the facts that are necessary at the time.

Using this text, students will find it easy to learn fundamentals and apply them. Important details are clarified with use of extra color. Numerous proven projects are included with construction procedures. These not only help motivate students but also build interest and confidence.

Suggested activities at the end of each chapter range from simple to demanding. These activities may be used to challenge the more advanced students or to diversify the program. An effort has been made to relate all activities and projects to actual applications in home and industry.

Stephen R. Matt

About the Author

Stephen R. Matt received his BSEd and MEd from Wayne State University and his PhD from Michigan State University. He has taught various courses in electricity and electronics at Michigan State University and the University of Georgia, as well as curriculum design courses at the University of Manitoba. He has also served as the Vocational Director of Barrow County Schools, in Winder, Georgia.

He has worked in industry as a technician for a tubing manufacturer, a training specialist at Ford Motor Company, and a training director for Philco-Ford Corporation. He is currently working as a consultant, serving companies such as A2D Technologies and Generate Technologies. He is married to Theresa; has two daughters, Tracy and Kristin; and a stepdaughter, Dorian. Dr. Matt has been living in Athens, Georgia for the past 30 years.

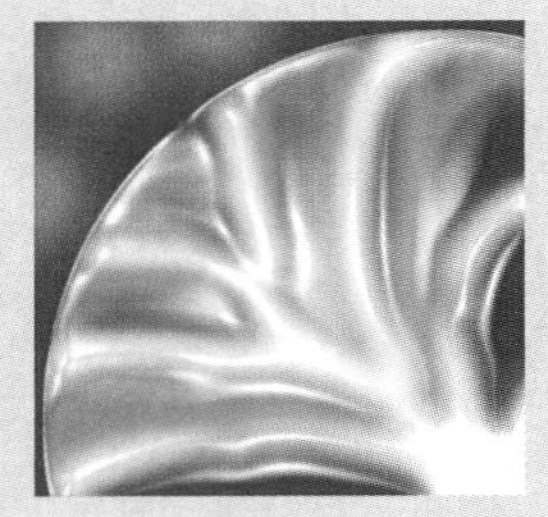

Chapter Listing

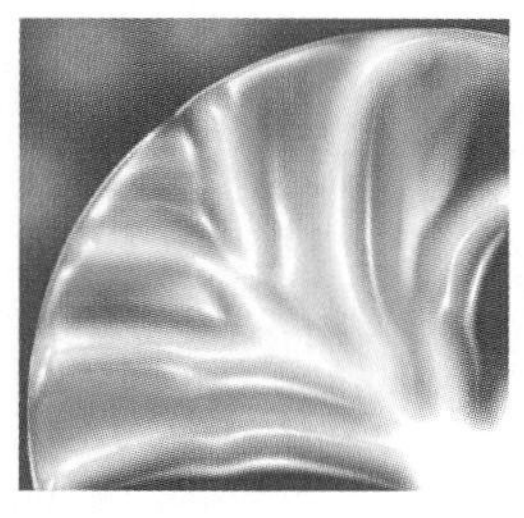

Contents

Chapter 21

Fiber Optics 351

Chapter 22

Switches 361

Chapter 23

Computers 377

Chapter 24

Energy Conservation 393

Chapter 25

Career Opportunities 409

Reference Section 419

Glossary 429

Index 443

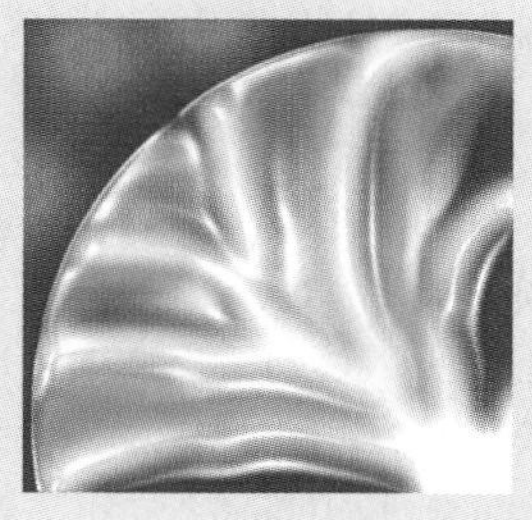

Features

Practical Application

Project

Math Focus

Safety Precautions for the Electricity-Electronics Shop

There is always an element of danger when working with electricity. Observe all safety rules that concern each project and be particularly careful not to contact any live wire or terminal, regardless of whether it is connected to either a low voltage or a high voltage. Projects do not specify dangerous voltage levels. However, keep in mind at all times that it is possible to experience a surprising electric shock under certain circumstances. Even a normal, healthy person can be injured or seriously hurt by the shock or what happens as a result of it. Do not fool around. The lab is no place for horseplay.

Learning and Applying the Fundamentals

Learning Objectives

After studying this chapter, you will be able to do the following:

- Name the three things required in any completed circuit.
- Explain what is meant by open and closed circuit.
- Explain what is happening inside a wire when electricity is flowing.
- List reasons for using an electrical schematic.
- Draw the symbols for a lamp, battery, wire, and speaker.
- Describe the direction and speed of electron flow in a completed circuit.

Technical Terms

battery
cell
closed circuit
current
electron
load
open circuit
schematic

Think for a minute what life would be like without electricity. There would be no electric lights, no radio or CDs, and no television or movies. Stoplights would not work, which would not matter since cars could not start. You would have to do your homework by candlelight, and you could not keyboard your papers on a computer or do research on the Internet.

Electricity is used in the production and operation of most things that you use. It helps heat your home in the winter and cool it in the summer. Newspapers, magazines, and books are printed on presses operated by electricity. Electricity powers the equipment used to monitor and gather information in hospitals, laboratories, and research centers. Consider all of the electricity needed to light the harbor in the city of San Diego, California, **Figure 1-1.**

In this chapter, you will be introduced to a whole new language. This language includes both terms and symbols. The easiest way to learn about anything new is to apply what you are being taught. Do not just read about electricity and electronics; build the projects, make your own drawings, and work on the problems. Do not just watch someone else build a project; do it yourself. The more hands-on activities you do, the more you will learn.

Use of Electricity

Using electricity is a lot like driving a car. It is not necessary for you to be able to repair everything that goes wrong with it to enjoy using it. To turn on a light at home, for example, all you have to do is flip a light switch. To start the car, you flip the switch by turning an ignition key.

The light in your home may not come on, or the car may fail to start. You may have to call in a specialist to make the repair. However, most of the time, everything will work correctly. You do not have to be an expert service technician to use and enjoy the benefits of the light or the car.

Knowing more about a given subject does make it easier to perform your own service work. However, a word of warning is given to do-it-yourselfers. Many people who know a lot about a given subject are injured because they become careless. As you learn more about electricity in this course and begin doing the projects, be careful. Always follow all of the safety tips and information.

Basic Circuit Concepts

How do you turn on a light? As discussed earlier, you flip a switch. But what does flipping a switch do that results in the light coming on? Before that question can be answered, you need to understand a few things about a circuit.

Look at **Figure 1-2.** It shows a battery, wires, and a test lamp (lightbulb and socket). In Figure 1-2A, one wire connects the test lamp to the battery. The other wire is connected to

Figure 1-1. Every large city is loaded with electrical wiring, lights, cables, and technology-related equipment. Looking at the lights of the San Diego harbor makes you realize how many electricians it took to complete this huge task.

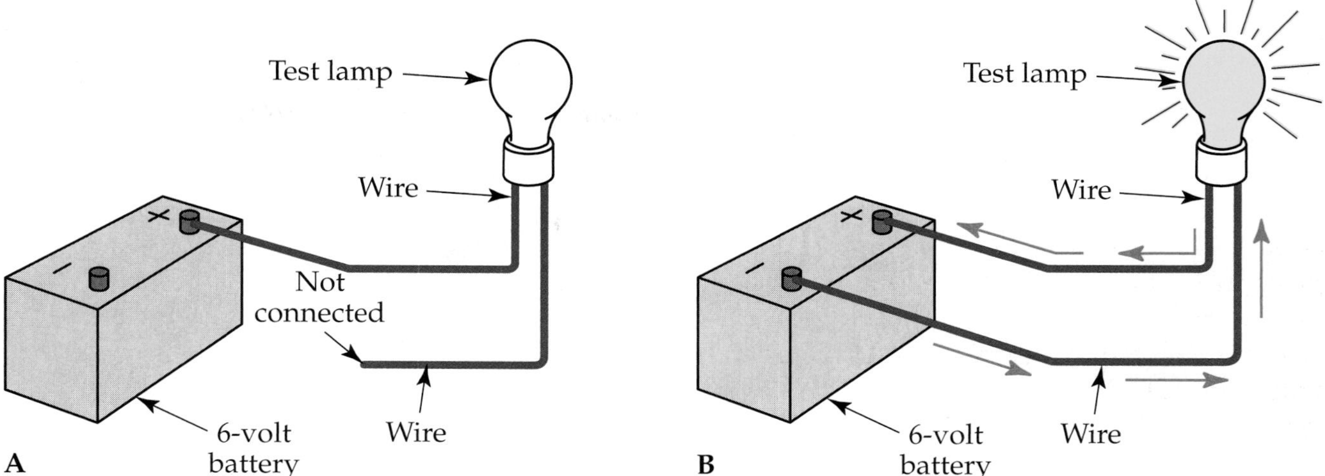

Figure 1-2. A battery, test lamp, and wires are used in this figure to form a simple circuit. A—A wire is disconnected; therefore, this circuit is not completed. B—This circuit has the same parts as in part A. The test lamp is on because the wire is connected to the battery.

the lamp, but not to the battery. Since a wire is disconnected, the test lamp is not lit. However, when both lamp wires are connected to the battery, Figure 1-2B, the lamp will light. When the wires are connected, electricity is allowed to flow through them. This is called a ***closed circuit*** or ***completed circuit.*** The electricity that flows is called ***current.***

Any completed circuit requires three things: First, it needs a source of electricity, or energy. In Figure 1-2, the source of electricity is the battery. Second, it needs a ***load,*** which is something that uses the energy provided by the battery. The load in Figure 1-2 is the test lamp. Third, it needs a path to get the energy from the source to the load and back. As you may have guessed, this path is formed by the wires.

Electrons

Suppose you have a very powerful microscope and can take a look at what is happening inside electrical wires. When there is no electricity flowing through the wires, you would see extremely small particles called *electrons* moving randomly, **Figure 1-3.** An ***electron*** is so small that millions of them would fit on the point of a pin.

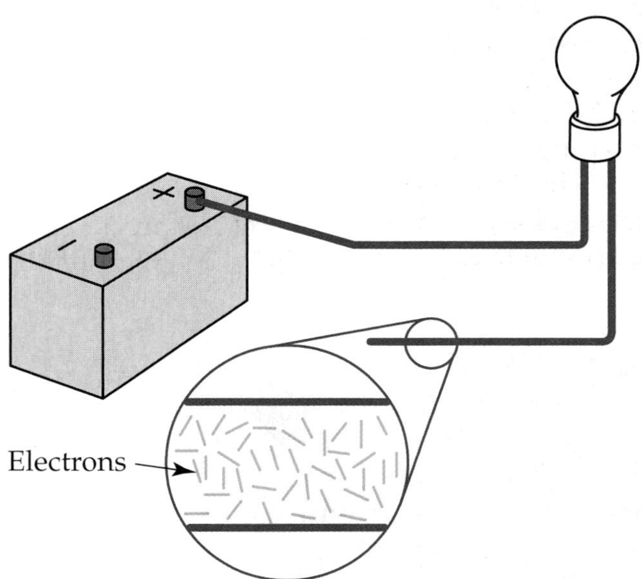

Figure 1-3. Using a very powerful microscope to look inside the wires of a circuit with no electricity flowing through it shows that the electrons are not flowing.

When the wire has electricity flowing through it, **Figure 1-4,** you would see that the electrons all move through the wire in the same direction. They always flow through the wires from negative (–) to positive (+). This will be explained in Chapter 3.

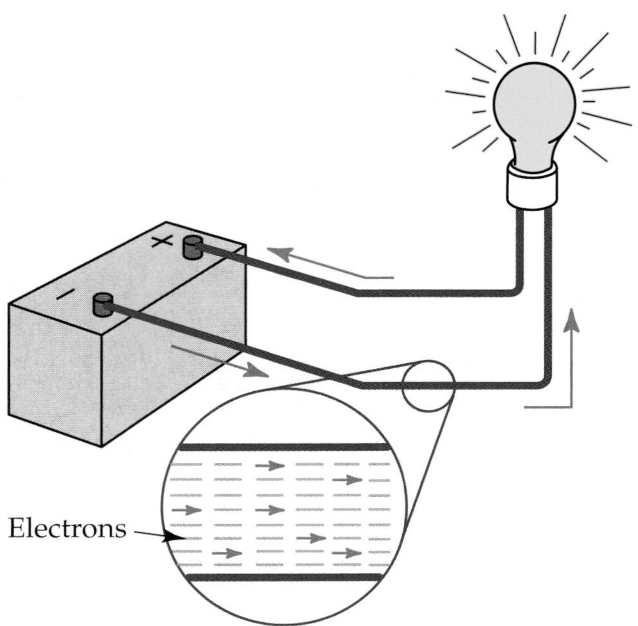

Figure 1-4. Using a very powerful microscope to look inside the wires of a circuit with electricity flowing through it shows that the electrons are moving rapidly in one direction.

Switches

A switch is a device used to open or close a circuit, **Figure 1-5.** Closing the switch allows the electrons to flow. When the switch is opened, none of the electrons can jump across the gap. This is called an ***open circuit.***

A switch provides a simple and convenient way of opening and closing the circuit without having to disconnect and reconnect wires to the battery. A push button can be connected to the circuit to do the same thing at your car or house. The light inside your car or your refrigerator is controlled by a push button, **Figure 1-6.** However, in these examples, you must open the door to turn on the light. When the door opens, the circuit is closed, allowing electrons to flow to the light as seen in **Figure 1-7.**

Switches and push buttons are not the only way to open and close circuits. Several other methods are explained in later chapters. By then, you will have a better understanding of electricity and electrical circuits.

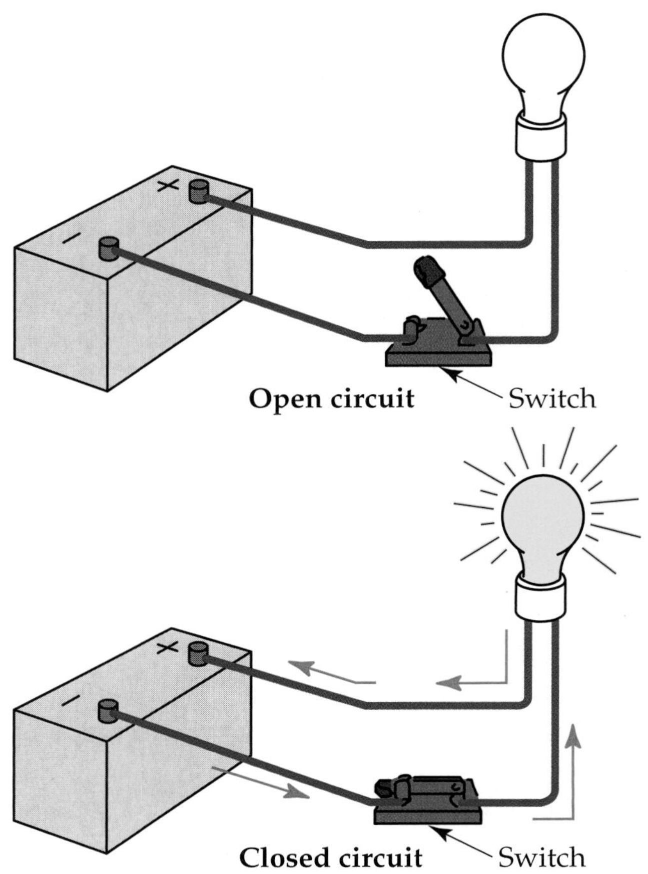

Figure 1-5. A switch installed in the simple circuit allows easier opening and closing of the circuit.

Batteries

You probably have noticed that the batteries shown in some of the examples are labeled 6 volts. Your flashlight uses two or more 1 1/2-volt cells like the one shown in **Figure 1-8.** A ***cell*** is a single power unit that produces electrical energy that can be used to power a circuit. When two or more cells are combined, they form a ***battery.*** **Figure 1-9** shows a 12-volt battery, which is used in cars, golf carts, and electric vans.

Symbols and Schematics

Take a closer look at the drawing of the power source, wires, and load in Figure 1-3. Note that the elements of the circuit are drawn in some detail. To make electrical drawings

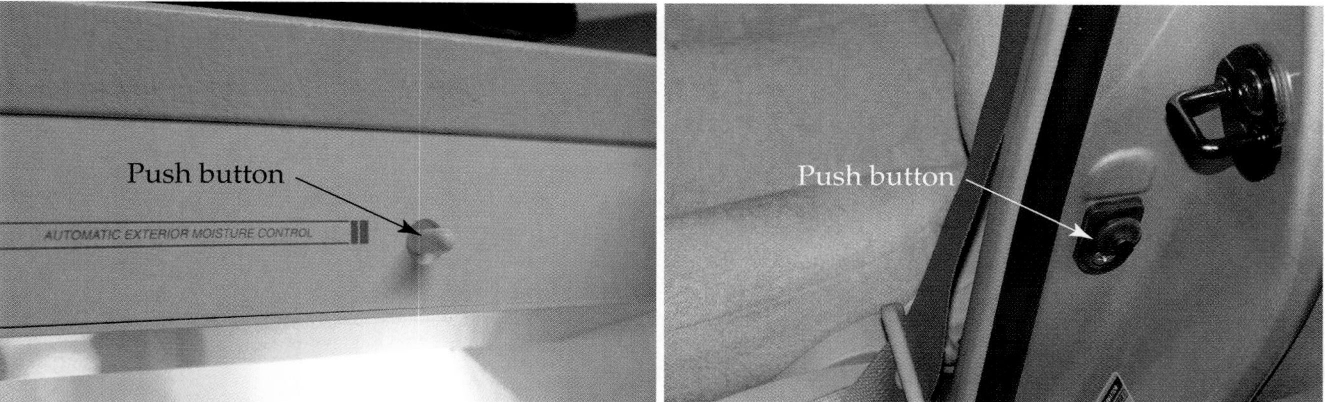

Figure 1-6. The light inside your refrigerator and car can be turned on and off with a door and a push-button switch.

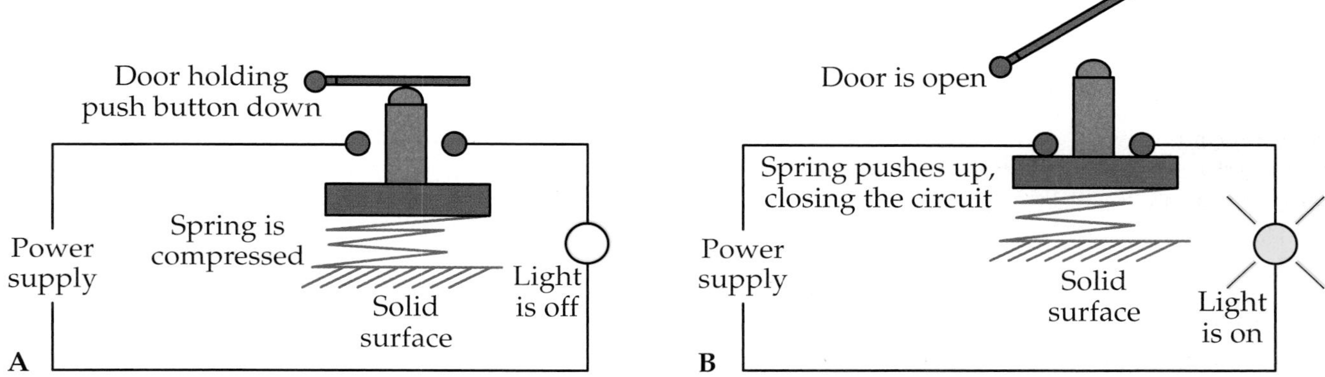

Figure 1-7. Wiring diagram of a door and push-button circuit. A—When the door is closed, the push-button switch does not complete the circuit. B—An open door allows the switch to complete the circuit.

Figure 1-8. The familiar flashlight battery is not a battery. Proper terminology for the items shown is "1 1/2-volt cell."

Figure 1-9. A 12-volt battery for automotive use is shown. It is an assembly of six 2-volt cells.

faster to produce and less complex, another method of showing circuits is needed. For this reason, symbols are used to represent the elements of a circuit. For example, instead of drawing a cell, the symbol for a cell shown in **Figure 1-10** is used.

Since a battery is a combination of two or more cells, its symbol should be drawn as a group of cell symbols, as shown in **Figure 1-11.** Note that the long line in the battery symbol is the positive terminal, and the short line is the negative terminal.

When using symbols, a wire is simply drawn as a line. The symbol for the load depends on the item you are using for a load. **Figure 1-12** shows the symbols of some common loads.

Knowing symbols allows us to redraw some of the circuits we saw earlier in this chapter. **Figure 1-13** shows another way to draw the circuit shown in Figure 1-2 using the symbols for a battery, wires, and a lamp. To add another lamp to the circuit, just put a large dot where the wires connect to one another. **Figure 1-14** shows how this is done.

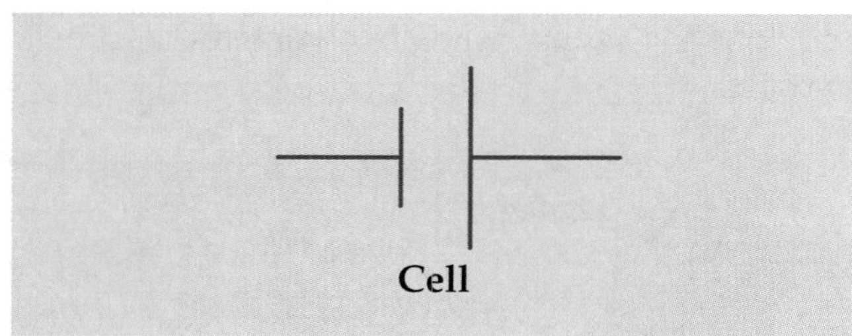

Figure 1-10. Symbols are used to identify electrical elements or components in circuit drawings. The symbol for a cell is shown.

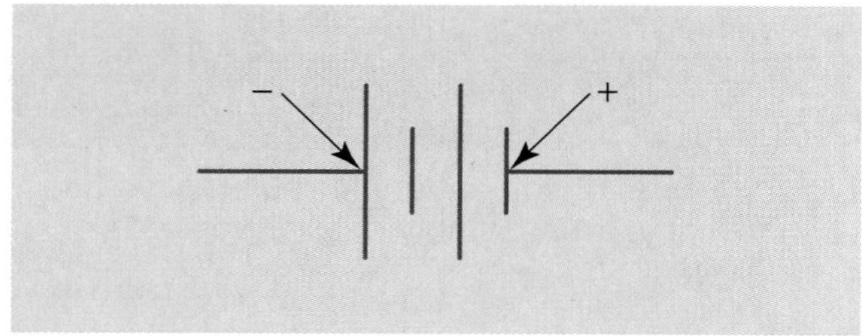

Figure 1-11. The symbol for a battery is a combination of several cell symbols.

If two circuits are shown near each other, or if the wires do not connect, leave off the dot, **Figure 1-15.** This tells the person reading

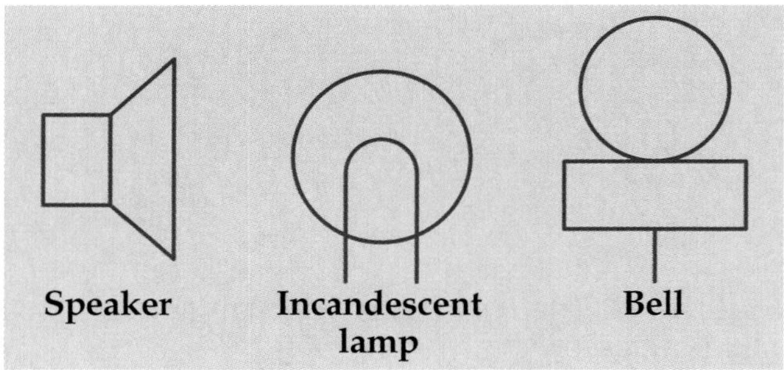

Figure 1-12. Symbols shown are, from left to right, a speaker, an incandescent lamp, and a bell.

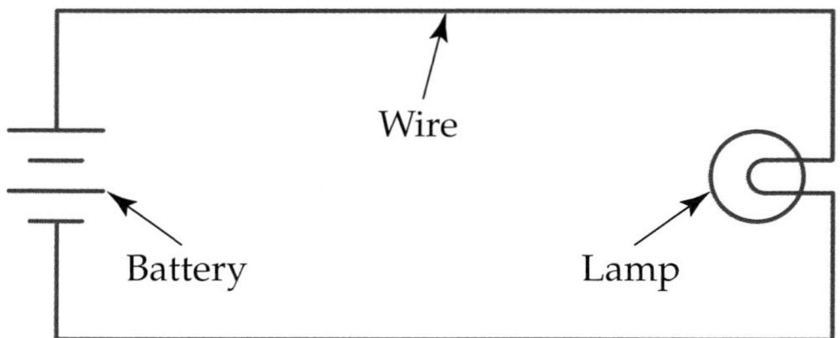

Figure 1-13. Use of symbols permits us to simplify the pictorial drawing of electrical circuits. Compare this schematic with the drawing in Figure 1-2.

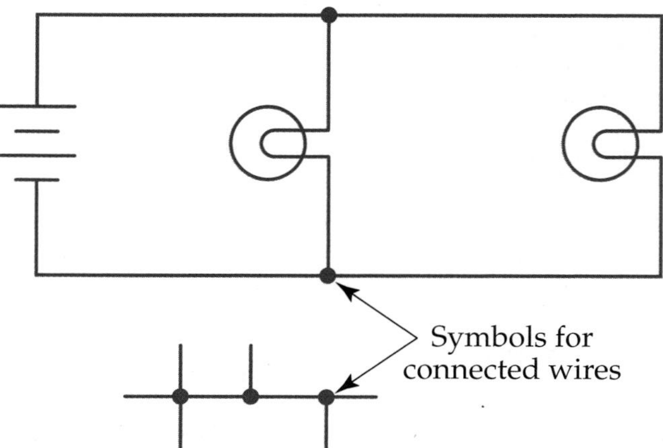

Figure 1-14. Another lamp has been added to the circuit in Figure 1-13. Note that a large dot is the symbol for "wires connected."

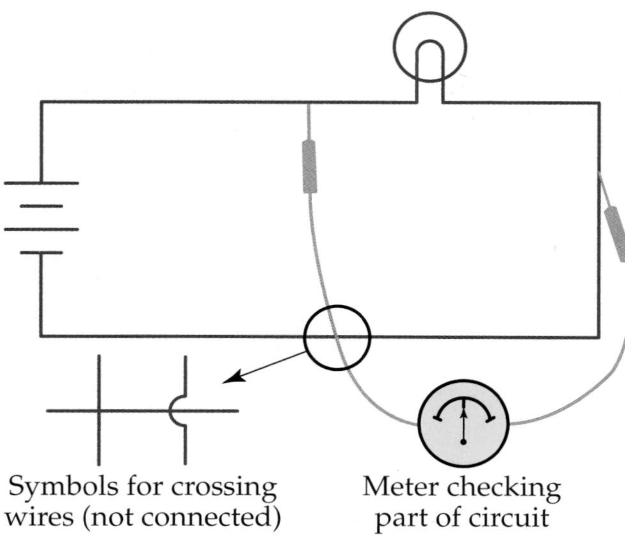

Figure 1-15. If the wires of a circuit are near each other, or if the wires cross but do not connect, either one of the two symbols shown may be used.

the drawing that the wires just pass over each other and are not connected.

A drawing that uses symbols is called a ***schematic***. **Figure 1-16** shows what a finished industrial schematic looks like. Schematics like these are frequently used in industrial applications. For example, an industrial schematic for a complex machine is essential to a service technician. Not all machines have the same circuits. Even the same type of machine has models that change from year to year. Automobile circuits have more than one hundred schematics. Can you imagine the trouble it would be to build and equip the control room in **Figure 1-17** without a large number of schematics?

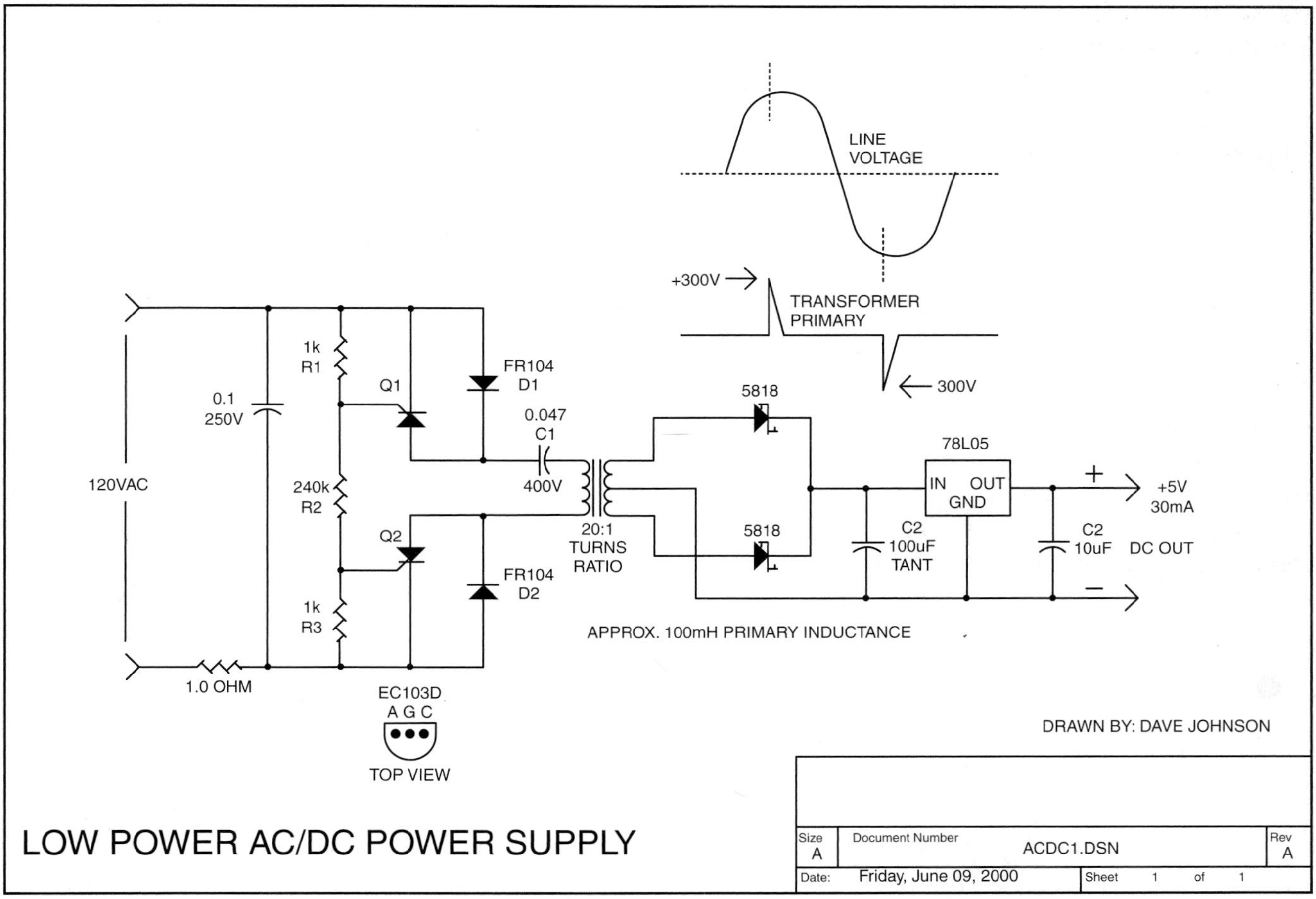

Figure 1-16. Schematic of a low-power AC/DC power supply illustrating its various circuits and symbols for its electrical components. (David Johnson and Associates)

Figure 1-17. This complex control room is loaded with robotic arms, switches, and buttons, which were wired with the help of a schematic.

The Chemistry of Electricity

As we discussed earlier, electrons are tiny particles that can be found inside the wires of circuits. You might wonder, "How small is an electron?" Well, if you could count the number of electrons on the tip of a ballpoint pen, you would count about one billion (1,000,000,000). In reality, you would need a powerful electron microscope to see the atoms they belong to. Electrons are parts of atoms. All matter is made of atoms.

The center of the atom is called the nucleus. It is made of neutrons, which have no charge, and protons, which have a positive charge. Electrons are negatively charged. They orbit the nucleus in specific patterns. The number of neutrons, protons, and electrons that make up an atom varies from element to element. Elements are made of atoms of the same type. For example, the element hydrogen consists of hydrogen atoms.

A periodic table of the elements, such as the one shown in **Figure 1-18,** lists all known elements and their atomic structure. To learn more about the names and atomic structure of the elements, visit the Los Alamos National Laboratory's Chemistry Division Web site: http://periodic.lanl.gov.

One of the simplest atoms has only two electrons (negative charge), two protons (positive charge), and two neutrons (no charge), **Figure 1-19.** The protons and neutrons make up the nucleus of the atom. Each electron orbits the nucleus in a separate path within a single layer (also called a *shell* or *energy level*). The first shell of an atom can have a maximum of two electrons. If an atom has more than two electrons in its first shell, other shells are filled with electrons.

Figure 1-20 shows an atom with two shells. This atom is neon. The first shell has two electrons and the second shell has eight. The second shell of an atom can have up to eight electrons in its orbit.

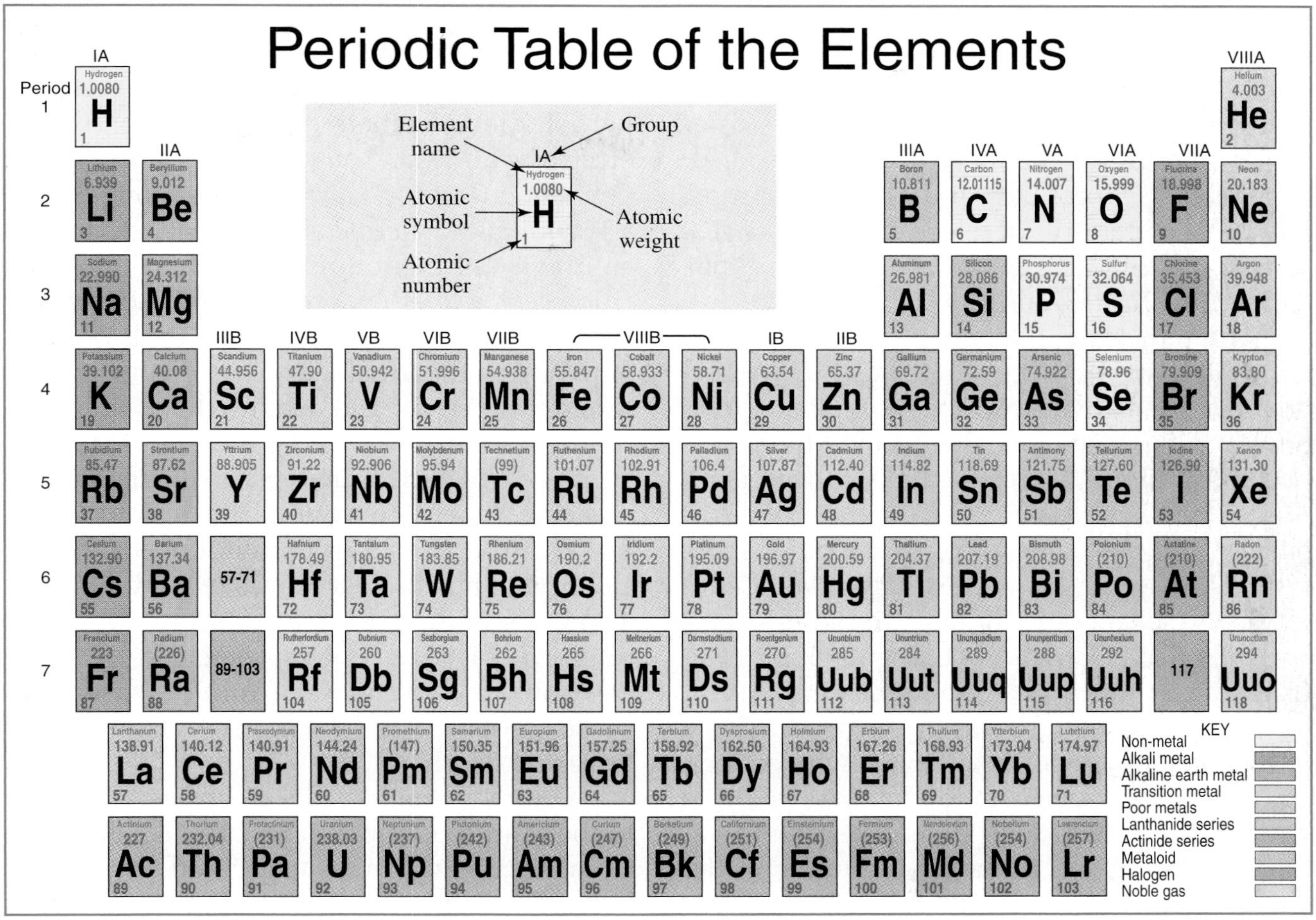

Periodic Table of the Elements

Element name
Group
IA
Hydrogen 1.0080 H 1
Atomic symbol
Atomic weight
Atomic number

Period	IA	IIA	IIIB	IVB	VB	VIB	VIIB	VIIIB	VIIIB	VIIIB	IB	IIB	IIIA	IVA	VA	VIA	VIIA	VIIIA
1	Hydrogen 1.0080 H 1																	Helium 4.003 He 2
2	Lithium 6.939 Li 3	Beryllium 9.012 Be 4											Boron 10.811 B 5	Carbon 12.01115 C 6	Nitrogen 14.007 N 7	Oxygen 15.999 O 8	Fluorine 18.998 F 9	Neon 20.183 Ne 10
3	Sodium 22.990 Na 11	Magnesium 24.312 Mg 12											Aluminum 26.981 Al 13	Silicon 28.086 Si 14	Phosphorus 30.974 P 15	Sulfur 32.064 S 16	Chlorine 35.453 Cl 17	Argon 39.948 Ar 18
4	Potassium 39.102 K 19	Calcium 40.08 Ca 20	Scandium 44.956 Sc 21	Titanium 47.90 Ti 22	Vanadium 50.942 V 23	Chromium 51.996 Cr 24	Manganese 54.938 Mn 25	Iron 55.847 Fe 26	Cobalt 58.933 Co 27	Nickel 58.71 Ni 28	Copper 63.54 Cu 29	Zinc 65.37 Zn 30	Gallium 69.72 Ga 31	Germanium 72.59 Ge 32	Arsenic 74.922 As 33	Selenium 78.96 Se 34	Bromine 79.909 Br 35	Krypton 83.80 Kr 36
5	Rubidium 85.47 Rb 37	Strontium 87.62 Sr 38	Yttrium 88.905 Y 39	Zirconium 91.22 Zr 40	Niobium 92.906 Nb 41	Molybdenum 95.94 Mo 42	Technetium (99) Tc 43	Ruthenium 101.07 Ru 44	Rhodium 102.91 Rh 45	Palladium 106.4 Pd 46	Silver 107.87 Ag 47	Cadmium 112.40 Cd 48	Indium 114.82 In 49	Tin 118.69 Sn 50	Antimony 121.75 Sb 51	Tellurium 127.60 Te 52	Iodine 126.90 I 53	Xenon 131.30 Xe 54
6	Cesium 132.90 Cs 55	Barium 137.34 Ba 56	57-71	Hafnium 178.49 Hf 72	Tantalum 180.95 Ta 73	Tungsten 183.85 W 74	Rhenium 186.21 Re 75	Osmium 190.2 Os 76	Iridium 192.2 Ir 77	Platinum 195.09 Pt 78	Gold 196.97 Au 79	Mercury 200.59 Hg 80	Thallium 204.37 Tl 81	Lead 207.19 Pb 82	Bismuth 208.98 Bi 83	Polonium (210) Po 84	Astatine (210) At 85	Radon (222) Rn 86
7	Francium 223 Fr 87	Radium (226) Ra 88	89-103	Rutherfordium 257 Rf 104	Dubnium 260 Db 105	Seaborgium 263 Sg 106	Bohrium 262 Bh 107	Hassium 265 Hs 108	Meitnerium 266 Mt 109	Darmstadtium 271 Ds 110	Roentgenium 270 Rg 111	Ununbium 285 Uub 112	Ununtrium 284 Uut 113	Ununquadium 289 Uuq 114	Ununpentium 288 Uup 115	Ununhexium 292 Uuh 116	117	Ununoctium 294 Uuo 118

Lanthanum 138.91 La 57	Cerium 140.12 Ce 58	Praseodymium 140.91 Pr 59	Neodymium 144.24 Nd 60	Promethium (147) Pm 61	Samarium 150.35 Sm 62	Europium 151.96 Eu 63	Gadolinium 157.25 Gd 64	Terbium 158.92 Tb 65	Dysprosium 162.50 Dy 66	Holmium 164.93 Ho 67	Erbium 167.26 Er 68	Thulium 168.93 Tm 69	Ytterbium 173.04 Yb 70	Lutetium 174.97 Lu 71
Actinium 227 Ac 89	Thorium 232.04 Th 90	Protactinium (231) Pa 91	Uranium 238.03 U 92	Neptunium (237) Np 93	Plutonium (242) Pu 94	Americium (243) Am 95	Curium (247) Cm 96	Berkelium (249) Bk 97	Californium (251) Cf 98	Einsteinium (254) Es 99	Fermium (253) Fm 100	Mendelevium (256) Md 101	Nobelium (254) No 102	Lawrencium (257) Lr 103

KEY
Non-metal
Alkali metal
Alkaline earth metal
Transition metal
Poor metals
Lanthanide series
Actinide series
Metaloid
Halogen
Noble gas

Figure 1-18. Periodic table of the elements. (Courtesy of Los Alamos National Laboratory)

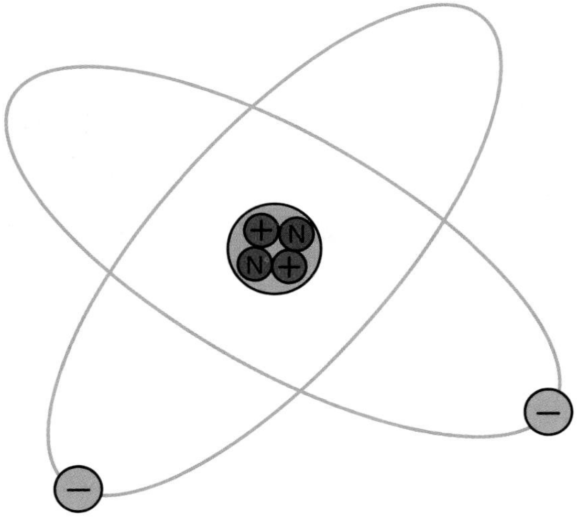

Figure 1-19. A helium atom is made of two electrons (negative charge) and a nucleus composed of neutrons (no charge) and protons (positive charge).

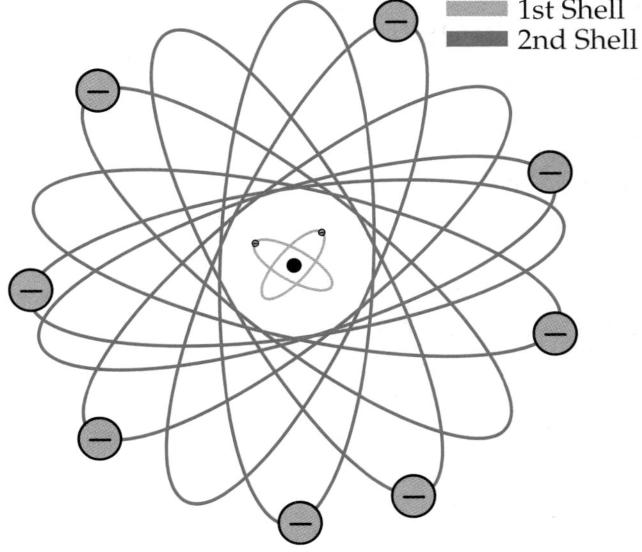

Figure 1-20. Neon atom. Note that electrons orbit the nucleus in different shells.

Rather than drawing all the orbits for each shell, shells are usually shown as a circle of electrons like those shown in **Figure 1-21.** Drawing them in this way makes it easier to see and count the electrons. The atom in Figure 1-21 is copper. It has 29 electrons. Notice the pattern of electrons and shells. The first shell has 2 electrons, the second has 8, the third has 18, and the last has 1.

The third shell of an atom can have a maximum of 18 electrons. The fourth shell of an atom can have 32. The copper atom, however, has only one electron in the fourth shell. This single electron can be moved from the outer shell easier than if there were a greater number of electrons in the outer shell. That is why copper is used to make wire found in most of our circuits. Copper is a good conductor of electricity because its outer electron can be moved easily. As the number of electrons in a shell increases, the conductivity of that element decreases. That is why copper, gold, and silver are good conductors. All three have a single electron in their outer shells.

How many electrons are in motion in the electrical circuits in your school, city, and state every second of the day? You just can't count such a large number. No doubt you have heard the term *ampere* or *amps* used to describe the amount of current flow. One amp is equal to the flow of over 6,241,000,000,000,000,000 electrons moving past some point in one second. Rather than counting billions of electrons, we call the movement of over 6 billion-billion electrons an *amp*. You will learn more about amps or current in later chapters. You will see why your knowledge about electron shells is so important in studying electricity and electronics. In a later chapter, you will also study insulators and semiconductors. Each of them depends on the understanding of what it takes to stop electrons and what it takes to make them flow.

As each electron moves, it sets up a chain reaction from atom to atom. The movement of one electron to the next atom forces a second electron to move. As it moves to another atom, it forces still another electron to move and so on down the line. The movement of all these electrons sets up a wave motion similar to those waves you see at the ocean as they move toward a beach.

The movement of the electrons is similar to the example shown in **Figure 1-22.** Imagine a plastic tube ten miles long filled with tennis balls. Once it is filled, force one more tennis ball into the tube. At the same instant that you force a new one in, another tennis ball falls out the other end. This is similar to how electrons force a flow from atom to atom.

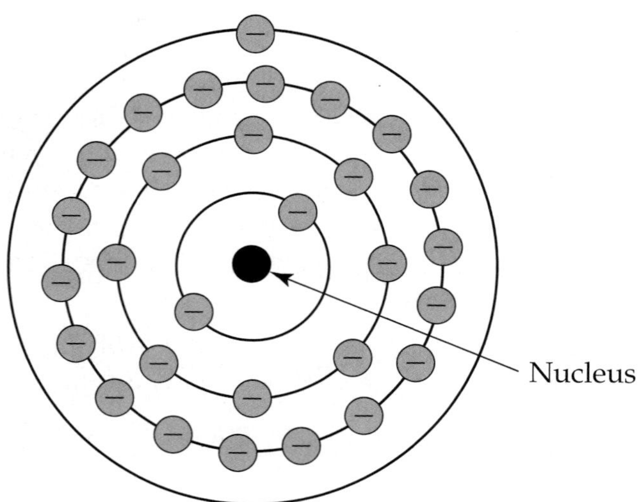

Figure 1-21. Copper atom.

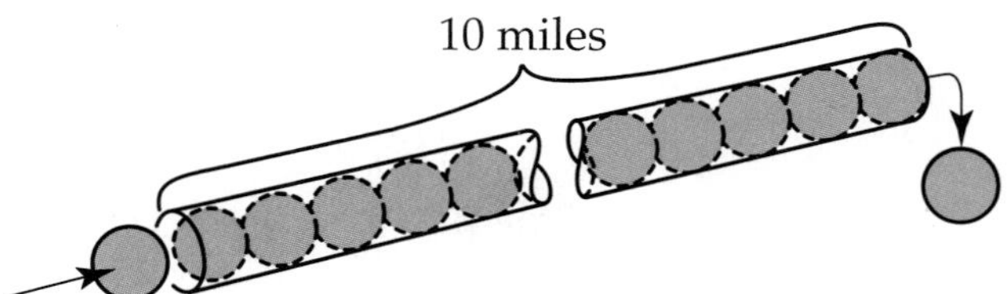

Figure 1-22. Another tennis ball forced into a full plastic tube causes one to fall out 10 miles away.

Practical Application 1-1: Electronic Kits

Electronic kits are an easy way to start working with circuits. There are many kits sold in stores. Some of them require a lot of knowledge about electricity to use them successfully. Other kits are loaded with inexpensive parts and components that will not stand up to much use. Select a kit that will stand up to the wear and tear from constantly putting pieces together and taking them apart. If you want to learn the concepts behind many electronic components, be selective in which kit you purchase. Some good kits containing about fifty components attached by connectors allow you to put together hundreds of different experiments.

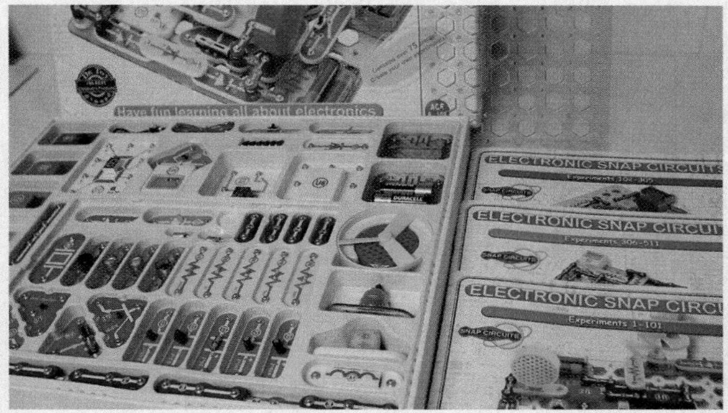

(Elenco Electronics Inc.)

Practical Application 1-2: Breadboarding

"Breadboarding" is the process of using a breadboard to construct circuits. A breadboard is a circuit board that allows components and wires to be easily inserted and removed. It is often used to quickly assemble a circuit for the purpose of testing a circuit's design. Breadboards are often used in the classroom because of their ease of use. Students can quickly assemble and disassemble circuits without the need for soldering and desoldering.

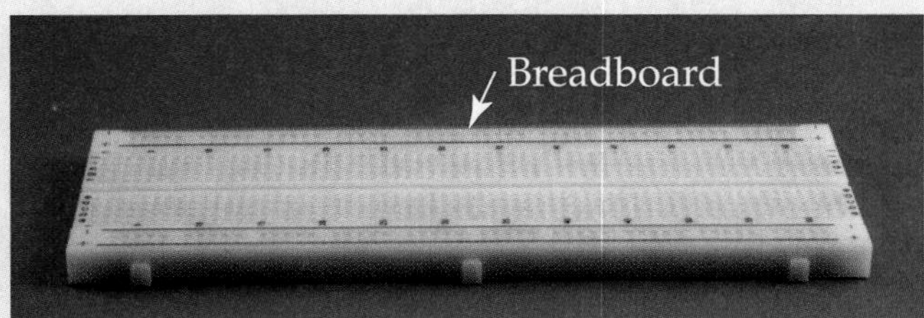

A breadboard is typically constructed of plastic. It contains rows of holes into which components and wires are inserted. Beneath its surface are metal strips. The metal strips run beneath the holes in a certain pattern. It is important to be familiar with this pattern to properly assemble a functioning circuit. You must be careful not to insert both or all of the leads of a component so that they contact a single strip, as this will create a short across the component.

There are typically one or two horizontal runs of metal strips on the top and bottom of the breadboard. These metal strips are used for the power source. The rest of the metal strips run vertically. The horizontal gap in the center of the breadboard makes a convenient location in which to install ICs, since each IC pin can make contact with its own metal strip. IC stands for integrated circuit, a device that is explained in Chapter 20.

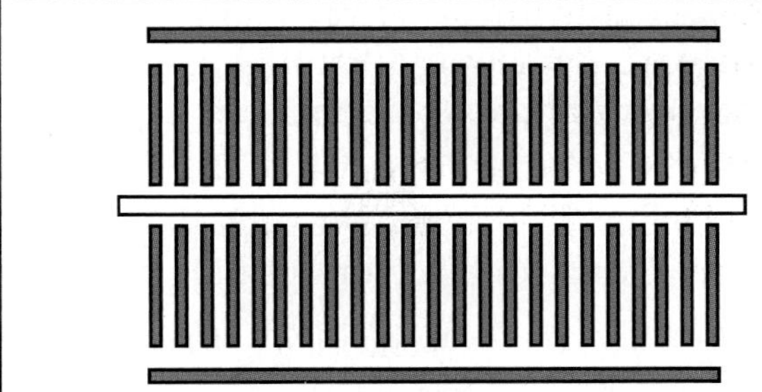

To make connections between runs, short wires called jumper wires are used. The following picture shows a simple circuit like the one presented earlier in this chapter. It consists of a power source (battery), lightbulb, and switch. Notice the way the components are assembled in the breadboard.

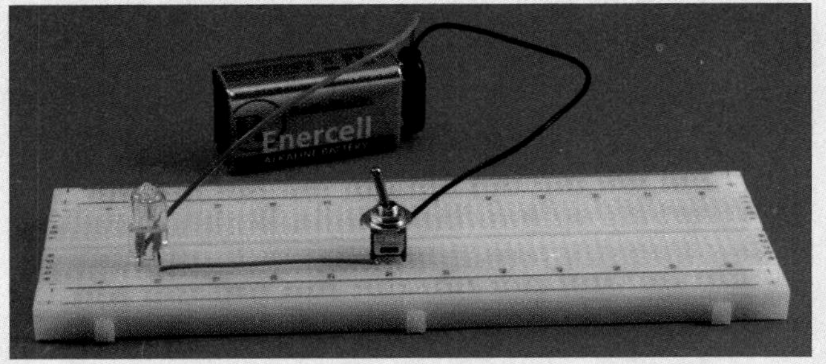

Advances in Electricity

Since Benjamin Franklin flew his now famous kite, many important discoveries have been made in the electrical field.

Danger
In attempting to duplicate Franklin's kite flying experiment in a thunderstorm, some people have been killed or seriously injured. As you will see in studying this text, there are safer ways to experiment with electricity.

Over the years, electrical discoveries have led to many things that have made work easier and life more comfortable. To understand these remarkable advances, it is necessary to examine many of the early discoveries. Each one has served as a building block in the construction of electrical knowledge.

1752: Benjamin Franklin (1706–1790) used his famous kite experiment in an effort to prove that lightning was a form of electricity.

1780: Luigi Galvani (1737–1798) discovered that moisture in frog legs caused them to twitch when they came into contact with two different metals. The electricity created was the beginning of the science of electrophysiology.

1800: Alessandra Volta (1745–1827) built a device (electrochemical cell) that produced a flow of current. He used stacks of copper, zinc, and cardboard moistened with saltwater. These stacks formed the voltaic pile. When a wire was attached at each end, it formed what was called an electric battery.

1820: Hans Christian Oersted (1777–1851) discovered that the magnetic field formed when electricity flows through a wire affects the needle of a compass.

1820: Andre-Marie Ampere (1775–1836) established the relationship between electricity and magnetism and developed the science of electromagnetism. The unit of electric current, the ampere, was named in his honor.

1821: Michael Faraday (1791–1867) discovered that moving a magnet through a coil of wire causes an electric current to flow through the wire. This led to his discovery of electromagnetic induction and his invention of the electric motor.

1826: George Ohm (1789–1854), using information from Volta's discovery of a cell, conducted experiments that uncovered the relationship between voltage, current, and resistance. This relationship is known as *Ohm's Law.* The unit of resistance, the ohm, was named in his honor.

1829: Joseph Henry (1797–1878) discovered that wrapping insulated wire around an iron core produced a strong electromagnet when connected to a battery. This concept also led to the discovery of how to generate an alternating current. The unit of electrical inductance, the henry, was named in his honor.

1838: Samuel Morse (1791–1872) demonstrated at a New York exhibition his new telegraph machine that used a system of dots and dashes to send 10 words per minute. His system became known as Morse code.

1843: Charles Wheatstone (1802–1875) made popular an instrument invented by Samuel Christie that measures unknown electrical resistances.

1876: Alexander Graham Bell (1847–1922) used electricity to transmit speech. He is considered to be the inventor of the telephone. He spoke the words "Mr. Watson, come here. I need you" into the telephone. However, the United States Congress officially recognized Antonio Meucci as the inventor of the telephone because of his work on a "telephone-like device" around 1857.

1879: Thomas Edison (1847–1931) developed an electrical distribution system that used direct current (dc) electric generators. His intention was to power America. He lost that battle when Tesla developed the alternating current(ac) system we use today. Edison was issued over one thousand patents. Some of the better-known ones are the incandescent lamp, electric locomotive, alkaline storage battery, microphone, and the phonograph.

1886: Heinrich Hertz (1857–1894) produced and detected electric waves in the atmosphere. This concept led to the invention of the wireless telegraph, a device credited to Marconi. The measure of electrical frequency, the hertz, was named in his honor.

1888: Nikola Tesla (1856–1943) developed an alternating current (ac) motor and a system of ac power generation. Seeing Tesla's system as a threat to his own, Edison maintained that Tesla's system was unsafe. When Tesla's system powered 1,000,000 electric lights at the 1893 Chicago World Fair, ac became the standard power supply in the United States. Tesla also designed the first hydroelectric plant.

1896: Guglielmo Marconi (1874–1937) was granted the world's first patent for a system of wireless telegraphy. Although he didn't invent radio, he is credited as being the "father of radio" because he did more than any of the early inventors to make radio popular.

1905: Albert Einstein (1879–1955) demonstrated that light energy could be used to produce electricity. He also wrote his theory of relativity. These were the beginning of the ideas behind photovoltaic cells and theories of light energy. The latter earned him the Nobel Prize in 1921.

1947: John Bardeen (1908–1991), Walter Brattain (1902–1987), and William Shockley (1910–1989) invented the transistor while working at AT&T Bell Telephone Laboratories. This invention would change the world of electronics from tube technology into the solid-state systems that we know today. For this discovery, the three shared the 1956 Nobel Prize in physics.

Summary

- A circuit must always have a source, a load, and a path.
- When electricity flows through a wire, the electrons in the wire move through it from the negative to the positive terminal of the power source.
- In a schematic drawing, the components of a circuit are replaced with symbols, which makes producing and reading the drawing easier.
- An atom is made of electrons, protons, and neutrons.
- Electrons orbit the nucleus of the atom, which is composed of protons and neutrons.
- The term *ampere* is used to describe the amount of current flow.
- One ampere is equal to the flow of over 6,241,000,000,000,000,000 electrons moving past some point in one second.

Test Your Knowledge

Do not write in this book. Write your answers on a separate sheet of paper.

1. The three requirements for a complete circuit are ______, ______, and ______.
2. In a closed circuit, electrons always flow from (negative, positive) to (negative, positive).
3. One way to stop the flow of electrons is to ______ the circuit.
4. Copper is a good conductor of electricity because ______.
   a. the protons orbit the outer shell at great speed
   b. there is only one electron in its outer shell
   c. its protons outnumber both the electrons and neutrons
   d. it is a pure metal, which is both easy to refine and cast into wire
5. A(n) ______ is used instead of a complex, detailed drawing of an electrical circuit.
6. Draw the symbols for a cell and battery.
7. When electricity is flowing through a wire, ______ are forced from one atom to the next.

Activities

1. Explain why there are so many different symbols for an electrical load.
2. Explain why a load is so important to an electrical current.
3. Conduct an Internet search on additional inventors in the electricity/electronics field. For example, you may want to research who invented the following items:
   - Television.
   - Compact disc.
   - DVD.
   - iPOD.
   - Walkman.
   - Satellite radio.
   - Digital camera.
   - MP3 player.
4. Visit the Los Alamos National Laboratory's Chemistry Division Web site: http://periodic.lanl.gov. Find which metals are the best conductors of electricity.

Sources of Electricity

Learning Objectives

After studying this chapter, you will be able to do the following:

- Identify five ways to produce electricity.
- Give examples of devices that produce electricity.
- Describe various devices that produce electricity.

Technical Terms

dead cell
direct current (dc)
dry cell
piezoelectricity
piezoelectric effect
primary cell
secondary cell
specific gravity
thermocouple

After reading the first chapter, you probably have several questions about how electricity is produced. This chapter provides the answers to those questions. There are many ways to generate electricity. They basically fall into the following five methods:

- Chemical energy.
- Light energy.
- Pressure.
- Heat.
- Magnetism.

These methods are explained in some detail in the order given. Related projects are included in this chapter.

Electricity from Chemical Energy

Chemical energy means that electricity is generated through the use of chemicals and their reaction with other materials. The simple cell mentioned in Chapter 1 is a good example of a device that produces electricity through a chemical reaction. Cells and batteries produce a stream of electrons that flow from negative (–) to positive (+). Remember, this was mentioned in Chapter 1. This constant flow of electrons in one direction in the electrical circuit is called ***direct current (dc).***

How many things can you think of that get their power from a dc source? Flashlights, emergency lamps, portable radios, and children's toys, **Figure 2-1,** all use dc as a source of electricity.

The capacity of a battery, or the amount of energy it can provide for a period of time, is labeled in ampere-hours (Ah). For example, a 1 Ah battery can provide a current of one ampere for one hour. A battery with a rating of 50 Ah can provide 1 ampere for 50 hours. Batteries with higher ratings usually have more electrolyte or electrode material in their cells. That is why AAA cells are rated with a smaller Ah capacity than C and D cells, even though they have the same voltage capacity.

Primary Dry Cells

If you were to cut a cell in half, you would see something like what is shown in **Figure 2-2.** This type of cell is often called a ***dry cell,*** although the chemicals used inside the cell are in the form of a paste. The paste is an electrolyte material that provides the path for electricity inside a cell. It is a mixture of sal ammoniac (ammonium chloride), manganese dioxide, and carbon with water. These chemicals react with the zinc can and the carbon rod to produce electricity. When the zinc can is finally eaten away, the cell no longer produces electricity. It is called a ***dead cell.*** This type of dry cell cannot be recharged. A cell that cannot be recharged is called a ***primary cell.***

Caution

Never leave a dead cell inside a device The chemicals can leak out of the cell and damage the device in which it is installed. The chemicals are corrosive and will eat away connectors, switches, and other electrical components.

Figure 2-1. Children's toys and remote controls get their power from dc sources.

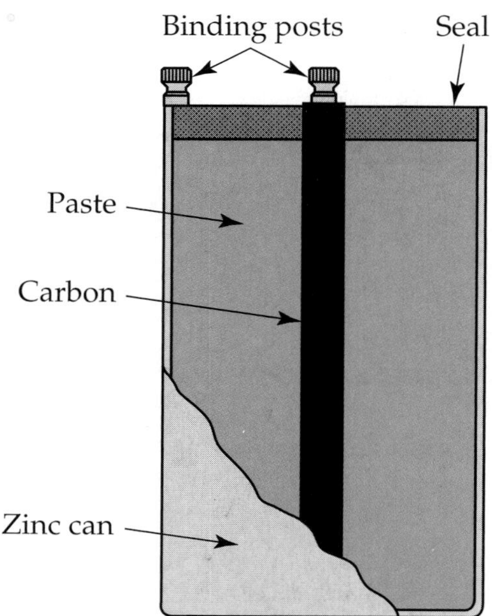

Figure 2-2. In a dry cell, chemical energy produces electricity. The electrolyte (paste) in the cell reacts with the zinc can and the carbon rod.

Some dry cells are referred to as acidic dry cell batteries. Although it is not totally correct, most people refer to cells as batteries or as dry cell batteries. Batteries are really made up of two or more cells.

Another type of battery, called alkaline dry cell battery, contains sodium hydroxide and potassium hydroxide. Acidic dry cell batteries and alkaline dry cell batteries contain chemicals that can harm equipment and people.

Dry cells are usually found in small, household items. The most common types of dry cells are found in the following sizes:

- AAA.
- AA.
- C.
- D.
- 9 V.

The *V* in 9 V refers to voltage. You can think of voltage as the pressure used to force the electrons through the circuit. More will be said about voltage in later chapters.

Secondary Dry Cells

With the rapid development in battery technology, you have probably heard the names mentioned of dry cells called nickel-cadmium (Ni-Cd), lithium ion (Li-Ion), and nickel-metal hydride (Ni-MH). All of these cells are rechargeable. A cell that is rechargeable is referred to as a ***secondary cell.***

Nickel-cadmium cells are typically found in small appliances such as digital cameras, electric shavers, calculators, and cordless power tools. This type of cell has a voltage of 1.2 V. The cadmium inside this cell is highly toxic and requires special handling during the battery's manufacture.

Nickel-metal hydride cells have the same voltage as nickel-cadmium cells but have a greater Ah rating. This means that they provide energy for a longer time than nickel-cadmium cells. Nickel-metal hydride cells are constructed similarly to the nickel-cadmium cell except that they use metal hydride instead of cadmium. They are typically used in digital cameras and portable devices, such as cell phones, laptops, and cordless power tools. Nickel-metal hydride and nickel-cadmium cells come in standard battery sizes AAA, AA, B, C, and D.

Lithium ion cells weigh less than nickel cadmium and have voltages of 3.6 V and 3.7 V. Their light weight and high power capacity make them ideal for use in laptops, cell phones, and other mobile devices. Lithium ion cells come in a variety of shapes and sizes. **Figure 2-3** shows two lithium ion cells.

Figure 2-3. Lithium ion batteries come in many shapes and sizes.

Project 2-1: Continuity Tester

The schematic for a continuity tester, a handy item you can make for yourself, is shown here. With it, you can test for broken wires or burned out (open) fuses. Another good use is finding the matching wire ends in a wiring harness, which is a group of wires routed together.

No.	Item
1	1.5-V cell
1	1.5-V lamp
1	Battery holder
1	Lamp holder
2	4′ to 6′ lengths of 18 or smaller gauge wire
2	Probes

To make the continuity tester, connect a 1.5-V dry cell and a 1.5-V test lamp, and then connect leads to each of those as shown in the schematic. If soldering is required, consult with your instructor. Soldering involves melting a tin-based metal to the wire joints. Keep the circuit small so you can put the dry cell and lamp in almost anything you want to, such as a hollow plastic tube. You can also mount the components on a small board or similar material.

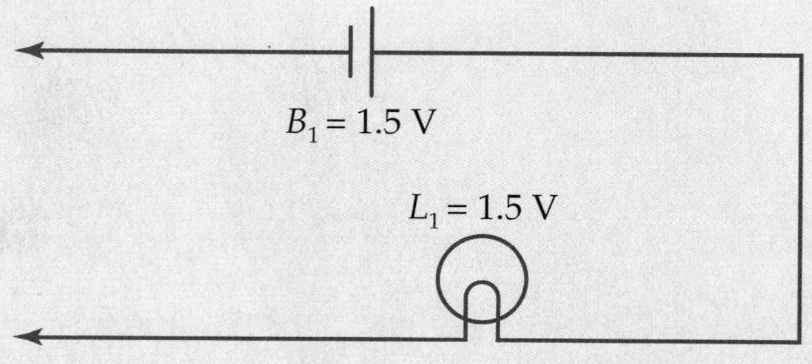

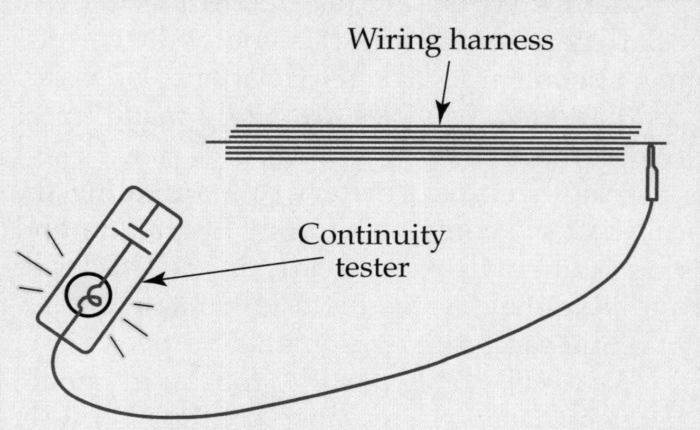

To use the continuity tester, touch the leads to the item under test. If the light comes on, the item has continuity (will conduct electricity). If the light stays off, the item lacks continuity (will not conduct electricity).

You cannot use this continuity tester to check everything. The coil in a transformer or a large inductor, for example, offers too much resistance for the low-voltage tester. A dc cell of 1.5 V cannot do the job. Later, you will learn why this is so.

If you want to test a wiring harness, hook one lead of the tester to one end of a wire. To find the other end of the wire, touch each of the other wire ends with the other test lead. When the light comes on, you have continuity. Now you have both ends of the same wire. Sometimes, one end of the wiring harness is hard to reach. For this reason, wires are often color coded, which means that each wire has a specific color of insulation to aid in identification.

Caution

Do not use the continuity tester on live circuits. Doing so can damage the continuity tester. Make sure all wires are disconnected before testing them.

Wet Cells

Wet cells contain electrolytes other than the paste type. A car battery, for example, is a wet cell battery that uses an acid as its electrolyte, **Figure 2-4.** It is a battery because it is made of several wet cells. This battery has an advantage over the primary dry cell because it can be recharged. It is therefore a secondary cell.

When a wet cell battery is used, the metal plates inside the battery are still eaten or dissolved like the zinc can in the primary dry cell. The difference, however, is that the metal in the wet cell battery can be made to deposit itself onto its plates. This is called *charging the battery.*

Danger

Do not charge a non-rechargeable battery. It can explode.

Danger

Do not charge a frozen battery. It can explode due to the mixture of hydrogen and oxygen trapped inside the battery's cells.

Figure 2-4. A car battery is made of six cells filled to a prescribed level of electrolytic material.

In the case of an auto battery, the alternator, which is a type of generator, does the recharging. However, it is possible to drain the battery of its charge. If, for example, you leave the headlights on when the car is not running, it may be necessary to use a battery charger, **Figure 2-5,** to recharge the battery. You have probably seen people use two heavy "jumper cables" to start an engine from the battery of a second vehicle. Once the engine in the first vehicle is started, the alternator begins to recharge the battery.

Danger

An explosion can occur due to sparks from connecting and disconnecting jumper cables. To ensure sparks occur away from the battery, always connect one end of the negative jumper cable to the chassis of the vehicle with the dead battery. This connection should be made last.

Even though they can be recharged, auto batteries do not last forever. When they get old, it is harder to keep them fully charged. How the vehicle owner cares for the battery also affects how long it lasts. Repeatedly allowing an auto battery to discharge shortens its life. Older batteries or those that have been misused do not hold their charge because their chemicals become weak. As the battery is used, the electrolyte reacts with the metal

Figure 2-5. A portable battery charger keeps the car battery fully charged.

plates as in parts 3 and 4 in **Figure 2-6.** This causes them to be eaten or dissolved like the zinc can in the dry cell.

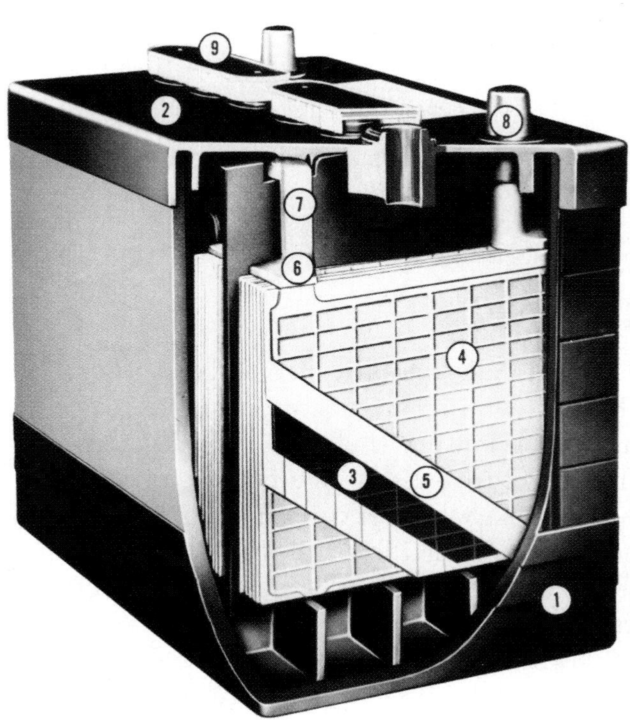

Figure 2-6. Cutaway battery reveals the relationship of its parts. 1—Case. 2—Cover. 3—Positive plates. 4—Negative plates. 5—Separators. 6—Cell. 7—Cell connector. 8—Terminal post. 9—Vent cap.

Hydrometer Test

The acid or electrolyte in a battery can be used to tell you whether or not the battery is fully charged. In **Figure 2-7,** a hydrometer is used to check an airplane battery.

The hydrometer bulb is squeezed. Then, the end is inserted into the electrolyte, and the bulb is slowly released to fill the hydrometer tube with acid from the battery. By reading the scale on the float inside, it is possible to tell how much charge the battery has available. Many hydrometers are marked to tell you whether the battery is fully charged, half-charged, or discharged. This is possible because the acid changes in strength. This strength is measured in terms of ***specific gravity.***

> **Note**
> Although many batteries are called *maintenance free,* the hydrometer is still used by technicians to measure acid strength.

Project 2-2: Buzzer

A schematic is shown for a buzzer you can build. This circuit is part of other circuits that are to follow. In some cases, the buzzer is replaced with a light. You could also use a bell or siren.

No.	Item
1	1.5-V cell
1	1.5-V buzzer
1	Battery holder
1	Single-pole single-throw (SPST) switch
1	18 or smaller gauge wire

The parts needed to complete this project include a 1.5-V cell, a 1.5-V buzzer, a single-pole single-throw (SPST) switch, and sufficient wire to make the connections shown. The purpose of the switch is to allow you to open and close the circuit without removing wires. Each time you close the switch, the buzzer makes noise. When you open the switch, the buzzer stops making noise. No electrons can flow when you have an open circuit. Electrons can flow once you close the circuit. The switch can be replaced with a doorbell pushbutton or a number of other devices that allow you to open or close the circuit, as you will see when you study switching with transistors and relays.

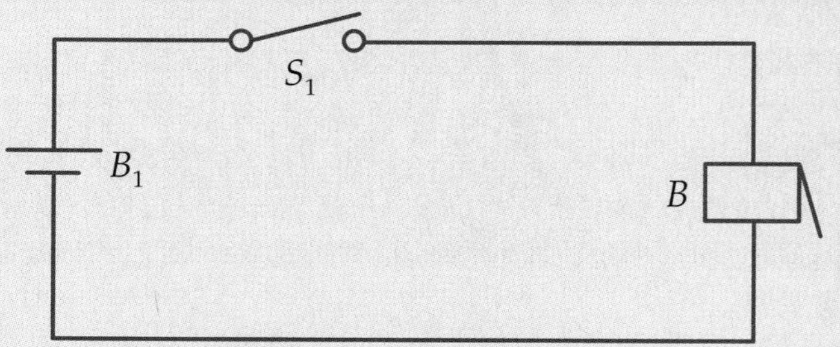

Specific gravity is a comparison of the density of the acid with the density of water. Water has a specific gravity of 1.000. As the battery discharges, the specific gravity of the acid gets closer to the 1.000 mark. In some colder climates, this charge loss can allow a battery to freeze in the winter. Note the table in **Figure 2-8.**

Figure 2-7. A hydrometer is an instrument used to measure the specific gravity of electrolyte in each cell to determine the battery's state of the charge.

Freezing Point of Battery Electrolyte		
	Specific Gravity	Freezing Point
Fully Charged	1.275	–85°F (–65°C)
	1.250	–62°F (–52°C)
Half-Charged	1.225	–35°F (–37°C)
	1.200	–16°F (–27°C)
	1.175	–4°F (–20°C)
Discharged	1.150	5°F (–15°C)
(Water)	1.000	32°F (0°C)

Figure 2-8. Freezing points of battery electrolyte at various specific gravity readings.

Practical Application 2-1: Material Safety Data Sheets

A Material Safety Data Sheet, or MSDS, is an important document required for products that contain hazardous materials. An MSDS lists information such as a product's ingredients, physical data, safe handling procedures, health hazards, procedures for recycling and disposal, and procedures for spill and leak clean up.

This information allows the user of the product to take the proper handling precautions and to immediately treat affected areas, such as skin and eyes, or to treat ingestion. It informs emergency workers, such as firefighters, hazardous material personnel, and medics, of how to treat those affected by the product's hazardous materials and how to clean and dispose of these materials.

Many products in the field of electricity and electronics have an MSDS. It is important for you to be familiar with the MSDS of the products you work with day to day and to have them on hand for reference. In fact, employers are required to distribute the MSDS of any hazardous materials to the employees working with them. They must also have the MSDS on hand in the room that contains the hazardous material.

Some of the products you may come into contact with in the field that have an MSDS are batteries, solder, contact cleaner, insulating varnish, and heat shrink tubing. To see a sample MSDS for any of these products, do an Internet search. For example, type into an Internet search engine "insulating varnish MSDS." As you read through the MSDS, note the various sections and identify any information that would be useful to you in the field.

Electricity from Light Energy

Light energy is another means of producing electricity. In recent years, more and more research has been done on photovoltaics, or the process of converting light energy into electricity. Light energy is especially important in space travel. **Figure 2-9** shows a simple photovoltaic (PV) system used for recharging cells. **Figure 2-10** shows solar cells being used with the Hubble Space Telescope. In space, liquid chemicals like the electrolyte used in batteries are of limited use. However, the sun is a huge source of power. By using solar cells, it is possible to convert the energy in the sun's rays into enough electricity to power all the electronic devices in the satellite.

There are a number of other examples of converting light energy into electricity. Some houses have installed special types of shingles on the roof for this task. One shingle can produce about 17 watts. This number will continue to increase as the ability of our technology increases. The shingles use a thin-film PV material which converts light into electricity. Examples of this material can be seen in Chapter 24.

Figure 2-10. The solar cells on the Hubble Space Telescope convert light into energy. (NASA/AURA/STSCI)

Figure 2-9. Photovoltaic system. A—Front of photovoltaic system for use with rechargeable, or secondary, cells. B—Back of the photovoltaic system showing where four cells can be mounted for charging.

Light-powered watches and calculators have been used for a number of years. The batteries in these devices are charged by light energy to provide the small amount of current needed to make them operate.

Sunlight is the key ingredient in PV modules for power stations that produce over 400-megawatts of electricity. A megawatt is equal to a million watts. If you compare this to your 100-watt lightbulb, you will have a rough idea what this represents. Each kilowatt of electricity annually produced by PV is roughly equal to eliminating the following pollutants from our environment:

- 9 kilograms of sulfur oxides.
- 16 kilograms of nitrogen oxides.
- 2,300 kilograms of carbon dioxide (CO_2).

The diagram in **Figure 2-11** shows how a PV cell works, without getting into details. The operation of a PV cell will be made clear to you when you study transistors in Chapter 18.

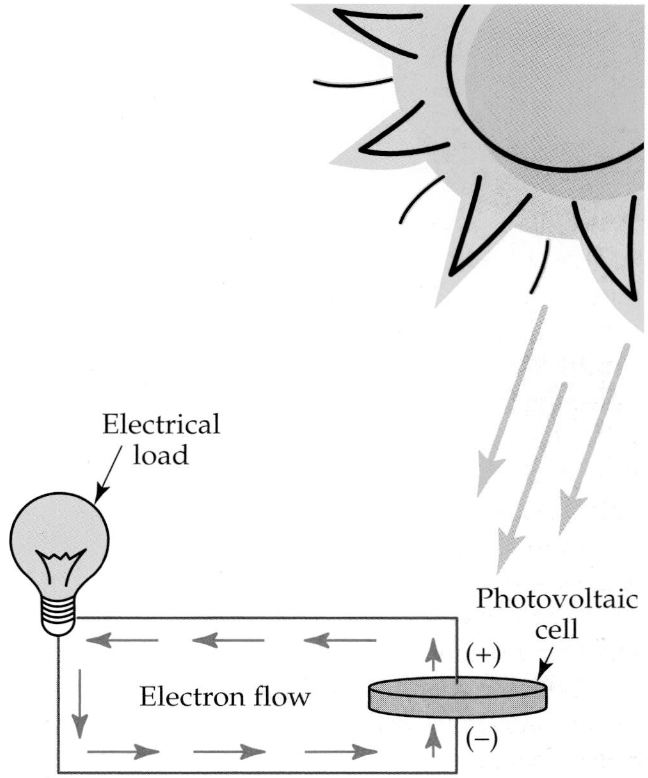

Figure 2-11. A simple circuit using a photovoltaic cell to power a lightbulb. When light strikes the photovoltaic cell, the cell becomes conductive and electrons flow from the photovoltaic cell through the circuit.

Electricity from Pressure

Pressure is another method used to generate electricity. *Piezo* is the Greek word for pressure. Therefore, ***piezoelectricity,*** or the ***piezoelectric effect,*** is the name often given to the creation of electricity from pressure. A number of crystals, including quartz, produce electricity when subjected to high pressure, **Figure 2-12.** You can see examples of this effect when pushing a button to create the spark used to light a gas grill or gas kitchen range. Two different types of lighters that use this effect are shown in **Figure 2-13.**

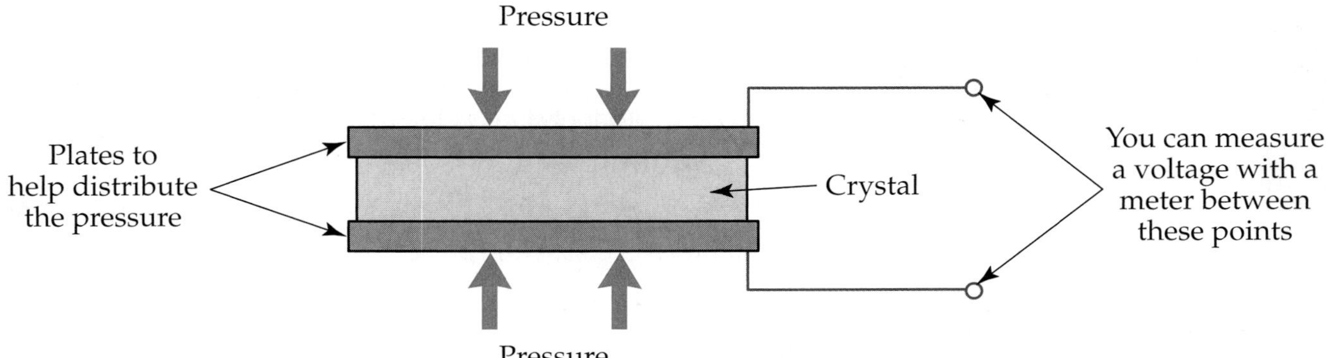

Figure 2-12. Electricity by pressure can be demonstrated by placing quartz crystal between pressure plates. When pressure is applied, electricity flows.

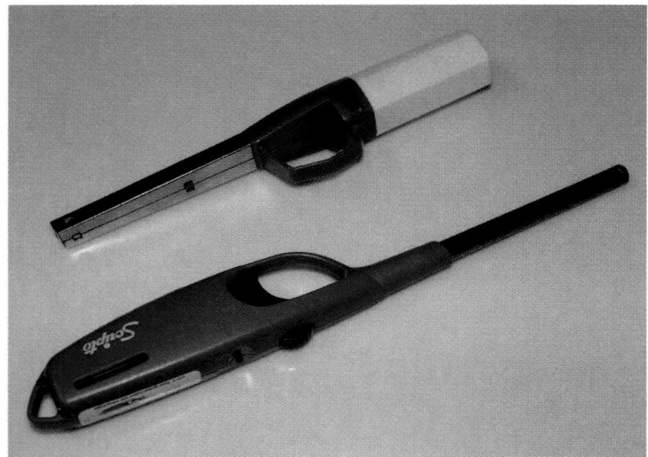

Figure 2-13. Lighters used to light a gas grill or kitchen range do so through piezoelectricity.

A microphone is a common application of pressure used to generate electricity. When pressure is applied to the microphone, an electric current is produced. The pressure in this case is in the form of sound waves, such as those produced by our voice. The voice entering the microphone, in turn, can be made much louder (amplified). If you ever attended a concert you know that sound waves can be made so loud they almost deafen you.

An underwater microphone, a hydrophone, has made it possible for us to hear the sounds of many of our sea creatures. The sound produced by a whale is converted into underwater pressure waves, which are picked up by the hydrophone. Hydrophones were once used by submarines to detect enemy ships and submarines. This, in turn, resulted in a number of other electronic devices that were much more sensitive in detecting sound. For example, it led to the development of sonar (**SO**und **NA**vigation and **R**anging). As you have read in Chapter 1, many new inventions are the result of improving previous inventions.

Now you know that electricity can be produced by sound waves. There are a number of experiments underway that use a large number of microphones to create electricity. These experiments are taking place along major highways and other locations that have a high level of noise pollution.

Electricity from Heat

Heat can generate electricity through the use of a device called a ***thermocouple***, **Figure 2-14.** A thermocouple consists of two dissimilar metals, such as copper and zinc or constantan and iron, bonded tightly together. Two of its symbols are shown in **Figure 2-15.** When the thermocouple is heated, the movement of the electrons between the two metals produces voltage, or electricity.

Gas furnaces, which can be found in most homes, use a thermocouple. The thermocouple is located by the pilot light. If the pilot light is lit, the thermocouple conducts electricity and the gas valve opens. The open gas valve allows the gas to fuel the fire that heats the home. This fire can only be started if the pilot light is lit. If it is not lit, the thermocouple does not conduct electricity and the gas valve remains closed. Gas

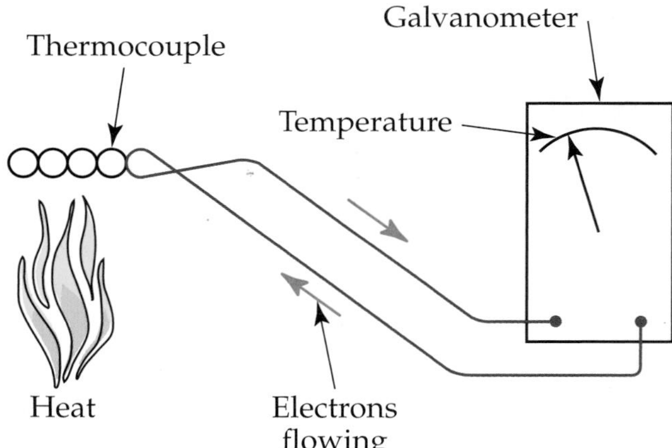

Figure 2-14. A thermocouple generates electricity when heat is applied to its two dissimilar metals. If the ends of the metals are connected to a meter, the meter needles will indicate the temperature of the applied heat.

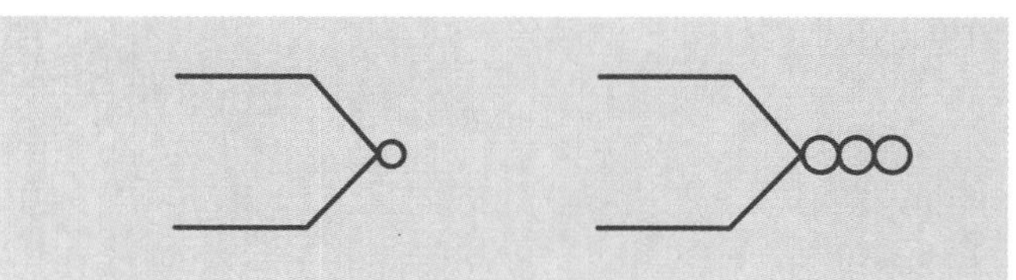

Figure 2-15. Electrical symbols for the thermocouple.

is not allowed to escape from the furnace into the home. This keeps the people who live there safe from the gas being lit by another source outside the furnace causing an explosion.

Thermocouples are used to measure temperature. Each metal is connected to one end of a sensitive meter, such as a galvanometer, to check the output voltage. The hotter the thermocouple becomes, the higher its voltage output. Since it will always have the same output at a given temperature, the size of the output can then be used to find the temperature.

A thermocouple can be used to determine how cold an item is. In **Figure 2-16,** a technician is using a thermometer to measure the temperature after a nitrogen release. The thermometer uses a thermocouple to take a temperature reading. Other uses of the thermocouple include thermometers to measure the temperature of meat, **Figure 2-17,** and to control the temperature inside an oven.

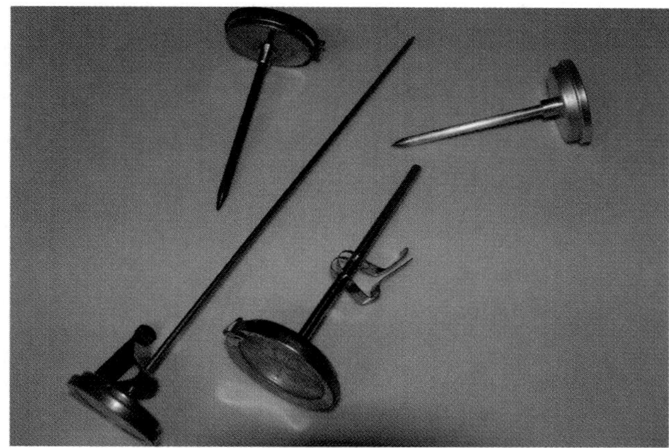

Figure 2-17. Meat thermometers use a thermocouple to measure the temperature of meat.

Electricity from Magnetism

Using magnetism to generate electricity is the most well-known and most frequently used method of all. Electricity from magnetism

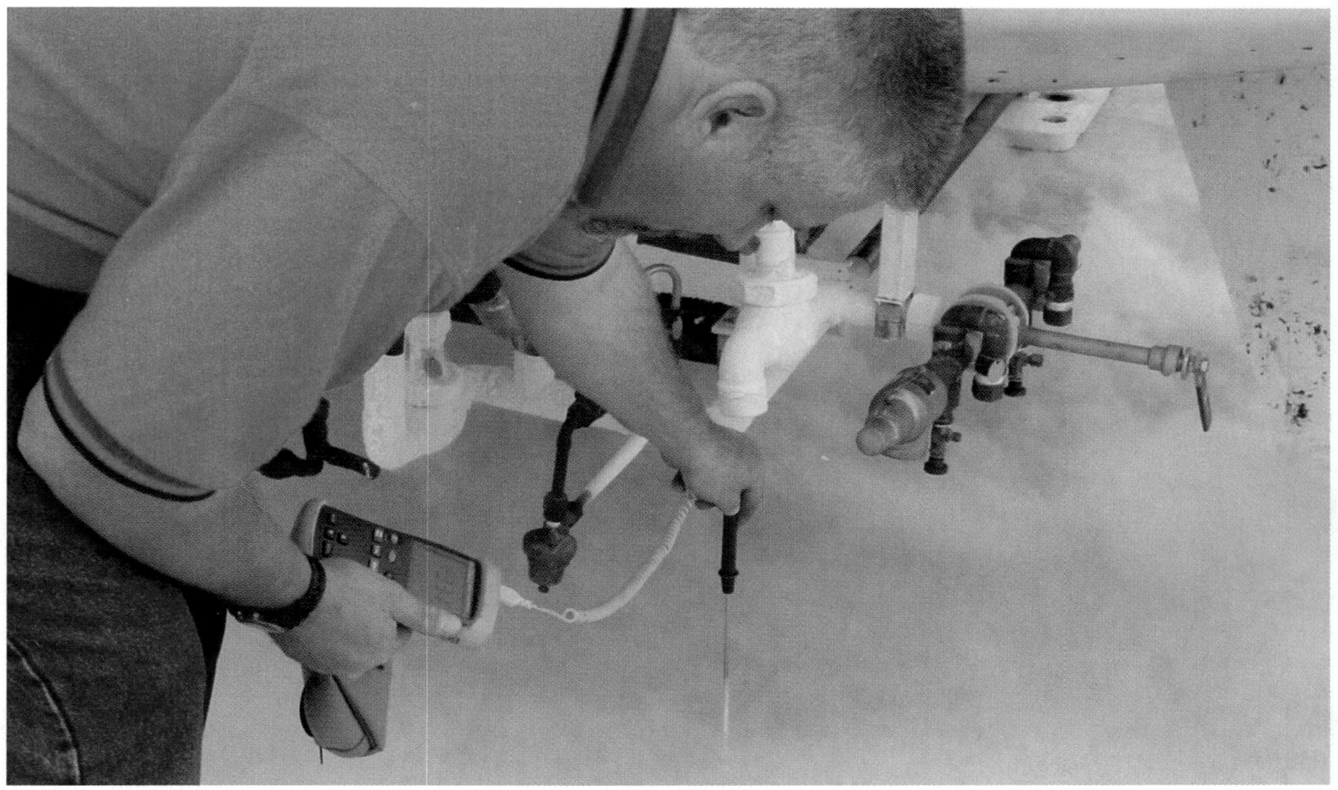

Figure 2-16. A technician is using a meter called a *thermometer* to measure the temperature after a nitrogen release.

is covered in greater detail in Chapter 10. Here we will provide a brief explanation so you will understand how the basic principles of magnetism are applied to generate electricity.

Passing a magnet by a wire coil or passing a coil by a magnet produces a flow of electrons. The power companies that supply the electricity for our communities use this principle. To produce electricity, they use a device called a *generator,* **Figure 2-18.** The armature, which is a conductive part of a generator, spins inside a magnetic field. You probably know some of the common methods used to make the armature turn. Waterfalls or water under pressure from dams and reservoirs, **Figure 2-19;** steam from burning coal or oil; or steam from nuclear reactors are all used to spin generators' armatures. The magnetic field then produces an electric current inside the moving armature. We will go into more detail about generators in Chapter 11. For now, remember that electricity from magnetism is what powers most of your electrical devices.

Figure 2-19. The high water pressure seen in this spray is used to spin the generators. In a hydroelectric power plant, water under pressure from the high side of the dam is applied to turbine wheels that power electric generators. (Bureau of Reclamation)

Figure 2-18. Electricity generated by magnetism is the method employed by huge generators installed in a power station. A—The outside of a generator. B—The inner parts of a generator. (Bureau of Reclamation)

Summary

- Electricity can be generated by a chemical reaction, as it is in cells and batteries.
- Electricity can be generated by using energy from light.
- Electricity can be generated through pressure on certain substances, such as quartz.
- Electricity can be generated through heat using a thermocouple.
- Electricity can be generated with magnetism, which is how most electricity that we use is produced.

Test Your Knowledge

Do not write in this book. Write your answers on a separate sheet of paper.

1. Name five ways to produce electricity.
2. Which of the following is a type of chemical energy used to produce electricity?
   a. Nuclear power plants.
   b. Burning coal to produce heat.
   c. The battery in a car.
   d. Lights in your house.
3. A(n) ______ is used to measure the specific gravity of the acid used in a storage battery.
4. Why would a discharged battery freeze in the winter?
5. One of the materials used to produce electricity by pressure is ______.
6. A thermocouple is made from ______ dissimilar metals.
7. Name two uses for a thermocouple.
8. Passing a magnet by a wire coil or passing a coil by a magnet produces a flow of ______.
9. A device called a(n) ______ uses the principles of magnetism to produce electricity.
10. The ______ of a generator spins inside a magnetic field.

Activities

1. Conduct an Internet search for Material Safety Data Sheets for lithium ion, nickel-cadmium, and nickel-metal hydride batteries. Note the possible hazards, such as inhaling its fumes if it is incinerated and coming into contact with the battery's chemicals. Record this information in a table that compares all three batteries.
2. Look up the specifications of several portable devices, such as digital cameras and PDAs, and note the types of batteries they use.
3. Research the ampere-hours of several types of batteries. Record this information in a table, listing them from highest ampere-hour capacity to lowest.
4. Thermocouples come in various types, such as type K and type J. Research these types and note their application and temperature range.

A wind turbine uses wind to generate electricity. The wind spins the blades of the turbine, which turns the armature of the generator.

Conductors and Insulators

Learning Objectives

After studying this chapter, you will be able to do the following:

- Identify items that are conductors and insulators.
- Describe the process of stripping the insulation from wire.
- Explain how wire size is measured.
- Draw the symbol for a fuse.
- Explain why fuses are used to protect circuits.
- Demonstrate the process of soldering wire and the use of a heat sink.

Technical Terms

arc
American Wire Gauge (AWG)
ampere (A)
circuit breaker
circular mils
cold solder joint
conductor
conduit
double-pole double-throw
double-pole single-throw
flux
fuse
heat sink
heat shrink
insulator
insulating spaghetti
poor solder joint
printed circuit boards
rosin core solder
short circuit
single-pole double-throw
single-pole single-throw
solder
soldering iron
superconductivity
wiring harness

As discussed in Chapter 1, electrons must have a path from the power source to the load. This path usually is in the form of a wire, although it could be any kind of conductor of electricity. A ***conductor*** is any material that easily passes an electric current. If it is difficult to pass an electric current through a material, the material is an ***insulator.*** Insulators are used to cover wire, which keeps the flow of electricity inside the wire.

In some cases, the path for the electrons is provided by printed wiring on specially made ***printed circuit boards,*** or PC boards. Many radios, television sets, and intercoms have circuits of this type, **Figure 3-1.** A PC board's components are connected with thin strips of copper or silver. Both of these metals are good conductors and serve as printed wiring. While PC boards are used extensively in certain applications, they are not practical for many electric circuits.

Figure 3-1. The PC board is connected to a speaker for an intercom system being used in a house. (Home Store, Atlanta)

Wire

Wire used in electric circuits can be made from many different materials. Common conductors are shown in the table in **Figure 3-2.**

Conductor	Approximate Resistance*	Typical Application
Silver	9.80	Electronic circuits
Copper	10.35	Solid and stranded wire
Gold	14.60	Specialized electrical instruments
Aluminum	17.00	Lightweight wire
Nickel	52.00	Rechargeable batteries
Steel	100.00	Telephone lines
Constantan	295.00	Heat-measuring devices
Nichrome®	676.00	Heater elements
Carbon	≥ 2000.00	Automobile wiring

*Resistance is given in ohms per circular mil-foot at 25° on the Celsius scale. These terms will be explained as you progress through this and the next two chapters.

Figure 3-2. Electrical conductors are listed in order of rising resistance. Typical applications are given for each conductor.

A material is considered a good conductor when the electrons can flow easily through it.

Some good conductors of electricity are copper, silver, and aluminum. Most wire is made from copper because it is a good conductor and can be purchased at a reasonable price. Copper wire is widely used in homes and appliances. Silver is actually a better conductor than copper, but it is more expensive and not as readily available. There are times when silver is used to make wire, such as the printed wiring mentioned earlier. Aluminum wire is used for high-voltage lines because it is so much lighter than copper, and sags less when used for this purpose.

Solid Wire and Stranded Wire

Not all wire is the same. Some wire is a solid, single strand (solid core wire), while other wire is made of many fine strands woven together into a braid (stranded wire), **Figure 3-3.** Solid wire does not flex easily and cannot withstand repeated bending. Therefore, it is used in places where the wire does not need to move. Stranded wire, on the other hand, is flexible, so it is used when wire must flex or easily bend.

Wire is used in most electrical circuits. Solid core wire completes the circuit to your lights, washer, dryer, and most circuits in your home. Stranded wire is used in the electric cords of many household appliances, such as lamps, because they are moved frequently. If solid wire were used, the constant flexing would cause the wire to break.

Wire Covering

Wire is made in many different ways. In addition to being either solid or stranded, wire can be coated with varnish, left bare, or covered with insulation or a braided covering.

Coated wire usually is referred to as magnet wire. It can be found in electromagnets, coils for transformers (see Chapter 12), and the antennas of portable radios. Generally, the varnish coating must be removed when this type of wire is to be soldered. However, there is a coated wire available that does not require the protective coating to be stripped before the wire can be soldered.

Insulated wire is used for lamp cords, house wiring, and telephone cables. A cable is an assembly of two or more wires inside a common covering. Cables used in telephone

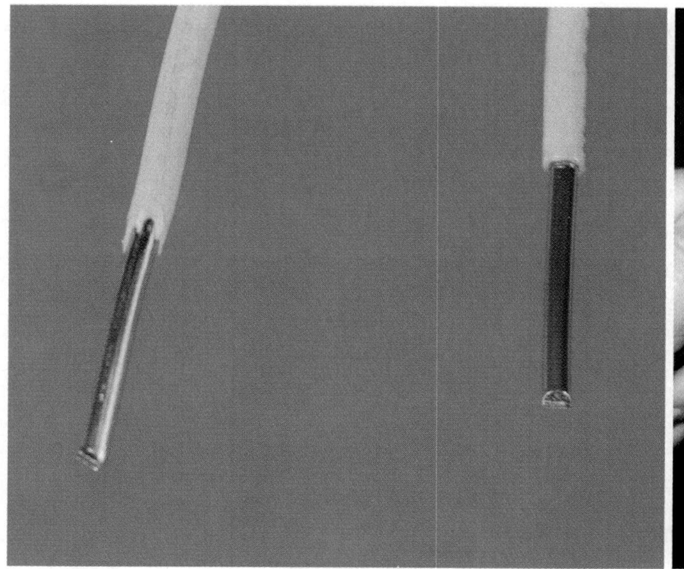

Solid Core

Stranded

Figure 3-3. Solid core and stranded wire.

circuits have color-coded insulation on the individual wires to make connecting or tracing wires easier, **Figure 3-4.**

Since many people are color blind, it is important to remember that certain electrical jobs require working with color-coded wires. Color coding is one reason why some companies insist on giving a color blindness test when they interview people for new job openings. Color-blind persons could have trouble working as telephone or fiber optics installers because some cables have 20 or more color-coded conductors.

Wire with braided covering is used extensively in communications work. Applications include lead-in cables that connect to TV sets, leads for electrical instruments, and transmission line cables. This type of cable is used because the braided covering reduces interference problems in radio and television reception. It is much more expensive than other wire, but it is used because of these special shielding properties. For this reason, wire with braided covering is commonly called shielded cable. A coaxial cable, which is a type of shielded cable and is shown in **Figure 3-5,** has a single conductor, insulator, shield, and an insulating jacket. The shield protects the inner conductor from magnetic fields. In some situations, copper wire has been replaced by fiber optics. See Chapter 21 for more information on fiber optics.

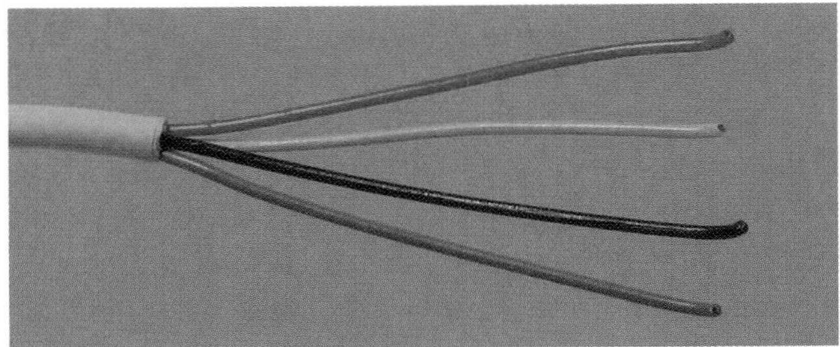

Figure 3-4. Telephone cable contains numerous color-coded wires inside a plastic jacket.

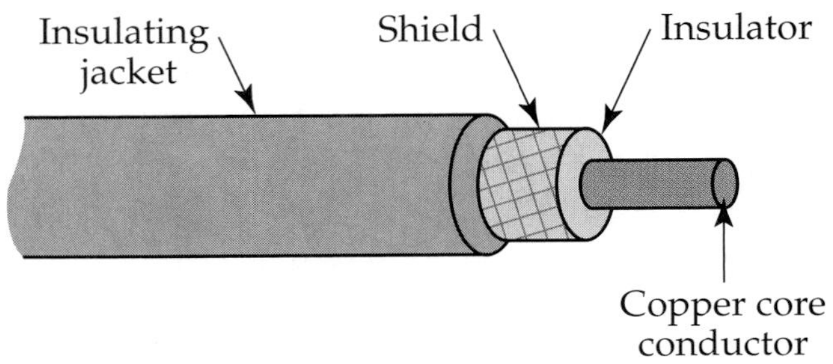

Figure 3-5. Coaxial cable layers.

Stripping Wire

Sometimes two pieces of insulated wire need to be connected in an electric circuit, and it is necessary to remove the insulation before they can be connected. Removing the insulation from wire is called *stripping wire*. It usually is done with wire strippers, **Figure 3-6,** but it can be done with a knife, **Figure 3-7.** If you use a knife, be careful that the wire is not nicked and that strands are not broken during the stripping. When the wire is damaged, the size of the wire is reduced as well as the amount of electricity it can carry.

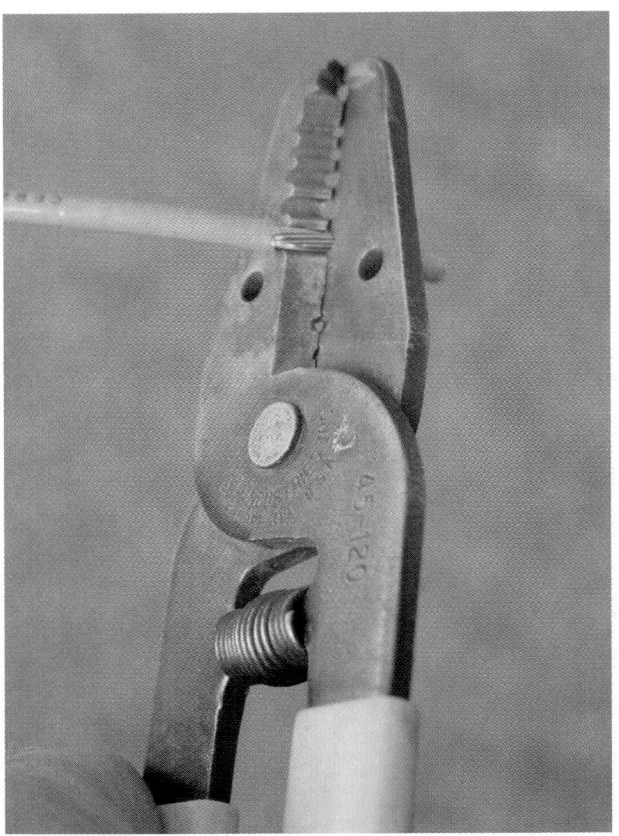

Figure 3-6. Spring-loaded wire stripper opens after each strip.

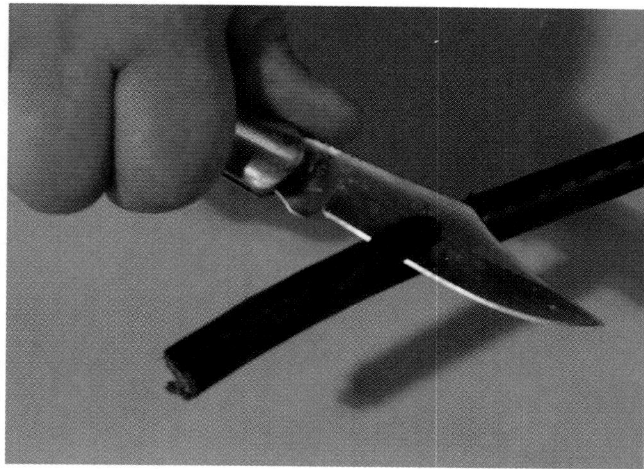

Figure 3-7. When using a knife to strip the insulation from a conductor, you must be careful not to knick the wire or cut off strands of stranded wire.

Figure 3-8 shows by comparing the wire to garden hoses what happens when broken strands reduce the size of a wire. In Figure 3-8A, water hoses of two different sizes are shown. The smaller hose allows less water to pass through it.

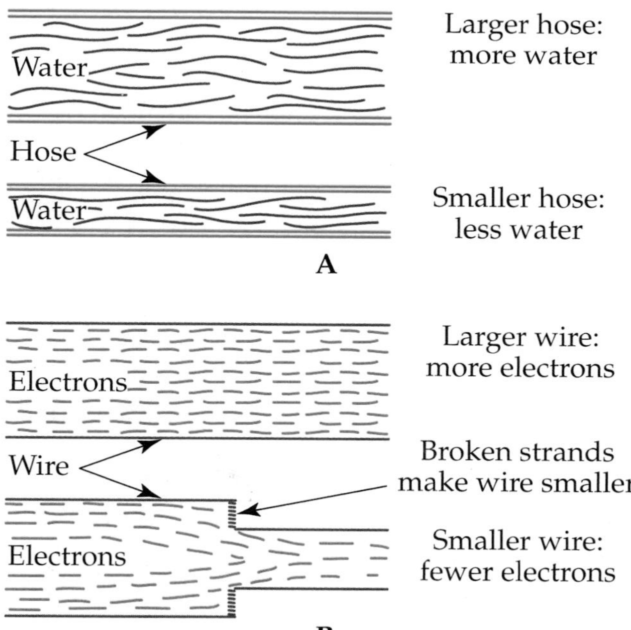

Figure 3-8. The size of the conductor makes a difference in volume of flow, whether the product is water or electricity. A—Large and small hoses. B—Whole and damaged wires.

Electron flow through two wires is shown in Figure 3-8B. One wire is whole and undamaged. The other is smaller because many strands are broken. Note that, like the hose, the smaller wire allows fewer electrons to pass through it.

Wire Size

As you can see from the garden hose example, size impacts the amount of electricity that can flow through a wire. The larger the wire, the more current it can carry. Different circuits require different amounts of current, and the wire used must be able to carry the current required.

Figure 3-9 lists various sizes of wire. Note that each size is given a number. The numbers range from 0000 to 40. The larger numbers refer to wires with small diameters, and the smaller numbers refer to wires with larger diameters. This numbering system is known as the ***American Wire Gauge (AWG).*** The AWG makes it easy to communicate which wire size is needed for a project. It is important to select the proper size of wire for the job. Relying on a small wire to carry too much current might cause a fire due to overheating.

Warning

Never use water on an electrical fire. Use the correct type of fire extinguisher or common baking soda. Also, be sure to disconnect the source of electricity powering the device that caused the fire.

There are other problems that result from the use of undersized wire. Toasters will not brown bread, and electric motors will run too slow. In general, the circuit will not work properly if wire of the proper gauge is not used.

Figure 3-10 shows how a wire gauge is used to determine the size of the wire. The gauge is easy to use. First remove the insulation from the wire. Then, hold the gauge firmly and slide the wire being measured in

Dimensions and Resistances of Copper Wire				
Gauge No. AWG or B&S	Diameter of Bare Wire in Inches	Diameter of Bare Wire in Mils	Area in Circular Mils	~Resistance in Ft./1 Ω at 68°F
0000	0.460	460.00	211,600	20,400
000	0.496	409.60	167,800	16,180
00	0.365	364.80	133,100	12,830
0	0.325	324.90	105,500	10,180
1	0.289	289.30	83,690	8070
10	0.109	101.90	10,380	1001
12	0.081	80.81	6530	630
14	0.064	64.08	4107	396
18	0.040	40.30	1624	157
20	0.036	31.96	1022	99
30	0.010	10.03	101	10
40	0.003	3.15	10	1

Figure 3-9. Copper wire is charted by gauge number, by diameter in inches and in mils, and by area in circular mils. Resistance is shown in feet per ohm at 68°F (20°C). See the Reference Section for a complete table.

the numbered slot where it just fits. Check the size, and then use the opening at the bottom of the slot for easy removal of the wire.

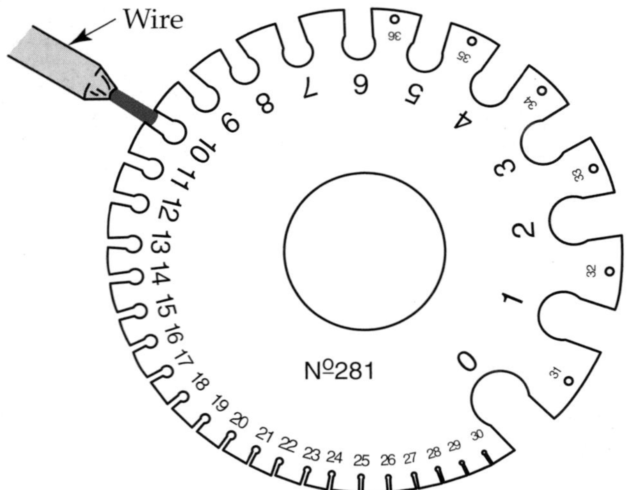

Figure 3-10. To use a wire gauge, insert bare wire in different slots until it fits. Then, note the gauge number.

Project 3-1: Intercom

You can build a simple intercom to use in your home. Use two speakers about 4″ in diameter that have large permanent magnets. Place the speakers in separate rooms and connect them with two wires as shown.

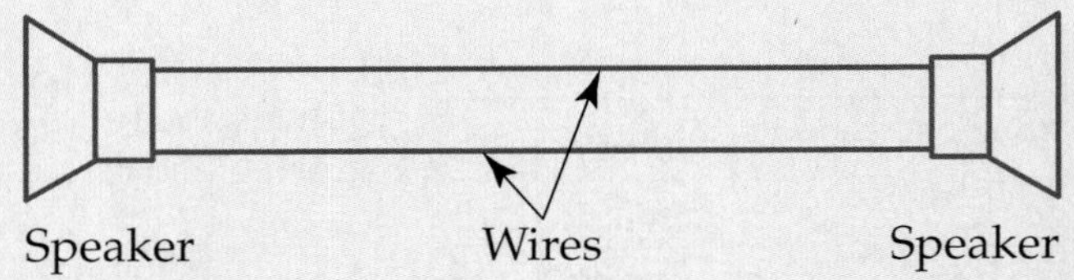

Unfortunately, this intercom will not work in a noisy building, so make sure you are in a relatively quiet place. Have one person stand by each speaker, and take turns talking and listening. Both parties cannot speak at once. The volume from this simple intercom setup may not be very loud, but the results will surprise you.

Amperes

Earlier in this chapter, we stressed the importance of wire size as it relates to the number of electrons that can flow through the wire. Since the number of electrons flowing in a circuit is so large, it is measured in amperes, or amps, instead. A current of one ***ampere (A)*** means that at a given point in a wire, 6,240,000,000,000,000,000 (6.24×10^{18}) electrons are flowing past it every second, **Figure 3-11.** The term *amp* is explained in detail in Chapter 5.

6,240,000,000,000,000,000 electrons
Past this point in 1 second = 1 amp
Wire

Figure 3-11. The term *amperes*, or *amps*, refers to how many electrons are flowing through a circuit.

The expression 6.24×10^{18} is written in scientific notation. It is a way people working with extremely large numbers can shorten the numbers to make them easier to use. The given value for one amp is a standard issued by the National Bureau of Standards and Technology (formally National Bureau of Standards). Remember, the larger the wire, the smaller the AWG number.

Math Focus 3-1: Exponents

An exponent is a number that tells how many times to multiply another number by itself. It is written as a superscript, like this: 2^3. The 3 is called the *exponent*. The 2 is called the *base*. We say this is "two to the third power." It tells you to multiply three 2s together.

$$2^3 = 2 \times 2 \times 2$$
$$= 4 \times 2$$
$$= 8$$

Practice Problems

Solve each of these expressions.

1. 2^2
2. 10^2
3. 4^4
4. 5^3
5. 9^2
6. 2^6
7. 3^3
8. 20^1

Math Focus 3-2: Scientific Notation

Scientific notation is a shorthand way to write very large and very small numbers. Many scientists work with these sorts of numbers every day. For example, the speed of light is about 300,000,000 meters per second. The average diameter of an atom is around 0.00000000025 meters. It takes time and space to write all of the zeros in these numbers, so most scientists put these numbers in scientific notation.

Scientific notation for the speed of light looks like this: 3.0×10^8. The number *3.0* is called the *coefficient*. Scientific notation for the diameter of an atom looks like this: 2.5×10^{-10}. The coefficient is 2.5. To go from scientific notation to a normal number, you can multiply it like any other equation.

$$
\begin{aligned}
3.0 \times 10^8 &= 3.0 \times (10 \times 10 \times 10 \times 10 \times 10 \times 10 \times 10 \times 10) \\
&= 3.0 \times (100{,}000{,}000) \\
&= 300{,}000{,}000
\end{aligned}
$$

Since the exponent is *8*, multiply 3.0 by eight 10s. Note that 300,000,000 has eight zeros. For each ten, you simply move the decimal point in the coefficient one spot to the right, adding zeros as needed. In this case, we moved the decimal point eight spaces to the right:

3.00,000,000.
1 2 3 4 5 6 7 8

If the exponent is negative, move the decimal to the left.

Resistance Wire

If you watch a toaster while it is turned on, you will notice that the inside of it gets so hot that it glows red. If you look closely, you will see that the part that turned red is made from wire. This type of wire, called resistance wire, is designed to heat up. It is used inside lightbulbs, in heating elements for hair dryers, and in electric space heaters.

Because resistance wire is designed to heat up, it is not made from the same material as most wire. Instead, it is made from special metals like nichrome® and constantan. Although most wire is round, resistance wire can also be purchased in a ribbon shape, the type commonly used in toasters.

Circular Mils

Round wire is measured in ***circular mils.*** A mil is one thousandth of an inch (0.001″). A circular mil is the cross-sectional area of a wire that has a diameter of one mil. Note in **Figure 3-12** that as the cross-sectional area gets larger, the number of circular mils increases by the square of the diameter. When you square a number, it means you must multiply the number by itself. For example, 5 squared (also written 5^2) is 5×5, which is equal to 25. Therefore, if you had a wire with a diameter of 5 mils, the cross-sectional area of the wire contains 25 circular mils. A wire with a diameter of 8 mils has the cross-sectional area of 64 circular mils.

Usually a coefficient in scientific notation is between 1 and 10. To convert a number, such as 0.000000284 to scientific notation, place the decimal point after the first non-zero digit and drop the zeros.

~~0.0000002~~.84

Since this number was less than one, the exponent is negative. Count the number of digits the decimal point moved.

~~0.0000002~~.84
1 2 3 4 5 6 7

In this case, the decimal moved seven places. This means the exponent is negative seven.

$0.000000284 = 2.84 \times 10^{-7}$

When the number is greater than one, you can follow the same steps.

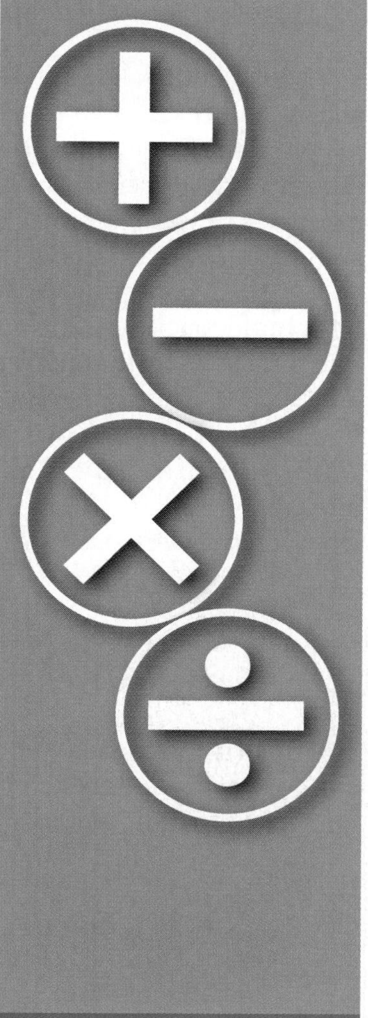

Practice Problems

Convert these normal numbers into scientific notation.

1. 540,000,000
2. 0.0000024
3. 37,000
4. 0.0000623

Convert these numbers in scientific notation to normal numbers.

5. 6.72×10^5
6. 7.5×10^7
7. 2.01×10^{-3}
8. 5.6×10^{-6}

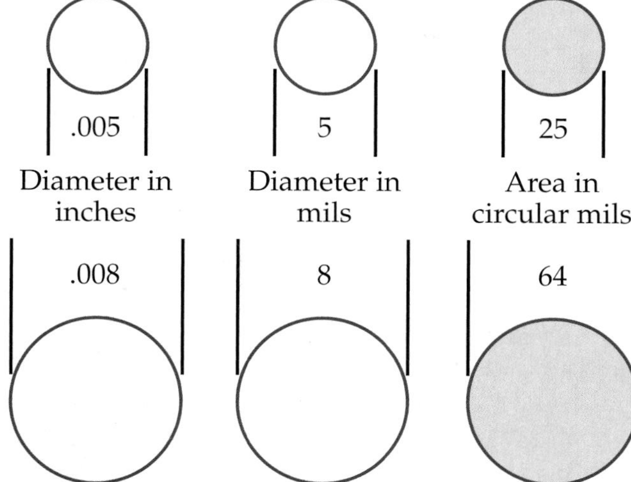

Figure 3-12. A given size of wire is expressed in inches, mils, and circular mils.

There is a good reason for converting the diameter of wire into circular mils. Knowing the cross-sectional area of the wire lets you know how much current it can carry. This makes it possible to select wire of the right size to safely carry the number of amps in the circuit.

A rough rule of thumb states that 700 circular mils of copper wire is rated to carry up to 1 amp. However, according to the National Electric Code, 14 AWG may carry a maximum of 20 amps in free air. Since cables that consist of two or more wires bound together lack air circulation, its current carrying ability is reduced. Therefore, a wire with a rating of 14 AWG as part of a three-conductor cable, can only carry 15 A.

Fuses

Most electric circuits are equipped with a safety device called a ***fuse.*** Fuses are designed to prevent fires caused by overloading the circuit with too many amps. The fuse symbols most commonly found on schematics are shown in **Figure 3-13.**

A fuse contains a thin element through which electricity runs. This element is made of a mixture of metals, including aluminum, and it gets hot when an electrical current passes through it. When too many amps pass through the fuse, the thin aluminum element overheats and melts. This opens the circuit, stopping the electrons from flowing, **Figure 3-14.**

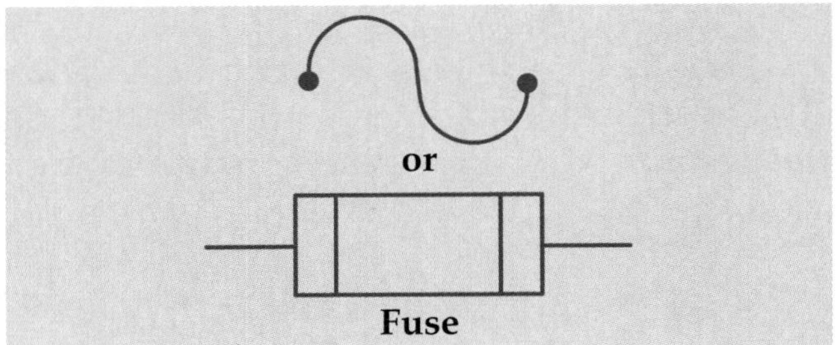

Figure 3-13. Two different wiring diagrams and schematic symbols for fuses are shown.

Each fuse has a maximum number of amps it can carry. Fuses range in size from 1/100 A to heavy-duty units able to carry several hundred amps.

When the element does melt, most people say that the fuse has blown. If this happens,

Project 3-2: Fuse Tester

Build the fuse tester shown here by obtaining the following materials and assembling them according to the construction procedure given.

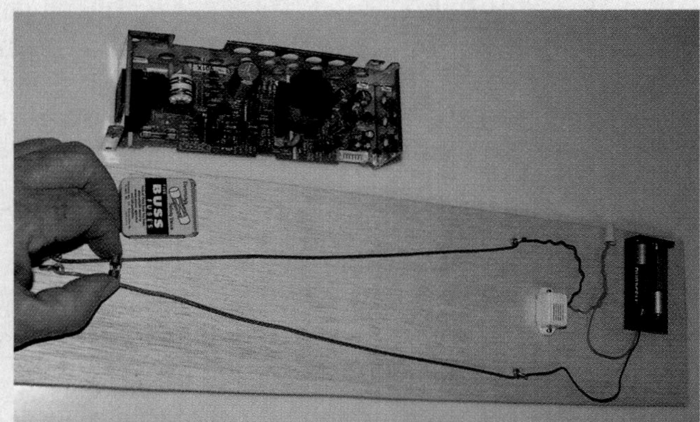

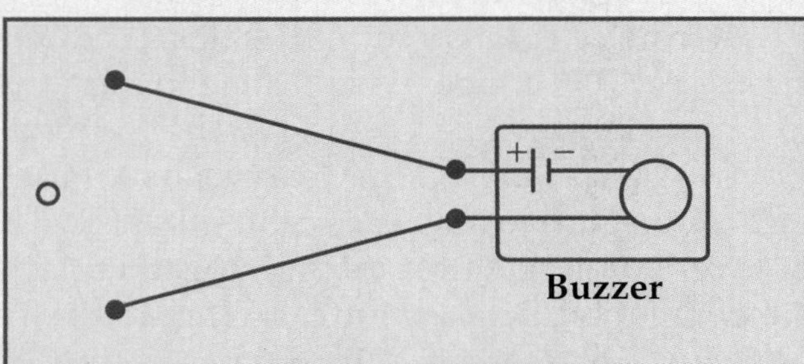

(Project by Allan Witherspoon)

No.	Item
1	1.5 V to 3 V dc alarm buzzer
1	1″ × 3″ × 2″ chassis box
1	Penlight battery (cell)
1	1/4″ × 6″ × 14″ plywood
4	1/2″ screw eyes
1	Battery holder
2	1′ lengths of stranded copper wire (uncoated)

Procedure

1. Cut a hole in the chassis box and mount the buzzer.
2. Install the battery, and then mount the chassis box at the bottom of the plywood.
3. Position the screw eyes on the plywood in a triangular shape.
4. Connect wires to the screw eyes as shown in the diagram.
5. Drill a 1/4″ hole in the plywood for hanging the fuse tester on the wall.

To use the tester, place a fuse between the two wires. Move the fuse inward until it contacts both wires. The buzzer will sound if the fuse is good. If the buzzer fails to sound, the fuse is blown.

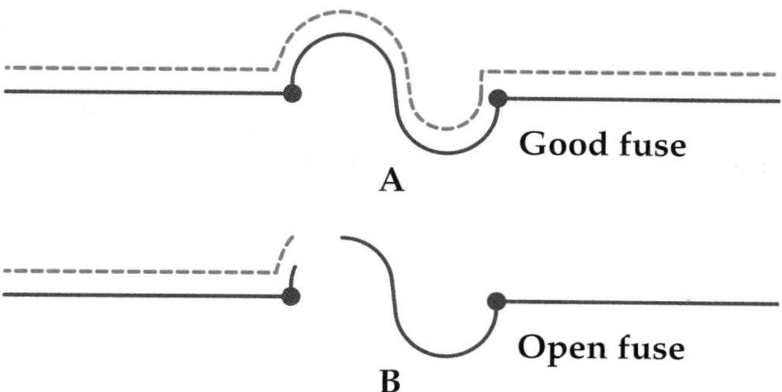

Figure 3-14. The most popular fuse symbol is used to indicate a good fuse. An open fuse is shown with a break in the symbol. A—Electrons can flow. B—Electrons are stopped.

the bad fuse must be replaced with a new one. It is very important to find out why the fuse blew. Typically, a blown fuse means that something is wrong with the circuit or the circuit is overloaded. The problem must be fixed so that replacement fuses do not continue to blow. A blown fuse has done its job of protecting the circuit. It has also prevented overheating wire and stopping a possible fire.

Blown Fuses

Blown fuses are shown in **Figure 3-15.** The blackened section is the result of a ***short circuit,*** which happens when a large number of amps surge through the circuit. When the thin metal strip heats up and melts, it turns into a vapor and discolors the fuse window.

A short circuit is shown in **Figure 3-16.** The load usually offers resistance to the flow

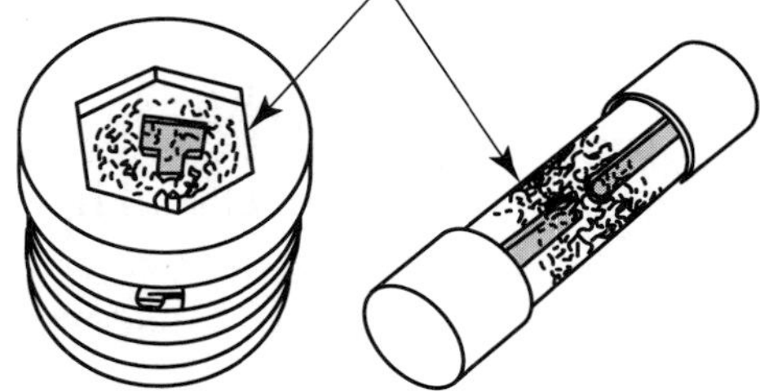

Figure 3-15. When fuses of the snap-in or screw-in type blow, the element vaporizes.

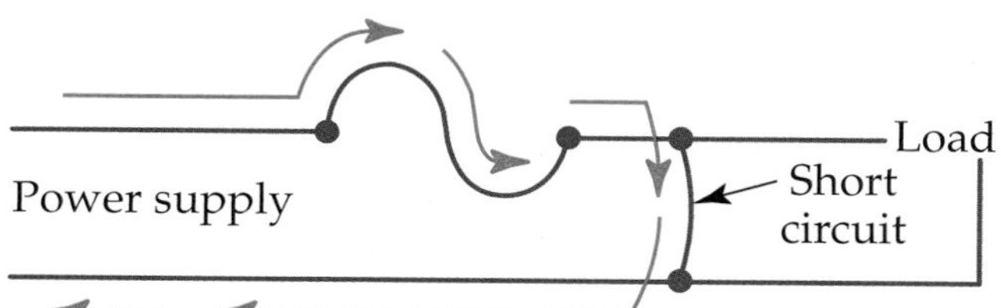

Figure 3-16. A short circuit occurs when two bare wires come into contact and bypass the load, allowing a large number of amps to take a shortcut.

of electrons in a circuit, but when there is a short circuit, the flow of electricity bypasses the load. Because there is nothing to resist the flow of electrons, a short circuit allows an extremely high number of amps to flow through it, which is what the fuse is designed to stop.

Figure 3-17 shows a fuse holder with a fuse in place to protect the circuit from being overloaded. This unit makes it easy to change a blown fuse.

Some circuits normally have a large number of amps flowing for short periods of time. You probably have seen some lights in your home dim when certain appliances, such as the refrigerator, first turn on. If you have an electric saw, you might notice the same dimming effect when the saw is turned on, or if the saw slows down while cutting a large

Figure 3-17. Fuse holders usually are placed for easy access and designed for quick fuse replacement.

piece of wood. At times like this, you do not want the fuse to blow if the overload is only going to last a few seconds. A slow-blowing fuse, **Figure 3-18,** can prevent this. Note that the dual element is heavier at one end. This makes it melt more slowly. Some types of slow-blowing fuses are designed to withstand an overload for 12 to 25 seconds without blowing.

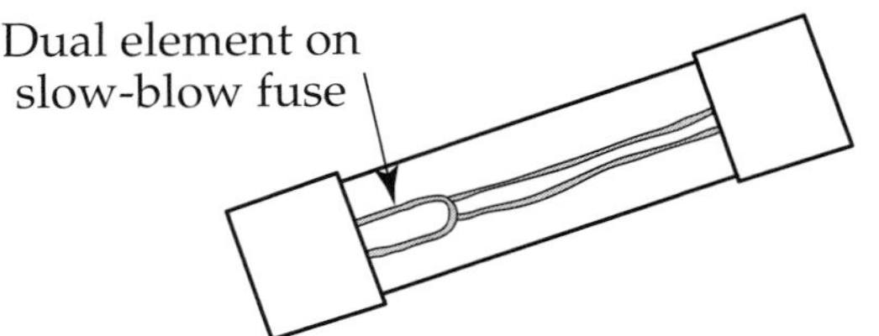

Figure 3-18. Slow-blowing fuses generally have a dual element to help withstand a brief overload.

Circuit Breakers

The electrical circuits in many older homes are protected with fuses. In newer homes, however, the circuits are protected with circuit breakers, **Figure 3-19.** Unlike a fuse, the ***circuit breaker*** does not have to be replaced after an overload. It works like a switch. When an overload occurs, the circuit breaker opens, which is generally called *tripping* the breaker, and the flow of electricity is stopped. If an overload trips a circuit breaker, the circuit breaker needs only to be closed again to complete the circuit.

Figure 3-19. A circuit breaker opens automatically when a current overload flows through it.

Switches

As mentioned in Chapter 1, switches are used to open and close circuits without the inconvenience of disconnecting and reconnecting wires. The symbols for single-pole and the double-pole switches are shown in **Figure 3-20.** The letters SPST stand for ***single-pole single-throw.*** SPDT stands for ***single-pole double-throw,*** DPST for ***double-pole single-throw,*** and DPDT for ***double-pole double-throw.*** Each switch is shown in both the normal position and in the second position after the switch is thrown. Switches are wired into circuits so that one or two circuits are completed by closing the switch. When the circuit is open, no electrons can flow and the switch is said to be "off." When the circuit is closed, electrons can flow and the circuit is said to be "on."

The number of switches available ranges in the hundreds, and many come with extra features. For example, magnet-actuated proximity switches indicate that a piece of magnetic material is close to it. Types of switches include slide switches; rotary switches, **Figure 3-21;** switches with a built-in light so the user knows when they are plugged in to a receptacle, **Figure 3-22;** plunger switches; toggle switches, **Figure 3-23;** push button switches; and switches with built-in safety features, **Figure 3-24.** The cover on the switch in Figure 3-24 can be removed, which locks it in the off position. This protects people from injury because the machine cannot be turned on accidentally.

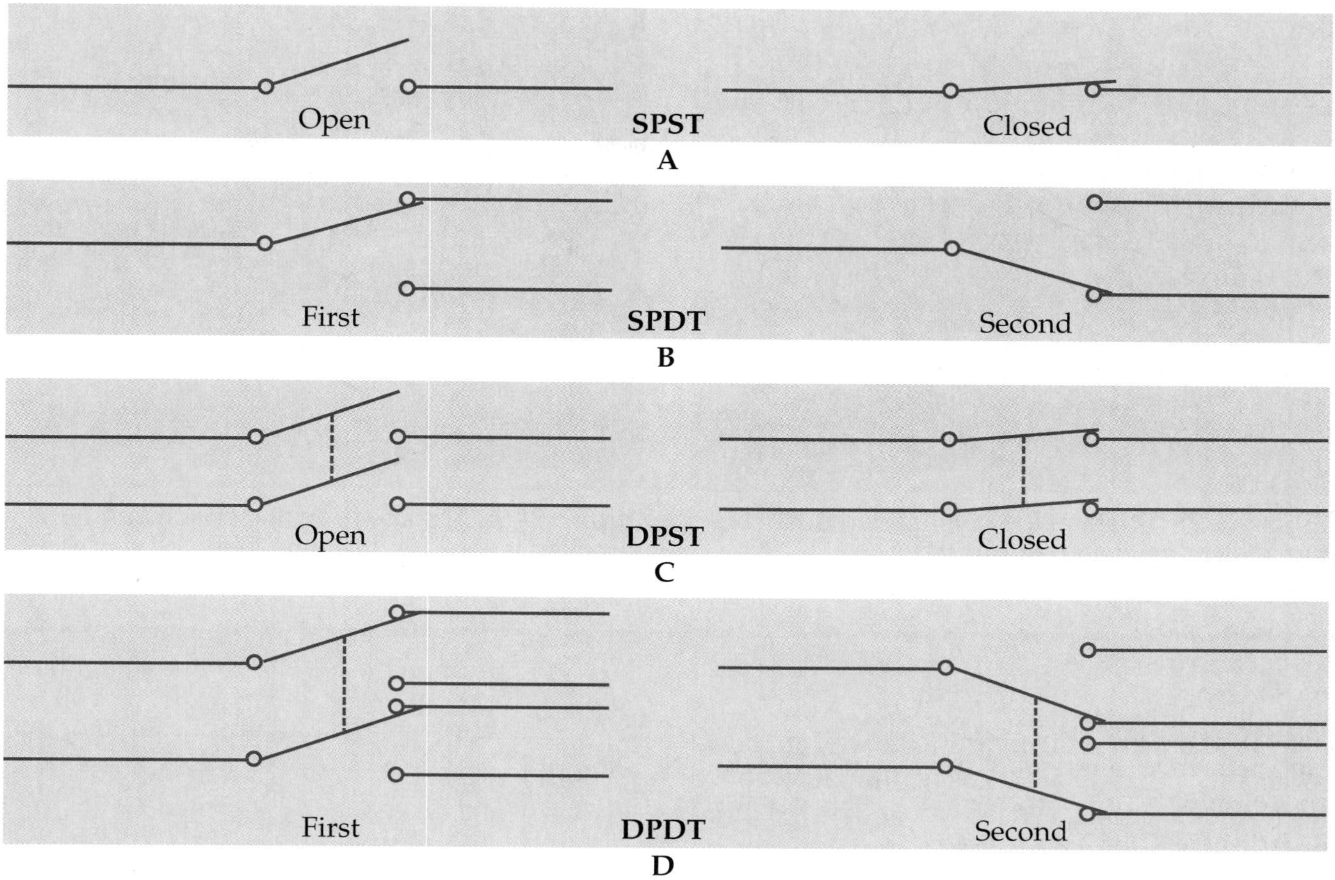

Open SPST Closed

A

First SPDT Second

B

Open DPST Closed

C

First DPDT Second

D

Figure 3-20. Symbols for switches. A—Single-pole single-throw switch in open and closed position. B—Single-pole double-throw switch in first and second positions. C—Double-pole single-throw switch in open and closed positions. D—Double-pole double-throw switch in first and second positions.

Figure 3-21. The rotary switch on this meter allows a technician to change the meter's settings.

Figure 3-22. The toggle switch used on this drill press has a built-in indicator light. The light turns on as soon as the machine is plugged into a receptacle.

Figure 3-23. Pushing the color-coded buttons turns this machine on or off.

Figure 3-24. The cover on this switch can be removed to lock the machine in the off position.

Project 3-3: Auto Alarm

Building an auto alarm is a way to put a switch to good use. Look at the automotive horn circuit shown. When you push the horn button, the horn blows. When you release the pressure, the noise stops. You can use the same sound to scare away a car thief. Do this by wiring a switch into the horn circuit as shown in the diagram.

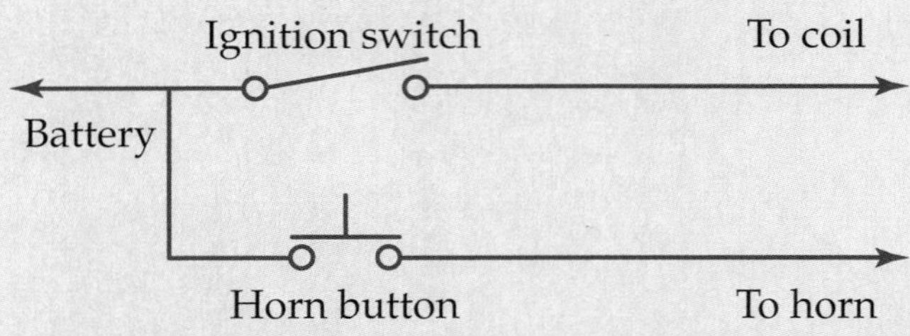

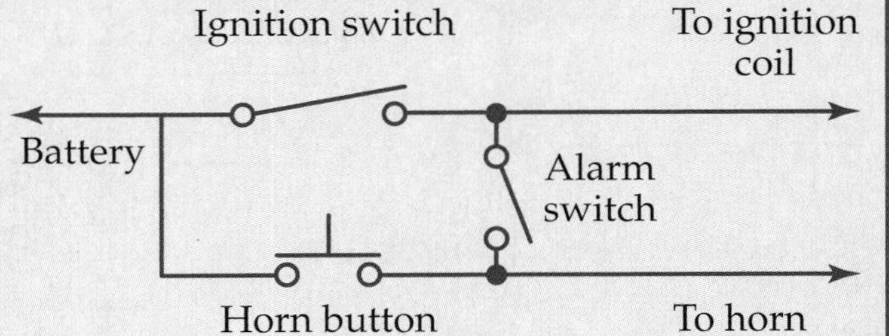

When the switch is installed and the car is in use, keep the car alarm switch in the open position. When you leave your car, close the switch and hide it. If someone starts the engine, the horn will start blowing and frighten away the would-be thief.

When you want to start the engine of your car, turn off the alarm switch. Always keep this in mind or the blasting sound of the horn will surprise you too.

Wire Insulation

As mentioned earlier, insulation is a material used to cover wire. Wires are insulated to keep them from touching each other or other conductors. If they did touch, it would cause a bright spark called an ***arc.*** Anytime there is an arc, there is danger a fire will start or a fuse will blow. Insulation also keeps anyone who touches the wire from getting shocked. Therefore, for protection of the circuit and safety of the user, wire is manufactured with an insulating cover. Also, plugs and connectors are molded from an insulating material. Both types of insulation offer a high resistance to the flow of electrons.

Insulation comes in many types and colors. The best insulators are rubber, shellac, glass, mica, plastics, and paper. In **Figure 3-25,**

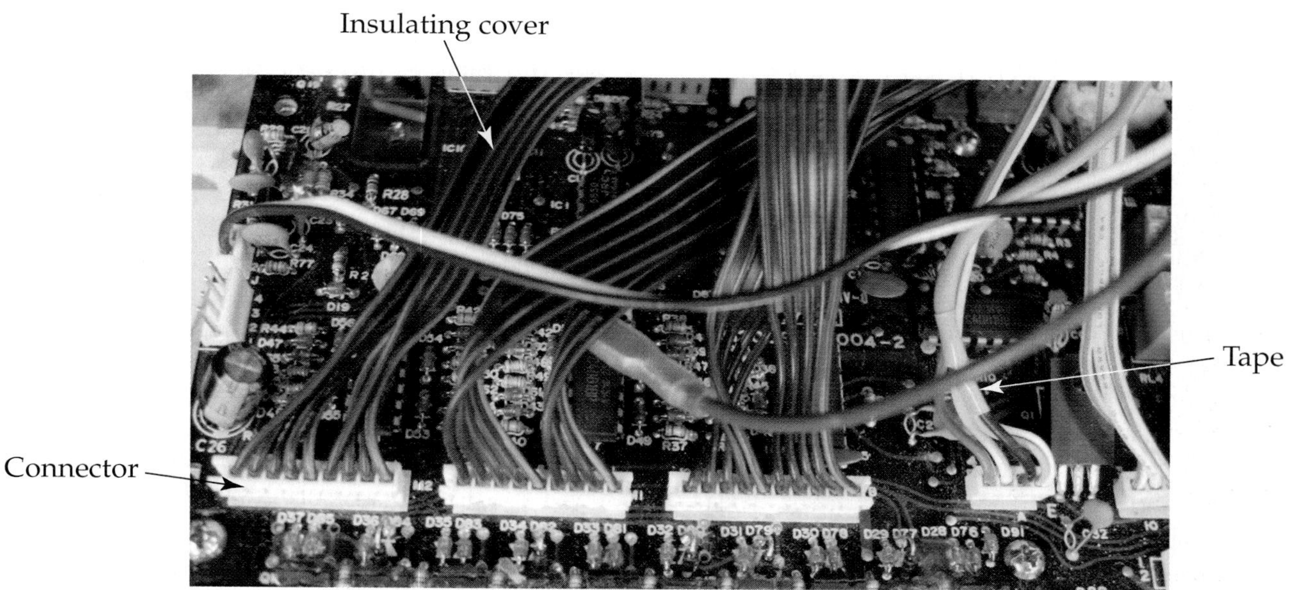

Figure 3-25. The molded plugs and color-coded wiring are insulators designed to stop a short circuit.

the plastic ends (connectors) on the wires of this intercom PC board, insulating cover, and the electrical tape are insulators. When a large number of wires are taped together or encased in protective tubing like in **Figure 3-26,** the assembly is called a ***wiring harness.***

Soldering

Many times, it will be necessary to connect two wires or attach a wire to an electrical part.

Figure 3-26. The wiring harness on this spiral freezer is designed to keep the circuit wiring neat and easy to troubleshoot. (Bravo Foods)

This is usually done by soldering the connection. ***Solder*** is a metal mixture, usually tin and lead, used to permanently connect metal parts or components. The solder is heated with a ***soldering iron,*** **Figure 3-27.** This causes the solder to change into a liquid and flow around

Figure 3-27. This soldering iron kit provides a built-in heat source. Note the heat control knob.

the electrical connection. When the soldering iron is removed, the solder returns to a solid.

A mixture of 60 percent tin and 40 percent lead, for example, is known as 60/40 solder. Most people use this 60/40 mixture because it melts at a lower temperature than other mixtures, such as 70/30 or 40/60. Due to health concerns, lead-free solder is now available. This type of solder is typically made of a mixture of 90% or more of tin and up to 10% of another metal, such as antimony, silver, and copper.

Lead-free solder has a higher melting point than lead-based solder. This means that when using lead-free solder, higher levels of heat are required to make a good solder joint. Extra care is also required to prevent heat damage to the circuit board, electronic components, and insulation materials. When working with lead-free solder, keep the soldering iron in place long enough to complete the process. The longer the soldering iron is in contact with the circuit, the greater the danger of damaging the circuit board, electronic components, and insulation materials.

Warning

Solder will melt at 370°F (188°C) or higher. At these temperatures, you can get a bad burn. Do not let any of the melted solder touch your skin. Also, do not touch the soldering iron or gun used to melt the solder.

Poor solder joints and cold solder joints are a major cause of projects that don't work. A ***poor solder joint*** is any combination of the following characteristics:

- Poor appearance.
- Does not conduct.
- Burned solder joint.
- Too much solder on joint.
- Loose wires at joint.

A solder joint that is dull or has lumps or small beads of solder is known as a ***cold solder joint.*** The following are some common reasons why some solder joints are unsatisfactory:

- **Solder was not heated to the proper temperature.** Too much heat can cause a dull appearance, and too little heat will not melt the solder properly.
- **Solder was applied to the iron instead of to the joint opposite the iron.** If this happens, the solder is melted by the soldering iron and does not stick to the joint. The joint has to be heated to attract the solder.
- **Too much solder was used.** This may look messy. It can also accidentally form connections between parts that are not supposed to be connected if the solder flows into places where it is not meant to be.

When you know how to solder, you will even be able to repair printed wiring with solder. Heating the bad spot and applying a small amount of solder can fix small breaks or cracks.

Types of Solder

Most hardware and building supply stores carry two types of solder: rosin core and acid core. *Use only rosin core solder when doing electrical work.* Acid core solder is used for sheet metal work. If it is used for electrical work, it will eat through connections and damage parts.

Rosin core solder means that the solder has a small opening in the center that is filled with special material called ***flux.*** Flux is a general term for materials used to remove impurities (called oxides) from the surfaces being joined by the solder. The clean surfaces permit the solder to stick to the connection when it reaches the correct temperature.

Heat Sinks

Many sensitive parts of a circuit can be damaged from the heat created by the soldering iron. Therefore, when soldering must be done near these delicate electrical parts, they should be protected with a ***heat sink.*** A heat sink is a better conductor of heat than the electrical part, so it will draw the heat and stop it from being transferred to the nearby part.

When preparing to solder electrical work, attach the heat sink to the circuit between the electrical part to be protected and the joint to be soldered, **Figure 3-28.** Needle nose pliers and other metal objects also can be used as heat sinks.

Protecting the Wire

Sometimes cables and wires have to be protected at a job site. Often, where there is danger of having the insulation worn away, cut, or scraped, the wire is protected by placing it in a hollow pipe called ***conduit.*** In some applications, it is protected by flexible steel conduit, also called *Greenfield*. Both types of protection are shown in **Figure 3-29.** In **Figure 3-30,** an electrician is bringing some heavy wire through a wall. The conduit protects the wire from being cut by any sharp object along the wire's path.

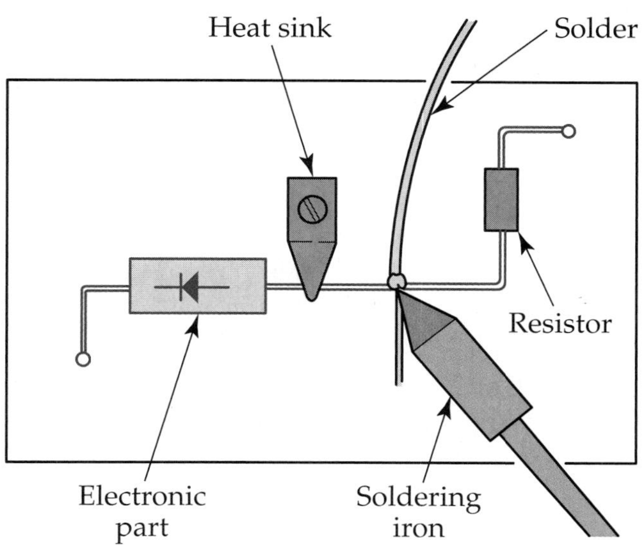

Figure 3-28. The heat sink is a means of attracting and giving off heat to protect delicate electrical parts from heat damage.

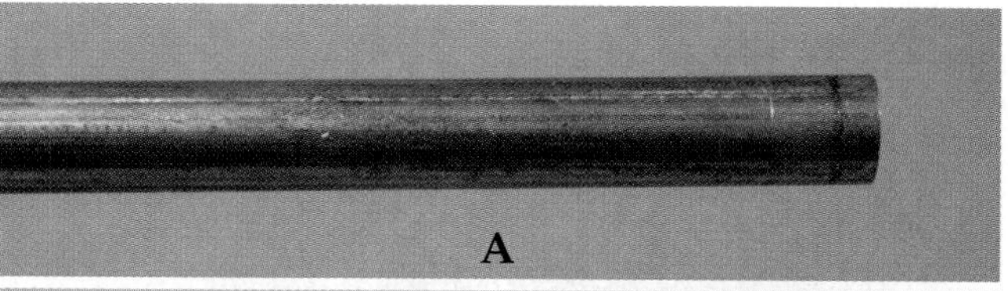

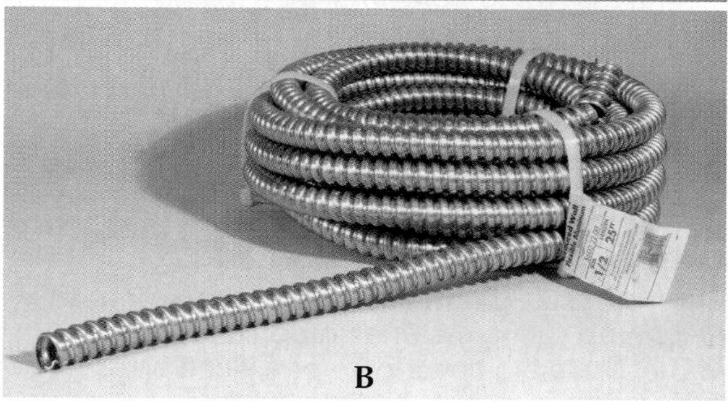

Figure 3-29. Wires between switches, receptacles, and electrical loads must be installed in protective tubing. A—Rigid conduit. B—Flexible steel conduit.

Figure 3-30. An electrician is pulling heavy gauge wire through some flexible steel conduit.

Another covering used to protect wire is called ***insulating spaghetti,*** **Figure 3-31.** It is a hollow insulating material similar to a plastic straw that is useful when splicing two pieces of wire. To use it, remove the insulation from both ends of the wire to be connected.

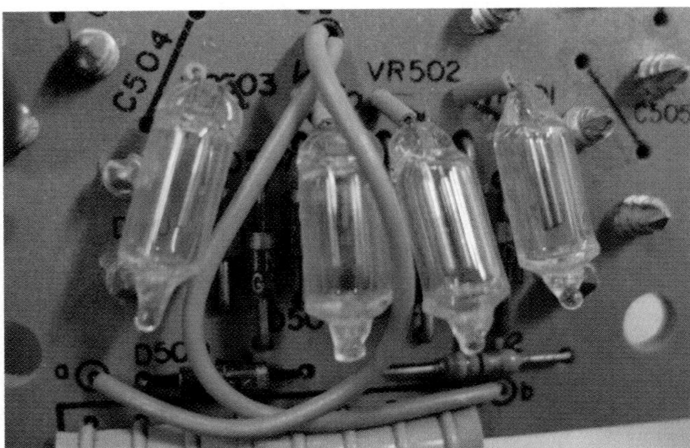

Figure 3-31. *Spaghetti* is a term used for a hollow insulating material (green, blue, and orange) used to cover bare or spliced wires. It protects components so their leads do not short out by touching one another.

Figure 3-32. Like spaghetti, heat shrink is a protective covering for wires. Heat shrink differs from spaghetti in that by applying heat to its surface, it shrinks to about one-half its size to fit more tightly around the wire.

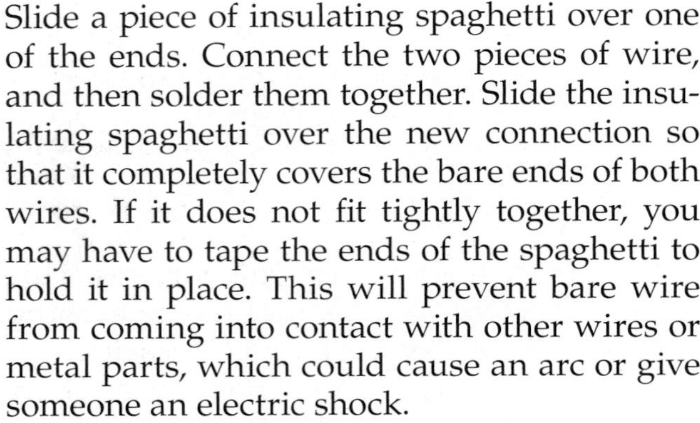

Slide a piece of insulating spaghetti over one of the ends. Connect the two pieces of wire, and then solder them together. Slide the insulating spaghetti over the new connection so that it completely covers the bare ends of both wires. If it does not fit tightly together, you may have to tape the ends of the spaghetti to hold it in place. This will prevent bare wire from coming into contact with other wires or metal parts, which could cause an arc or give someone an electric shock.

Another material, called ***heat shrink,*** also can be used to protect wires. It looks much like spaghetti. With heat shrink, however, heat is applied to it once it is covering the wires, **Figure 3-32.** The heat causes the material to shrink to about one-half its size. This makes the covering fit more tightly on the wire. Heat shrink is used for waterproofing, wire marking, color coding, and as an electrical insulation for the wire. The heat shrink material comes in a wide variety of colors and sizes, **Figure 3-33.**

Figure 3-33. Heat shrink comes in various sizes and colors.

Electrical Codes

Guidelines for properly routing wires and for other safe practices vary from city to city, depending on local electrical codes. These codes generally require that electrical work be completed in compliance with the National Electric Code (NEC). The NEC contains the general rules and regulations for electrical contractors, builders, and manufacturers. The codes have been established to ensure safe electrical installation. City and state building inspectors visit new buildings under construction to make sure that codes have been followed.

Practical Application 3-1: Soldering

The only way to learn to solder is to do it. Read through the following steps, and then practice, practice, practice!

1. If soldering wire, strip the insulation or coating from the ends of the wire.
2. Make a strong mechanical connection between the pieces to be soldered to ensure good electrical flow after solder has been placed on the joint. Twisting wires together tightly or bending the end of a lead against a circuit board can make a good connection. If the joint is not mechanically tight, you will get poor electrical flow because solder does not conduct electricity as well as copper.
3. Use a soldering iron to heat the mechanical connection on one side while applying the solder on the other side. Do not put the solder directly on the soldering iron. A poor solder joint may result. By heating one side of the connection, it will get hot enough to melt the solder on the other side. Cover all the surfaces and wires at the connection with solder, but do not use any more than is necessary.
4. Remove the iron when enough solder has melted on the connection. Do not move the parts being soldered until the solder has cooled. If you accidentally move the parts while the connection is cooling, apply the iron and reheat the joint. This technique ensures a good solder connection.

The finished soldering joint should be completely covered by the flowing solder and should be bright and smooth with all wires firmly held in place.

There are a number of books written to explain and expand on the NEC. These books cover topics such as residential wiring, electrical raceways, electrical grounding, and electrical bonding. These and many other books will help you to learn and understand the rules dealing with electricity.

Superconductors

Superconductivity is the point at which a conductor loses its resistance to electricity. Scientists are continually experimenting with new methods of producing superconductive materials. One way to reach the point of superconductivity is with very cold temperatures. In the past, the point of superconductivity was thought to be around –452° Kelvin (–725°C). However, scientists have now discovered that some special materials can be manufactured to have little or no resistance to the flow of electrons at even warmer temperatures. This type of research will result in some exciting new possibilities for the use of electricity.

For example, superconductivity may finally allow electricity to be stored by methods other than batteries. Electricity could be sent around a loop of superconductive material until it is ready to use. Generating stations could be made smaller because their output could be stored for later use. Generators designed for peak demands would no longer be needed; a reserve supply of electricity could be built up by running a smaller generator overnight.

Superconductivity may also affect the computer industry. With no resistance to electron flow, computer circuits could operate at blinding speeds. This could result in supercomputers small enough to fit on a desktop.

Another possibility lies in the distribution of electricity. If electric lines have no resistance, electricity can be sent great distances without loss of energy. This means hydroelectric plants located in remote mountain areas could ship electricity to distant urban areas. Electricity produced in desert areas using photoelectric cells could then be carried where needed, all without energy loss.

Summary

- A conductor allows electricity to flow through it easily.
- An insulator is any material through which it is difficult to pass electricity.
- The insulation on a wire can be removed using wire strippers or a knife.
- Wire size is measured according to the American Wire Gauge (AWG).
- There are many different types of wire, but the most common is copper wire.
- Resistance wire is designed to heat up and is used in toasters, hair dryers, and space heaters.
- A fuse is a device put in a circuit to prevent a surge of electricity from destroying the circuit's components or starting a fire.
- Circuit breakers serve the same purpose as fuses, but they do not need to be replaced each time there is an overload.
- Soldering is a way to permanently connect a joint in a circuit.
- Heat sinks are used during soldering to protect the delicate electrical components in the circuit from overheating.
- Codes for properly routing wires and for other safe practices generally require that electrical work be completed in compliance with the National Electric Code (NEC).
- A superconductor is a conductor that offers little or no resistance to electricity.

Test Your Knowledge

Do not write in this book. Write your answers on a separate sheet of paper.

1. Three good materials for conducting electricity are ______, ______, and ______.
2. Name three commonly used insulator materials.
3. Name two tools used to remove insulation from wires.
4. Which size wire is larger, No. 12 or No. 16?
5. Describe how a wire should fit into the slot on the American Wire Gauge to determine its size.
6. What kind of wire is used in heating elements?
7. How is a circuit breaker reset after it has opened the circuit?
8. Draw the schematic for a circuit to show how a single-pole, single-throw switch can be used to control a 6-V lightbulb connected to a storage battery.
9. The purpose of ______ in electrical circuits is to attach two wires or metal pieces together.
10. *True or False?* Rosin-core solder should be used for all electrical work.
11. Acid core solder should not be used for ______.
12. What two precautions should you observe when soldering?

Activities

1. Why are the ends of stranded wire twisted together and tinned before it is used?
2. List the different mixtures of solder and find the temperature where each will melt.

3. Many companies use a method of soldering which does not use soldering irons or guns. Find out what they call this type of soldering and how it is done.
4. There are many different types of insulation used on wire. Make a list of different kinds and their colors.
5. How is insulation put on wires?
6. How do companies make the solder with a hollow center? How is the rosin flux put into the core of the solder?
7. List the different types of fuses you can find and determine how much current they will carry without blowing.
8. How is stranded wire made?
9. Find out what types of tests are given to see if a person is color blind.
10. What is flux and how is it made?
11. Where does copper come from? Make a map of the different places where it is found.

Close-up of circuit board conductors.

Resistors and Capacitors

Learning Objectives

After studying this chapter, you will be able to do the following:

- Explain how a resistor works and what it does.
- Draw the symbols for resistors and capacitors.
- Describe some of the uses for thermistors.
- Compute the value of a resistor by its color code.
- Contrast the different types of capacitors.
- Explain the difference between a farad, microfarad, and picofarad.
- Demonstrate the proper way to handle and discharge capacitors.

Technical Terms

capacitance
capacitor
carbon resistor
coulomb
dielectric
electrolyte
electrolytic capacitor
farad (F)
fixed resistor
microfarad (μF)
negative temperature coefficient (NTC) thermistor
ohm
picofarad (pF)
polarized
positive temperature coefficient (PTC) thermistor
potentiometer
resistor
rheostat
thermistor
thin film resistor
trimmer resistor
variable resistor
wirewound resistor

In Chapter 3, you studied electrical conductors and insulators. You found that good conductors are made from materials that allow electrons to flow easily and that insulators, on the other hand, are poor conductors of electricity. There are electrical components that fall somewhere between these two extremes. Two such components are resistors and capacitors.

Resistors

Between conductors and insulators, there is another material that conducts electricity but makes it difficult for electrons to flow. This material is used to make the electrical component called the ***resistor.*** One familiar type of resistor is resistance wire. A resistance wire offers so much resistance that it turns bright red and gives off enough heat to toast bread, heat rooms, and dry hair.

You might recall from Chapter 3 that carbon is a poor conductor. For this reason, carbon is often used to make resistors. The material, size, and makeup of the resistor determine its resistance value, which is measured in ohms. The symbol for the ***ohm*** is the Greek letter omega (Ω). Resistors can be purchased in hundreds of different values, ranging from about one ohm to over twenty-two million ohms.

On schematics and circuit boards, resistors are often identified by the letter *R* and a number. This letter helps identify the component. In **Figure 4-1,** for example, note resistor R27. The number *27* refers to a parts list, which gives the actual value of the resistor.

The symbols for a resistor (or resistance) are shown in **Figure 4-2.** Do not confuse the second symbol in the illustration with the symbol for a fuse. Unlike the fuse symbol, there are no lines in the box in the resistor symbol.

Resistors are of two types: fixed and variable. A ***fixed resistor*** has one specific value, such as 150 Ω. A ***variable resistor*** has a range of values. Some can be preadjusted to the resistance needed for a specific circuit.

Figure 4-1. Several resistors are used in this electrical circuit.

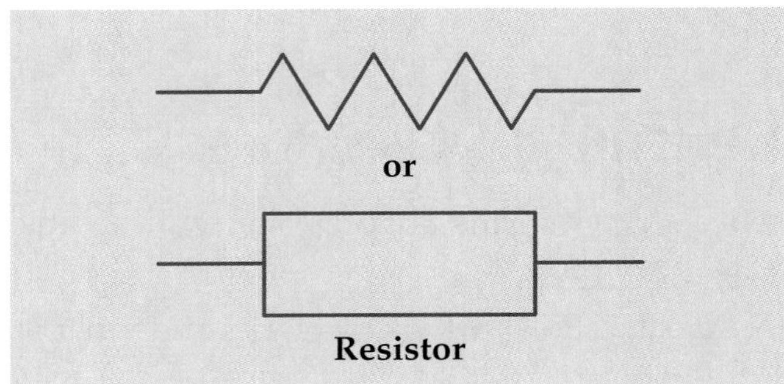

Figure 4-2. These two symbols are the most popular resistor symbols used in schematics.

Fixed Resistors

Fixed resistors vary in makeup. They may be molded composition, thin film, or wire-wound. The main differences among resistors are the heat they create and the heat they can safely dissipate (give off). When specifying a resistor for a circuit, the designer must consider the amount of heat and humidity in the area where the resistors will be used.

Molded composition resistors

A molded composition resistor is often referred to as a *carbon resistor.* A ***carbon resistor*** is made of a carbon material molded together with a binder, **Figure 4-3.** The wire leads at each end of the resistor do not go all the way through. They stop soon after entering the

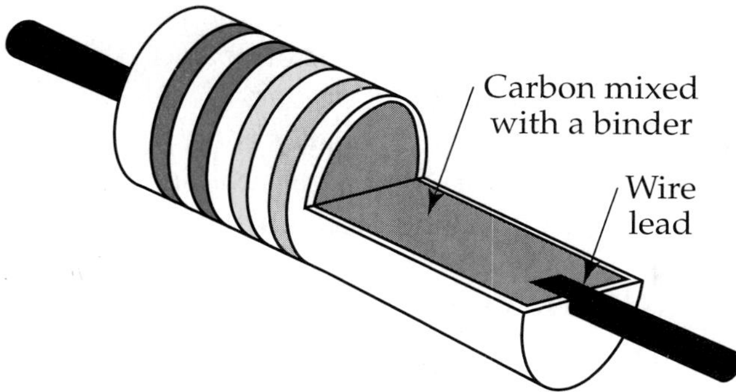

Figure 4-3. A cutaway view of a carbon resistor.

body of the resistor so that the current must flow through the carbon. The different numerical value of a resistor depends on the amount of carbon and binder used in the resistor's construction.

Carbon resistors come in many different sizes. The most commonly used carbon resistors have a diameter between 3/32″ and 5/16″.

The different sizes of carbon resistors, **Figure 4-4,** also tell the amount of heat that a given resistor can dissipate into the air. Carbon resistors are rated at 1/8 watt, 1/4 watt, 1/2 watt, 1 watt, and 2 watts. The term *watts* will be described in detail in Chapter 5. For now, just think of this term as the amount of heat the resistor can withstand without being destroyed.

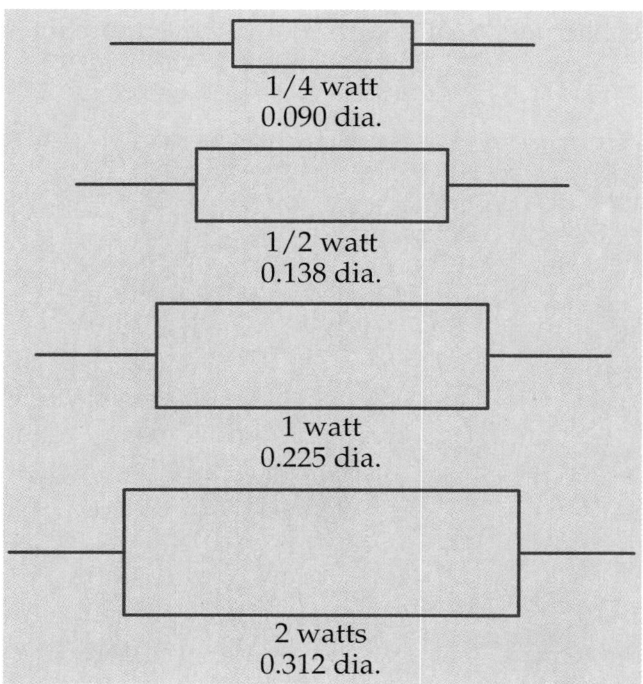

Figure 4-4. Resistors of four of the most common sizes are shown by wattage and diameter.

Thin film resistors

A ***thin film resistor,*** or ***chip resistor,*** can be considered a choice between carbon resistors and wirewound resistors. The primary differences are in size and cost. **Figure 4-5** shows how thin film resistors are constructed. The substrate (base material) shown is an insulating material similar to ceramic.

The materials used to make the thin film include nichrome and tantalum nitride. Thin film resistors come in values ranging in size from 4.7 Ω to 10 MΩ (megohms). Some of their features specify low noise. This is important in applications such as hearing aids and wireless devices used by hospitals for 24-hour patient monitoring.

Wirewound resistors

A ***wirewound resistor,*** **Figure 4-6,** is constructed by winding a long length of wire around a core. The length and diameter of the

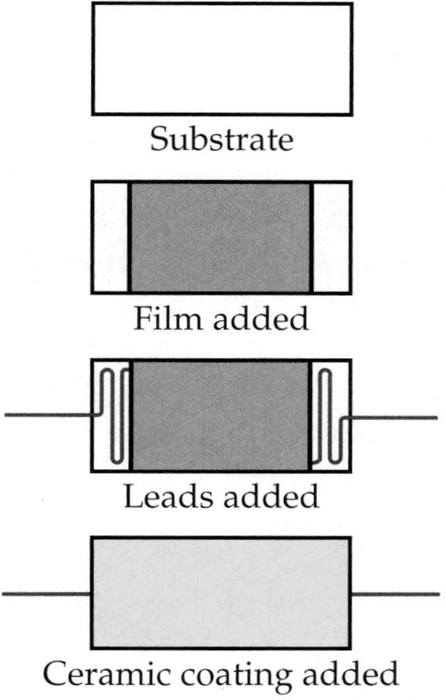

Figure 4-5. Four steps in the construction of thin film resistors.

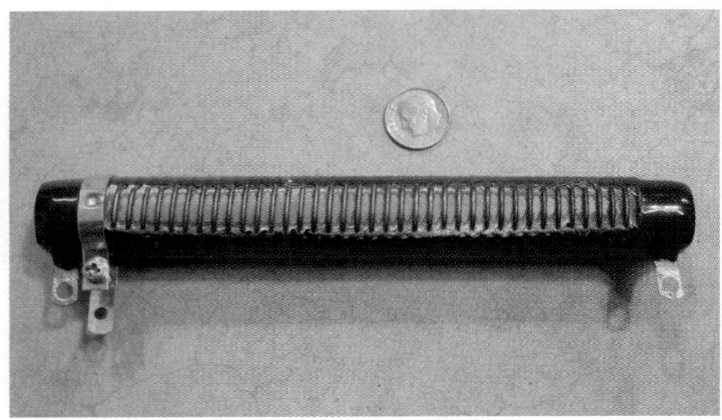

Figure 4-6. This wirewound resistor is made of a core wound with a continuous piece of wire and coated with ceramic insulation.

wire and the material from which it is made determine how much resistance it offers.

After the wire is wound around the core, an insulating ceramic coating is applied to the body of the resistor. This coating protects the wires when the resistor is used. Each end of the winding is attached to a terminal used to install the resistor in the circuit.

Wirewound resistors are used in circuits where large amounts of heat must be dissipated. As a result, they usually are much larger than carbon resistors. Some wirewound resistors are over 1″ in diameter and 10″ long. Power ratings range from about 3 watts to over 200 watts—much higher than the ratings of carbon resistors. For this reason, they are much more expensive than carbon resistors.

Variable Resistors

Variable resistors are used in applications where it is necessary to frequently adjust resistance values. Many people use variable resistors every day. Some common uses include adjusting the volume of a stereo or television set, changing the speed of a vehicle's windshield wipers, dimming the lights in a house, and changing the speed setting on a machine. Adjusting a variable resistor can change speed or volume because the amount of its resistance in the circuit is changed.

There are three different kinds of variable resistors: potentiometers, rheostats, and trimmer resistors. A ***potentiometer*** is also called a *pot*. The resistance value of a pot is changed by moving the wiper along the resistive element. **Figure 4-7** shows the inner construction of a pot. Notice that the center

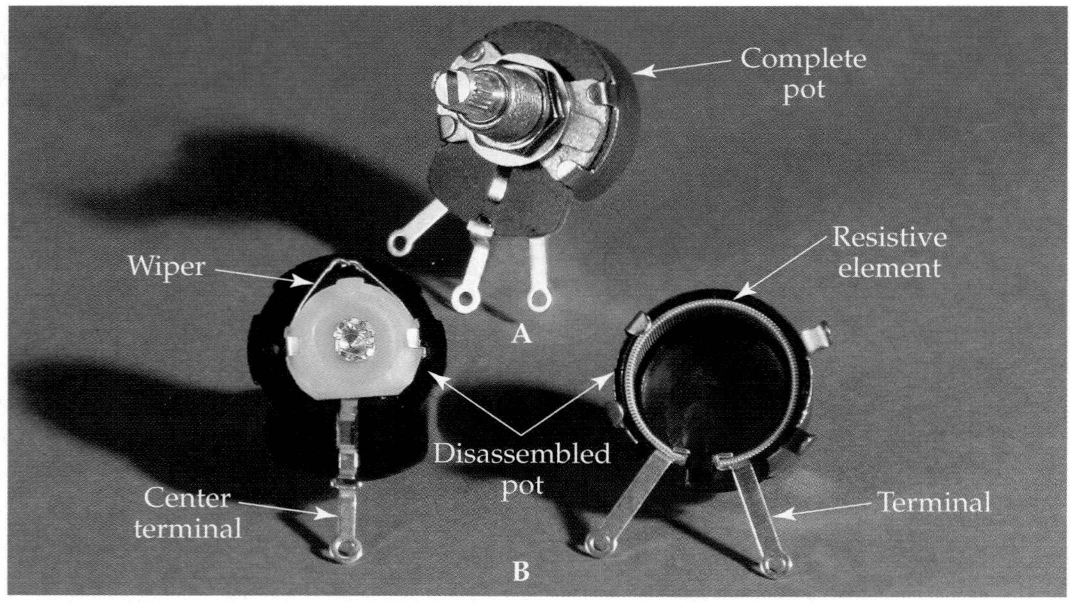

Figure 4-7. A potentiometer is used to vary the resistance of a circuit. A–Complete potentiometer. B–Disassembled potentiometer.

terminal is connected to the wiper and the outside terminals are connected to the resistive element.

Because of their construction, pots allow electrons to have two paths. One path goes through part of the resistance and then to the load. The other path goes only through the resistance, **Figure 4-8.** When there is more than one path for electrons, you have a parallel circuit. This will be explained more fully when you get to Chapter 7. At this point, you should know that some electrons travel each path that is present in a circuit. Much like water flowing through multiple branches of a stream, **Figure 4-9,** the number of electrons on a given path depends on the size of that path and how much resistance it offers.

Pots are added to circuits to make small changes in resistance. They have power ratings from 1 watt to 5 watts. There are some power potentiometers that can handle a lot more wattage, but the majority of what you will see will be the smaller sizes. Some applications of pots include power supply voltage adjustment and setting the operating point for temperature sensors.

When higher current adjustments are required, a ***rheostat*** is used. Rheostats usually have heavy wire windings for the resistive element. There are only two terminals on a

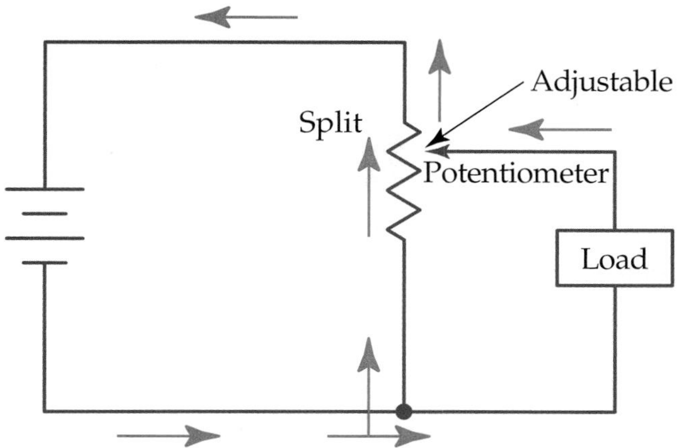

Figure 4-8. The arrows in this potentiometer circuit indicate the path of electron movement. Notice that there are two paths in this circuit.

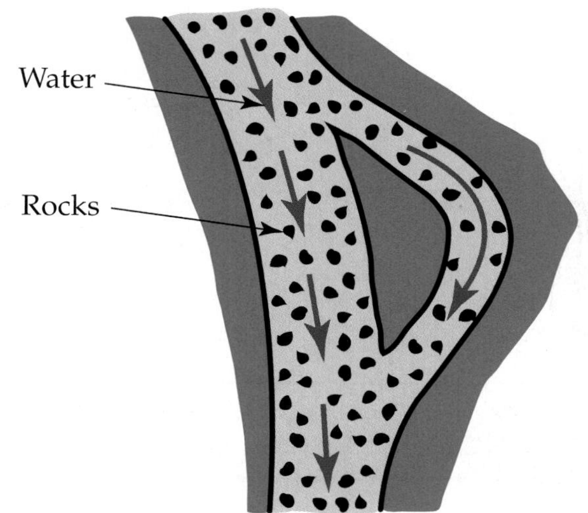

Figure 4-9. The principle of splitting the path of electron flow is illustrated by water flowing in the mainstream and in a branch. Rocks show resistance to flow.

rheostat. As you turn the rheostat knob, you increase the amount of resistance it controls. In this way, it acts as large variable resistors.

The rheostat adds resistance to a circuit. Its purpose is simply to further restrict the path of the electrons. Unlike the potentiometer, the rheostat allows only one path for the electrons, **Figure 4-10.** The farther the electrons must travel through the rheostat's resistance before getting to the load, the more resistance the electrons must overcome. This means that the more resistance you want in the circuit, the farther you must turn the wiper. There are times when rheostat-type control is needed with some small circuits. In many instances, a potentiometer is used by attaching wires to only two of its terminals.

A ***trimmer resistor,*** **Figure 4-11,** is a variable resistor that is similar to a potentiometer in construction. It is used to make very small changes in the resistance of a circuit. Fine-tuning certain circuits in television sets, stereos, and other communication devices are typical applications of trimmer resistors.

Some wirewound resistors are adjustable. In this type of resistor, the windings are not

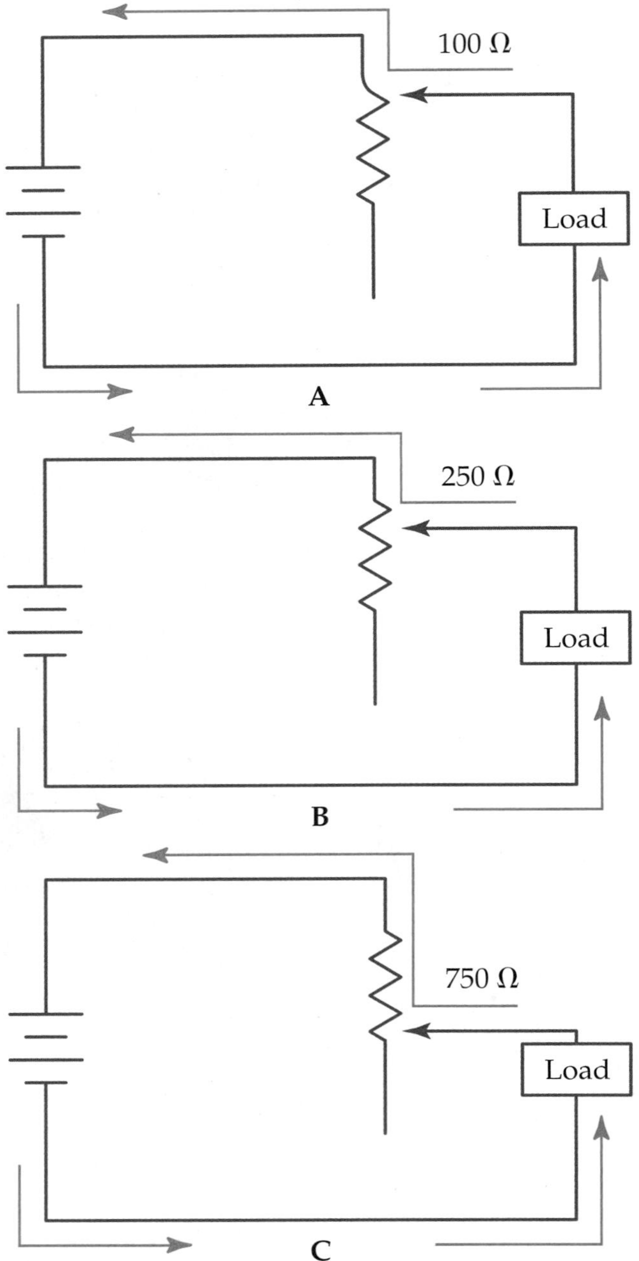

Figure 4-10. Note that there is only one path for electrons in this rheostat circuit. The rheostat adds resistance to the circuit. A—Easy flow of electrons. B—More difficult flow of electrons. C—Very difficult flow of electrons.

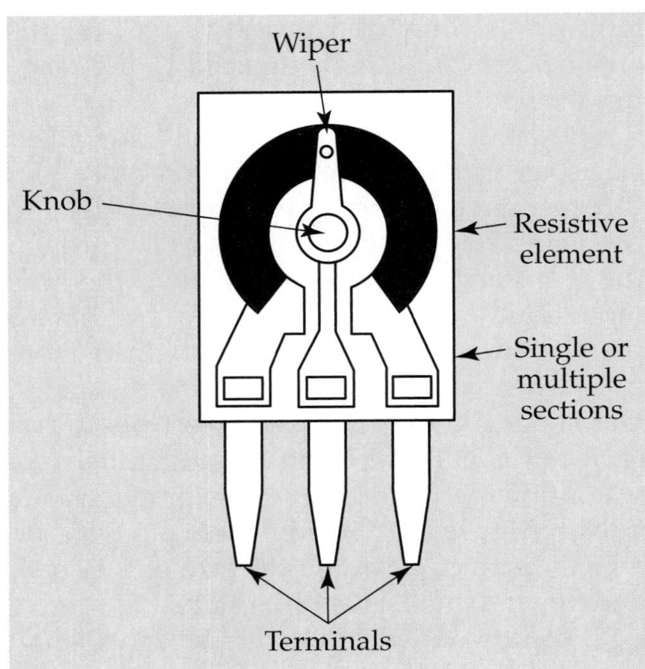

Figure 4-11. Trimmer resistors are used in fine-tuning applications.

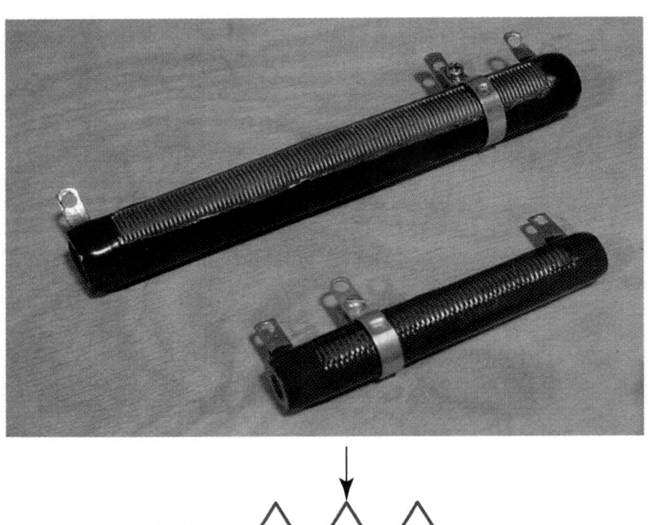

Figure 4-12. Adjustable wirewound resistors and their symbol.

completely covered with the protective ceramic material. As you can see in **Figure 4-12,** these resistors have a small portion of one side exposed. A slider arm is mounted across the exposed portion as a means of adjustment.

To make a resistance adjustment, the slider arm is moved along the resistor and clamped in a given position. Different amounts of resistance can be obtained by changing the slider arm location, but this type of resistor is designed to not be adjusted very often. If repeated adjustments are made, the resistor will wear out from the movement of the slider arm on the exposed copper.

Project 4-1: Probes

Some of the projects in this textbook require you to use probes to make tests. You can buy a set of probes or make your own. The probes you can make are nails with wires soldered to them. Once it is soldered, the whole thing must be cast in plastic to insulate the tips. This is done so you can hold the probes in your hand without fear of electric shock.

Some people use plastic tubing for insulation instead of casting in plastic. Simple ideas like this make projects more economical to build than if you bought the materials from an electronics store. Also, making your own materials is more fun and allows you to learn how the project works. If something happens to your project, it will be easier to fix.

Project 4-2: Variable Neon Voltmeter

You can use this simple voltmeter to find the exact circuit voltage. Gather the materials in the parts list and follow the procedure listed below:

No.	Item
1	NE-2 bulb
1	200 k resistor (R_1)
1	300 k potentiometer (R_2)
2	Probes
1	Mounting box
1	4′ of AWG 18 wire
	Grommets (insulators) for wires and bulb
	Heat shrink or spagetti

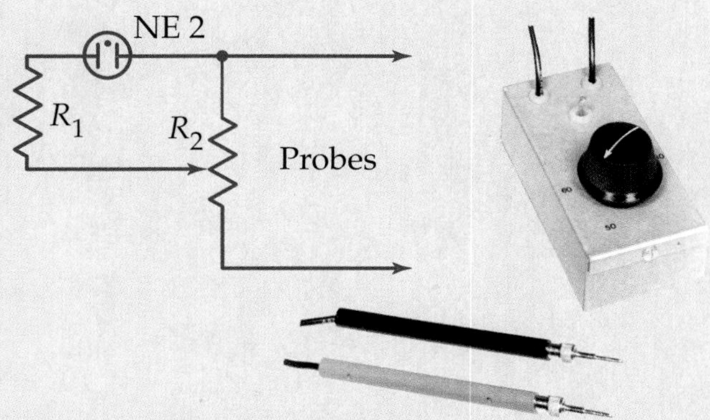

(Project by Allan Witherspoon)

Procedure

1. Solder one end of the resistor to the neon lamp and the other end to the center of the potentiometer (wiper terminal).
2. Solder the other end of the neon lamp to the first terminal of the potentiometer and to one of the probes.
3. Solder the third terminal of the potentiometer to the second probe. Probes serve as insulated handles so you can safely touch the exposed metal end to the voltage source tested.
4. After soldering the components together, mount them in a box. Be sure to insulate all the soldered connections with heat shrink or spaghetti. Do not allow any bare wires to touch the inside of the box.

After you have built the circuit and mounted it in a box, you can calibrate the instrument by doing the following:

1. Connect the test leads to a variable power supply. Do not exceed 600 volts.
2. Turn the dial on the potentiometer to the point where the light goes out.
3. Mark this voltage value on the front of your voltmeter box.
4. Keep changing the variable power supply and making this test until you have all of the values you want marked on the front of your voltmeter. Be sure to use grommets where the wires come through the metal box.

Once you have calibrated the tester, you can use it to test unknown voltages. Always start testing with the potentiometer set to the highest setting. Turn the knob on the potentiometer down until the light comes on. This will give you the value for your unknown voltage. Always work your way down to the unknown voltage. Remember, this is a simple voltmeter. It is not intended to be an expensive, highly-calibrated meter.

Thermistors

One special type of resistor, called a ***thermistor,*** deserves to be discussed, **Figure 4-13.** The resistance of a thermistor changes according to its temperature. Its resistance cannot be adjusted the way variable resistors can. Thermistors are used mostly in the electric circuits of heating and cooling systems. The name *thermistor* was derived from the phrase *therm*ally sensitive res*istor.*

Most thermistors are of the ***negative temperature coefficient (NTC) thermistor*** variety. NTC thermistors differ from conventional resistors in how they react to temperature changes. All resistors start to produce heat when electrons begin to flow in the circuit. As the temperature rises, resistance in most resistors increases. NTC thermistors, on the other hand, have an opposite reaction. As the temperature of an NTC thermistor rises, the resistance of an NTC thermistor decreases.

A ***positive temperature coefficient (PTC) thermistor*** behaves as conventional resistors. Their resistance increases as temperature rises.

Since thermistors have this special ability to react to temperature changes, they are used to measure temperatures, protect electric motors from overheating, serve as safety switches in fire or frost alarms, and sense liquid levels.

The frost alarm application is important to farmers and those in the agriculture industry. A frost could destroy hundreds of thousands of dollars worth of crops in just one night. When the air becomes too cold, thermistors can trigger water sprays, fans, or other protective devices.

Most thermistors have two terminals. This permits them to be installed into circuits the same way as resistors. However, the thermistor is placed so that it is on the outside of the unit where the air temperature can be measured. When the temperature drops to the thermistor setting, for example 33°F (0.55°C), its resistance triggers the circuit to the "on" state. For water spray, the "on" signal electrically opens a valve to allow water flow.

To get a better understanding of the use of resistors and thermistors, read Chapter 5. It will clear up questions about the kinds and sizes to be used, their wattage value, the types of circuits in which they are found, and their value in ohms.

Resistor Values

The design of most circuits includes one or more resistors. Resistors come in many different shapes and sizes. Most resistors use color-coded bands to help you identify

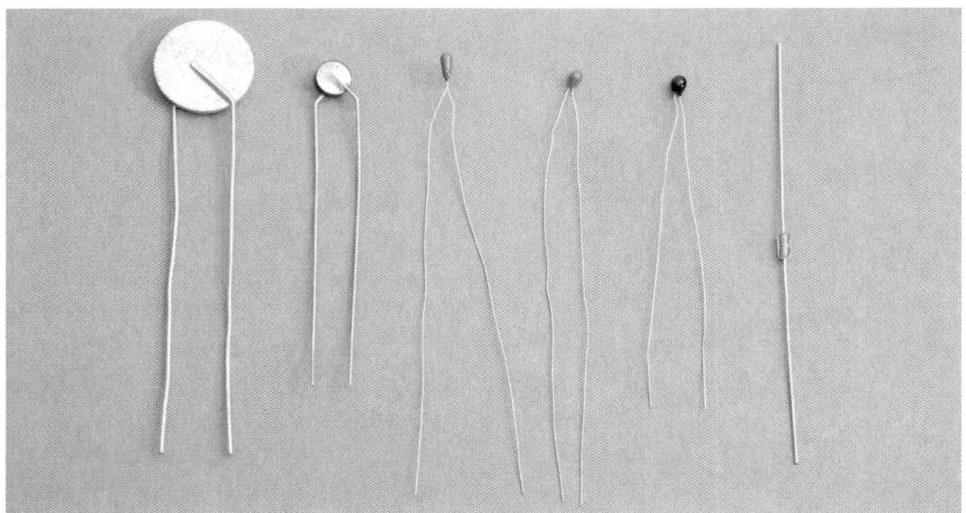

Figure 4-13. Thermistors, which come in many shapes and sizes, are thermally sensitive resistors. Since thermistors can detect slight temperature changes, they often are used as safety switches.

their values. Some of the larger sized resistors have their values printed on the outside of the resistor. You will see numbers like 5 Ω or 450 Ω. Values such as 3.8 kΩ (kilohms) or 3 MΩ (megohms) are a short method of labeling resistors with large values without having to write out all the zeros.

The letter *k* stands for kilo, which is the same as *thousand*. A 3.8 kΩ resistor, for example, has a value of 3800 Ω. Meg is another way of writing *million*. Therefore, a 3 MΩ resistor has the value of 3,000,000 Ω. The terms *meg* and *kilo* are commonly used with resistors and for sizing other electrical components, which you will learn about later.

Color Code—First Four Bands

You have probably noticed that resistors have three, four, or five bands of color on them. These bands indicate the value of the resistor, **Figure 4-14.** The bands are always closer to one end of the resistor. When you read a resistor's value, always place the resistor so the bands closest to the end of the resistor are to your left, **Figure 4-15.** The order in which you read the colors is important in determining the value of the resistor.

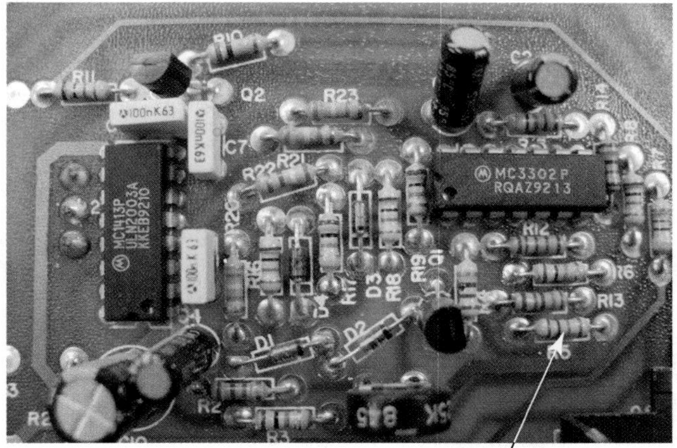

Figure 4-14. Resistors (*R* in photo) have colored bands. Just by looking at them, you will know their values once you understand how to read their color code.

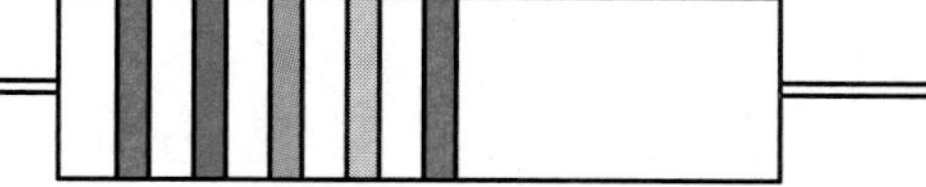

Figure 4-15. The correct way to read the color bands on a resistor is to position the resistor with the bands at your left.

Each color represents a number. The first three bands indicate the value, while the fourth band indicates a tolerance. The following colors are used for the first three bands.

Color	Number It Represents
Black	0
Brown	1
Red	2
Orange	3
Yellow	4
Green	5
Blue	6
Violet	7
Gray	8
White	9

To make a reading, list each of the colors used in the three bands. Then, list their respective numbers as follows:

Red—2 Green—5 Orange—3

The first two numbers, in order, are the base value of the resistor. The third is the number of zeros to add to the base value. Therefore, the resistor has a base value of 25 with 3 zeros. This is equal to 25,000 Ω, or 25 kΩ.

Example 4-1:

What is the value of this resistor?

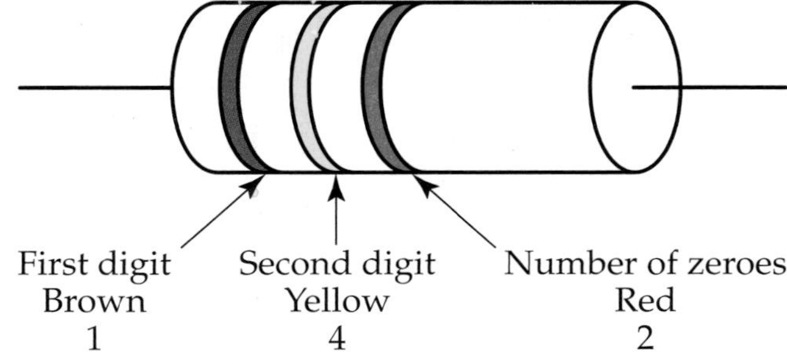

The resistor's value is 1400 Ω, or 1.4 kΩ.

Math Focus 4-1: Metric and Electrical Prefixes

The metric system (SI) is a measurement system based on powers of 10. This system uses a set of prefixes to show the size of the measurement. The following table shows some of the most commonly used metric prefixes and their meanings.

Prefix	Abbreviation	Exponential Equivalent	Decimal Equivalent
Tera	T	10^{12}	1,000,000,000,000
Giga-	G	10^{9}	1,000,000,000
Mega-	M	10^{6}	1,000,000
Kilo-	k	10^{3}	1000
Base unit	—	10^{0}	1
Milli-	m	10^{-3}	0.001
Micro-	m	10^{-6}	0.000001
Nano-	n	10^{-9}	0.000000001
Pico-	p	10^{-12}	0.000000000001

Although there are other combinations, the prefixes you will see used most in electricity and electronics are as follows:

Volts			
Gigavolts	GV	10^{9}	1,000,000,000 V
Megavolts	MV	10^{6}	1,000,000 V
Kilovolts	kV	10^{3}	1000 V
Ohms			
Gigaohms	GΩ	10^{9}	1,000,000,000 Ω
Megohms	MΩ	10^{6}	1,000,000 Ω
Kilohms	kΩ	10^{3}	1000 Ω
Farads			
Microfarads	μF	10^{-6}	0.000001 F
Picofarads	pF	10^{-12}	0.000000000001 F
Amperes			
Miliamps	mA	10^{-3}	0.001 A
Microamps	μA	10^{-6}	0.000001 A
Nanoamps	nA	10^{-9}	0.000000001 A

In the field of electricity and electronics, you will often need to add a prefix to a measurement and convert between electrical prefixes. When a prefix is added to a measurement, it is typically done to express the measurement as briefly as possible and without having to write out all the zeros. For example, say the color code of a resistor reads 100,000. To express this in a simpler way, you could change this value to kilohms. The prefix kilo is equal to 10^3. However, since the decimal needs to move to the left to make this conversion, you simply multiply 100,000 ohms by the negative exponential equivalent of 10^3:

$$100{,}000 \text{ ohms} \times 10^{-3} = 100 \text{ kilohm (k}\Omega\text{)}$$

Suppose a resistor reads 320,000,000 ohms. To express this measurement in a simpler way, you could change this value to megaohms. The prefix mega is equal to 10^6. Since the decimal needs to move to the left to make this conversion, multiply 320,000,000 ohms by the negative exponential equivalent of 10^6:

$$320{,}000{,}000 \text{ ohms} \times 10^{-6} = 320 \text{ megaohms (M}\Omega\text{)}$$

This method works the same for values that are less than one. For example, a capacitor with a value of 0.00000000032 farads could be expressed in picofarads. The exponential equivalent of a pico is 10^{-12}. However, since the decimal point will move to the right, the positive exponential equivalent of 10^{-12} must be used in the equation:

$$0.00000000032 \text{ F} \times 10^{12} = 320 \text{ picofarads (pF)}$$

To convert from one measurement prefix to another, you must first find the difference between their exponents. For example, to convert 7 kilovolts (10^3) to gigavolts (10^9), you must find the difference between the exponents *3* and *9*:

$$3 - 9 = -6$$

Use this difference as the new exponent:

$$7 \text{ V} \times 10^{-6} = 0.000007 \text{ gigavolts (GV)}$$

To convert from 7 megavolts (10^6) to kilovolts (10^3), you would again need to find the difference between the exponents. This difference is

$$6 - 3 = 3$$

Therefore, the exponential equivalent that you would need to multiply by is 10^3.

$$7 \text{ V} \times 10^3 = 700 \text{ kilovolts (kV)}$$

Practice Problems

Convert the following measurements.

1. 1 kV to V
2. 300 mA to A
3. 10,000 Ω to kΩ
4. 132 μF to pF
5. 0.000000000320 pF to μF
6. 8.41 GV to kV
7. 89 MV to V
8. 3.72 kΩ to Ω

Example 4-2:

What is the value of this resistor?

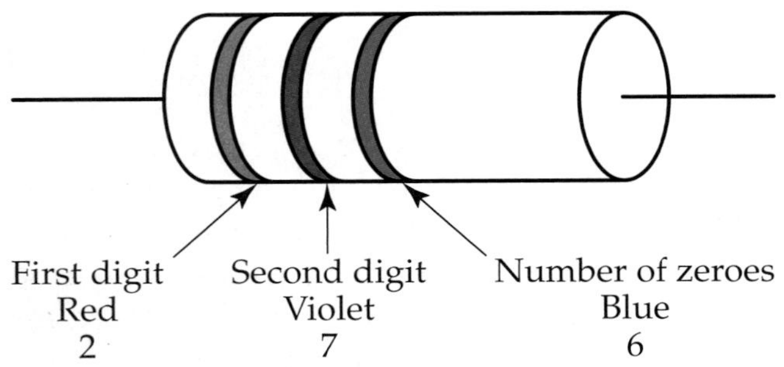

The resistor's value is 27,000,000 Ω, or 27 MΩ.

The fourth band, if used, is either gold or silver. This band tells how close the rated value of the resistor is to the actual value. Usually, the resistor is coded within some percentage of the value determined by the first three bands. The range of values is specified by the manufacturer. This is known as its *tolerance.*

Resistors have a tolerance because it is difficult to control the exact amount of carbon and binder used to make each resistor. When making a batch of resistors, it is impossible to use the same amount of carbon and binder for each one. A very small change in the amount of material used changes the value of the resistor. For this reason, the values of the resistors vary within a group of resistors produced to have the same value. Therefore, the manufacturer gives a resistor's value within a range or tolerance from one value. The colors and meanings of the fourth band are as follows:

Gold	Plus or minus 5 percent	(±5%)
Silver	Plus or minus 10 percent	(±10%)
No 4th band	Plus or minus 20 percent	(±20%)

Using our previous examples, assume that the 1400-Ω resistor had a gold fourth band. Its value and tolerance are 1400 Ω plus or minus 5 percent (±5% or ±0.05). Five percent of 1400 ohms is

$$1400\ \Omega \times 0.05 = 70\ \Omega$$

Therefore, this resistor's tolerance is ±70 Ω.

To calculate the maximum and minimum values that are possible for the actual resistor, add and subtract the tolerance from the resistor's value:

Maximum value:

$$1400\ \Omega + 70\ \Omega = 1470\ \Omega$$

Minimum value:

$$1400\ \Omega - 70\ \Omega = 1330\ \Omega$$

This means that if you measure the resistor with a meter that shows values in ohms, the value could be as high as 1470 Ω or as low as 1330 Ω and still be within the manufacturer's tolerance.

Example 4-3:

What is the range of tolerance for a 25,000-Ω resistor with a silver fourth band?

A silver fourth band means the resistor has a tolerance of ±10%.

Ten percent of 25,000 Ω is

$$25{,}000\ \Omega \times 0.10 = 2500\ \Omega.$$

Therefore, its range would be 25,000 Ω ±2500 Ω.

Maximum value:

$$25{,}000\ \Omega + 2500\ \Omega = 27{,}500\ \Omega$$

Minimum value:

$$25{,}000\ \Omega - 2500\ \Omega = 22{,}500\ \Omega$$

This resistor's range of tolerance is from 27,500 Ω to 22,500 Ω.

Example 4-4:

What is the range of tolerance for a 27 MΩ resistor with no fourth band?

No fourth band means the resistor has a tolerance of ±20%.

27 MΩ is 27,000,000 Ω. Twenty percent of 27,000,000 Ω is

$$27{,}000{,}000\ \Omega \times 0.20 = 5{,}400{,}000\ \Omega.$$

Therefore, its range would be 27,000,000 Ω ±5,400,000 Ω, which is the same as 27 MΩ ±5.4 MΩ.

Maximum value:

$$27\ M\Omega + 5.4\ M\Omega = 32.4\ M\Omega$$

Minimum value:

$$27\ M\Omega - 5.4\ M\Omega = 21.6\ M\Omega$$

This resistor's range of tolerance is from 21.6 MΩ to 32.4 MΩ .

Occasionally, you will find a gold or silver color on the third band of a carbon resistor. When you do, write down the first two numbers based on the color code, and then multiply by 0.1 for a gold band or by 0.01 for a silver band. For example, consider a resistor with bands of red, green, and gold:

Red—2 Green—5 Gold—Multiply by 0.1

This resistor has a value of 25×0.1, which equals 2.5 Ω. Resistors with these values are not found in many circuits.

Example 4-5:

What is the value of a resistor with the first three bands of red, green, and silver?

Write the code as follows:

Red—2 Green—5 Silver—Multiply by 0.01

This resistor has a value of 25×0.01.

$$25 \times 0.01 = 0.25\ \Omega$$

The resistor's value is 0.25 Ω.

Color Code—Fifth Band

Earlier, the possibility of a resistor having a fifth band was mentioned. When such a resistor is found, the fifth band is also color coded. The fifth band is used to predict the percentage of failure per thousand hours of use. It is based on tests that the manufacturer has conducted on similar resistors. The coding is as follows:

Color	Percentage of Failures/1000 hrs
Brown	1.0%
Red	0.1%
Orange	0.01%
Yellow	0.001%

For example, if a group of 100 resistors had a brown fifth band, only 1 out of the 100 (1.0%) would be likely to fail during 1000 hours of use. Do not worry if this color code seems difficult or confusing. The more you practice reading resistors, the easier it becomes. To sum up the entire resistor code, look at **Figure 4-16.**

Color	1st and 2nd Band Digits	3rd Band Multiply by	4th Band Tolerance ±%	5th Band Failure Rate %/1000 hrs
Black	0	1	–	–
Brown	1	10	–	1.0
Red	2	100	–	0.1
Orange	3	1000	–	0.01
Yellow	4	10,000	–	0.001
Green	5	100,000	–	–
Blue	6	1,000,000	–	–
Violet	7	10,000,000	–	–
Gray	8	100,000,000	–	–
White	9	1,000,000,000	–	–
Gold	–	0.1	5	–
Silver	–	0.01	10	–
No Color	–	–	20	–

Figure 4-16. Resistor color codes and tolerances.

Shelf Life

Resistors tend to heat up when they are used in a circuit. Therefore, when reusing a resistor, make sure it is still good. If a resistor has been overheated, it may be over or under its tolerance. A circuit design with a bad or overheated resistor may not work as intended. You can test a resistor using an ohmmeter, which is used to measure resistance. This type of measuring instrument is described in detail in Chapter 9.

Resistors can be stored for long periods of time without going bad. Although some electrical devices, such as batteries, should be used as soon as possible after they have been purchased, a resistor that has been on a shelf and never used in a circuit should be as good as new.

Capacitors

Another key component found in many electric circuits is a ***capacitor*, Figure 4-17.** A capacitor's main function is to store electrons until they are needed. **Figure 4-18** shows two symbols for a capacitor.

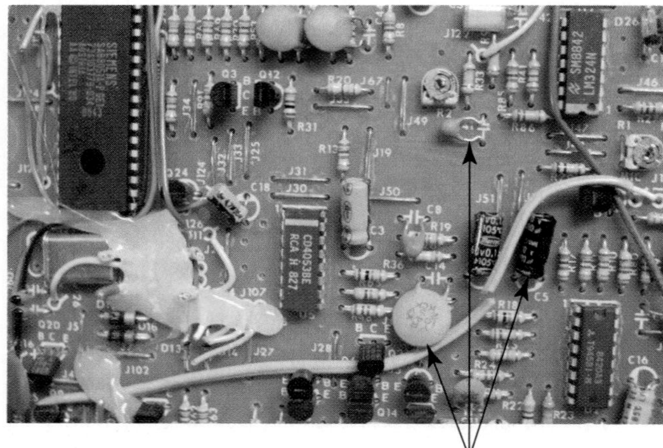

Figure 4-17. Capacitors store electrons, holding the electrical charge until the circuit needs it. Look for its symbol or the letter *C* on the circuit board to help identify the capacitors. How many capacitors can you find in the photo?

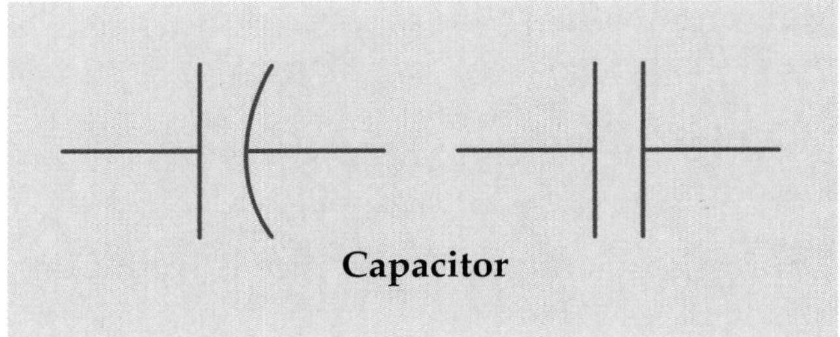

Figure 4-18. Commonly used symbols for a capacitor.

To understand how a capacitor works in a circuit, try relating it to a water jug placed in a cooler in an office. It must be filled when it is first installed. People take water by the cupful until the jug is empty. Once all of the water is gone, it must be refilled.

With a capacitor, the sequence is the same. It must first be charged with electrons. It then gives up the electrons as needed. Once it has no more electrons stored, the capacitor must be recharged so it can give off more electrons when called on to do so.

In Chapter 14, we will go into details of capacitor operation for some circuits. At this point, however, all you need to know is that a capacitor can be charged and discharged.

A capacitor consists of two or more plates and a dielectric. The plates are made from material that can be charged, such as aluminum foil. To hold this charge, the ***dielectric***, which is an insulator, is placed between the plates, **Figure 4-19.** If the plates touched one another, the charge would be

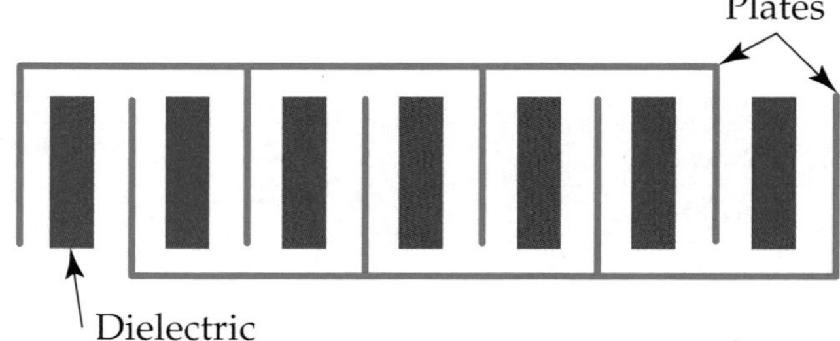

Figure 4-19. The dielectric in a capacitor is the insulator placed between the plates that store the electrical charge.

lost. The job of the dielectric is to keep the plates separated so they do not touch one another. If they were to touch, the capacitor would be shorted. This is an indicator that the capacitor is ruined.

The purpose of the capacitor is to store a charge until needed by the circuit. One good example of this is the capacitor in the flash unit of a camera. The camera battery is used to charge the capacitor. Some cameras have a light on them to indicate that the flash unit is charged. You aim the camera and push the button to take the picture. Pushing the button discharges the capacitor in the flash unit giving you the extra light needed to take the picture. The battery repeats the charging process to charge the capacitor for the next photo.

The more rapidly a capacitor is charged and discharged, the more heat that is created. If the amount of heat produced is not held to a reasonable level, it may destroy the capacitor.

Capacitors are designed to do specific jobs. Capacitors are usually named for the material from which they are made. They have names like *ceramic disc*, *polyester film*, *tantalum*, and *ceramic chip*, shown in **Figure 4-20.**

Tantalum and ceramic capacitors are small and have a rating between 0.1 μF to 22 μF. They are useful in tasks such as providing a steady dc output from a power supply. Tantalum capacitors have polarity and are referred to as an *electrolytic capacitor*. One lead is marked with a negative sign (–) to indicate it should be connected to the negative side of the

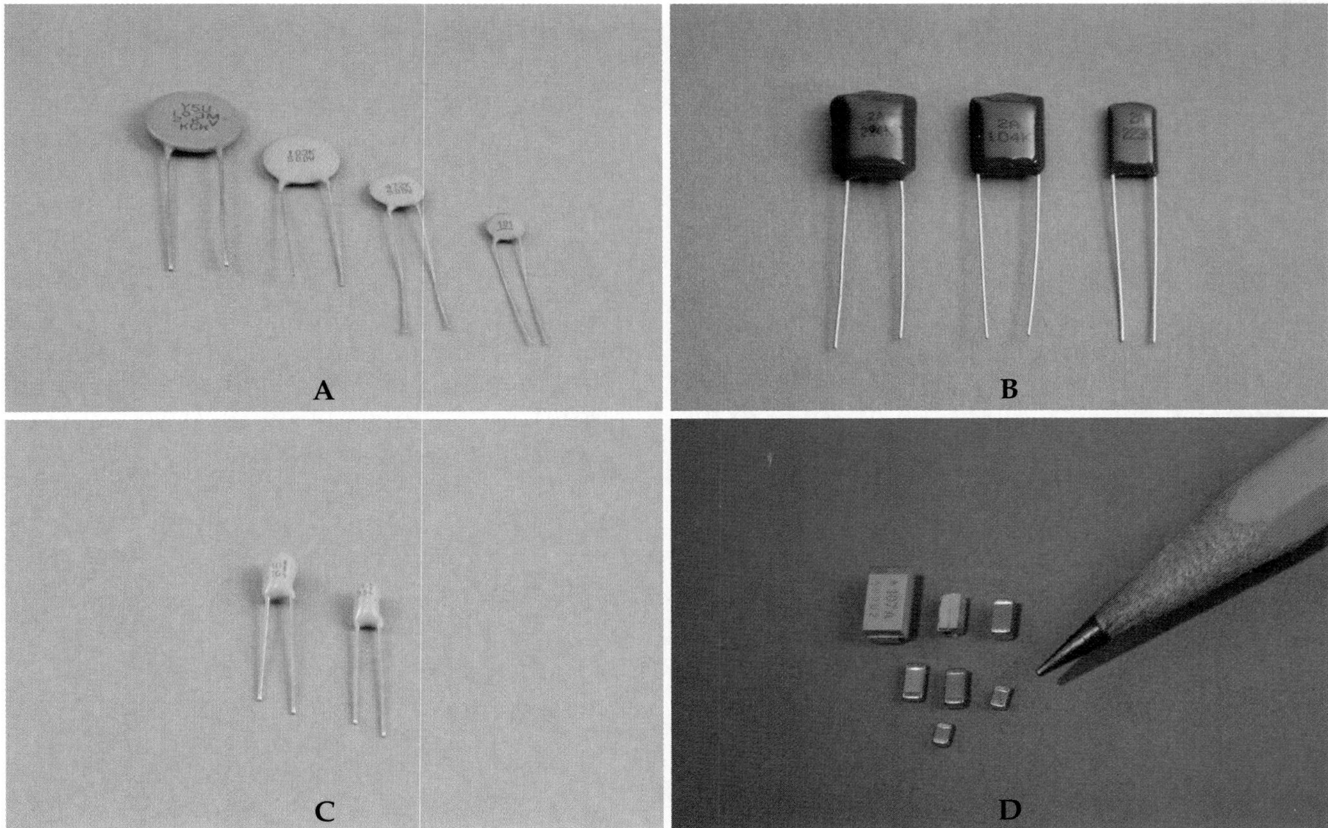

Figure 4-20. Variety of capacitors. A—Ceramic capacitors use a ceramic material as an insulator between the two conductors. B—Polyester film capacitors usually are made of two conductors separated by a polyester film. C—Tantalum capacitors use tantalum as an insulator. D—Ceramic chip capacitors pack a high level of capacitance in a tiny package. They are constructed of alternating layers of ceramic insulation and electrodes.

Project 4-3: Flasher

This is a good project you might make for a child. Gather the parts in the parts list and then perform the following procedure.

No.	Item
5	NE-2 bulbs
1	0.25 A fuse
1	1000 k resistor (R_1)
1	390 k resistor (R_2)
1	330 k resistor (R_3)
1	300 k resistor (R_4)
1	270 k resistor (R_5)
5	0.1 μF capacitors (C_1–C_5)
1	0.5 A, 200 PIV diode (D_1)
1	Plug
1	Grommets
	Connecting wire

(Project by Don Winchell)

Procedure

1. Paint a picture of a rabbit or some other animal against a flowered background.
2. Drill small holes at the center of the flowers for five NE-2 bulbs
3. Position the NE-2 bulbs in position to determine the length of wire and circuit assembly. Use grommets for the bulbs.
4. Install the rest of the components as shown in the schematic. When installing the fuse and plug, be sure to secure them so they cannot be pulled out when disconnecting the circuit from the wall outlet.

After you have assembled this project, plug it into the wall outlet. Once electrons fully charge the capacitors, the capacitors will discharge, making the lamp flash. Then, the capacitors will recharge and the process is repeated. Changing the value of the resistors changes how often the lights flash. Large ohm resistors increase the time needed to charge the capacitor. Smaller valued resistors decrease the charging time, making the lamps flash more often.

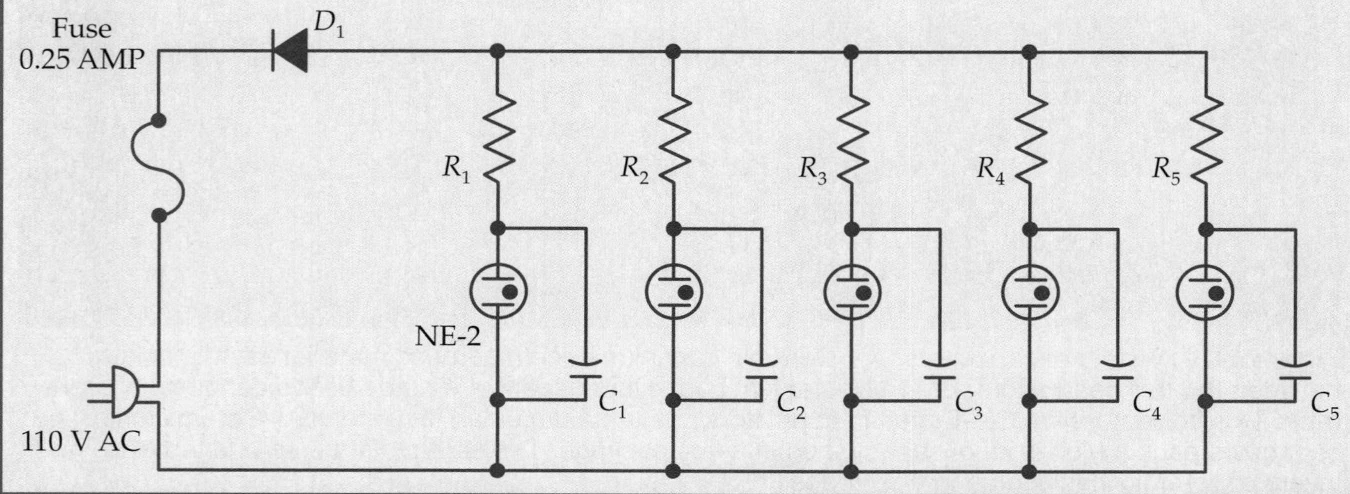

Project 4-4: Project Tester

If a project is not properly assembled, you can blow fuses and destroy a circuit that has taken hours to build when you plug it in to use it the first time. However, with the project tester shown, you can save yourself a lot of trouble after building a project.

No.	Item
1	Lightbulb
1	Lightbulb socket
1	Receptacle
1	SPST switch
1	Plug
	Connecting wire

The project tester is just what its name implies. It will test your circuit and serve as an advance guard against project failure when you put the project into operation for the first time.

Assemble the project tester and plug it into a wall receptacle. You can check a new 110-volt project for short circuits (unwanted electron flow) or opens by plugging it into the test receptacle with the switch open. The tester will prevent blown fuses and avoid the possibility of destroying the circuit.

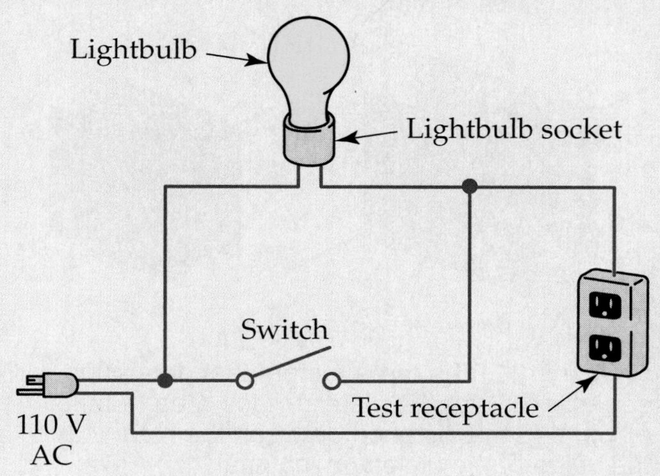

To use the tester, be sure the tester switch is open. Screw in a test lamp of about the same wattage as the project you are building. Plug the power lead of your tester into a 110-volt wall outlet, and plug your 110-volt project into the test receptacle. There are three possible results:

- The circuit is shorted; the test lamp burns brightly.
- The circuit is open; the test lamp does not light.
- The circuit is good; the test lamp is dim.

The brightness of the bulb will tell you what is wrong with your circuit, if anything. You can then fix the circuit and test it again. Once you get the desired result with your tester, your project is ready to be put to use.

circuit. Ceramics are useful in high-frequency applications. One type of film capacitor uses a polyester film as its dielectric. Their tolerance values might be as high as +/–10% but film capacitors are very inexpensive to produce. Mica capacitors are used in applications where it is necessary to charge and discharge the capacitor billions of times in a second. There are hundreds of types and uses for capacitors.

Capacitance

Capacitance is the amount of charge a capacitor can store. Although the capacitor was discovered by Pieter van Musschenbroek, capacitance is measured in farads (F), which are named for Michael Faraday, the scientist who discovered the principle of induction. The most common units you will be using when working with capacitors are the ***microfarad*** (μF), which is 1×10^{-6} F, and the ***picofarad*** (pF), which is 1×10^{-12} F.

The amount of charge that can be stored varies greatly among capacitors. It is affected by three important factors: the surface area of the plates, the distance between the plates, and the dielectric, **Figure 4-21.**

The distance between the plates is the same for both of the capacitors in Figure 4-21. However, the surface area of the plates and

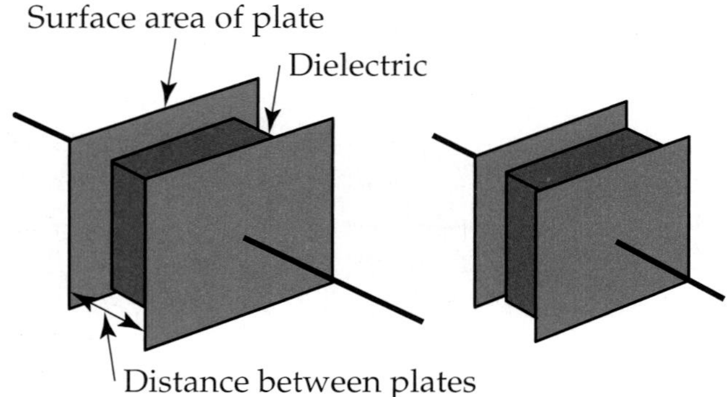

Figure 4-21. The three factors that determine capacitance are the plate surface area (total of all plates), the distance between the plates, and the value of the dielectric material.

dielectric is different. As you can see, the larger capacitor (the capacitor with the most plate surface area) has the greatest capacitance. It has more space to store an electrical charge. As the surface area increases, the capacitance increases as well. This is only true, however, if the other variables are kept constant and the plates are perfectly parallel to each other.

The second thing that affects capacitance is the distance between the plates. Again, we will keep the other variables—the dielectric and surface areas—constant. If the two charged plates are moved apart, the strength of the capacitor's charge will be lowered. Capacitance is inversely proportional to the distance between the plates.

Math Focus 4-2: Order of Operations

In mathematics, there is a specific order for doing operations in a complex equation. If the equation is not worked in this order, it may not yield the correct answer. Consider the equation $7 + 3^2 \times 5 = ?$. Three students doing this problem may come up with three different answers:

Student 1		Student 2		Student 3	
$7 + 3^2 \times 5$	= ?	$7 + 3^2 \times 5$	= ?	$7 + 3^2 \times 5$	= ?
$10^2 \times 5$	= ?	$7 + 9 \times 5$	= ?	$7 + 9 \times 5$	= ?
100×5	= 500	16×5	= 80	$7 + 45$	= 52

All three students performed the mathematical operations correctly, but each did these operations in a different order. This is why there is a prescribed order to follow. The following lists the order in which an equation should be worked.

1. Operations contained within parentheses( ), brackets [ ], and braces { }. If all three are used in an equation, the set of parentheses are contained within brackets [ ] and the brackets are contained within braces:{[( )]}. Perform the operation in this order: parentheses first, brackets second, and braces third. If more sets are needed, the pattern repeats {[({[( )]})]}. In this case, start with the innermost parentheses and work your way outward.

One way manufacturers change the distance between the plates is by putting a thicker dielectric between them. Another way is by making the plates adjustable. In later chapters, we will be working with an adjustable unit known as a variable capacitor. By adjustment, you can either increase or decrease the distance between plates. This will lower or raise the capacitance of the capacitor.

Suppose, for example, the distance between the plates is increased by one thousandth of an inch (0.001″). Will the capacitance increase or decrease? Since the relationship is inversely proportional, capacitance will decrease. If the distance between the plates is decreased, capacitance will increase.

The third effect on capacitance concerns the type of dielectric used in the capacitor. The dielectric is chosen based on the types of capacitors manufactured. Certain dielectric materials work best for specific capacitors used in specific kinds of circuits. For example, as mentioned earlier, ceramics are useful in high-frequency applications. The dielectric constant (K) is used in the formula to determine capacitance. Remember from Chapter 3 that an insulator resists the flow of electrons. See **Figure 4-22** for dielectrics used in capacitors. Approximate values are given. You can use this information to build your own capacitor.

2. Exponents and radicals (roots).
3. Multiplication and division, working left to right.
4. Addition and subtraction, working left to right.

Only student 3 has worked the equation correctly:

$$7 + 3^2 \times 5 = 52$$

Start with operations in parentheses. There are no parentheses, so move on to exponents and radicals. There is one exponent.

$$7 + 3^2 \times 5 = ?$$

$$7 + 9 \times 5 = ?$$

Now do multiplication and division, working left to right:

$$7 + 9 \times 5 = ?$$

$$7 + 45 = ?$$

Finally, do addition and subtraction, working left to right:

$$7 + 45 = 52$$

Practice Problems

Solve each of these equations using the proper order of operations:

1. $2 \times (3 + 7)$
2. $2 \times 3 + 7$
3. $[4^3 + (7^2 - 5)] \div 2$
4. $8 - 1 + 5 \times 2$
5. $(3^2 - 5 \times 2)^2$
6. $(2^3)^2$
7. $1 + \{2 \times [3 + (4 \times 5) + 6] \times 7\} - 8$
8. $1 + 2 \times 3 + 4 \times 5 + 6 \times 7 - 8$

Dielectric Constants of Capacitor Materials	
Material	**Dielectric Constant (K)**
Air	1.0
Aluminum Oxide	7.0
Beeswax	2.4–2.8
Cambric (varnished)	4.0
Celluloid	4.0
Glass	4.0–14.5
Mica	4.0–9.0
Mylar	3.0
Paper	1.5–4.0
Porcelain	5.0–6.5
Quartz	5.0
Tantalum Oxide	11.0

Figure 4-22. Dielectric materials are listed on the left. Dielectric constants for calculating the capacitance are listed on the right.

Formula

To better understand the three major factors that affect capacitance and how they relate to one another, examine the following formula:

$$C = 0.2235\frac{KA}{d}(N-1) \tag{4-1}$$

where

C is the capacitance in picofarads (pF)

K is the dielectric constant

A is the area of one plate in square inches

d is the distance between the plates in inches

N is the number of plates

Example 4-6:

What is the capacitance (*C*) of four 1″ square plates separated by quartz that is 0.001″ thick?

$$C = 0.2235\frac{KA}{d}(N-1)$$

$$= 0.2235 \times \frac{5 \times 1 \text{ in}^2}{0.001}(4-1)$$

$$= 0.2235 \times \frac{5}{0.001} \times 3$$

$$= 0.2235 \times 5000 \times 3$$

$$= 3352.5 \text{ pF}$$

The capacitance is 3352.5 pF.

Example 4-7:

What would be the value of the capacitor from Example 4-6 if we double the distance between the plates by increasing the thickness of the quartz to 0.002″ thick?

$$C = 0.2235\frac{KA}{d}(N-1)$$

$$= 0.2235 \times \frac{5 \times 1 \text{ in}^2}{0.002}(4-1)$$

$$= 0.2235 \times \frac{5}{0.002} \times 3$$

$$= 0.2235 \times 2500 \times 3$$

$$= 1676.25 \text{ pF}$$

The capacitance is 1676.25 pF.

As you can see, doubling the distance cuts the capacitance to one half of the capacitance in the previous example.

Terms

A capacitor has a capacitance of one ***farad*** **(*F*)** when, with one volt applied to its plates, it stores one ***coulomb*** of charge on each of its plates. Earlier you learned that one coulomb is the amount of electric charge carried by a current of one ampere flowing for one second. The terms *coulomb* and *ampere* seem to be interchangeable. This is because the base unit was changed from coulomb to ampere almost 50 years ago. Therefore, today we say that the ampere is equal to one coulomb (6.24×10^{18}) of electrons flowing past a point in one second. You may be familiar with the term *volt,* having seen it marked on batteries. We will cover what a volt is and how it works with other units in the next chapter.

The farad is a very large unit and is not used much. Most capacitors are measured in picofarads and microfarads. A microfarad is one-millionth of a farad, while a picofarad is one-millionth of a microfarad. The following will give you an idea of the relationships among these units:

1 farad (F) = 1,000,000 microfarads (μF)

1 microfarad (μF) = 1,000,000 picofarads (pF)

1 microfarad (μF) = 0.000001 farad (F)

1 picofarad (pF) = 0.000001 microfarad (μF)

1 picofarad (pF) = 0.000000000001 farad (F)

It is common to use the term *microfarad* down to about 0.0001 μF. The term *picofarad* is used up to about 1000 pF (0.001 μF). Even though the same number can be referred to in microfarads or picofarads within these limits, the idea is to try to use the smallest number of digits.

Discharging Capacitors

At this point, you may be thinking that these are pretty small numbers. As small as they may seem, they carry enough electrical charge to kill someone.

Project 4-5: Capacitor Discharger

There are times when you have to discharge a large capacitor and do not want to be too close to it. Here is a simple design that can be adjusted and will keep you away from arcs. Capacitor discharging devices can be useful when working on radios, television sets, and other electrical circuits involving capacitors. However, the right kind of equipment is needed to do the job properly. A schematic is shown, revealing a resistor built into the probe. How the boot fits over the alligator clip is also shown.

No.	Item
2	Probes (one probe can be exchanged for an alligator clip)
1	9 k, 1-watt resistor (or larger, depending on the capacitor to be discharged)
3′	14-gauge wire

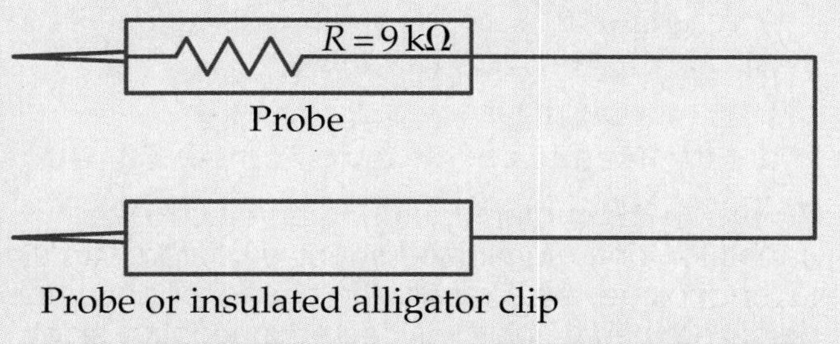

Procedure

1. Obtain parts and check resistor ratings.
2. Drill the inside of one probe to accommodate the resistor or resistors.
3. Strip the wire on both ends and make good mechanical joints.
4. Connect the wire to one end of the resistor and connect the other end of the resistor to the metal part of the probe. Connect the other end of the wire to the second probe or alligator clip, if used.
5. Screw the handles on the metal part of the probes.

To use the capacitor discharger, attach the alligator clip (or probe) to one terminal of the capacitor and touch a probe to the other terminal for at least five seconds. This gives the capacitor time to discharge slowly.

The principle of operation of the discharger is simple. The wire acts as a path for the electrons to flow from one set of capacitor plates to the second set of plates. The resistor slows the discharging process and stops arcing by preventing a sudden buildup of heat.

Use care when working with electrical circuits that include capacitors. Always assume that any capacitor is charged. Therefore, use some special means of discharging it before attempting to handle it. **Figure 4-23** shows a discharging device that is safe and easy to use. A capacitor is discharged simply by touching a probe to each terminal.

Some people make the mistake of using a screwdriver or a short section of insulated wire to discharge a capacitor. This could ruin the capacitor, or, if the capacitor is very large, it may arc. With some capacitors, such as the capacitors in an electronic flash for a camera, the end of the screwdriver could melt.

Danger

Remember that capacitors can hold a charge for many hours and, in some cases, for several days. Before you handle any capacitor, make sure it is discharged.

Electrolytic Capacitors

The ***electrolytic capacitor*** deserves special attention, **Figure 4-24.** It is made differently and provides more capacitance for its size than any other type of capacitor.

Instead of having its plates separated by a dielectric, the electrolytic capacitor consists of two aluminum sheets separated by a layer of paper soaked with a liquid chemical called ***electrolyte.*** All three sheets are rolled together and sealed in a container, and dc voltage is applied between the two aluminum sheets, **Figure 4-25.**

The voltage causes a thin oxide layer to be formed on one of the aluminum sheets. As a result, the capacitor is ***polarized.*** This means that one side of the capacitor is positive, and the other side is negative. The oxide acts as the

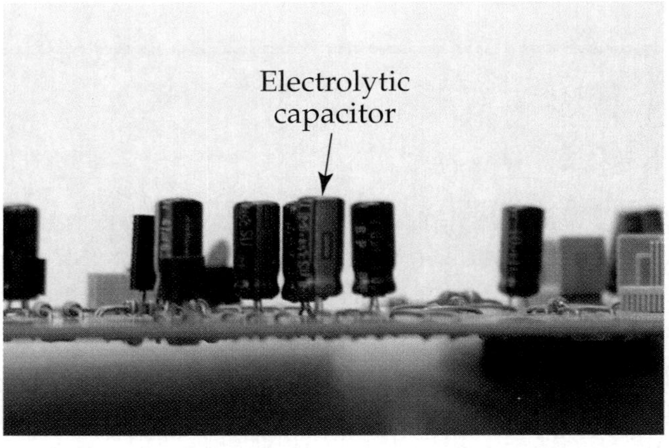

Figure 4-24. How many electrolytic capacitors can you identify on this circuit board?

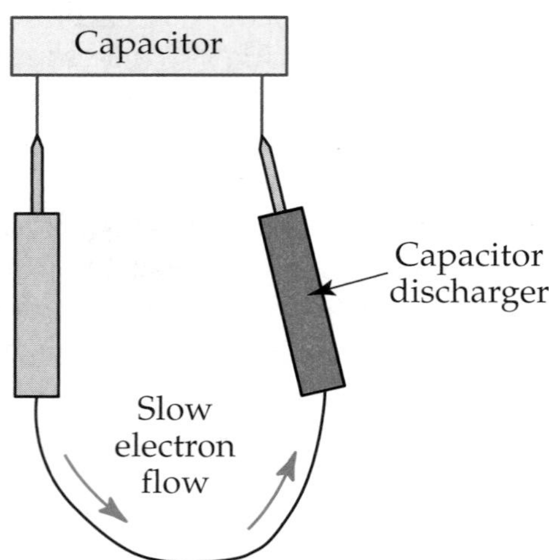

Figure 4-23. The resistor built into one probe of the capacitor discharger slows the flow of electrons to avoid damaging the capacitor through a sudden discharge.

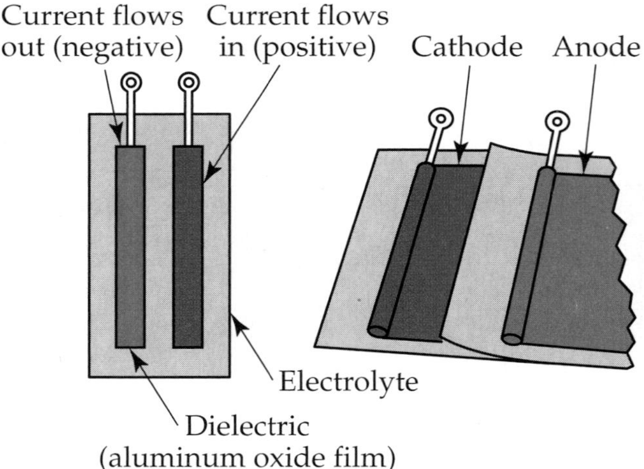

Figure 4-25. In manufacturing an electrolytic capacitor, one aluminized sheet is coated with a layer of oxide. As a result, the capacitor becomes polarized (has a positive side and a negative side).

dielectric. It has a high resistance to the flow of electrons in one direction and very low resistance in the opposite direction.

> **Caution**
>
> Electrolytic capacitors are marked with a positive (+) end and a negative (–) end. If they are not connected into circuits positive to positive and negative to negative, the oxide film will be ruined. In some cases, the capacitor may explode.

An electrolytic capacitor will slowly deform (decrease in capacitance) if it is not used for a while. Once an electrolytic capacitor begins to deform, a sudden surge of electrons may ruin it. Just installing the electrolytic capacitor in a circuit can cause this surge. When electrolytic capacitors are purchased from a store or taken off a shelf after being stored for a long time, they can still be used, but they must be charged slowly.

Many capacitors have voltage ratings marked on their outside cover. Some electrolytic capacitors are marked WVDC, which stands for *working voltage direct current*. To reform an electrolytic, hook up the circuit shown in **Figure 4-26.** It is important that the voltage output of the power source is lower than the voltage marked on the outside of the capacitor.

Substituting Capacitors

When repairing a circuit or building a new one, install capacitors of the specified voltage rating. If you cannot find a capacitor of the correct voltage rating, substitute one that has a higher voltage rating. If you install one with a lower rating you might get an arc across the plates and destroy the capacitor. This would be similar to putting in a resistor with too small a value for the circuit, causing it to burn up.

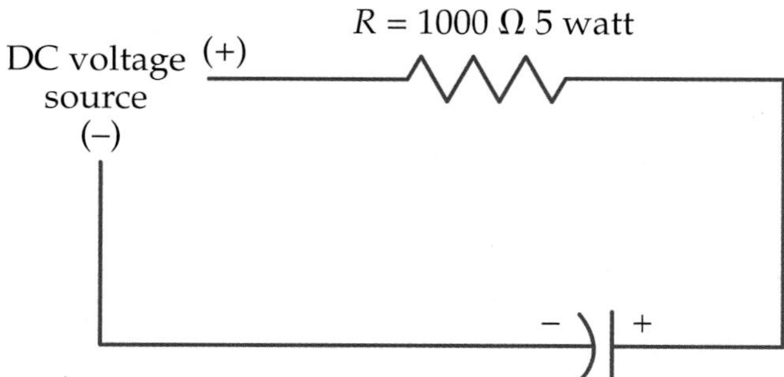

Figure 4-26. This schematic shows an electrolytic capacitor being reformed (charged) to build its capacitance. Note that it is installed positive-to-positive and negative-to-negative.

Soldering Capacitors

There are some precautions you should take if you have to replace a capacitor in a circuit. Make sure the new capacitor is soldered in place the same way as the old one when it was installed. If you get the capacitor leads too close or too far away from the chassis (the metal box on which electronic components are mounted and wired), it may shift the capacitance. Even a small shift during the replacement process could, for example, cause the repaired radio to hum. Try to keep distances, lead length, polarity, physical size, and capacitance the same size as that of the replaced capacitor. A difference in lead length could affect the resonant frequency (described in Chapter 15). Use a heat sink between the capacitor and the solder joint, as explained in Chapter 3. Some capacitors have a dielectric that could be melted by the heat of a soldering iron.

Summary

- Resistors are made of a material that makes it difficult for electrons to flow.
- A fixed resistor has a constant value, while a variable resistor can be adjusted within a range of values.
- A thermistor's resistance changes with temperature change.
- You can tell the resistance of a resistor by decoding its color bands.
- Capacitors are devices that hold a charge until those electrons are needed in the circuit.
- Different types of capacitors include film capacitors, mica capacitors, and electrolytic capacitors.
- Capacitance is the amount of charge a capacitor can hold; its basic unit is the farad (F).
- A microfarad (μF) is 1×10^{-6} farad (F), a picofarad (pF) is equal to 1×10^{-12} farad (F).
- Three factors affect capacitance: the surface area of the plates, the distance between the plates, and the material of the dielectric.
- Always make sure a capacitor is discharged before handling it; a charged capacitor could store enough electricity to kill some one.

Test Your Knowledge

Do not write in this book. Write your answers on a separate sheet of paper.

1. A potentiometer is another name for a ______.
   a. variable capacitor
   b. polarized capacitor
   c. fixed resistor
   d. variable resistor
2. Which of the following is *not* a common use for a thermistor?
   a. Protecting motors from overheating.
   b. Frost alarms.
   c. Predicting weather.
   d. Measuring temperature.
3. Copy these colors on a separate paper and place the number alongside each color that corresponds to the color code found on the resistor.
   ______ White
   ______ Orange
   ______ Gray
   ______ Red
   ______ Yellow
   ______ Blue
   ______ Black
   ______ Green
   ______ Violet
   ______ Brown
4. Draw the symbols for a resistor and a capacitor.
5. List two different types of capacitors.
6. Name at least two things that affect the amount of capacitance.
7. What is the basic unit of measure for capacitance?

8. Change 0.0001 μF into pF.
9. List the following readings from highest to lowest capacitance.
   3 F
   700 pF
   600 μF
   0.03 F
10. Never touch a capacitor without first ______ it.
11. Since a(n) ______ capacitor is made in a special way, you must match its polarity when installing it into a circuit.
12. How long can a capacitor hold its charge?

Activities

1. How do you know which tolerance resistor to use in a circuit?
2. Find out how a company makes carbon resistors.
3. Visit a store that sells electrical components and see how many different sizes of resistors they sell. Make a list of the sizes.
4. Make a list of the items in your car or home that use variable resistors.
5. Make up some memory device to help you remember the first letters in the resistor color code.
6. Why is it important to know the failure rate of resistors? Why do some resistors lack a fifth band that shows its failure rate?
7. Make your own capacitor using knowledge that you have gained about things used in their construction.
8. See if you can find a capacitor with a rating over one farad. Where are these types of capacitors used?
9. On the left side of Figure 4-17, there are two patches or globs of material that cover some of the components on the circuit board. What is the material? Why was it placed there?
10. What kind of capacitor holds its charge for the longest period of time? Where is it used?
11. Should capacitors and resistors be soldered into circuits the same way? Why should you pay attention to the direction in which each lead is installed in a circuit?
12. Find an easy way to store resistors and capacitors when you are laying them out for soldering into a circuit.

Capacitors come in various sizes and ratings. This super large capacitor has a 250 kV, 0.01 μF rating.

Ohm's Law

Learning Objectives

After studying this chapter, you will be able to do the following:

- Calculate the value of voltage, current, and resistance using Ohm's law.
- Apply the power formula to determine wattage.
- Read a kilowatt-hour meter.
- Determine the amount and cost of the electricity used in your home.
- Measure your body resistance with an ohmmeter.

Technical Terms

electromotive force
hot wire
kilowatt-hour (kWh)
kilowatt-hour meter
multimeter
Ohm's law
power
voltage
watt

George Simon Ohm (1789–1854), a German scientist, discovered how current, resistance, and voltage are related mathematically. Ohm found a simple equation that allows us to calculate a third value when only two of the values are known. This equation is called ***Ohm's law.***

Three Variables of Ohm's Law

The variables of Ohm's law are voltage, current, and resistance. If two of these values are known, the third can be calculated. Ohm's law states that

$$\text{Voltage} = \text{Current} \times \text{Resistance} \qquad (5\text{-}1)$$

It is common to substitute letters for words in a formula. The letter *E* is used to represent voltage, the letter *I* is used to represent current, and the letter *R* is used to represent resistance. When these have been substituted, the formula reads

$$E = I \times R \tag{5-2}$$

However, it still means the same thing: voltage is equal to current times resistance. You will sometimes see Ohm's law written as $E = IR$, which means the same thing as $E = I \times R$.

Electromotive Force

The letter *E* in the formula stands for ***electromotive force***, or EMF. Electromotive force is the pressure that is required to force the electrons through the circuit. It is also called ***voltage*** and is measured in volts (V). **Figure 5-1** helps explain voltage. The flow of electrons in an electrical circuit is compared with the flow of water from a squirt gun. If no pressure is exerted on the gun, no water flows. As more pressure is exerted on the gun, the greater the amount of water that flows and the farther it travels. This same principle applies to the pressure in an electrical circuit.

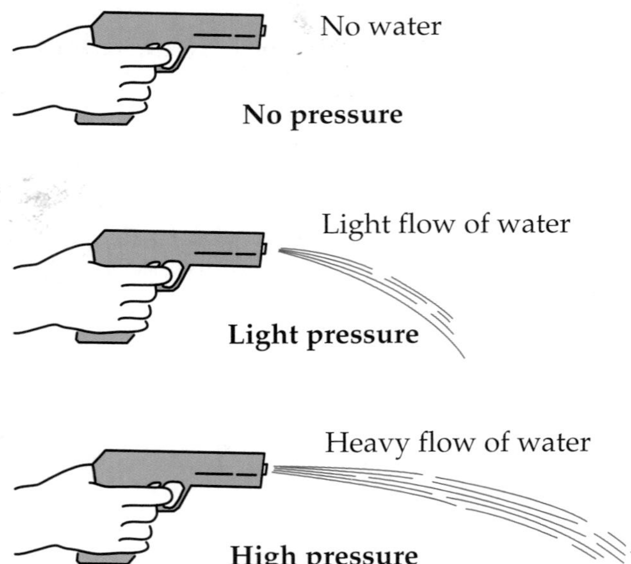

Figure 5-1. Just as increased pressure on a squirt gun forces more water to squirt farther, increased electrical pressure (voltage) forces more electrons to flow.

The pressure, or voltage, in electrical circuits can come from many different sources. At this point, we are discussing only direct current (dc) sources, such as dry cells and batteries. Other types of voltage sources produce alternating current (ac). These sources are described in Chapter 11.

Current

The letter *I* in Ohm's law stands for the intensity of current flow. Earlier, we discussed the billions of electrons that flow past a given point in the circuit in one second. Numbers this large are hard to work with. So, instead of measuring the electrons, the current flow rate is shortened to read in amps or amperes. One amp flowing through a circuit means that 624 billion or 6.24×10^{18} electrons are flowing past a given point in one second.

Amperes were named in honor of Andre Marie Ampere (1775–1836). As you continue reading, you will find that many laws—and even some circuits—have names that honor the people who first discovered them.

Resistance

The letter *R* stands for the resistance to current flow. This resistance is measured in ohms. Earlier, we saw in the table of wire sizes that resistance increases as the length of wire increases. We also saw that resistance decreases as the wire's diameter increases. This means we can get a longer piece of wire and keep resistance the same by increasing the diameter. All three elements, *E*, *I*, and *R*, can be measured with meters similar to the one shown in **Figure 5-2.**

- Voltage is measured with a voltmeter.
- Current is measured with an ammeter.
- Resistance is measured with an ohmmeter.

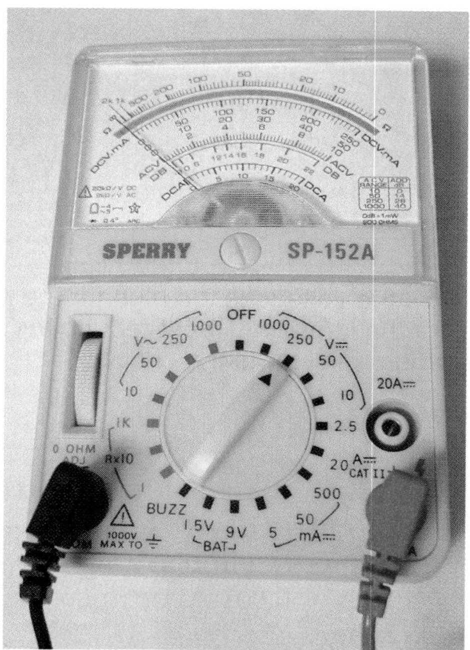

Figure 5-2. A multimeter is used to measure voltage, amperage, and ohms.

You do not have to use three separate meters to take measurements. One combined unit, called a ***multimeter,*** can be used to measure volts, amperes, and ohms. Many times, meters can be used to help find some unknown values. It is also possible to find the value of *E*, *I*, or *R* by using Ohm's law.

Figure 5-3 can help you remember the forms of Ohm's law. By covering the unknown letter with your finger, it is possible to see how you get the different forms of Ohm's

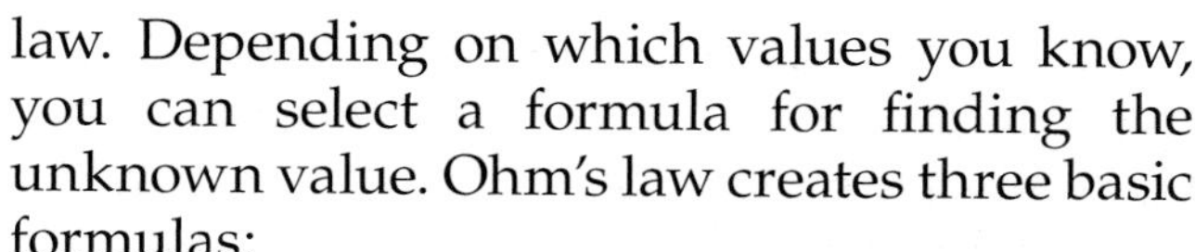

law. Depending on which values you know, you can select a formula for finding the unknown value. Ohm's law creates three basic formulas:

- $E = I \times R$
- $I = \dfrac{E}{R}$
- $R = \dfrac{E}{I}$

Using Ohm's Law

The simple circuit shown in **Figure 5-4** offers a problem that you can solve by applying Ohm's law. In this circuit, the symbol for a battery represents the voltage source, and the bulb symbol represents the load, or resistance. The wires are shown as lines, and the circle with a large *A* inside is an ammeter. The voltage, in this case, is unknown.

Looking at the known values tells us that the amount of current flowing in the circuit is 3 A, and the resistance is 2 Ω. To determine the voltage (*E*) provided by the battery, use Ohm's law as follows:

$$E = IR$$

By substituting known values for the letters, we have the following:

$$\begin{aligned} E &= 3\ \text{A} \times 2\ \Omega \\ &= 6\ \text{V} \end{aligned}$$

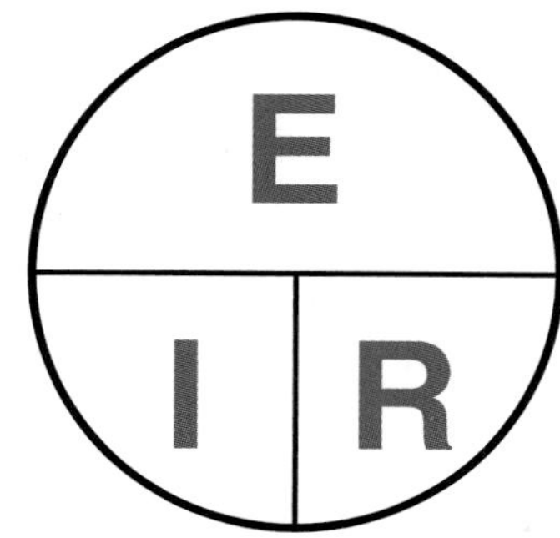

Figure 5-3. As a visual aid to learning Ohm's law, cover the letter for the unknown value. For example, cover *I* and the formula becomes *I* = *E*/*R*.

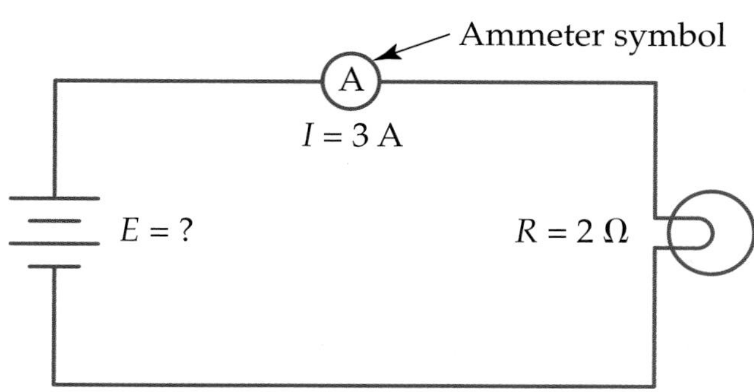

Figure 5-4. This simple electrical circuit shows a problem that can be solved by applying Ohm's law, *E* = *IR*.

The circuit has a 6-V battery, which provides the proper amount of current to light this bulb. If you mistakenly use a 3-V battery, **Figure 5-5,** the bulb will be dim or not light at all. A 3-V battery does not provide enough voltage to push the right amount of current through the resistance of the bulb.

$$I = \frac{E}{R}$$

$$= \frac{3\text{ V}}{2\ \Omega}$$

$$= 1.5\text{ A}$$

This voltage only provides 1.5 A of current, which is not enough to light the bulb. On the other hand, if you had selected a 12-V battery, **Figure 5-6,** it would produce too much pressure. Battery voltage would push so much current through the bulb that the filament would get hot and burn out.

$$I = \frac{E}{R}$$

$$= \frac{12\text{ V}}{2\ \Omega}$$

$$= 6\text{ A}$$

This voltage provides 6 A of current, which causes the bulb to burn out. Replacing the bulb would not solve the problem. No matter how many new 6-V bulbs are placed in this circuit, each would burn out. You must reduce the voltage of the battery or add resistance to the circuit to reduce the current flow to 3 A.

The values of current in these examples are used to keep the math easy to understand as you start off in your studies. In reality, the amount of current with low voltage dc devices is much less than one amp. It would be in the milliamp (0.001 A) or microampere (0.000001 A) range.

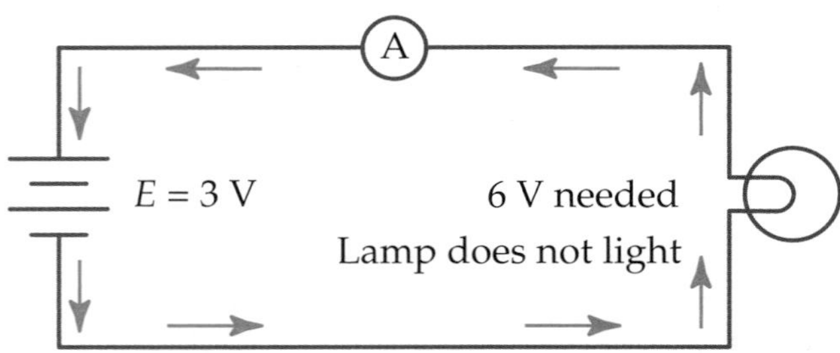

Figure 5-5. The voltage is too low in this circuit to force electrons through the resistance of the lamp.

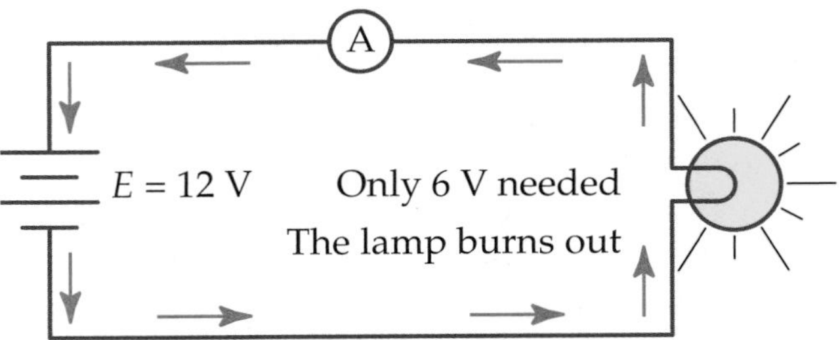

Figure 5-6. If the voltage is too high for the low resistance of the lamp, you must reduce the voltage or add a resistor to the circuit.

> **Danger**
> When starting out in the study of electricity, high-amp projects can be dangerous. Choose projects that allow you to work in the milliamp or microampere range.

Example 5-1:

How much current (I) would flow through a 36-Ω resistor in a car radio powered by a 12-V battery? Start by listing what is known:

$E = 12\text{ V}$

$R = 36\ \Omega$

$I = ?$

Since current is unknown, use the following formula:

$$I = \frac{E}{R}$$

Now insert the known values and solve for the unknown value:

$$I = \frac{12\text{ V}}{36\ \Omega}$$

$$= 0.333\text{ A}, \textit{ or } 333\text{ mA}$$

The completed formula tells you that 0.333 A of current would flow through the circuit.

Example 5-2:

How much current (I) would flow through a circuit in which the voltage is 12 V and the resistance is 12,000 Ω? Since current is again the unknown value, use the same formula as in Example 5-1.

$$I = \frac{E}{R}$$

Now insert the known values, and solve for the unknown value:

$$I = \frac{12 \text{ V}}{12{,}000 \ \Omega}$$
$$= 0.001 \text{ A, } \textit{or } 1 \text{ mA}$$

The current in this case would be equal to 1 mA.

Example 5-3:

What would happen to the current if the resistance in the previous example had been 12 MΩ? Use the same equation as in Example 5-2, but change 12,000 Ω to 12,000,000 Ω:

$$I = \frac{E}{R}$$
$$= \frac{12 \text{ V}}{12{,}000{,}000 \ \Omega}$$
$$= 0.000001 \text{ A, } \textit{or } 1 \ \mu\text{A}$$

The current in this case would be equal to 1 μA.

Example 5-4:

Study **Figure 5-7.** What size resistor would you pick to keep the current 0.003 A or less?

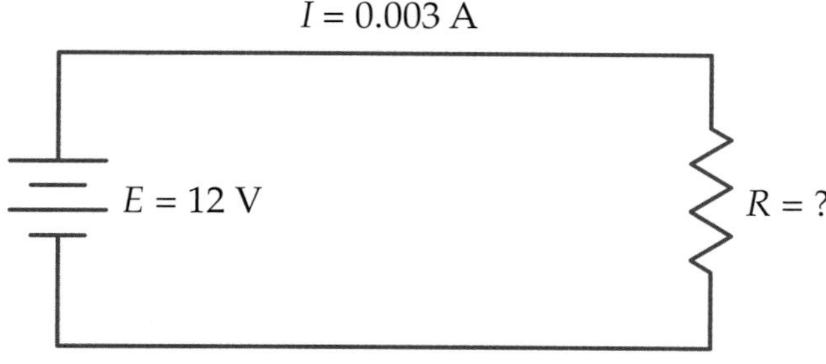

Figure 5-7. A circuit having an unknown resistance value is shown. Ohm's formula $R = E/I$ will help supply the answer.

Start by listing what is known:

$$E = 12 \text{ V}$$
$$I = 0.003 \text{ A}$$
$$R = ?$$

Since resistance (R) is unknown, use the following formula:

$$R = \frac{E}{I}$$

Now insert the known values, and solve for the unknown value:

$$R = \frac{12 \text{ V}}{0.003 \text{ A}}$$
$$= 4000 \ \Omega, \textit{ or } 4 \text{ k}\Omega$$

This means that a resistor of at least 4 kΩ should be used.

Example 5-5:

If you picked a 6-kΩ resistor for the circuit in Figure 5-7, would the current (I) be kept below 0.003 A? Voltage and resistance are known, so current is the unknown value. Using the following formula, insert the known values, and solve for the unknown value:

$$I = \frac{E}{R}$$
$$= \frac{12 \text{ V}}{6000 \ \Omega}$$
$$= 0.002 \text{ A, } \textit{or } 2 \text{ mA}$$

This is lower than the high limit you set (0.003 A), so it is acceptable.

Example 5-6:

What if you picked a 2-kΩ resistor to complete the circuit in Figure 5-7? Again, current (I) is unknown. Follow the same steps as in earlier examples:

$$I = \frac{E}{R}$$
$$= \frac{12 \text{ V}}{2000 \ \Omega}$$
$$= 0.006 \text{ A, } \textit{or } 6 \text{ mA}$$

There would be more than 3 mA of current in this circuit. This setup would result in a blown fuse (if the circuit is fused) or an overheated resistor (if the circuit is not fused).

Examples 5-4, 5-5, and 5-6 show the importance of choosing a resistor with the correct value. They also give you the tools to calculate what resistance is needed to limit the current in a circuit.

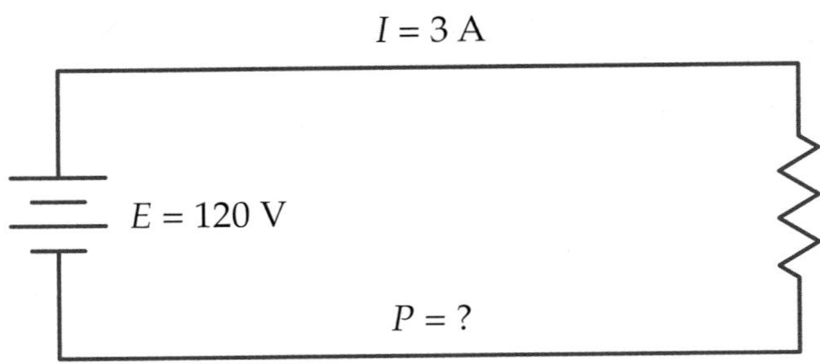

Figure 5-8. What amount of power (wattage) is used in this circuit? Power formula $P = IE$ will help solve this problem.

Power

Power is measure of the amount of current that moves through a load at any given instant. The unit of power is the ***watt*** (**W**), named after James Watt (1736–1819). A watt is the power produced when one ampere of current flows through an electrical component under a potential of one volt.

The formula for power uses many of the same terms as Ohm's law. By putting down known values, the formula will permit you to figure out the correct wattage of resistor to use. The wattage of a resistor tells how much heat it will dissipate. As long as the resistor can transfer heat into the air around it, the resistor will not overheat. If it cannot transfer heat, the resistor will be ruined.

The formula used to calculate the wattage is known as Watt's law. The formula is

$$P = EI \tag{5-3}$$

where P is power, and E and I are voltage and current, respectively, as they are in Ohm's law. You know that $3 \times 2 = 6$ and that $2 \times 3 = 6$. It does not matter which number comes first in multiplying numbers. Therefore, an easy way to remember the formula for power is to think of "PIE" ($P = IE$).

A simple circuit is shown in **Figure 5-8.** To find the number of watts used, use the power formula. In our problem, the circuit has a voltage of 120 V and a current of 3 A. Therefore, substituting values for letters, we would have the following:

$$\begin{aligned} P &= IE \\ &= 3\text{ A} \times 120\text{ V} \\ &= 360\text{ W} \end{aligned}$$

The power formula is just as flexible as Ohm's law. The variations of Ohm's law and the power formula have been put into table form in **Figure 5-9.**

Example 5-7:

Find the watts used if a 6-V power supply were used in the circuit in Figure 5-8. Start by listing what is known:

$$E = 6\text{ V}$$
$$I = 3\text{ A}$$
$$P = ?$$

Since power (P) is unknown, use the following formula:

$$\begin{aligned} P &= IE \\ &= 3\text{ A} \times 6\text{ V} \\ &= 18\text{ W} \end{aligned}$$

The number of watts used would therefore be 18 W.

Variations of Ohm's Law and the Power Formula			
E =	IR	$\frac{P}{I}$	$\sqrt{PR}$
I =	$\frac{E}{R}$	$\frac{P}{E}$	$\sqrt{\frac{P}{R}}$
R =	$\frac{E}{I}$	$\frac{P}{I^2}$	$\frac{E^2}{P}$
P =	IE	I^2R	$\frac{E^2}{R}$

Figure 5-9. All of Ohm's formulas and all power formulas are arranged in this table.

Example 5-8:

What if the amount of power used in Figure 5-8 was known to be 15 W, and the amount of voltage was 120 V? You would, therefore, have to arrange the Watt's law formula to solve for current (*I*). The formula used would be

$$I = \frac{P}{E}$$

$$= \frac{15\text{ W}}{120\text{ V}}$$

$$= 0.125\text{ A, } or \text{ } 125\text{ mA}$$

The amount of current (*I*) flow would therefore be 125 mA.

Example 5-8 uses only one form of the Watt's law formula. In the following chapter, you will use some of the other forms to determine the proper wattage resistance to use in a given circuit.

Kilowatt-Hours

A ***kilowatt-hour* (*kWh*)** is 1000 W of electricity used in one hour. Power companies use this unit when measuring amounts of power consumed. The power company charges about eight cents for one kilowatt-hour. **Figure 5-10** gives the average wattage of many common appliances. The cost of operation shown in this table is based on the number of kilowatt-hours used.

Approximate Cost of Operation				
Appliance	**Average Wattage**	**Estimated Hours of Use/Month**	**Approximate kWh/Month**	**Estimated Cost/ Month ($0.08/kWh)**
Clock	2.5	730	1.825	$0.15
Clothes dryer	4900	17	83.300	$6.66
Central a/c (30,000 btu)	3500	200	700.000	$56.00
Coffeemaker	1100	4	4.400	$0.35
Computer and printer	350	75	26.250	$2.10
Dishwasher	1200	25	30.000	$2.40
Hair dryer	750	10	7.500	$0.60
Radio	70	90	6.300	$0.50
Range with oven	12,200	8	97.600	$7.81
Refrigerator (auto defrost)	500	300	150.000	$12.00
Shaver	15	5	0.075	$0.01
Television—plasma	300	180	54.000	$4.32
Toaster	1100	3	3.300	$0.26
Washing machine, energy saver	150	20	3.000	$0.24
Water heater	4500	90	405.000	$32.40
Vacuum cleaner	650	10	6.500	$0.52

Figure 5-10. Wattages and cost of operation are given for many popular appliances. Note that higher cost is in direct proportion to higher wattages. The cost of operation is based on the Energy Information Administration's Monthly Energy Review. (Edison Electric Institute)

Math Focus 5-1: Algebraic Equations

Through algebra, you can come up with different forms of Ohm's law and the power formula for use in finding unknown values in electrical circuits. For example, suppose you knew the voltage and current for a circuit and needed to find the value of resistance, but you could only recall the formula for Ohm's law as

$$E = I \times R$$

$$12\text{ V} = 0.025\text{ A} \times R$$

Using the basic principles of algebraic equations, you could rearrange this formula so that the unknown value is on one side of the equal sign and the known values are on the other side. This would allow you to easily solve for the unknown value. When solving for an unknown in any formula, keep the following principles in mind:

- Values on both sides of the equal sign are equal.
- Values can be rearranged, as long as the equality of the formula is maintained.

For example, to arrange the previous formula so that the unknown value, *R*, is on one side and the known values are on the other, you must move 0.025 to the other side of the equal sign. In a multiplication equation like this one, you can move a value and maintain the equality of the formula by dividing each side by the value to be moved. For example, to move 0.025 A to the other side of the equal sign, you must divide each side by 0.025 A.

$$E = I \times R$$

$$12\text{ V} = 0.025\text{ A} \times R$$

$$\frac{12\text{ V}}{0.025\text{ A}} = \cancel{0.025\text{ A}} \times \frac{R}{\cancel{0.025\text{ A}}}$$

$$\frac{12\text{ V}}{0.025\text{ A}} = R$$

$$480\ \Omega = R$$

Suppose you knew the current and resistance values of the circuit but could only remember the formula

$$I = \frac{E}{R}$$

Using the values of current and resistance from the previous example, you would have the following:

$$0.025\text{ A} = \frac{E}{480\ \Omega}$$

In a division equation like this one, you can rearrange it so that the unknown value is on one side and the known values are on the other. You can do this by multiplying both sides of the formula by the value to be moved, such as by 480 Ω.

$$480\ \Omega \times 0.025\ \text{A} = \frac{E}{\cancel{480}\ \Omega} \times \cancel{480}\ \Omega$$

$$480\ \Omega \times 0.025\ \text{A} = E$$

$$12\ \text{V} = E$$

In the previous formula, the unknown value was a numerator (positioned above the division line). If the unknown value were a denominator (positioned below the division line), the operation would be different in order to maintain the equality of the equation.

$$0.025\ \text{A} = \frac{12\ \text{V}}{R}$$

First, you would need to invert both sides of the formula like this:

$$\frac{1}{0.025\ \text{A}} = \frac{R}{12\ \text{V}}$$

Then, you could multiply both sides by the value to be moved, such as by 12 V.

$$\frac{12\ \text{V} \times 1}{0.025\ \text{A}} = \frac{R}{\cancel{12}\ \text{V}} \times \cancel{12}\ \text{V}$$

$$\frac{12\ \text{V}}{0.025\ \text{A}} = R$$

$$480\ \Omega = R$$

Practice Problems

Find the unknown value for each of these equations using the basic principles of algebraic equations. Show your work and write your answers on a separate sheet of paper.

1. $\frac{36}{x} = 6$
2. $3y = 96$
3. $45 = \frac{x}{5}$
4. $25 = \frac{150}{a}$
5. $30 = 5x$
6. $\frac{80}{a} = 2$
7. $18 = 9y$
8. $\frac{64}{x} = 16$

To keep a record of how much electrical power you use every month, the power company installs a meter at your home, like that shown in **Figure 5-11.** This meter is called a ***kilowatt-hour meter.*** By recording meter readings at regular intervals, the electric company is able to bill you for the exact amount of electricity used.

When reading a kilowatt-hour meter, it is important to read the dials from right to left to get an accurate reading. This is because the rotation of a right-hand dial affects the movement of the dial to its immediate left. The pointer on a right-hand dial must make a full rotation before the pointer on the dial to its left can move to the next highest number. Therefore, the correct reading of a dial will often be determined by the reading of the right-hand dial. When reading a kilowatt-hour meter, keep the following guidelines in mind:

- If a pointer on the dial being read is between two numbers, you should select the lower number of the two.
- If the pointer appears to be on a number, check if the pointer on the dial to the right is on or has just past the *0.* If it is on the *0* or has just past the *0,* select the number on the dial being read that the pointer is on. Otherwise, select the preceding number.

Also, be aware that the numbers on each dial change directions. In **Figure 5-12,** look at the direction the numbers flow in each dial. Notice that the numbers on the first (far right), third, and fifth dials turn in a clockwise direction. The numbers on the second and fourth dials turn in a counterclockwise direction. It is easy to make a mistake when reading a kilowatt-hour meter because of this.

Now, suppose you want to read your electric meter to verify your monthly bill. To do this you will need to subtract this month's reading from last month's reading. Then, multiply the difference times the rate per kilowatt-hour your electric company charges. The following paragraphs explain the complete process of reading a kilowatt-meter to determine your monthly bill. Refer to **Figure 5-13.**

Suppose the meter in Figure 5-13A represents your last month's reading. You would read the meter starting with the dial on the

Figure 5-11. The electric company uses a kilowatt-hour meter like the one shown to keep track of how much power you use per month in your home.

Figure 5-12. The kilowatt-hour meter is read from right to left. Notice that the numbers on each dial change directions.

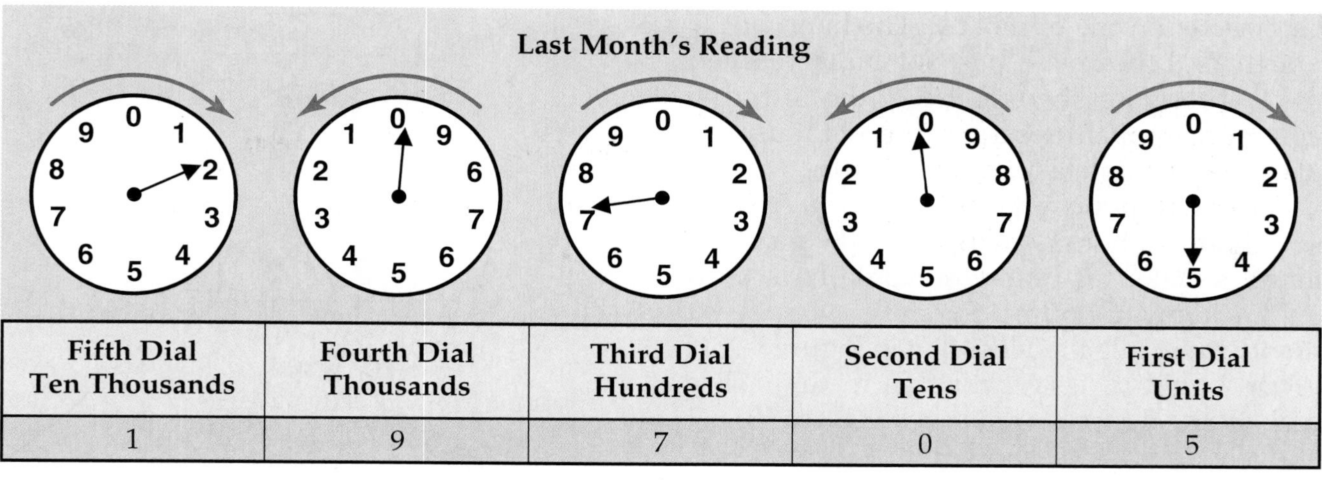

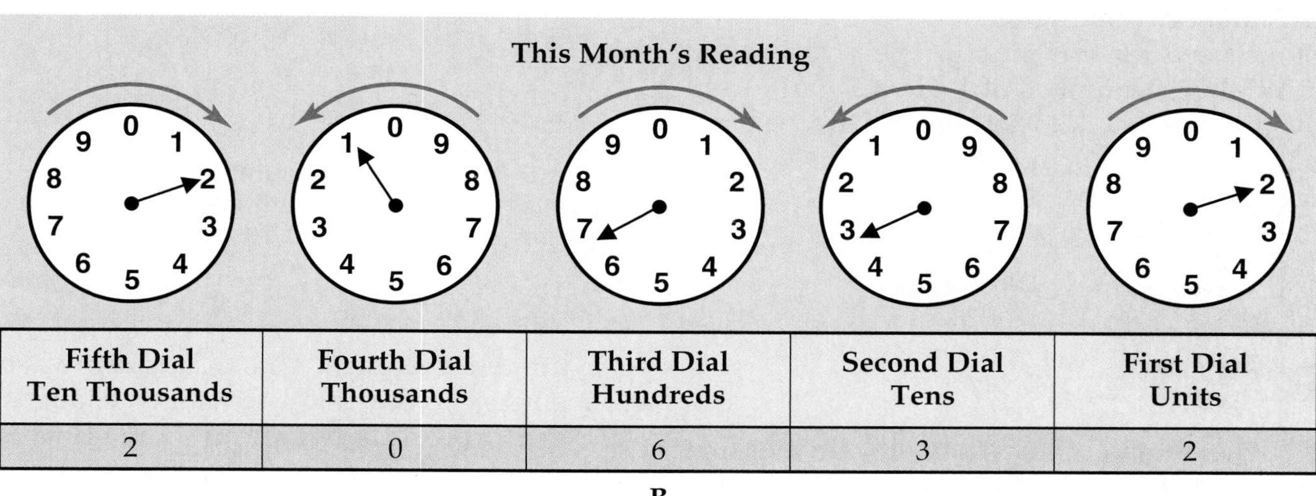

Kilowatt-Hours Used for This Month

This month's reading	20,632
Last month's reading	19,705
	927 Kilowatt-hours (kWh)

C

Figure 5-13. Reading a kilowatt-meter to determine your monthly bill. A—Last month's reading. B—This month's reading. C—The difference between last month's reading and this month's reading is the number of kilowatt-hours of electricity used.

right. The pointer on the first dial points to the number *5*. This dial represents the units place, so the number 5 is recorded in this location. The pointer on the second dial is between the *0* and *1*, so the lower of the two numbers, *0*, is selected. This dial represents the tens place. The pointer on the third dial appears to point directly to the *7*. Since the hand of the second dial is just past the *0*, the correct reading is *7* and represents the hundreds place. The hand on the fourth dial is between the *9* and *0*. The lower number of the two is selected. This number is *9* and represents the thousands place. The hand of the fifth dial appears to be directly on the number 2. The hand of the fourth dial, however, has not quite reached the *0*. Therefore, the reading of the fifth dial is *1*. This number represents the 10 thousands place. The complete reading is 19705.

You will notice that the face of the meter says kilowatt-hours. Had this been your last month's meter reading, you would have used 19705 kilowatt-hours. Suppose your next month's meter looked like that in Figure 5-13B. What would you have for this reading? The table in the figure summarizes this reading.

The next step, shown in Figure 5-13C, would be to subtract last month's reading from this month's reading to arrive at the kilowatt-hours used for this month. Then, you would multiply the number of kilowatt-hours used by the cost per kilowatt-hour. This cost can be found on your monthly bill. If the charge were $0.08 during that month, your cost for the reading shown would be:

$$927 \times 0.08 = \$74.16$$

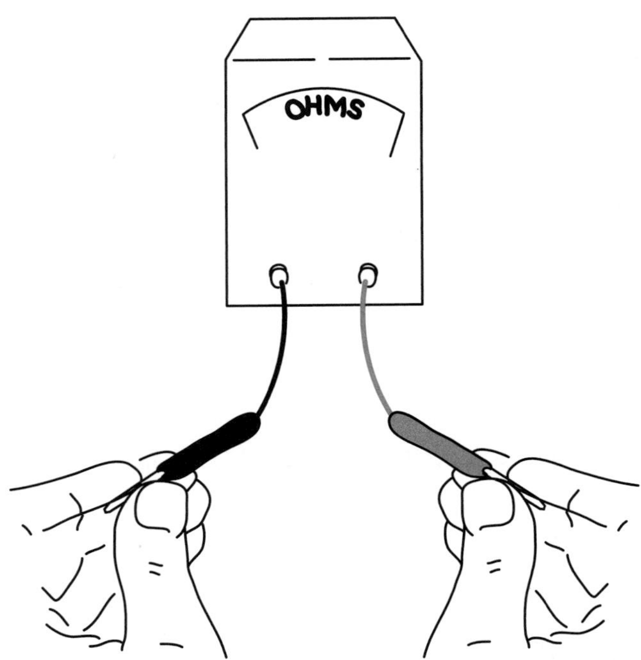

Figure 5-14. You can measure body resistance with an ohmmeter. Note that when you increase pressure on the probes, the resistance reading drops.

Body Resistance

The human body is a resistor. To measure your body resistance from hand to hand, hold the ohmmeter probes as shown in **Figure 5-14.** Compare your reading to your classmates' readings. You will find that everyone has a different amount of body resistance, and most will fall between 10 kΩ and 150 kΩ. The resistance differs because of weight, height, bone size, and many other body chemistry factors. In other words, because no two people are exactly alike, no two people's body resistances are the same.

> **Danger**
> If the conditions are right, it takes less than one amp to kill you!

Body differences are not the only things that can affect body resistance. If you measure your body resistance on a hot summer day and measure it again on a cool day, you will get different readings. This is because when you perspire, your body resistance changes. You can prove this by wetting your fingers and measuring your body resistance again. Note that this reading is lower.

Pressure also affects body resistance. Measure your body resistance while holding each probe loosely between your thumb and forefinger. Take a second reading while squeezing the probes as hard as you can. The second reading will be lower than the first. Squeezing increases the surface area of skin in contact with the probes. This means there is more area through which the current can flow.

All of these differences affect a person's level of danger of being shocked or even killed while working with electricity. Because every person has a different body resistance, no two people will be affected the same way by coming into contact with electricity. One person may be shocked while another is seriously injured by the same source. The result can also vary based on weather conditions and whether or not the victim is perspiring.

There are safe ways to work with electricity. Professional electrical technicians are taught how to safely work on equipment. They must be careful when working below ground, in tunnels, or in damp locations. Everyone, even those who have been trained, must be extremely careful when working with electricity. Unsafe work habits can be fatal.

Danger
Avoid touching anything electrical in a damp basement or an area with water on the floor. Never touch anything electrical soon after coming out of a tub or shower.

The most dangerous situation is when the path of electricity goes through the heart, **Figure 5-15.** When this happens, the heart seizes and stops beating. Some people think they can release their grip on a wire when they discover that it is carrying current. Electricity travels at 186,000 miles per second—the speed of light—and no human can react that fast. When you touch a current carrying wire, or ***hot wire,*** it is too late to change your mind.

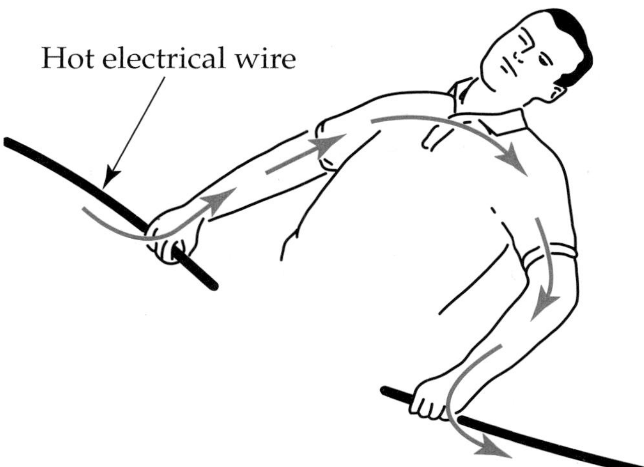

Figure 5-15. Picking up a hot electrical wire can cause electrocution because of the direct path of current through the heart.

Danger
A small shock, such as that from a charged capacitor, can knock the heartbeat out of rhythm, having fatal consequences.

The muscles in your hand and arm contract when you touch a hot wire. Instead of letting go of the wire, you are forced to grip it even harder. As you learned earlier, when this happens, the surface area in contact with the wire increases, and your body resistance goes down. If the current has a path through the heart, the shock can easily be fatal. If there is no path through the heart, the chances of survival are greatly increased. Therefore, keep one hand behind your back or in your pocket when working on a hot wire. Obviously, your best chance of survival would be to turn off the current by removing a fuse or opening a circuit breaker. This eliminates the source of power when working on a circuit.

Danger
When testing live circuits, keep one hand in your pocket. This is done to prevent accidental contact with current from finding a path through the heart.

To understand how body resistance affects your chances of survival when working with electricity, look at Ohm's law. Imagine that you are working on a 120-V circuit in your home. Since $E = IR$, you know that $I \times R$ must be equal to 120 V. Body resistance is estimated at 50,000 Ω. We know that

$$I = \frac{E}{R}$$
$$= \frac{120 \text{ V}}{\text{body resistance}}$$
$$= \frac{120 \text{ V}}{50{,}000 \ \Omega}$$
$$= 0.0024 \text{ A, } or \text{ } 2.4 \text{ mA}$$

As the body resistance falls, the current increases. The following table shows how the current goes up as your body resistance falls. When your resistance gets below 500 Ω, you enter the danger zone if the current path crosses your heart.

E	*R*	*I*
120 V	50,000 Ω	0.0024 A
120 V	25,000 Ω	0.0048 A
120 V	5000 Ω	0.024 A
120 V	500 Ω	0.24 A
120 V	100 Ω	1.2 A

Summary

- Ohm's law says that $E = IR$ where E is voltage, I is current, and R is resistance.
- One watt (W) is the result of one amp pushed through a circuit by one volt.
- Power (P) is found by multiplying the voltage (E) by the current (I), as expressed by the following formula:

  $P = EI$
- A milliamp (mA) is one-thousandth of an amp, and a microamp (μA) is one-thousandth of a milliamp and one-millionth of an amp.
- By using algebra, Ohm's law can be used to calculate E, I, or R when two of these values are known.
- Electric companies use kilowatt-hour meters to keep track of how much electricity is used in a home.
- A kilowatt-hour meter is read from right to left.
- You can measure your body resistance by holding onto both leads of an ohmmeter.
- Your body resistance decreases when you hold the leads of the ohmmeter more tightly.
- If electricity is allowed a path through the heart, it can kill a person instantly with less than one amp.

Test Your Knowledge

Do not write in this book. Write your answers on a separate sheet of paper.

1. Give another name for voltage.
2. List two devices that can produce voltage.
3. What is the unit of measure for current flow in a circuit?
4. The unit of measure for resistance is the ______.
5. Using the correct letters, write three different equations using Ohm's law.
6. If we increase the voltage in an electrical circuit containing a lightbulb, what might happen? Why?
7. An electrical circuit has a voltage of 10 V and the current flowing through it is 2 mA. What is the resistance (R) of the circuit?
8. The term "milliamp" means ______ of an amp.
9. A circuit has a current of 2.150 A flowing in it. How many mA does that equal?
10. The unit of measure for electrical power is the ______.
11. A circuit has 3.2 A and 110 V. Calculate the power.
12. The unit used to measure the quantity of electricity used in your home is the ______.
13. If you measured your body resistance while your palms were damp, your ohmmeter would get a (lower, higher) reading than if measured when your palms were dry.

14. Explain what is meant by "hot" in an electrical circuit.
15. *True or False?* It takes less than 1 A to kill you if it travels through your heart.

Activities

1. How did George Ohm discover the law named in his honor?
2. If you wanted to protect a circuit that carried 6 amps, what size fuse should you use?
3. If you wanted to install a fuse in a circuit carrying 1 amp, what size fuse should you use?
4. How much current can flow in a 1/2-watt resistor before it would burn up?
5. Take a reading of the meter that measures the amount of electricity you use in your home. Take another reading two weeks later. Find out from the power company what they charge for electricity and use this information to figure out what it cost you for those two weeks.
6. Take readings each morning and evening for two weeks on your home's kilowatt-hour meter. When do the peak readings take place?
7. Many of your home appliances are rated in watts. How much current flows in each of these appliances? Make a list of the appliances starting with the one that uses the most current and ending with the one that uses the least current.
8. Measure your body resistance on a cool day and again on a hot day. Is there a difference? Why? Do the same thing while lightly squeezing the probes and again by squeezing the probes as hard as you can. Is there a difference? Why?
9. Knowing the speed of electricity, how many times will it travel around the world in one minute? In 30 seconds? In 5 seconds? How long would it take for it to travel from the east coast to the west coast of the United States?
10. Describe the number of ways electricity can go through your heart while working with electrical tools or appliances.

Electricians use insulated hand tools to help protect them against electrical hazards.

Series Circuits

Learning Objectives

After studying this chapter, you will be able to do the following:

- Construct a series circuit using various components.
- Calculate the total resistance, voltage, and current in a circuit.
- Calculate the total capacitance in a series circuit.
- Determine the polarity of batteries and install them with correct polarity in a device.
- Use a continuity light to find the two ends of a wire in a cable.

Technical Terms

ammeter
equivalent circuit
Kirchhoff's voltage law
parallel circuit
series circuit
total capacitance
total resistance
total voltage drop
voltage drop
voltmeter

You will find two different types of circuits on most electrical prints. These are known as series circuits and parallel circuits, **Figure 6-1.** This chapter covers series circuits. The next chapter covers parallel circuits.

A ***series circuit*** is a circuit with only one path for the current. A ***parallel circuit*** has more than one path for the current. These paths are called *branches*. There are times when electrical circuits have both series and parallel branches (paths for current to flow) on the same schematic. Note that there is only one branch in the series circuit, while the parallel circuit has a number of branches.

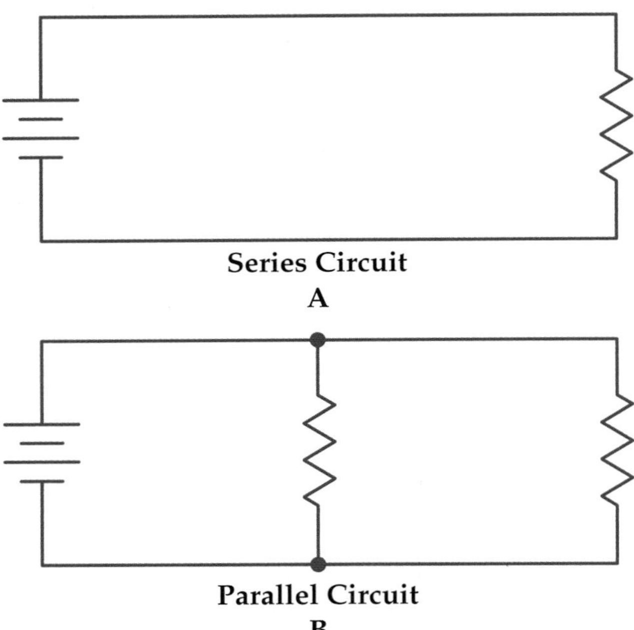

Figure 6-1. Two types of circuits. A—Series circuit. B—Parallel circuit.

Current in a Series Circuit

When current flows in a series circuit, the same number of electrons passes through each component. Therefore, every part of a series circuit carries the same amount of current. For example, **Figure 6-2** illustrates a circuit with four lamps connected in series. Electrons flowing in this circuit have only one path to follow. Starting at the source of voltage, all the electrons travel through each lamp and return to the source.

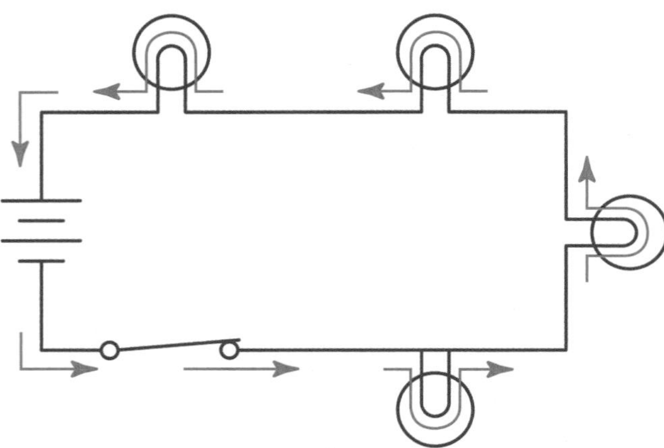

Figure 6-2. Path of electron flow in a completed series circuit.

You can prove that the entire circuit carries the same current by wiring a four-lamp circuit with an ammeter placed after each resistance so that the current flows through the ammeter after it goes through the resistance. An ***ammeter*** is a meter that measures current in amps and milliamps.

In **Figure 6-3,** all of the resistances are lamps. Each ammeter reads the same number of milliamps when the switch is closed. If you open the switch or remove one of the wires, **Figure 6-4,** the lamps go out and all the meter readings drop to zero. This shows that the current has stopped flowing.

Note

An ammeter is always connected in series with the components of any electrical circuit. When hooking up an ammeter to an existing circuit, some part of the circuit must be disconnected to permit installation of the meter. Usually, it is placed between the power source and the load.

Resistance in a Series Circuit

As you saw in Figure 6-3, every part of a series circuit carries the same amount of current. For further proof of this, we will examine three different test circuits, each with a total resistance of 450 Ω. The ***total resistance*** is the sum of the individual resistances in the circuit.

First, look at the series circuit in **Figure 6-5.** This series circuit has a 9-V power source, one 450-Ω resistor, and an ammeter. When this simple circuit is completed, the meter reads 0.02 A, or 20 mA. A second test circuit, shown in **Figure 6-6,** has a 9-V power source in series with three 150-Ω resistors and an ammeter. This ammeter also reads 20 mA. Finally, a third test circuit with a 9-V power source has three different resistors (100 Ω, 150 Ω, and 200 Ω) and an ammeter, **Figure 6-7.** Again, the ammeter reads 20 mA. Each of these three test circuits carries the same number of amps.

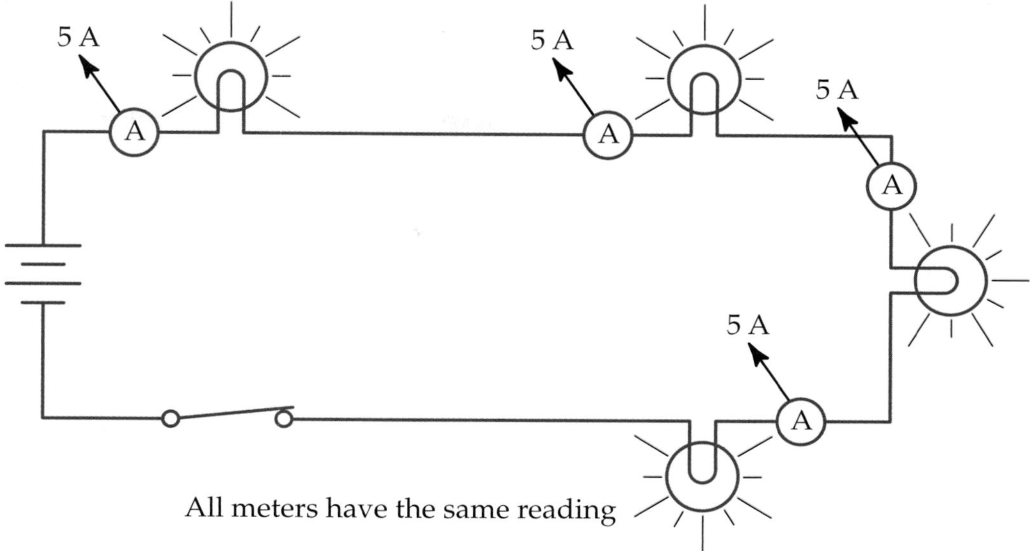

Figure 6-3. Ammeters placed after resistances in a completed series circuit will read the same amperage.

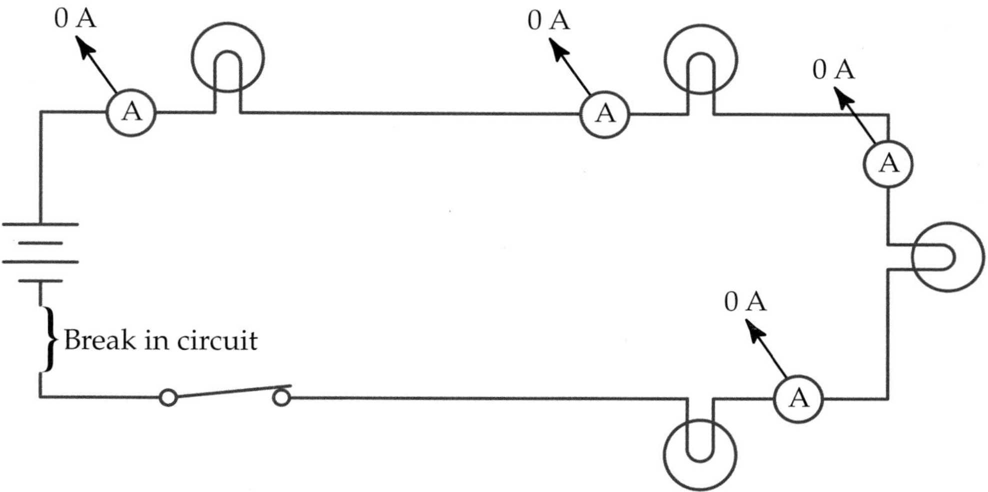

Figure 6-4. With the same circuit shown in Figure 6-3, a break in the circuit will cause ammeters to read zero.

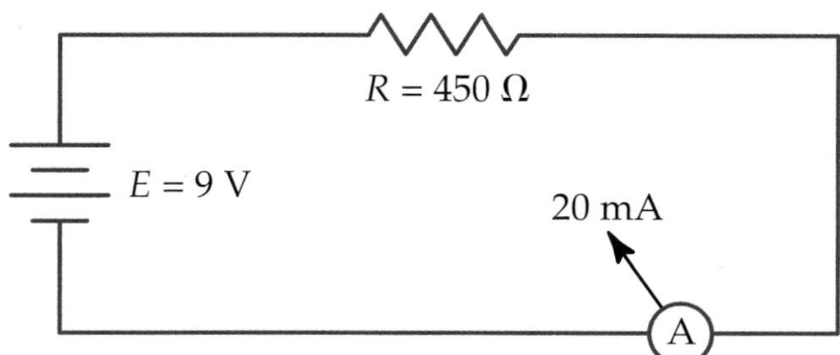

Figure 6-5. Simple series circuit with one resistor. Electrical values are given.

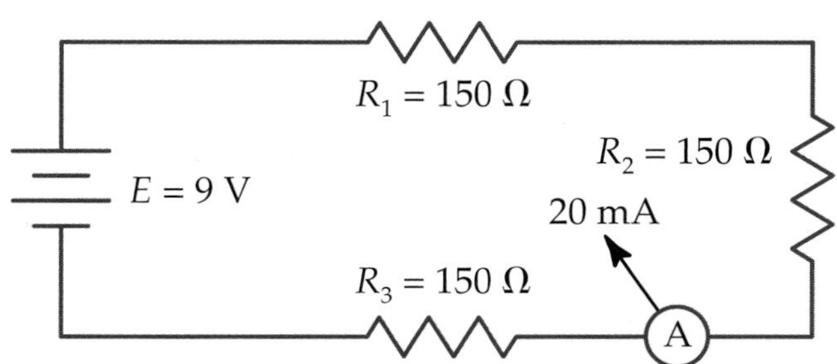

Figure 6-6. Three resistors of equal value in a series circuit have the same total resistance as the single resistor in Figure 6-5.

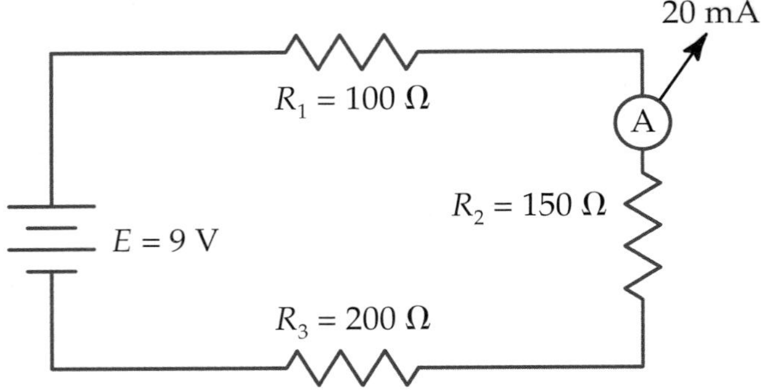

Figure 6-7. Three resistors of unequal values in a series circuit also have the same total resistance as the single resistor in Figure 6-5.

In Chapter 5, we found that it is possible to calculate the number of amps flowing in a given circuit by using Ohm's law. We can put this law to use when working with the test circuits just covered. In Figure 6-5, use Ohm's law as follows to calculate the current:

$$I = \frac{E}{R}$$
$$= \frac{9\ \text{V}}{450\ \Omega}$$
$$= 0.02\ \text{A}, \textit{or } 20\ \text{mA}$$

To apply this formula to the other two test circuits, you need to know the total resistance of the circuit. To find the total resistance (R_T) in a series circuit, simply add the values of the individual resistances:

$$R_T = R_1 + R_2 + R_3 + \ldots \qquad (6\text{-}1)$$

The ellipsis (…) means that something has been omitted. In this case, it means that you keep adding resistance values until you have accounted for all resistances in the circuit.

Example 6-1:

In the circuit shown in Figure 6-6, there are three resistors of equal value. Compute the total resistance (R_T) and current (I). Start with the formula for total resistance:

$$R_T = R_1 + R_2 + R_3$$

Insert the known resistance values for the variables:

$$R_T = 150\ \Omega + 150\ \Omega + 150\ \Omega$$

Solve the equation:

$$R_T = 450\ \Omega$$

Now that you know the total resistance, you can solve for current. Use the appropriate version of Ohm's law:

$$I = \frac{E}{R}$$

Insert the known values for voltage and resistance:

$$I = \frac{9\ \text{V}}{450\ \Omega}$$

Solve the equation:

$$I = 0.02\ \text{A}, \textit{or } 20\ \text{mA}$$

The resistance in this circuit is 450 Ω, and the current is 20 mA.

Example 6-2:

In the circuit shown in Figure 6-7, three resistors of different values are connected in series. Compute the total resistance (R_T) and current (I). Start by finding the total resistance:

$$R_T = R_1 + R_2 + R_3$$
$$= 100\ \Omega + 150\ \Omega + 200\ \Omega$$
$$= 450\ \Omega$$

Now that you know the resistance, solve for current using Ohm's law:

$$I = \frac{E}{R}$$
$$= \frac{9\ \text{V}}{450\ \Omega}$$
$$= 0.02\ \text{A}, \textit{or } 20\ \text{mA}$$

The resistance in this circuit is 450 Ω, and the current is 20 mA. These examples show that total resistance determines how much current is flowing in a series circuit with a given applied voltage.

When you find the total resistance, you are also finding an ***equivalent circuit.*** An equivalent circuit has the same amount of resistance as the original circuit, but with only one resistor. The equivalent circuit of the one shown in Figure 6-6 and Figure 6-7 is the circuit in

Figure 6-5. It is equivalent because it offers the same resistance as the other two circuits.

In Example 6-3, you are asked to find the total resistance of a circuit with six components. This circuit contains a 40-V power supply, a lamp with 15 Ω resistance, a fixed 45-Ω resistor, a variable resistor set at 100 Ω, a switch, and a fuse.

Although the switch, fuse, and wire have some resistance, it usually is overlooked in circuit calculations. This is because the amount of resistance is very small. Most circuits, for example, use wire in the range of 12 gauge to 22 gauge. It takes 629′ of 12-gauge wire or 61′ of 22-gauge wire to provide one ohm of resistance. Besides the wiring in buildings, few circuits have enough wire resistance to matter in circuit calculations. Most circuits you make in your laboratory at school involve a small amount of wire, so you can ignore the resistance from the wire.

Example 6-3:

Examine this circuit and compute the total resistance (R_T) and the current (I).

$R_T = ?$
$I = ?$
$R_1 = 15\ \Omega$
$R_2 = 45\ \Omega$
$E = 40$ V
$R_3 = 100\ \Omega$

$$R_T = R_1 + R_2 + R_3$$
$$= 15\ \Omega + 45\ \Omega + 100\ \Omega$$
$$= 160\ \Omega$$

Now that you know the total resistance, solve for current using Ohm's law:

$$I = \frac{E}{R}$$
$$= \frac{40\text{ V}}{160\ \Omega}$$
$$= 0.25\text{ A}, \textit{or}\ 250\text{ mA}$$

The total resistance in this circuit is 160 Ω, and the current is 250 mA.

Example 6-4:

Remember, in a series circuit you must use the total resistance to find the current. A bell, a resistor, and a lightbulb provide the resistance in this circuit.

$R_T = ?$
$I = ?$
$R_1 = 200\ \Omega$
Gen $E = 24$ V
$R_3 = 1200\ \Omega$
$R_2 = 1000\ \Omega$

What is the current through these components? Start by finding the total resistance (R_T):

$$R_T = R_1 + R_2 + R_3$$
$$= 200\ \Omega + 1000\ \Omega + 1200\ \Omega$$
$$= 2400\ \Omega$$

To compute the current (I), use Ohm's law:

$$I = \frac{E}{R}$$
$$= \frac{24\text{ V}}{2400\ \Omega}$$
$$= 0.01\text{ A}, \textit{or}\ 10\text{ mA}$$

The current in this circuit is 10 mA. This is a series circuit, so the current is the same in each component.

Note that the source of voltage in the last example is a generator. A generator is used in some circuits instead of, or together with, a battery. This is similar to running your battery-powered laptop while it is plugged into the wall outlet. The generator produces voltage needed to run the circuit, thus eliminating the need to use batteries.

Math Focus 6-1: Adding Fractions

To find total capacitance, you need to add fractions. You also need to add fractions when finding total resistance in a parallel circuit. (Finding total resistance in a parallel circuit is covered in Chapter 7.) You should recall from your early math studies that to add a fraction, you must follow two rules:

- For fractions with the same denominator (bottom number), add the numerators (top numbers) and keep the same denominator. For example:

$$\frac{1}{4}+\frac{2}{4}+\frac{3}{4}=\frac{1+2+3}{4}=\frac{6}{4}$$

- For fractions with different denominators (bottom numbers), you must find a common denominator. Then change all of them to a common denominator before adding the numerators (top numbers).

For example, to find the common denominator of 1/2 + 1/4 + 3/8, you need to find a common number into which all of the denominators can be evenly divided. For the fractions in this example, this number is *8*. All of the denominators can divide evenly into *8*.

For the first fraction, 1/2, *2* goes into *8* four times. For the other fraction, 1/4, the denominator *4* goes into *8* two times.

Change the fractions in this example so that they all have the common denominator *8*. You need to multiply the numerator and denominator of each fraction by the number that will provide the denominator of *8*. Therefore, for the fraction 1/2, you need to multiply both the numerator and denominator by 4:

$$\frac{1\times 4}{2\times 4}=\frac{4}{8}$$

Capacitance in a Series Circuit

Total capacitance (C_T) is the combined value of all the capacitors in a circuit. To find the total capacitance (C_T) of capacitors in a series circuit, use the following formula:

$$\frac{1}{C_T} = 1/C_1 + 1/C_2 + 1/C_3 + \cdots \qquad (6\text{-}2)$$

Example 6-5:

The circuit shown in the following schematic has two 20 μF capacitors in series.

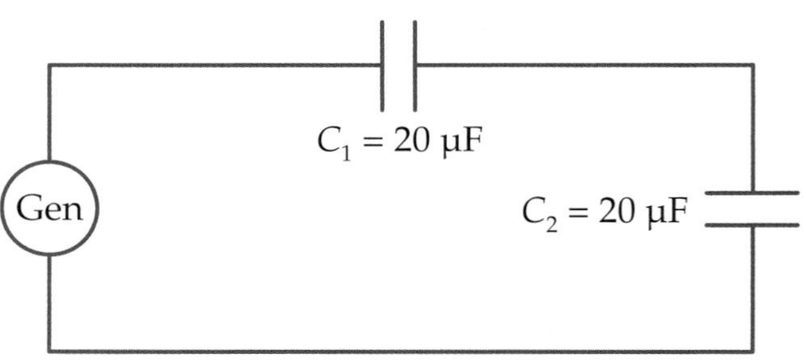

What is the total capacitance (C_T) in this series circuit? Start with the formula for total capacitance:

$$\frac{1}{C_T}=\frac{1}{C_1}+\frac{1}{C_2}$$

For the fraction 1/4, you would multiply the numerator and denominator by 2:

$$\frac{1 \times 2}{4 \times 2} = \frac{2}{8}$$

Once all fractions have a common denominator, you can add the numerators:

$$\frac{4}{8} + \frac{2}{8} + \frac{3}{8} = \frac{9}{8}$$

Notice that the numerator is greater than the denominator. When this happens, you have an *improper fraction*. To convert 9/8 to a proper fraction, divide the numerator by the denominator:

$$\frac{9}{8} = 1\frac{1}{8}$$

You have now reduced the fraction in your answer to its lowest term. This means that the numerator and denominator can no longer be divided by a common factor.

Practice Problems

Add the following fractions. Be sure your answer is a proper fraction and is reduced to its lowest term. Write your answers on a separate sheet of paper.

1. $\frac{1}{20} + \frac{1}{40} + \frac{1}{80} =$
2. $\frac{1}{6} + \frac{2}{24} + \frac{1}{4} =$
3. $\frac{1}{25} + \frac{7}{50} =$
4. $\frac{3}{75} + \frac{1}{150} =$
5. $\frac{2}{9} + \frac{1}{3} =$
6. $\frac{1}{10} + \frac{9}{60} + \frac{3}{120} =$
7. $\frac{6}{7} + \frac{1}{8} =$
8. $\frac{1}{4} + \frac{3}{8} =$

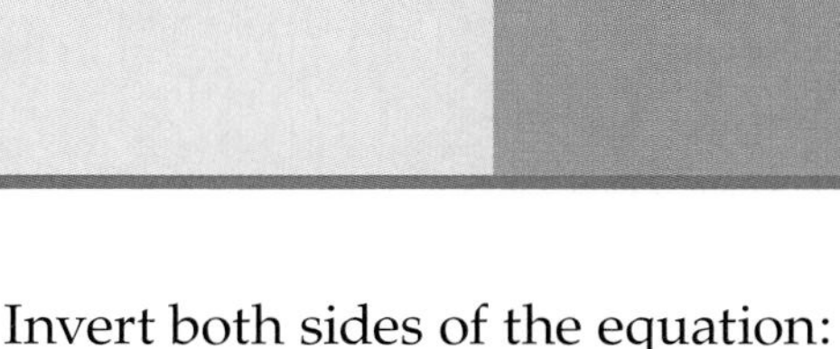

Insert the known capacitance values for the variables:

$$\frac{1}{C_T} = \frac{1}{20\ \mu F} + \frac{1}{20\ \mu F}$$

Add the right side of the equation and simplify. (For a review of adding and subtracting fractions, see Math Focus 6-1, included in this chapter.)

$$\frac{1}{C_T} = \frac{2}{20\ \mu F}$$

$$= \frac{1}{10\ \mu F}$$

Invert both sides of the equation:

$$C_T = 10\ \mu F$$

The total capacitance of this equivalent circuit, shown in the previous schematic, is 10 µF.

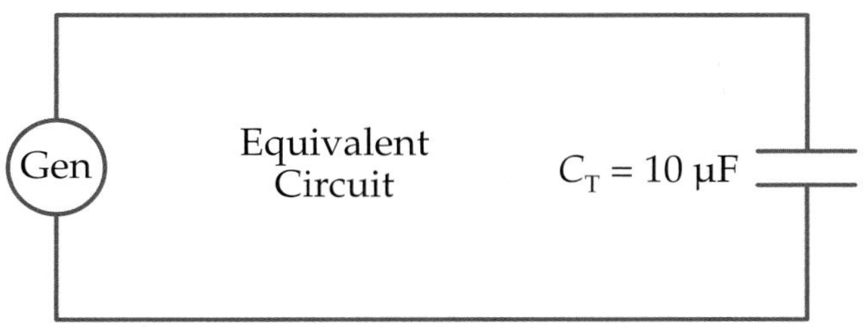

Project 6-1: Safety Switch

A safety switch is useful for anyone who operates a lathe, drill press, or other machine that uses a chuck to hold a tool. Too often, the chuck key is left in the chuck. Then, when the machine is started, the chuck key is thrown out with great force. If the key strikes someone, that person could be badly injured. A simple way to avoid this problem is to use a safety switch inserted in the power line to the machine. The switch will keep the machine from starting if the chuck key is in the chuck.

(Project by John Kleihege)

Some of the newer machines are built so you can tighten the tool without using a chuck key. This eliminates the safety problem of chuck keys that require this type of project.

To build the project, obtain the following materials. Assemble them according to the diagram.

No.	Item
1	Microswitch
1	Surface mounting outlet box and plain cover
1	Wood block
2	Wire nuts
1	110-V three-wire extension cord
1	Junction box grounding clip with wire
2	1/2″ connectors

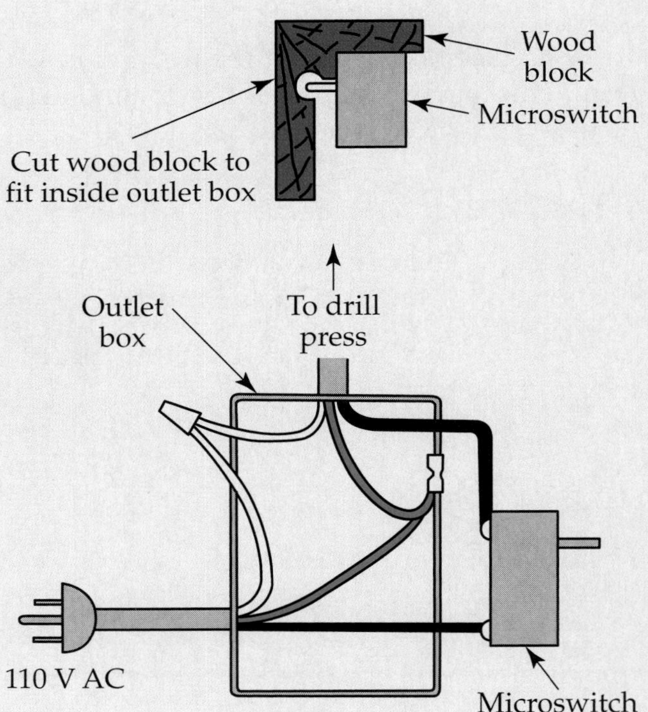

With the safety switch connected to the machine power cord and a 110-V receptacle, the machine will not operate unless the chuck key is pushed in the hole in the cover of the outlet box. In this position, the key will close the microswitch and allow current to pass to the on-off switch on the machine.

Many people hear the word microswitch and think that it could not carry much current. However, many of these switches are made to carry current in the 18 A to 25 A range. They are, therefore, ideal to use for this project.

As you can see, the total capacitance in a series circuit is smaller than any individual capacitor. This indicates that your answer is correct and provides an easy way to check your work. If your answer for total capacitance is larger than the individual capacitors, your answer is incorrect. You should check where you made the error.

You may wonder why capacitors placed in series offer less capacitance than a single capacitor. In **Figure 6-8,** you can see that the distance between the outer plates of C_1 and C_2 increases when placed in series. This lowers the capacitance of the circuit.

You should recall from Chapter 4 that the distance between the plates is one factor that affects capacitance. As the distance between the plates increases, capacitance decreases. The total width of the dielectric for the equivalent circuit is the sum of the distances between the plates of the individual capacitors. The length of wire between the capacitors does not affect capacitance. The thickness of the dielectric affects capacitance.

Polarity

There are many electrical devices that cannot be plugged into a wall outlet. These devices include flashlights and children's toys, all of which use dry cells to operate. Most use more than one cell because they need a higher voltage.

Have you ever noticed the (+) and (–) markings on a cell? There are similar markings on many of the devices that use cells. The markings show you the correct way to install the cells into the device.

Cells have polarity. This means they have a positive (+) and a negative (–) pole. To get the voltage needed to operate a device, you must face these poles in the proper direction. This is called *correct polarity*. If you face the cells in the wrong direction, you will have incorrect polarity.

Figure 6-9 illustrates why polarity is important. Note that when all cells face in the same direction, their individual voltages add:

$$1\frac{1}{2}+1\frac{1}{2}+1\frac{1}{2}+1\frac{1}{2}=6$$

If one cell faces the opposite way, it reduces total voltage:

$$1\frac{1}{2}+1\frac{1}{2}-1\frac{1}{2}+1\frac{1}{2}=3$$

The sum of voltages in a series circuit depends on the polarity of the voltage sources. If their polarity is correct, the voltages add. Any opposite polarity is subtracted from the total voltage of the circuit.

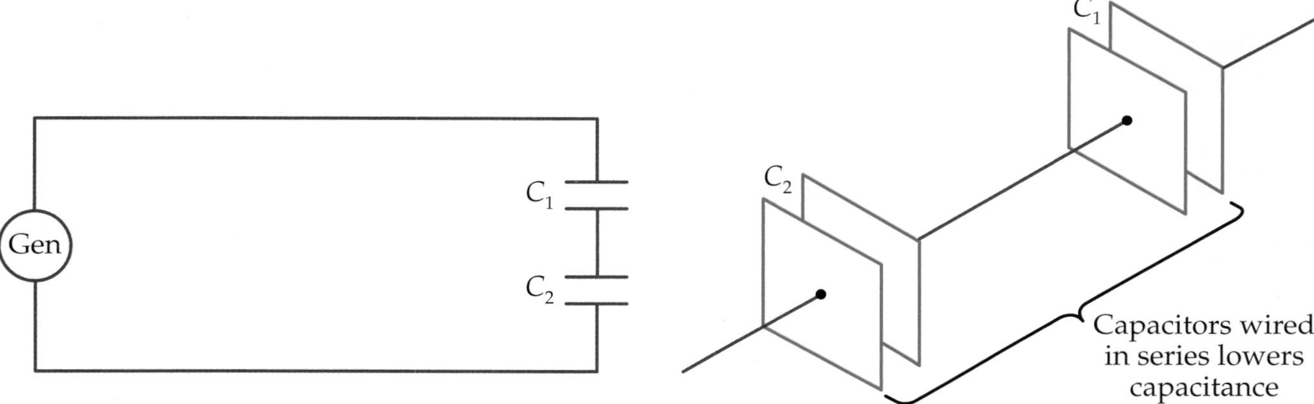

Figure 6-8. When capacitors are placed in series, the capacitance in the circuit decreases because the distance between plates increases.

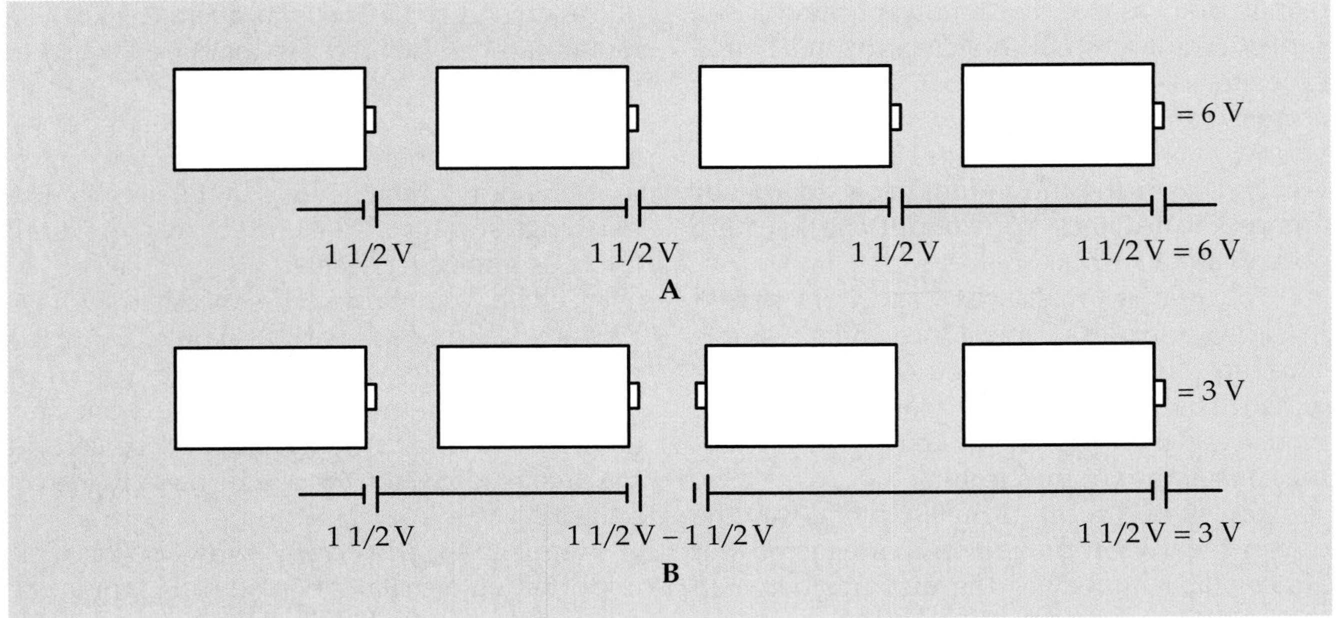

Figure 6-9. Polarity of cells connected in series. A—Correct polarity. B—Incorrect polarity, with resulting loss of voltage.

The polarity of electrical devices affects the direction in which a meter needle moves. It also affects the brightness of the flashlight or the direction a motor will turn. If installed with wrong polarity, some of the devices will not operate at all.

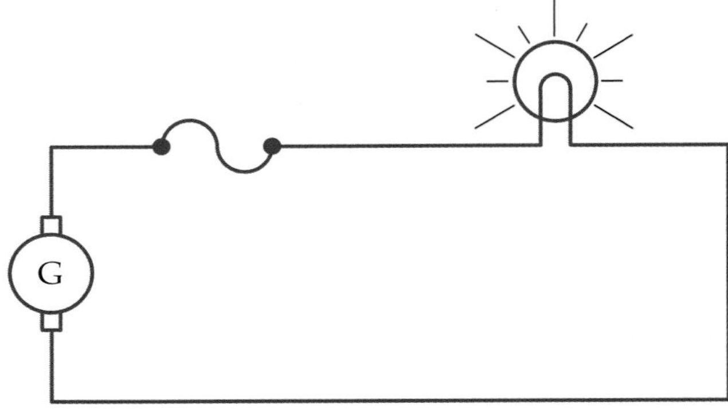

Figure 6-10. Fuse connected in series with a lamp.

Continuity Tester

There are times when you must install wire or components in a circuit so that they are in series with one another. In **Figure 6-10,** the fuse is in series with the lamp. After the circuit has been in operation for a while, something could happen that would create an overload and blow the fuse.

If the lamp in this circuit does not light because of a blown fuse, there are several ways of checking it. The simplest way is to see if the fuse element is burned through (appears to be broken or completely missing).

Another way of checking for a blown fuse is to use a continuity tester, **Figure 6-11.** When the fuse element is broken, the light of the tester will not come on. If the element is good

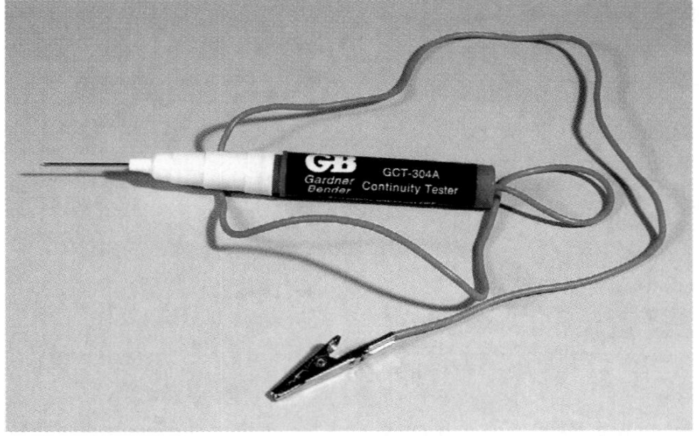

Figure 6-11. Continuity tester.

(solid), the light will come on, **Figure 6-12.** This tells you that the circuit is continuous.

Danger
Do not use a continuity tester on live circuits or to test batteries. This is dangerous to you and the continuity tester. Current flowing in the circuit could result in the tester being destroyed, the batteries exploding, and you being seriously injured.

A continuity tester also can be used to check wires to find out whether a break in a wire exists. If there is a break in the copper under the insulation, you cannot see it. A continuity tester will help pinpoint the location of the break. In another application, a continuity tester can help locate both ends of a given wire from among a group of wires bundled together. This is described in Practical Application 6-1.

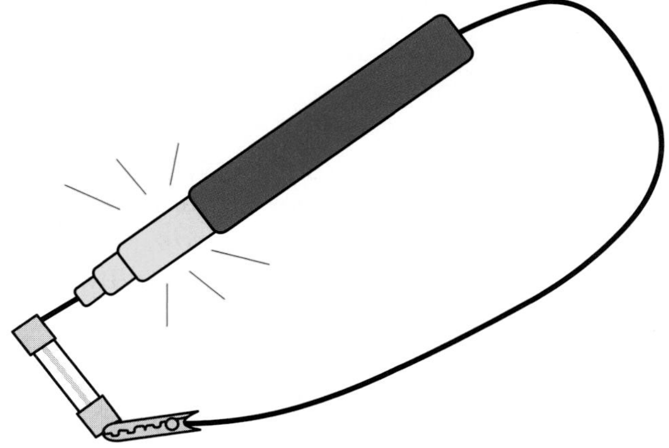

Figure 6-12. A good continuity tester will come on if its tip touches the alligator clip. It will also light if the tip and alligator clip are placed in contact with opposite ends of a good fuse.

Voltage Drop

You have learned that current is constant in a series circuit. No matter where in the

Practical Application 6-1: Identifying Cables

Many times wires in a cable are color coded to help you identity those you want to use. What happens if the wire is not color coded or you can only see the wire ends? If the wires are short and you have a continuity tester, you can identify the wire ends by doing the following:

1. Attach the alligator clip of the continuity tester to the wire in the cable bundle you wish to identify.
2. Touch each wire on the other end of the cable bundle with the tip of the continuity tester. When the light comes on, you have identified the two matching ends. These ends can be tagged for later use in a circuit.

If the wires are long, you may need to use a specialized meter that shows continuity. An ohmmeter is one of these specialized meters. It will be described in more detail in Chapter 9.

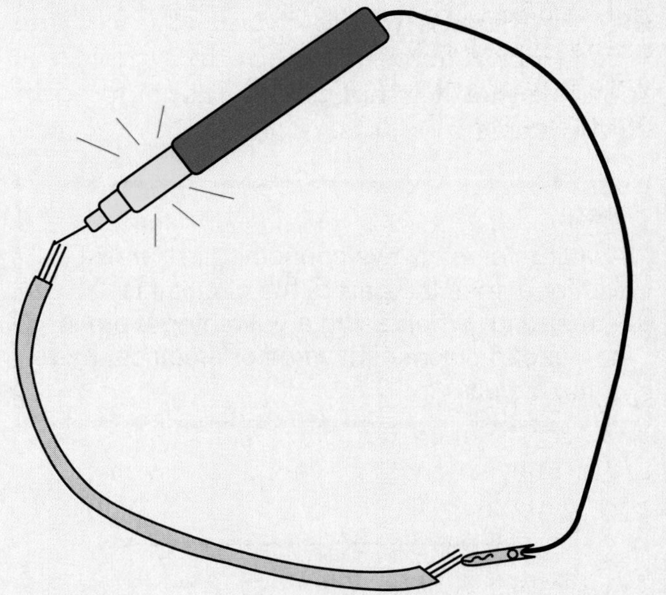

Depending on the length of the wire, it may have too much resistance for the continuity light to illuminate. In this case, most of the voltage will be applied to the wire and little or no voltage to the light. You will learn why this is so in the section on voltage drop.

circuit it is measured, it is always the same value. This is not true of voltage. If you measure the voltage across each component in a series circuit, you will find that the sum of the voltages is equal to the applied voltage. The amount of voltage measured across each component is called a ***voltage drop.***

Each component in a series circuit may have a different voltage drop. When the individual voltage drops are added, the sum is equal to the applied voltage. This discovery was named in honor of Gustav Kirchhoff. It is known as ***Kirchhoff's voltage law.***

Voltage drop can be measured with a ***voltmeter*** or calculated with Ohm's law. To measure it, the voltmeter must be connected across the component, **Figure 6-13.** This means that the two leads of the voltmeter are connected on opposite sides of the component. The reading on the voltmeter is the voltage drop. If you measure the voltage across the power source, the reading is the applied voltage. This is the same thing as measuring the voltage drop across all the components in the circuit at once. The voltage drop across the wire in a project is so small that it is not usually considered in circuit measurements.

Note

A voltmeter is always connected in parallel with or across the part of the circuit it is measuring. When using a voltmeter, there is no need to disconnect wires or electrical parts of the circuit.

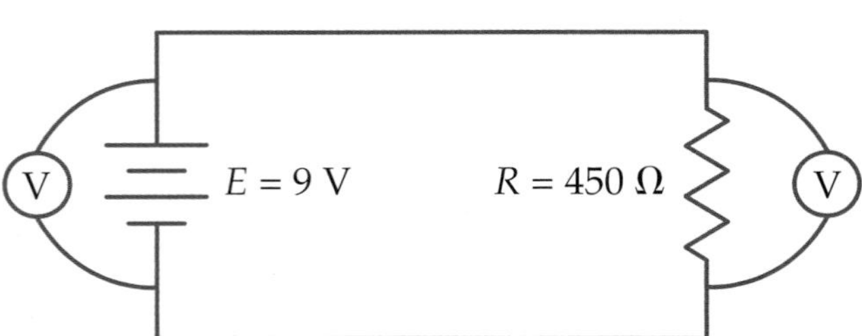

Figure 6-13. A voltmeter is always connected in parallel with an electrical component.

Finding Total Voltage Drop

Total voltage drop in a series circuit is always equal to the applied voltage. To find the total voltage drop in a circuit, use the following formula:

$$E_T = V_{R1} + V_{R2} + V_{R3} \ldots \tag{6-3}$$

To find the total voltage drop (E_T) in the circuit in **Figure 6-14,** use this formula, and substitute the individual voltage drop values shown by the voltmeters.

$$E_T = V_{R1} + V_{R2} + V_{R3}$$
$$= 3\text{ V} + 4\text{ V} + 2\text{ V}$$
$$= 9\text{ V}$$

The total voltage drop in this circuit is 9 V, which equals the source voltage.

Voltage Drop and Ohm's Law

If the current in the circuit and the resistance of the component are known, voltage drop can be calculated using Ohm's law. Look

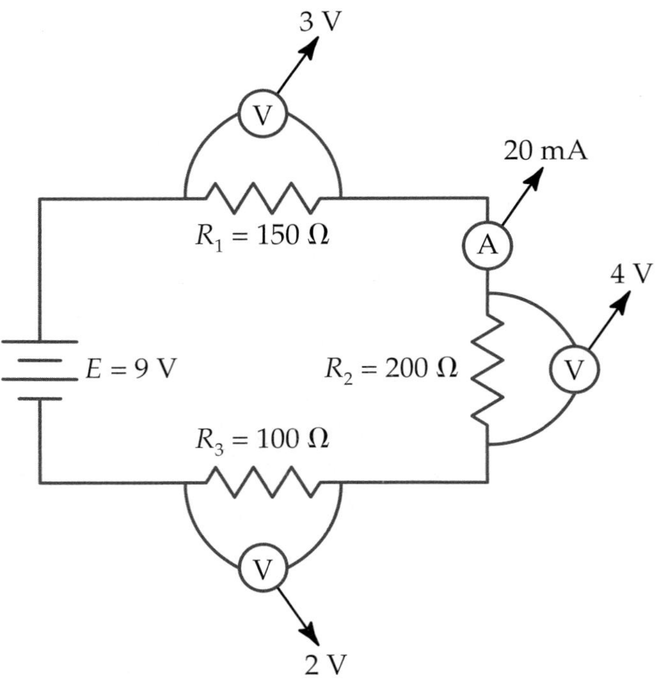

Figure 6-14. Voltmeter readings taken across each of three resistors with unequal values.

at the circuit in **Figure 6-15.** To find the voltage drop across each of the resistors, current must be calculated first. To do this, start by computing the total resistance (R_T).

$$R_T = R_1 + R_2 + R_3$$
$$= 10\ \Omega + 40\ \Omega + 30\ \Omega$$
$$= 80\ \Omega$$

Calculate the current (I) using Ohm's law:

$$I = \frac{E}{R_T}$$
$$= \frac{12\text{ V}}{80\ \Omega}$$
$$= 0.15\text{ A, } or \text{ 150 mA}$$

To find the voltage drop across an individual resistor, use Ohm's law. For the voltage drop (V) across R_1, we have

$$V_{R1} = I_{R1} \times R_1$$

Multiply the total current in the circuit by the individual resistance:

$$V_{R1} = 150\text{ mA} \times 10\ \Omega$$
$$= 1.5\text{ V}$$

Now, calculate the voltage drops for R_2 and R_3:

$$V_{R2} = I_{R2} \times R_2$$
$$= 150\text{ mA} \times 40\ \Omega$$
$$= 6\text{ V}$$

$$V_{R3} = I_{R3} \times R_2$$
$$= 150\text{ mA} \times 30\ \Omega$$
$$= 4.5\text{ V}$$

Add these voltage drop values together to check the calculations:

$$E_T = V_{R1} + V_{R2} + V_{R3}$$
$$= 1.5\text{ V} + 6\text{ V} + 4.5\text{ V}$$
$$= 12\text{ V}$$

The answer is 12 V, which is the same as the applied voltage. This means the calculations are correct. Let us examine what would happen if we moved to higher voltages and more current.

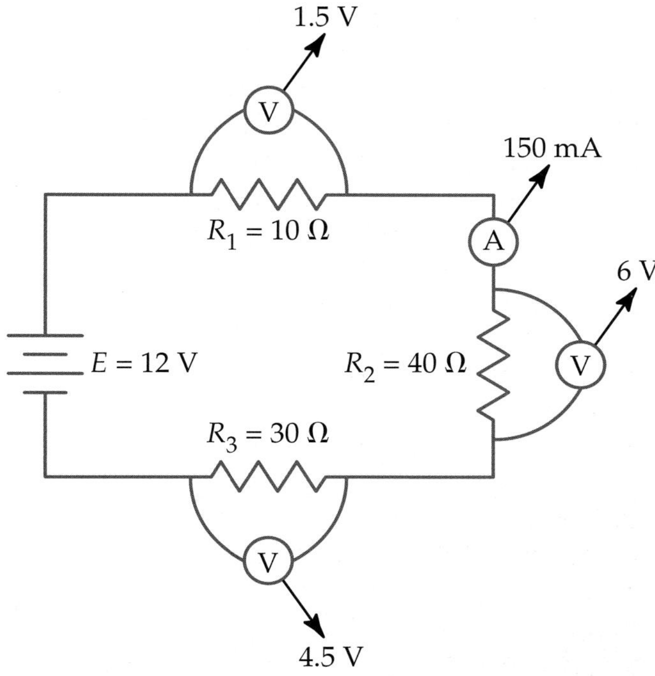

Figure 6-15. When calculating unknown voltage drops, determine total resistance and current. Then, using Ohm's law, calculate the individual voltage drop across each resistor.

Remember, for safety reasons we have been using low voltage and current readings until you have more knowledge. If you attempt to hook up any circuits with these higher readings, do not put yourself in the circuit. Do not touch any exposed wiring or the metal ends of any probes.

Example 6-6:

Find the voltage drop across each resistor and the total voltage drop (E_T) of this circuit.

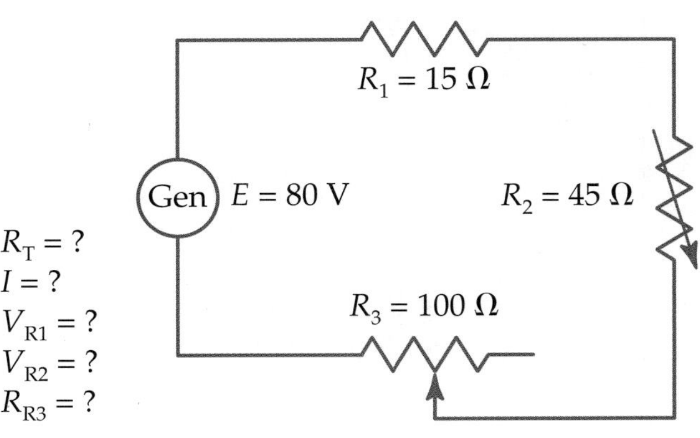

$R_T = ?$
$I = ?$
$V_{R1} = ?$
$V_{R2} = ?$
$R_{R3} = ?$

To start, find the total resistance (R_T):

$$R_T = R_1 + R_2 + R_3$$
$$= 15\ \Omega + 45\ \Omega + 100\ \Omega$$
$$= 160\ \Omega$$

Now calculate the current (I):

$$I = \frac{E}{R}$$
$$= \frac{80\ \text{V}}{160\ \Omega}$$
$$= 0.5\ \text{A}$$

Next, find the voltage drop across each resistor. The voltage drop across R_1 is

$$V_{R1} = I \times R_1$$
$$= 0.5\ \text{A} \times 15\ \Omega$$
$$= 7.5\ \text{V}$$

The voltage drop across R_2 is

$$V_{R2} = I \times R_2$$
$$= 0.5\ \text{A} \times 45\ \Omega$$
$$= 22.5\ \text{V}$$

Project 6-2: Foot Switch

There are many good uses for a foot switch, including woodcarving, sewing machines, and working in the area of graphics animation. Pushing down on the switch with your foot controls the flow of power to some electrical device. You can use the foot switch as an on-off control or to control the speed of the electrical device.

The chief advantage is that a foot switch leaves your hands free to do other work. In woodcarving, it leaves you free to hold the wood with one hand and move the carving tool with the other. There are many other good uses for a foot switch. Many machines found in industry are cycled with a combination of a foot switch along with safety buttons for the hands.

(Project by Dave Wightman)

To build this project, you will need the following items:

No.	Item
1	Foot switch
1	Double outlet receptacle
1	Male plug
2	Black lead wires
1	White lead wire

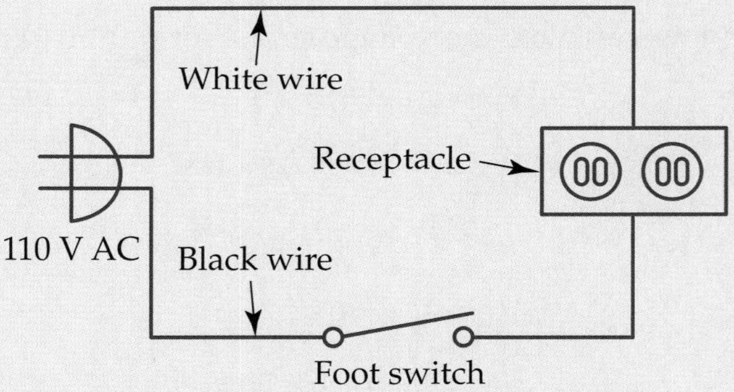

Procedure

1. Study the diagram.
2. Install the white wire and one black wire in the male plug.
3. Connect the other end of the white wire to the double outlet receptacle.
4. Connect the other end of the black wire to the foot switch.
5. Use the other black wire to connect the foot switch to the double outlet receptacle.

The voltage drop across R_3 is

$$V_{R3} = I \times R_3$$
$$= 0.5 \text{ A} \times 100\ \Omega$$
$$= 50 \text{ V}$$

To check your work, add the individual voltage drops to see if their sum equals the applied voltage:

$$E_T = V_{R1} + V_{R2} + V_{R3}$$
$$= 7.5 \text{ V} + 22.5 \text{ V} + 50 \text{ V}$$
$$= 80 \text{ V}$$

Since 80 V equals the applied voltage, you know that your calculations are correct.

Summary

- Electrons flowing in a series circuit have only one path in which they can flow.
- Every part of a series circuit has the same amount of current flowing through it.
- An ammeter is always connected in series with the components.
- The total resistance (R_T) of a series circuit can be found by adding the values of each resistance:

  $R_T = R_1 + R_2 + R_3 + \dots$
- If you know the voltage and total resistance of a series circuit, you can calculate total current using Ohm's law ($I = E/R$).
- The total capacitance (C_T) of a series circuit can be found by the following equation:

  $\frac{1}{C_T} = \frac{1}{C_1} + \frac{1}{C_2} + \frac{1}{C_3} + \dots$
- Batteries have polarity, which means they have a positive (+) and a negative (–) pole.
- A continuity tester is used to check if there is a continuous path through a wire. Do not use on a live circuit!
- Kirchhoff's voltage law states that the sum of the voltage drops across all the components in a series circuit is equal to the applied voltage.
- A voltmeter is connected across (parallel to) the component it is measuring.

Test Your Knowledge

Do not write in this book. Write your answers on a separate sheet of paper.

1. There is only one ______ in a series circuit, while a parallel circuit has several.
2. Every part of a series circuit carries the same amount of ______.
3. An ammeter is always connected in ______ with components of an electrical circuit.
4. A voltmeter is always connected in ______ to components of an electrical circuit.
5. What does R_T stand for?
6. Draw the symbol for ohms.
7. Give the Ohm's law formula for finding the amount of current flow or amperage in an electrical circuit.
8. If you had three resistors connected in series, each with a value of 30 Ω, how much total resistance would you have?
9. How much current will flow in a circuit having a 24-V power supply, a 200-Ω resistor, and a 2200-Ω resistor in series?
10. If a series circuit has four capacitors, two with a capacitance of 60 μF and two with a capacitance of 30 μF, what is the total capacitance of the circuit?
11. What is polarity?
12. To find the voltage drop across a resistor, we can use a voltmeter to measure its value, or we can use ______.
13. Why should you never use a continuity tester on a live circuit?

Activities

1. Why will several lights in your house go out if you open a circuit breaker?
2. If a continuity light will not light up when checking a coil, what is indicated?
3. Visit a new house under construction. What type and size of wire is being used? Where is the ground wire, and what color is it?
4. List some places or things where you will find three or more sources of resistance connected in series.
5. Explain why a continuity light should never be used on a circuit while it is connected to its source of power.
6. List Kirchhoff's laws and tell what they mean.
7. Make a list of the people who have electrical laws named after them. Show the dates when these people lived. Notice in which years most of their discoveries were made.
8. Where did omega, the name of the symbol used for resistance, first come from? How old is the term?
9. Which electrical devices in your home have the greatest resistance? If they are listed in watts, explain why your answer is correct.

Parallel Circuits

Learning Objectives

After studying this chapter, you will be able to:

- Construct a parallel circuit using various components.
- Calculate the total resistance, voltage, and current in a parallel circuit.
- Calculate the total capacitance in a parallel circuit.
- State a simple rule for calculating the total resistance of equal resistors in parallel.
- Measure the current flow and voltage drop in a parallel circuit.

Technical Terms

conductance (G)
parallel circuit

In Chapter 6, you learned that parallel circuits have more than one path for the current, **Figure 7-1.** Parallel circuits have broad applications, including radios, TVs, and the wiring in your home. As you work your way through this chapter, you will see why all circuits cannot be wired in series.

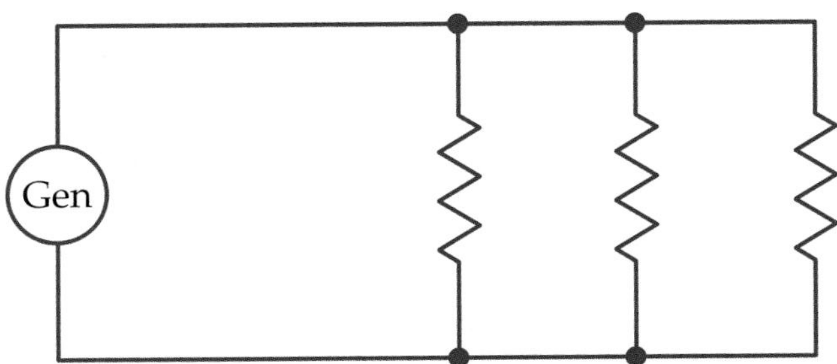

Figure 7-1. A parallel circuit provides two or more paths for the flow of electricity.

Current in a Parallel Circuit

A ***parallel circuit*** has two or more paths, or branches, through which current flows. Current in a parallel circuit is similar to water flowing in a river, **Figure 7-2.** Note the full flow down the main branch of the river. When the water reaches the two islands, the river splits and takes three paths. Since each path is the same size, the same amount of water flows in each.

The same thing is true of electrical circuits. The electrons have only one path to follow when they leave the power source, **Figure 7-3.** However, when three equal paths are provided, the electrons split evenly among those paths.

Again relating current to water flow, note that the river shown in **Figure 7-4** also flows down three paths. However, since these paths are unequal in size, the water does not divide equally. Most of the water flows through the largest path, which is the one that offers the least resistance. Less water flows through the smaller paths, or those offering more resistance.

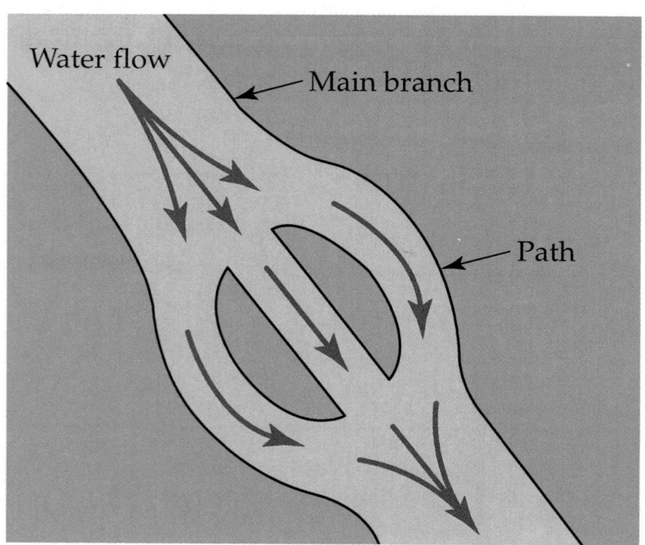

Figure 7-2. Water flowing in a river divides evenly when branching into three equal parts.

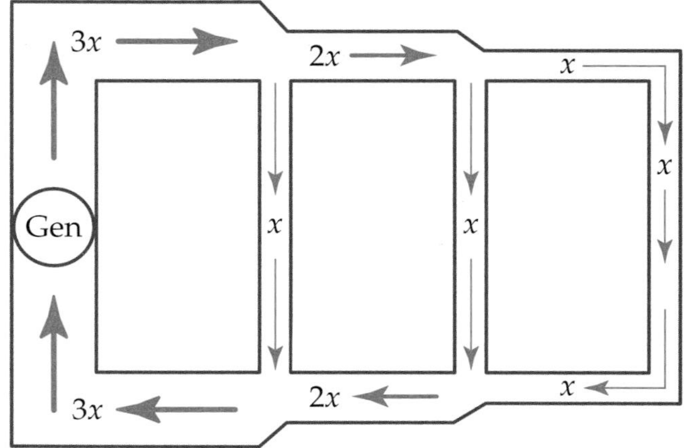

Figure 7-3. Electron flow in a parallel circuit divides evenly when branching into three equal electrical paths.

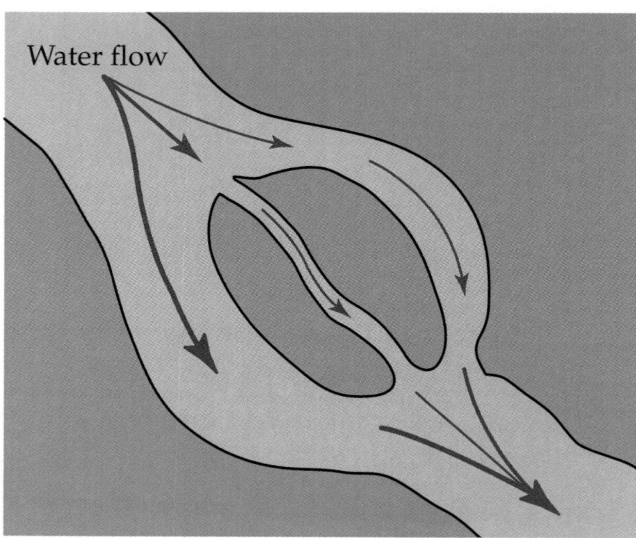

Figure 7-4. Water flowing in a river divides unevenly when branching into several unequal paths.

The same thing is true of electron flow in a parallel circuit. If the circuit has three unequal paths, the current divides unequally. The branch of a parallel circuit that offers the smallest amount of resistance provides the easiest path for current flow, **Figure 7-5.** Ohm's law can be used to determine the amount of current flowing in each branch of the circuit, but it is necessary to first find the total resistance of the circuit.

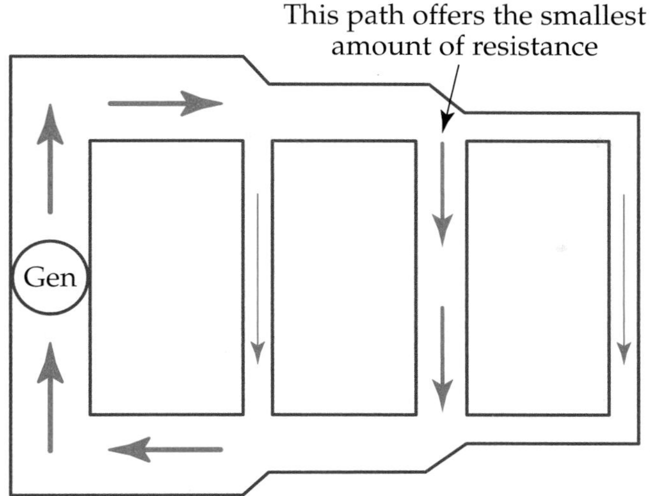

Figure 7-5. Electron flow in a parallel circuit also divides unevenly when branching into several unequal electrical paths.

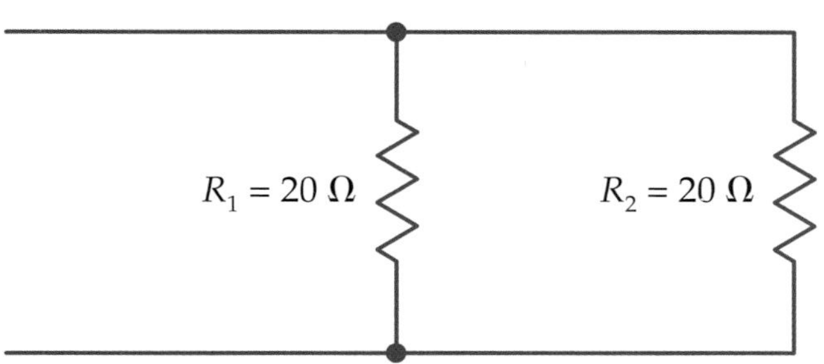

Figure 7-6. Parallel paths of equal resistance permit the same amount of current to flow in each electrical path. Total resistance in this circuit is 10 Ω (see text).

Resistance in a Parallel Circuit

Resistance is the measure of how difficult it is for electrons to flow through some component. This is just the opposite of the ability for electrons to flow in a circuit, which we call ***conductance (G).*** Therefore, total resistance in a parallel circuit can be found using the formula for conductance:

$$\frac{1}{R_T} = \frac{1}{R_1} + \frac{1}{R} + \frac{1}{R_3} + \dots \qquad (7\text{-}1)$$

Both sides of the equation are then inverted to find the total resistance:

$$R_T = \frac{R_T}{1} \qquad (7\text{-}2)$$

Using this as a formula, we would say that conductance is the reciprocal of resistance. It is written as

$$G = \frac{1}{R} \qquad (7\text{-}3)$$

This is why the inversion of $1/R_T$ is equal to the total resistance.

In the portion of a parallel circuit shown in **Figure 7-6,** the two resistors connected in parallel have values of 20 Ω each. To find the total resistance, substitute these values for the variables in the formula as follows:

$$\frac{1}{R_T} = \frac{1}{R_1} + \frac{1}{R_2}$$

$$= \frac{1}{20\ \Omega} + \frac{1}{20\ \Omega}$$

$$= \frac{2}{20\ \Omega}$$

$$= \frac{1}{10\ \Omega}$$

Invert both sides of the equation to find the total resistance:

$$\frac{R_T}{1} = \frac{10\ \Omega}{1}$$

$$R_T = 10\ \Omega$$

Example 7-1:

Find the total resistance of the following circuit. Use the parallel resistance formula:

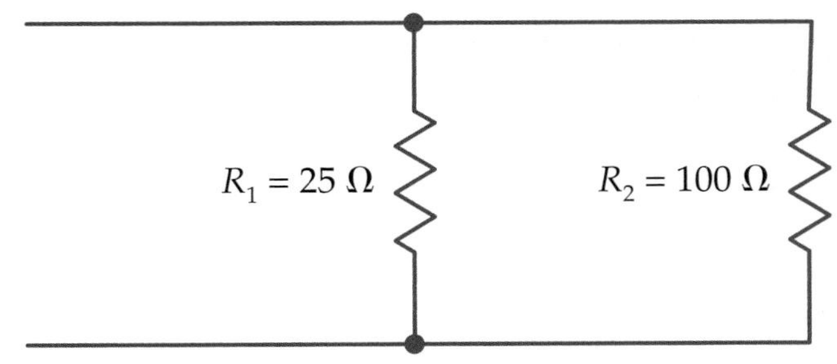

$$\frac{1}{R_T} = \frac{1}{R_1} + \frac{1}{R^2}$$

Insert the known values and add:

$$\frac{1}{R_T} = \frac{1}{25\ \Omega} + \frac{1}{100\ \Omega}$$

$$= \frac{4}{100\ \Omega} + \frac{1}{100\ \Omega}$$

$$= \frac{5}{100\ \Omega}$$

$$= \frac{1}{20\ \Omega}$$

Invert both sides of the equation to find the total resistance:

$$\frac{R_T}{1} = \frac{20\ \Omega}{1}$$

$$= 20\ \Omega$$

The equivalent circuit is shown. Notice that it contains one 20-Ω resistor.

$R_T = 20\ \Omega$

If your calculated R_T is larger than either of the resistors, the answer is wrong. Recheck your work. A parallel circuit, by adding at least one new path for current flow, widens the path through which current can move. As you learned in Chapter 3 while studying wire gauge, when the path is widened, the resistance decreases. Therefore, the total resistance in a parallel circuit is always less than or equal to the value of each of the resistors in the branches.

In the last example, if you were to replace all the branches with a single 20-Ω resistor, you would have an equivalent circuit. An equivalent circuit for parallel circuits replaces the parallel branches with a single resistor. The single resistor offers the same resistance as the total resistance of the parallel circuit. A parallel circuit's equivalent circuit is simpler and easier to work with for calculations. It is easier to find how much current the power supply is being asked to deliver. An equivalent circuit makes it easy to tell what effect the load has on the power supply.

Example 7-2:

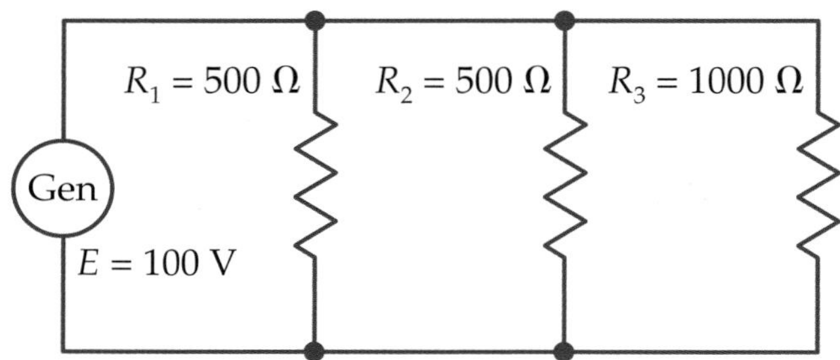

What is the total resistance of this circuit? Use the parallel resistance formula:

$$\frac{1}{R_T} = \frac{1}{R_1} + \frac{1}{R_2} + \frac{1}{R_3}$$

Substitute the known values of the resistors and solve the equation:

$$\frac{1}{R_T} = \frac{1}{500\ \Omega} + \frac{1}{500\ \Omega} + \frac{1}{1000\ \Omega}$$

$$= \frac{2}{1000\ \Omega} + \frac{2}{1000\ \Omega} + \frac{1}{1000\ \Omega}$$

$$= \frac{5}{1000\ \Omega}$$

$$= \frac{1}{200\ \Omega}$$

Invert both sides of the equation:

$$\frac{R_T}{1} = \frac{200\ \Omega}{1}$$

$$R_T = 200\ \Omega$$

Note again that the total resistance is smaller than any of the resistors in the circuit.

Voltage in a Parallel Circuit

The voltage drop in a parallel circuit is the same across each branch. The voltage drop across each branch is the same as the voltage drop across the entire parallel portion of the circuit. This can be calculated using Ohm's law with the total resistance and circuit current. Using these values, you do not need to know the amount of current in each branch. Look at the series-parallel circuit in **Figure 7-7.** (Series-parallel circuits are covered in detail in chapter 8.) The circuit current is 3 A. To find the voltage drop across the parallel portion of the circuit, we must first calculate the total resistance (R_T):

$$\frac{1}{R_T} = \frac{1}{R_1} + \frac{1}{R_2}$$

$$= \frac{1}{60\ \Omega} + \frac{1}{20\ \Omega}$$

$$= \frac{1}{60\ \Omega} + \frac{3}{60\ \Omega}$$

$$= \frac{4}{60\ \Omega}$$

$$= \frac{1}{15\ \Omega}$$

Invert both sides of the equation to find the total resistance:

$$\frac{R_T}{1} = \frac{15\ \Omega}{1}$$

$$R_T = 15\ \Omega$$

Now use R_T and Ohm's law to find the voltage drop across the parallel portion of the circuit.

$$V_{\text{Parallel}} = I \times R_T$$

$$= 3\text{ A} \times 15\ \Omega$$

$$= 45\text{ V}$$

The voltage drop across the parallel portion of the circuit is 45 V. This means the voltage drop across each branch is also 45 V. This information is needed to calculate the current in each branch of the parallel circuit.

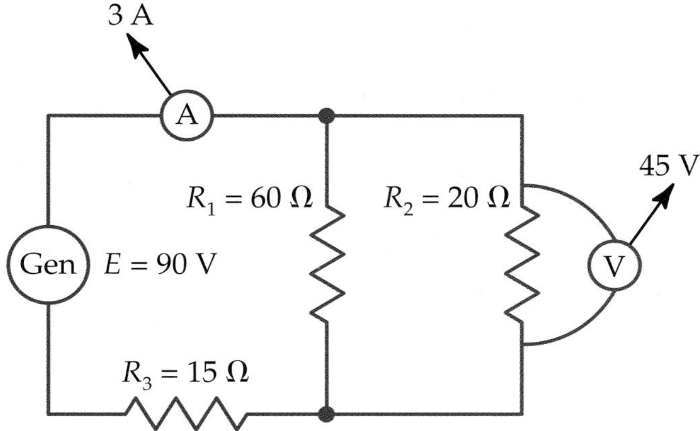

Figure 7-7. R_1 and R_2 form the parallel portion of this circuit, which is in series with R_3.

Calculating Current

Using the total resistance and voltage drop, we need to find the amount of current flowing in a circuit. We can use Ohm's law and **Figure 7-8:**

$$I_T = \frac{V}{R_T}$$

$$= \frac{100\text{ V}}{40\ \Omega}$$

$$= 2.5\text{ A}$$

Since R_T was used, not an individual resistor for a single branch, the value 2.5 A is the amount of current flowing in the total circuit.

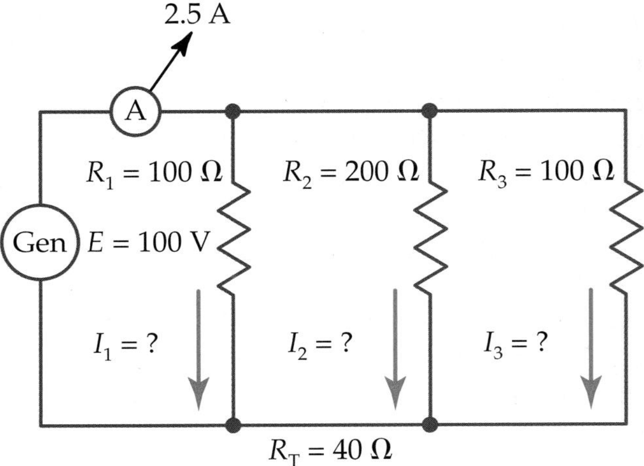

Figure 7-8. The total resistance (R_T) is given. Use this value to calculate the current in the circuit.

Project 7-1: Blown Fuse Indicator

A useful application of a parallel circuit is a blown fuse indicator. When a circuit is bad, many people spend a lot of time troubleshooting other parts of the circuit first. Sometimes the problem is as simple as a blown fuse. To avoid this problem, make the simple circuit shown.

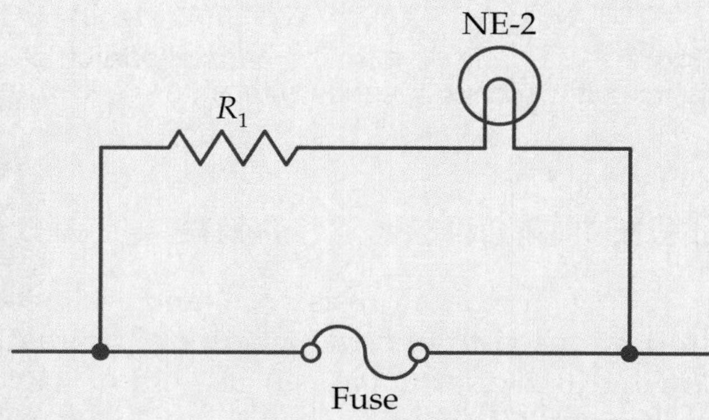

The circuit can be added across the fuse in the original circuit. When the light comes on, you know that the fuse has blown. Parts needed for the project include the following:

No.	Item
1	NE-2 neon lamp
1	Resistor (value depends on circuit voltage)
	Connecting wire

To assemble this project, you must know the voltage across the fuse. This information will allow you to select the right size resistor. The formula to find the correct size resistor (R_1) is

$$R_1 = \frac{E}{0.0025 \text{ A}}$$

The current required for an NE-2 bulb is 2.5 mA. Convert milliamperes to amperes. Therefore, substitute using Ohm's law $R = E / I$. If you use a different bulb, find its current rating and substitute that amount into your formula. For example, if the voltage drop across the fuse is 110 V, then

$$R_1 = \frac{110 \text{ V}}{0.0025 \text{ A}}$$

$$R_1 = 44{,}000\ \Omega$$

Based on this computation, a 110-V circuit requires a 44,000-Ω resistor. Since an NE-2 neon lamp requires between 60 V and 65 V to ionize, this blown fuse indicator is only good for sources greater than 65 V.

To find the amount of current flowing in each resistor, again use Ohm's law. The amount of current flowing through R_1 can be calculated as follows:

$$I_1 = \frac{E}{R_1}$$

$$= \frac{100 \text{ V}}{100\ \Omega}$$

$$= 1.0 \text{ A}$$

Since R_1 and R_3 are equal, I_3 is also 1.0 A. Use the same method to find the current in R_2:

$$I_2 = \frac{E}{R_2}$$

$$= \frac{100 \text{ V}}{200\ \Omega}$$

$$= 0.5 \text{ A}$$

The formula for total current (I_T) in a circuit is

$$I_T = I_1 + I_2 + I_3 + \ldots \qquad (7\text{-}4)$$

where I_1, I_2, I_3... represent the amount of current flowing through each resistor. Using this formula, we add the three current values from the example:

$$I_T = 1.0\text{ A} + 0.5\text{ A} + 1.0\text{ A}$$
$$= 2.5\text{ A}$$

This is equal to the total current flowing in the circuit, which means our calculations are correct.

Figure 7-9 shows how ammeters would be installed to measure the current in each branch. Notice that the ammeter is connected in series with its load. Again, adding these three current values gives 2.5 A, which is the total current flowing in the circuit.

We would expect these values if we built this circuit. However, for safety purposes, use a voltage less than 30 volts and current in the milliamp range. If you actually build this circuit, you may get different results. This difference is due to the tolerance of various resistors. The results may vary by ±5 percent if the resistors have a gold fourth band, ±10 percent if they have a silver fourth band, or ±20 percent if they have no fourth band.

Capacitance in a Parallel Circuit

The last chapter showed how to find total capacitance in a series circuit. The formula for finding total capacitance of capacitors in parallel is very similar to the formula you used for resistors in series.

Figure 7-10 shows a circuit with two capacitors in parallel and an equivalent circuit. The two capacitors in Figure 7-10A have ratings of 30 μF and 40 μF. To get the total capacitance of this circuit, use the following formula for capacitors in parallel:

$$C_T = C_1 + C_2 + C_3 + \ldots \quad (7\text{-}5)$$

Therefore, for our circuit

$$C_T = C_1 + C_2$$
$$= 30\ \mu\text{F} + 40\ \mu\text{F}$$
$$= 70\ \mu\text{F}$$

This would result in the equivalent circuit shown in Figure 7-10B. Total capacitance of this equivalent circuit is 70 μF.

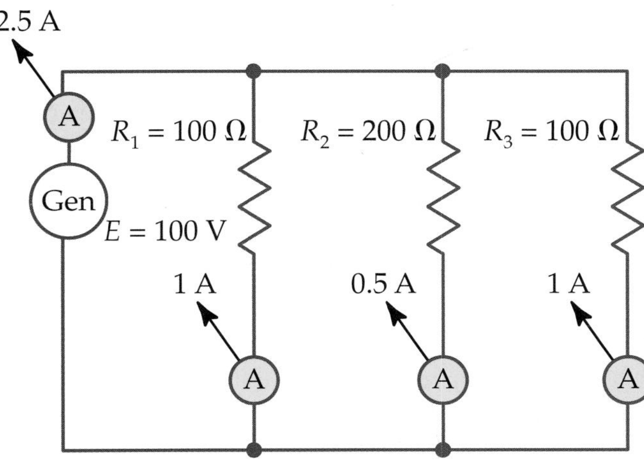

Figure 7-9. Ammeters in a parallel circuit show the different current values in the branches. The different resistors make the current vary among the branches.

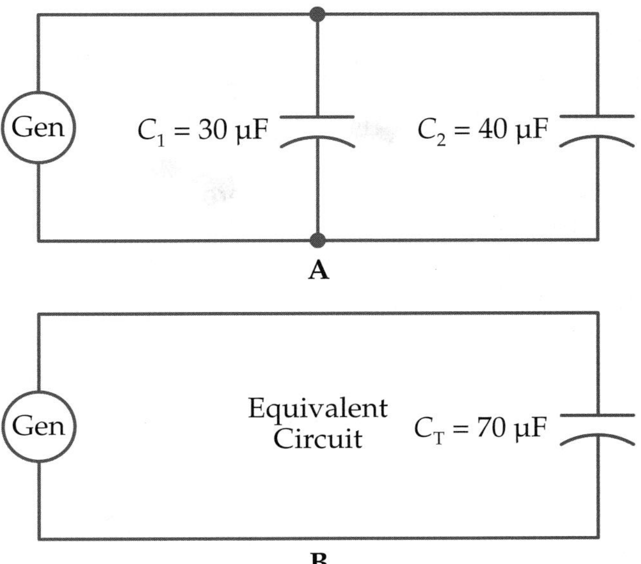

Figure 7-10. Find the total capacitance. A—Two capacitors of unequal value in parallel. B—An equivalent circuit.

The reason capacitors in parallel differ from resistors in parallel is shown in **Figure 7-11.** Remember, a larger plate area increases capacitance. When we have two capacitors in parallel, we have a greater number of plates and more electrons on the plates. Therefore, when you are finding total capacitance, you are adding the total plate area of all the capacitors in parallel.

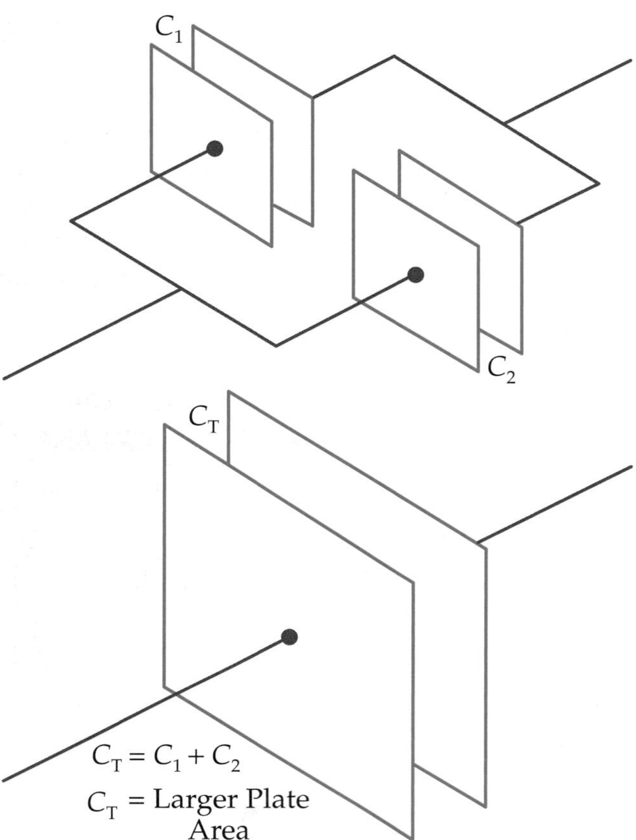

Figure 7-11. When capacitors are placed in parallel, capacitance in the circuit increases because the plate area is larger.

Example 7-3:

What is the total capacitance in this parallel circuit? Use the formula for total capacitance:

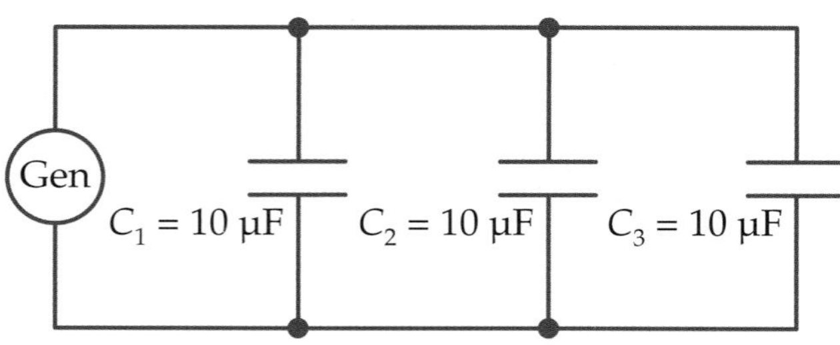

$$C_T = C_1 + C_2 + C_3$$
$$= 10\ \mu F + 10\ \mu F + 10\ \mu F$$
$$= 30\ \mu F$$

Therefore, the equivalent circuit shown has a total capacitance of 30 µF.

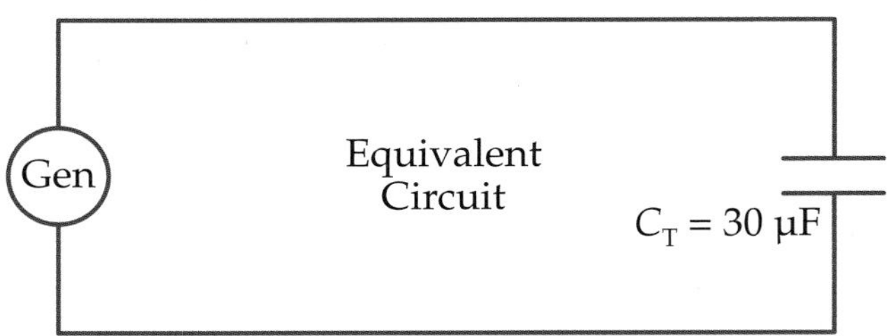

Uses for Parallel Circuits

Next, let us see how parallel circuits are valuable. Suppose you have a radio, and you want to attach two speakers to it. You find some speakers marked 4 Ω, some marked 8 Ω, and some marked 16 Ω. However, the back of the radio is marked 4 Ω, meaning it calls for one 4-Ω speaker. Since you wish to install two speakers, you have a problem.

One way to solve this problem is with a parallel circuit and two speakers marked 8 Ω each, **Figure 7-12.** The total resistance needs to be 4 Ω because of the radio's rating. Find the total resistance (R_T) of this circuit with the 8-Ω speakers.

$$\frac{1}{R_T} = \frac{1}{R_1} + \frac{1}{R_2}$$
$$= \frac{1}{8\ \Omega} + \frac{1}{8\ \Omega}$$
$$= \frac{2}{8\ \Omega}$$
$$= \frac{1}{4\ \Omega}$$

Invert both sides of the equation to find total resistance:

$$R_T = 4\ \Omega$$

The problem is solved. The total resistance of the two speakers matches the radio's 4-Ω specification.

Practical Application 7-1: Using a Parts Catalog

When building or troubleshooting circuits, you may find that the circuit does not work due to a faulty component. In this case, you will need to use a parts catalog to find a replacement part. If you do not have access to the original parts manufacturer catalog, you may have to substitute that component with a component from another manufacturer. When this happens, use a substitution manual. You can also use a catalog from a parts distributor. A parts distributor carries original parts from many manufacturers.

A substitution manual is typically arranged first by major part type and then by manufacturer. Then, the original part number is listed in the left column and the substitution part number is listed in the right column.

Once you have identified the substitution part number, you can obtain it from a parts supply house that carries those parts. Sometimes a circuit built with substitute parts will work poorly or not at all. Be sure to breadboard the circuit with the replacement parts first to see if the circuit will operate properly.

A catalog from a parts distributor is also typically arranged first by major part type and then by manufacturer. The original part number is listed in the left column; however, a stock number is listed in the right column. The stock number is the one the distributor assigns to the original part.

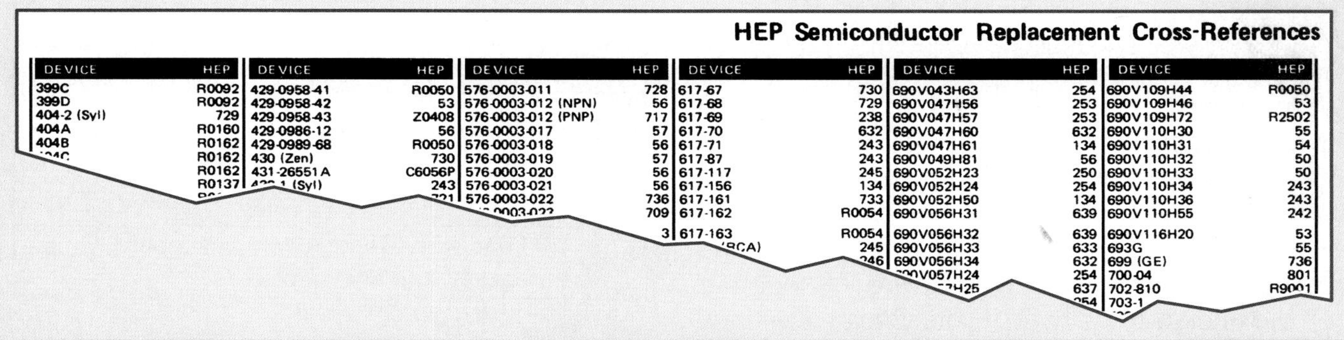

HEP Semiconductor Replacement Cross-References

DEVICE	HEP	DEVICE	HEP	DEVICE	HEP	DEVICE	HEP	DEVICE	HEP	DEVICE	HEP
399C	R0092	429-0958-41	R0050	576-0003-011	728	617-67	730	690V043H63	254	690V109H44	R0050
399D	R0092	429-0958-42	53	576-0003-012 (NPN)	56	617-68	729	690V047H56	253	690V109H46	53
404-2 (Syl)	729	429-0958-43	Z0408	576-0003-012 (PNP)	717	617-69	238	690V047H57	253	690V109H72	R2502
404A	R0160	429-0986-12	56	576-0003-017	57	617-70	632	690V047H60	632	690V110H30	55
404B	R0162	429-0989-68	R0050	576-0003-018	56	617-71	243	690V047H61	134	690V110H31	54
[illegible]	R0162	430 (Zen)	730	576-0003-019	57	617-87	243	690V049H81	56	690V110H32	50
	R0162	431-26551A	C6056P	576-0003-020	56	617-117	245	690V052H23	250	690V110H33	50
	R0137	[illegible] (Syl)	243	576-0003-021	56	617-156	134	690V052H24	254	690V110H34	243
	[illegible]		[illegible]	576-0003-022	736	617-161	733	690V052H50	134	690V110H36	243
				[illegible]	709	617-162	R0054	690V056H31	639	690V110H55	242
					3	617-163	R0054	690V056H32	639	690V116H20	53
						[illegible] (RCA)	245	690V056H33	633	693G	55
							[illegible]	690V056H34	632	699 (GE)	736
								[illegible]	254	700-04	801
								[illegible]	637	702-810	R9001
									[illegible]	703-1	
										[illegible]	

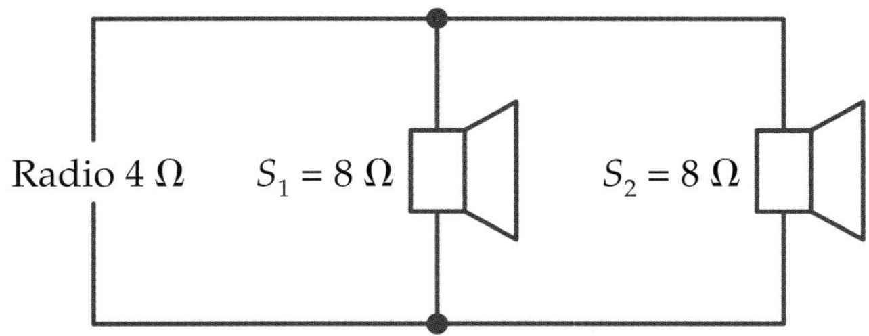

Figure 7-12. In this circuit, two speakers are connected in parallel, each with 8 Ω of resistance. This matches the radio's requirement for 4 Ω of total resistance.

Drawing Parallel Circuits

You will find parallel circuits drawn in many different ways. So far in this chapter, we have shown simple rectangular circuits with resistors placed parallel to each other in add-on fashion. **Figure 7-13** shows two other ways of drawing parallel circuits.

Note that in **Figure 7-14** the two resistors are connected to separate branches of the circuit. The resistors are in parallel electrically, but not in parallel position on the drawing. When resistors are connected in parallel, it is not necessary to draw them in positions parallel to each other.

Project 7-2: Electric Puzzle

Here is a puzzle that you can build with four switches, a lamp, and two dry cells. The idea is to wire the project so that the lamp will come on if a mistake is made while working the puzzle.

Each switch, in turn, represents a farmer, a chicken, corn, and a fox. The trick is to operate the four switches so that the farmer can get all three possessions safely across the river. On the shore is a rowboat that is used to carry the farmer and only one possession at a time. This is the catch: the fox will eat the chicken and the chicken will eat the corn if they are left together on either side of the river. If you wire the switches as shown, there is a way for the farmer to get all three possessions safely across the river.

Try your skill to see if you can figure out how to do it. Remember, the farmer can move only one possession per trip across the river. When working the puzzle, imagine that the switches span the river.

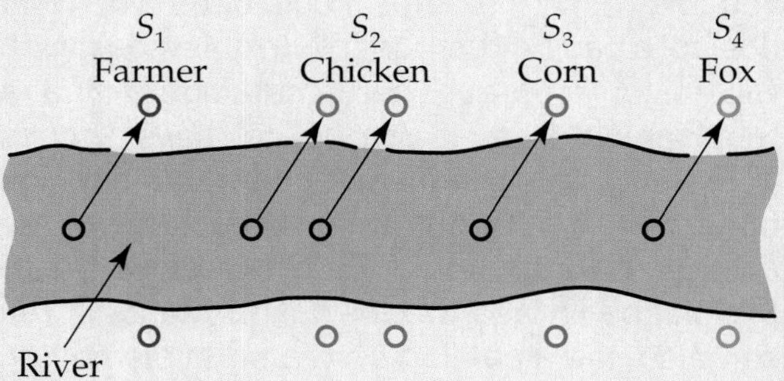

S_1 is for controlling the farmer's crossings. S_2 is for the chicken. S_3 is for the corn. S_4 is for the fox.

No.	Item
2	1.5-V dry cells (B_1)
1	3-V dc lamp (L_1)
3	SPDT knife switches (S_1, S_3, and S_4)
1	DPDT knife switch (S_2)
	Lead wire, wooden box, baseboard, lamp socket, and holder for cells

The answer follows the construction procedure.

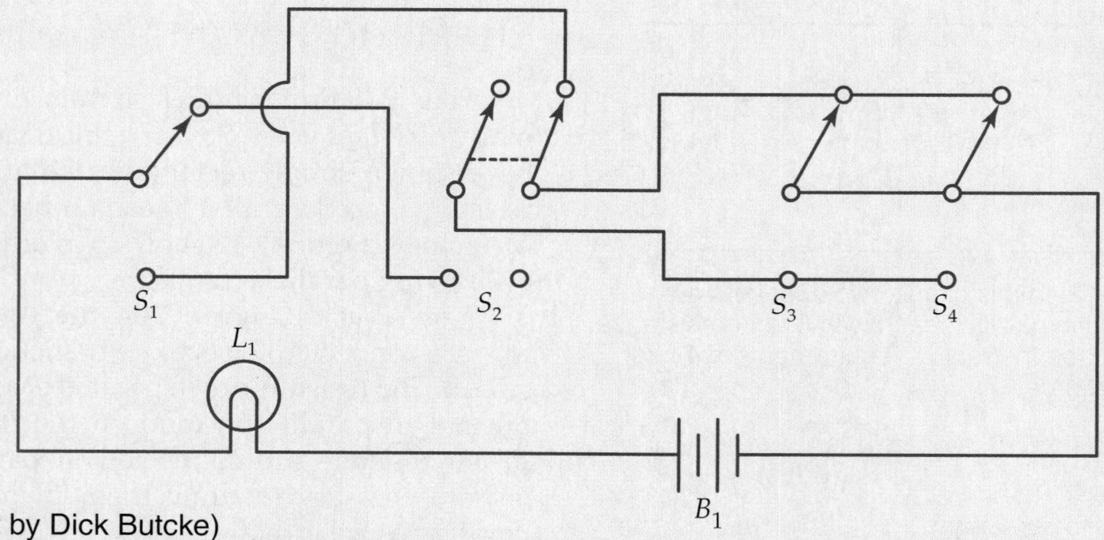

(Project by Dick Butcke)

Project 7-2: Electric Puzzle *(Continued)*

Procedure

1. Lay out the location of the switches and the lamp on the baseboard.
2. Drill all the necessary holes in the baseboard.
3. Wire and install all the switches and the lamp according to the schematic diagram.
4. Assemble the wooden box and install the switchboard on it.

Here is the correct switching sequence that will get the farmer and possessions to the opposite side of the river:

1st trip	Farmer (S_1) and chicken (S_2)
Return trip	Farmer (S_1) alone
2nd trip	Farmer (S_1) and corn (S_3)
Return trip	Farmer (S_1) and chicken (S_2)
3rd trip	Farmer (S_1) and fox (S_4)
Return trip	Farmer (S_1) alone
4th trip	Farmer (S_1) and chicken (S_2)

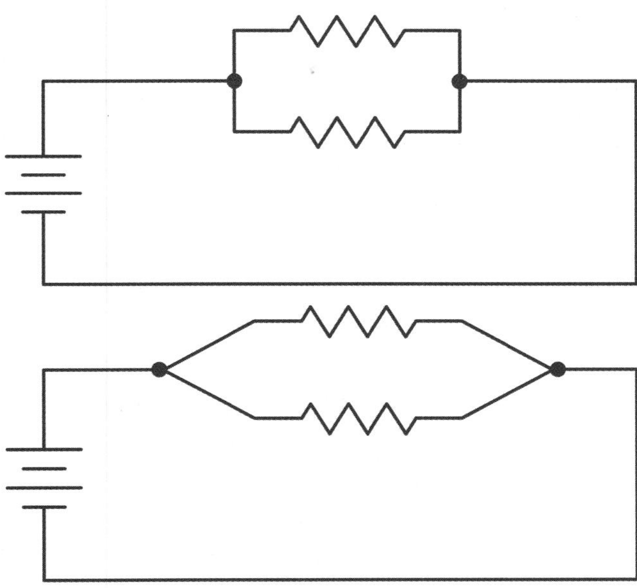

Figure 7-13. Parallel circuits can be drawn in many different ways. Drawings shown are typical arrangements of power source and resistors.

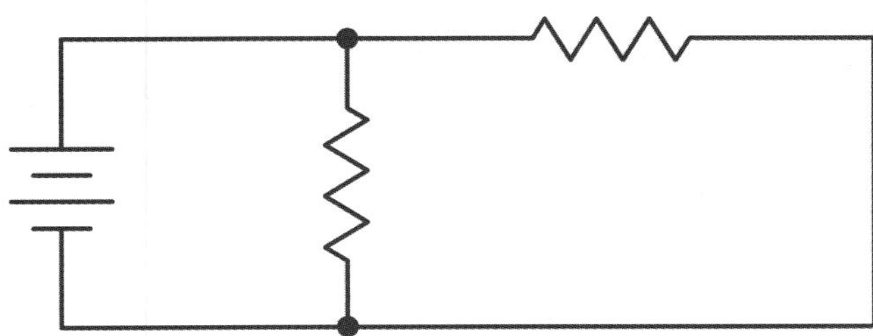

Figure 7-14. In this parallel circuit, the resistors are not actually drawn parallel. This is perfectly acceptable. They still work properly.

Parallel Power Sources

Power sources can also be connected in parallel. In **Figure 7-15,** note that three batteries are connected in parallel across the load. This arrangement is used when one power source simply cannot supply enough current for an extended time.

There are times when a load draws so much current it shortens the life of the power source. When this happens, it is necessary to install several power sources in parallel. This enables them to share the current draw by the load. This extends the life of the total power supply. Only batteries or power sources that have the same voltage should be connected in parallel.

Equal Resistors in Parallel

In **Figure 7-16,** two resistors are connected in parallel, each having a value of 100 Ω. When a parallel circuit has two resistors of the same value, another formula can be used to find total resistance. The resistance formula says that

$$\frac{1}{R_T} = \frac{1}{R_1} + \frac{1}{R_2} + \frac{1}{R_3} + \ldots$$

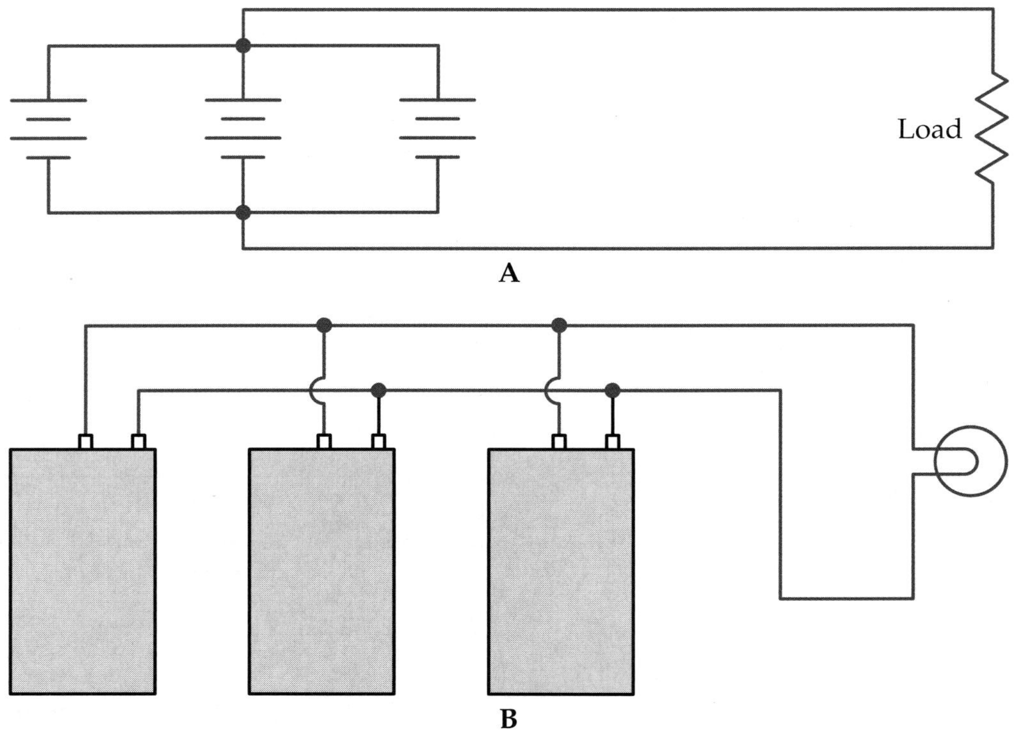

Figure 7-15. Multiple power sources connected in parallel share the electrical load and last longer.

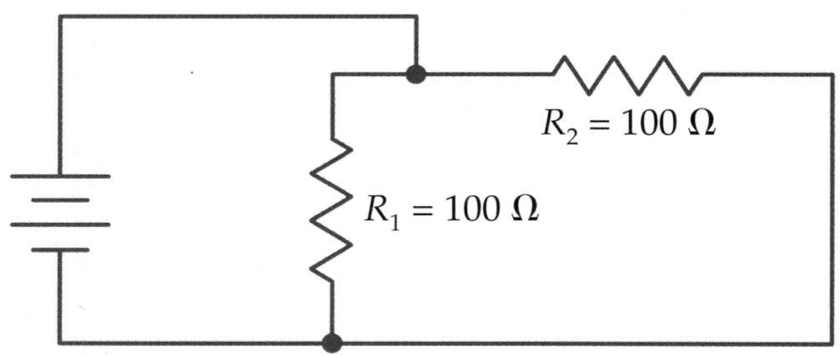

Figure 7-16. Two 100-Ω resistors connected in parallel offer 50 Ω of total resistance.

Suppose that all of the resistors in parallel have the same resistance. Each fraction in the resistance formula would have same denominator. The numerator is 1 for each resistor in the circuit. When they are added together, the numerator equals the number of resistors in the circuit. When we invert the fraction to find R_T, the numerator and denominator trade places. So, for equal resistors in parallel, the following formula can be used:

$$R_T = \frac{R}{N} \quad (7\text{-}6)$$

where R is the value of the resistors, and N is the number of resistors in the circuit.

The following example helps show why this is true. Suppose a circuit has three resistors ($N = 3$), and each has a value of 10 Ω ($R = 10\ \Omega$). We have the following:

$$\frac{1}{R_T} = \frac{1}{10\ \Omega} + \frac{1}{10\ \Omega} + \frac{1}{10\ \Omega}$$

$$= \frac{3}{10\ \Omega}$$

$$= \frac{10\ \Omega}{3} = \frac{R}{N}$$

Example 7-4:

Using the new formula, find the total resistance (R_T) of the circuit in Figure 7-16.

$$R_T = \frac{R}{N}$$

You know that $R = 100\ \Omega$ and that there are two resistors:

$$= \frac{100\ \Omega}{2}$$

$$= 50\ \Omega$$

The total resistance in this circuit is 50 Ω.

Example 7-5:

In the following circuit, three resistors are connected in parallel. Use the same formula to find total resistance (R_T):

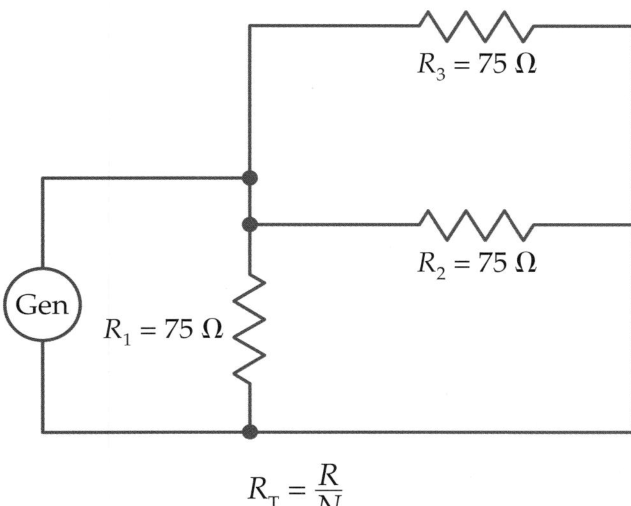

$$R_T = \frac{R}{N}$$

You know that R equals 75 Ω and there are three resistors:

$$R_T = \frac{75\ \Omega}{3}$$

$$= 25\ \Omega$$

The total resistance in this circuit is 25 Ω. This means that to get an equivalent circuit, the resistors should be replaced with one 25-Ω resistor.

Sometimes series circuits and parallel circuits are combined to make complex electrical equipment. For example, **Figure 7-17** shows the cockpit of a small airplane that has thousands of dollars worth of instruments. This equipment allows the pilot to fly safely through or around all kinds of potentially dangerous weather conditions. A complex electrical device such as this has many parallel circuits and series circuits. It also has series-parallel circuits, which are explained in Chapter 8.

Figure 7-17. The wiring for the instrument panel of this private airplane is made of many parallel circuits. It took many hours to assemble the meters, radios, and navigation gear. (Scott Murwin)

Measuring Voltage in a Parallel Circuit

Voltage in a parallel circuit can be measured by using a voltmeter. The voltmeter leads must be connected across the circuit. Look at **Figure 7-18.** Note that by connecting a voltmeter across each resistor and the source, all leads are connected to the same two lines. Therefore, all meters read the same voltage.

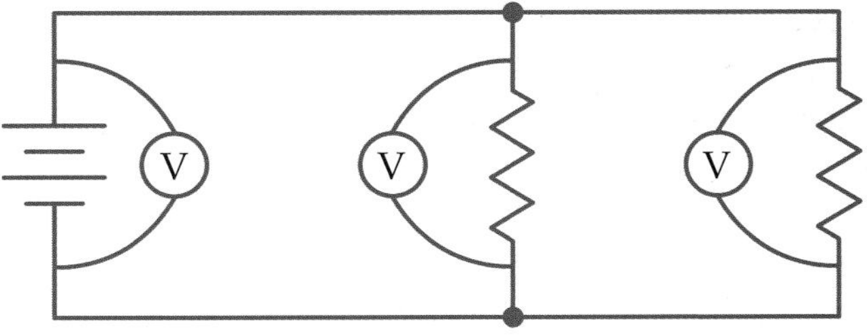

Figure 7-18. When measuring voltage in this parallel circuit, connect the voltmeter across the power source or load.

Consider another resistance experiment. If you installed a voltmeter across a short portion of the wire in a circuit, you would get a reading of zero. There is no reading because voltage drop across a short piece of wire is very small. It is so small, in fact, that you would need a highly sensitive meter to record the drop.

For example, consider the No. 14 wire in **Figure 7-19.** It is carrying 2 A of current. No. 14 wire has 1 Ω of resistance for 396′ of wire. The resistance for 1′ of the wire can be found as follows:

$$R = \frac{1'}{(396 \text{ ft.}/\Omega)}$$

$$= 0.00252\ \Omega$$

Therefore, one foot of wire between the clips of the voltmeter has a resistance of 0.00252 Ω. Using Ohm's law and this resistance, we can find the voltage drop:

$$E = IR$$

$$= 2\text{ A} \times 0.00252\ \Omega$$

$$= 0.00504\text{ V, } \textit{or } 5.04\text{ mV}$$

You can see that a very sensitive voltmeter would be needed to make this voltage measurement. Since there is practically no voltage drop across the wire in a circuit, we can ignore it when using meters. For all practical purposes, we will assume that there is no voltage drop across wire in our circuits.

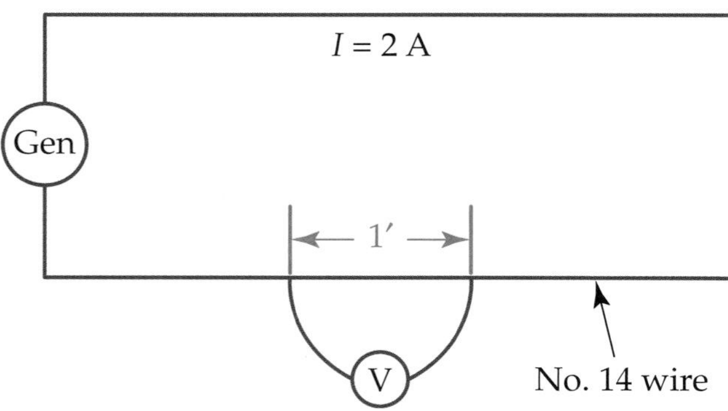

Figure 7-19. There is no voltage drop when measuring across a wire because the wire's resistance is so small.

Summary

- A parallel circuit has more than one branch.
- The voltage of each branch of a parallel circuit is the same as the source voltage.
- Conductance is the reciprocal of the total resistance in a parallel circuit.
- Total resistance is found with the following formula:

$$\frac{1}{R_T} = \frac{1}{R_1} + \frac{1}{R_2} + \frac{1}{R_3} + \ldots$$

- The voltage drop in a parallel circuit is the same across each branch.
- Total current (I_T) in a parallel circuit is equal to the sum of the current flowing in the branches:

$$I_T = I_1 + I_2 + I_3 + \ldots$$

- Current is measured by connecting an ammeter in series with each load.
- The total capacitance is found by adding the capacitance of the individual capacitors in the branches of the circuit:

$$C_T = C_1 + C_2 + C_3 + \ldots$$

- Power sources connected in parallel share the current draw of a load.
- Only power sources of the same voltage level should be connected in parallel.
- When resistors connected in a parallel circuit have the same value, the total resistance is equal to the value of one resistor divided by the number of resistors:

$$R_T = \frac{R}{N}$$

- Voltage drop is measured with a voltmeter in parallel with each resistor.

Test Your Knowledge

Do not write in this book. Write your answers on a separate sheet of paper.

1. How many branches are there in a parallel circuit?
2. In a parallel circuit, how much of the total current will flow in the path of least resistance?
   a. Less current than the branch with the most resistance.
   b. Most current of the total current.
   c. All of the total current.
   d. Exactly 2/3 of the total current.
3. What is the total resistance in a circuit having a 150-Ω, 50-Ω, and 75-Ω resistor connected in parallel?
4. Show the formulas you used to get the answer to question number 3.
5. How can you check if the answer to question number 3 is correct?
6. The voltage in a parallel circuit is ______ across all branches.
7. When you measure the current in a circuit, you may get different results than if you solved the problem using the formula. Why?
8. If a parallel circuit has three branches, each branch has a capacitor, and the capacitors' values are 30 μF, 50 μF, and 75 μF, what is the total capacitance of the circuit?
9. Three resistors are connected in parallel. If they are all of equal value, what equation would you use to find the equivalent circuit?
10. *True or False?* When two resistors are connected in parallel, they must be placed parallel to each other on the circuit drawing.
11. What is the total voltage of a circuit with three 6-V power sources connected in parallel?

Activities

1. Why do audio speakers have different ratings?
2. Why does a voltmeter read zero when it is connected across a piece of wire?
3. Why do you get a voltage drop across a piece of resistance wire and not across copper wire or aluminum wire?
4. Why do some circuits found in books and magazines fail to work after you build them?
5. Make a list of some of the places where you can find resistors connected in parallel.
6. Why can a bird land on electric wires between telephone poles without getting shocked or killed?

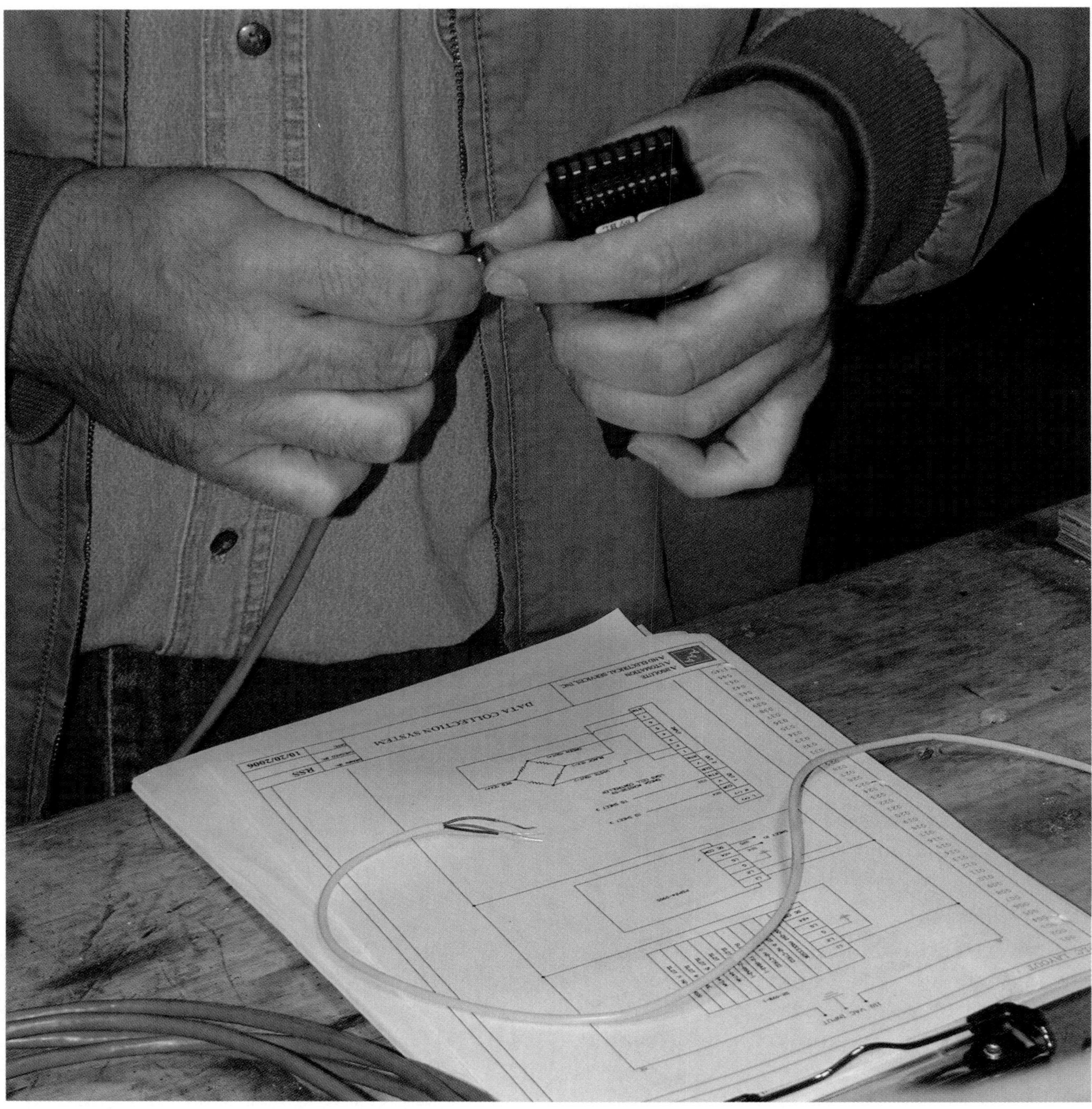

Reading schematics and other electrical diagrams is an important skill for the electronics technician and electrician.

Series-Parallel Circuits

Learning Objectives

After studying this chapter, you will be able to do the following:

- Construct a series-parallel circuit using various components.
- Calculate the total resistance, voltage, and current in a series-parallel circuit.
- Calculate the total capacitance of a series-parallel circuit with capacitors.
- Explain how an automobile can use a one-wire system to operate its electrical components.
- Design a voltage divider circuit.
- Explain the purpose of a Wheatstone bridge.
- Construct a Wheatstone bridge on a breadboard.

Technical Terms

chassis ground
circuit analysis
ground
one-wire electrical system
open circuit
series-parallel circuit
short circuit
voltage divider
Wheatstone bridge

When series circuits and parallel circuits are combined, the resulting single circuit is commonly called a ***series-parallel circuit*** or ***network.*** Most series-parallel circuits are complex. However, if you follow some simple guidelines, you will be able to reduce any series-parallel circuit to its simplest form to determine current and voltages.

Lightning

Perhaps the best way to describe a series-parallel circuit is to compare it to the many paths and branches of lightning. Look at the photo of lightning in **Figure 8-1.** The lightning

Figure 8-1. Lightning is similar to a series-parallel circuit in that it has many paths and branches. (NOAA Photo Library, NOAA Central Library; OAR/ERL/National Severe Storms Laboratory Collection)

strike captured in this photo shows the visible flow of high current values between the earth and the sky.

Note

A lightning bolt can carry 30,000 A and as much as 2,000,000 V.

You will notice that some of the lightning appears to discharge in a straight line with the earth. Other paths appear to split off in separate branches, while others form several parallel paths. These combinations of different paths are a visible example of a series-parallel circuit.

Resistance in a Series-Parallel Circuit

Now let's analyze a schematic diagram of a series-parallel circuit and its equivalent circuits, **Figure 8-2.** We will determine the total resistance of a series-parallel circuit. The series-parallel circuit in Figure 8-2A has three 60-Ω resistors. Two of the 60-Ω resistors are in parallel. The other 60-Ω resistor is in series with the two parallel resistors.

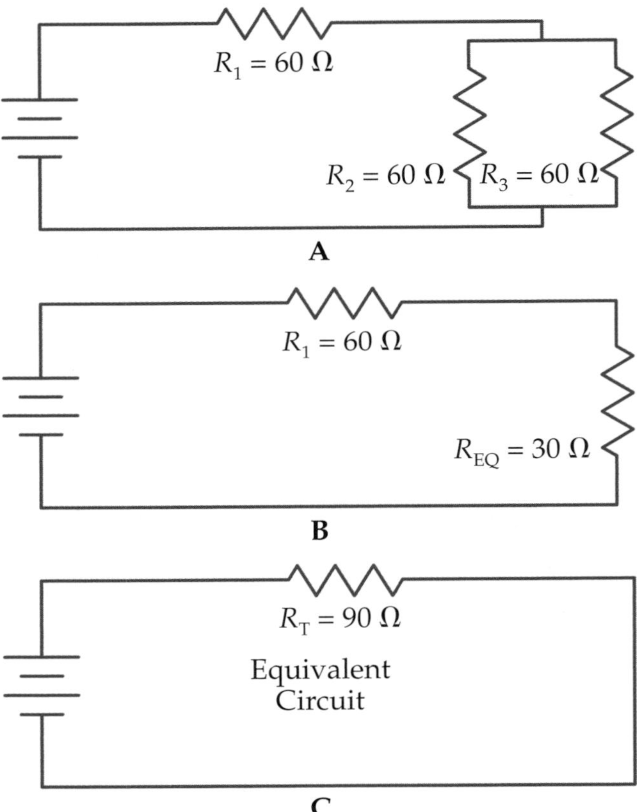

Figure 8-2. Series-parallel circuit and equivalent circuits. A—This series-parallel circuit has three resistors. R_1 is in series, and R_2 and R_3 are in parallel. B—Equivalent resistance (R_{EQ}) is 30 Ω. C—Total resistance (R_T) is 90 Ω.

The first step in calculating values in a series-parallel circuit is to determine which resistors are in series with the power supply and which resistors are in parallel. Note that R_1 is in series, while R_2 and R_3 are in parallel.

Use the following formula from Chapter 7 for finding the total resistance (R_T) of the two resistors of equal value in the parallel branch of the circuit:

$$R_T = \frac{R}{N}$$

For our analysis, we will call this value the equivalent resistance (R_{EQ}) to distinguish this value from the total resistance (R_T) of the series-parallel circuit. Therefore,

$$R_{EQ} = \frac{R}{N}$$

$$= \frac{60\ \Omega}{2}$$

$$= 30\ \Omega$$

Next, use the following formula for finding total resistance (R_T) in the newly formed series circuit, Figure 8-2B.

$$R_T = R_1 + R_{EQ}$$

$$= 60\ \Omega + 30\ \Omega$$

$$= 90\ \Omega$$

This gives you an equivalent circuit, Figure 8-2C, with a total resistance (R_T) of 90 Ω.

Example 8-1:

What is the total resistance (R_T) of the following circuit?

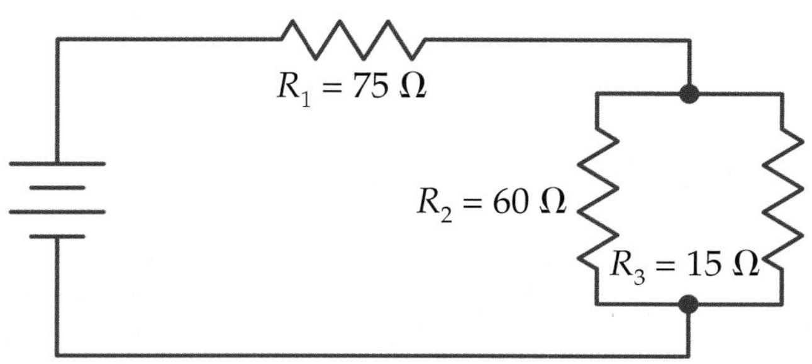

The resistors in the parallel branch are not equal, so we must use the basic formula for finding equivalent resistance (R_{EQ}) in a parallel circuit:

$$\frac{1}{R_{EQ}} = \frac{1}{R_2} + \frac{1}{R_3}$$

Insert the values of the resistors and solve the equation:

$$\frac{1}{R_{EQ}} = \frac{1}{60\ \Omega} + \frac{1}{15\ \Omega}$$

$$= \frac{1}{60\ \Omega} + \frac{4}{60\ \Omega}$$

$$= \frac{5}{60\ \Omega}$$

Invert the equation to find R_{EQ}.

$$R_{EQ} = \frac{60\ \Omega}{5}$$

$$= 12\ \Omega$$

The equivalent resistance for the parallel part of the circuit is 12 Ω. Now we can find the total resistance (R_T) by using the formula for resistances in series:

$$R_T = R_1 + R_{EQ}$$

$$= 75\ \Omega + 12\ \Omega$$

$$= 87\ \Omega$$

The total resistance in this circuit is 87 Ω.

Example 8-2:

What is the total resistance of the circuit shown?

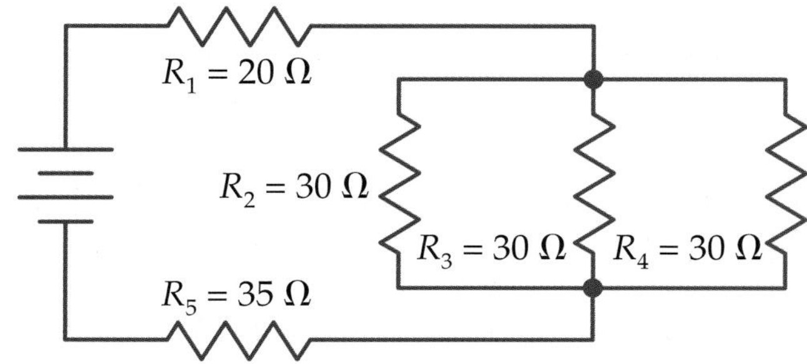

To start, find the equivalent resistance (R_{EQ}) of the parallel section of the circuit. Resistors R_2, R_3, and R_4 have the same value, so you can use the formula for equivalent resistors in parallel:

$$R_{EQ} = \frac{R}{N}$$
$$= \frac{30\ \Omega}{3}$$
$$= 10\ \Omega$$

The equivalent resistance of this part of the circuit is 10 Ω. Now, we can find the total resistance (R_T) by using the formula for resistances in series:

$$R_T = R_1 + R_{EQ} + R_5$$
$$= 20\ \Omega + 10\ \Omega + 35\ \Omega$$
$$= 65\ \Omega$$

The total resistance in this circuit is 65 Ω.

Location of Resistances

It does not make a difference when calculating total resistance where you place the series element of a series-parallel circuit. Switching the order of the series and parallel parts of a circuit changes the amount of current flow in various parts of the circuit, but it does not change the total resistance.

For example, **Figure 8-3** shows a circuit with the same values as the circuit in Figure 8-2A, but this circuit has the series element ahead of the parallel element. If you go through the same steps in calculating total resistance (R_T) that you did when analyzing Figure 8-2, you will get the same answer (90 Ω). In Figure 8-3B, the equivalent resistance (R_{EQ}) of R_1 and R_2 is 30 Ω. When this is added to the 60 Ω of R_3, Figure 8-3C, you would again have an equivalent circuit with a total resistance (R_T) of 90 Ω.

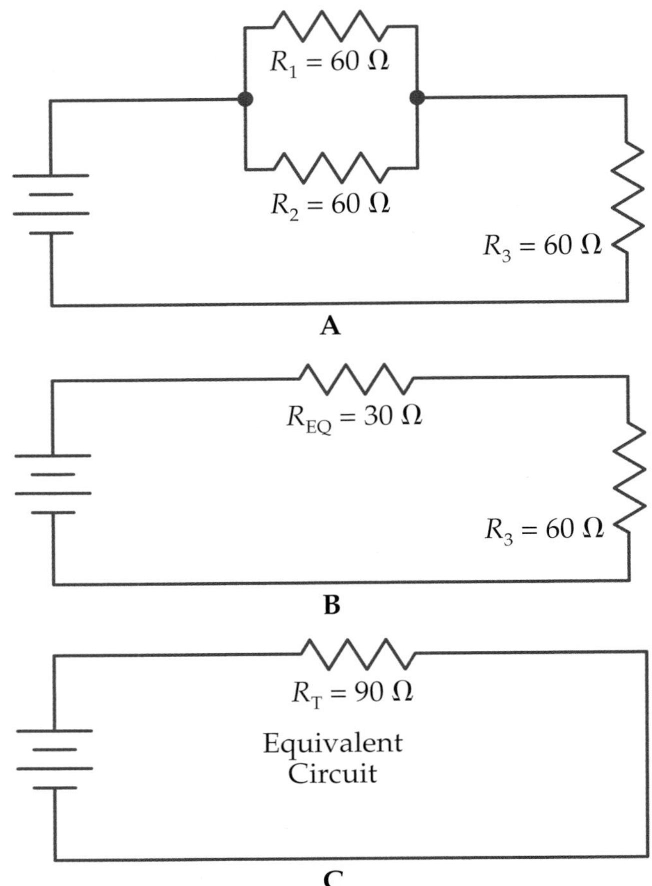

Figure 8-3. This circuit is similar to Figure 8-2, but with the series and parallel resistors reversed. A—Series resistor R_3 is ahead of parallel resistors R_1 and R_2. B—Equivalent resistance (R_{EQ}) is 30 Ω. C—Total resistance (R_T) is 90 Ω.

Order for Solving Total Resistance

We solved the previous examples by first finding the resistance in the parallel portion of the circuits. It may seem that total resistance for all series-parallel circuits is solved by first looking for a parallel circuit. This is not always true. Sometimes you must first find the equivalent resistance of a series portion before you can consider the parallel circuit.

Look at **Figure 8-4.** Notice that the schematic diagram in Figure 8-4A has a 10-Ω, 20-Ω, and 30-Ω resistor. The first step in solving this circuit is to find the equivalent resistance of the two resistors in series (R_2 and R_3). You must do this first because you need a single resistance for each branch of a parallel circuit before you can find its equivalent resistance. R_2 and R_3 are in series, but they form one branch of the parallel circuit. To find the equivalent

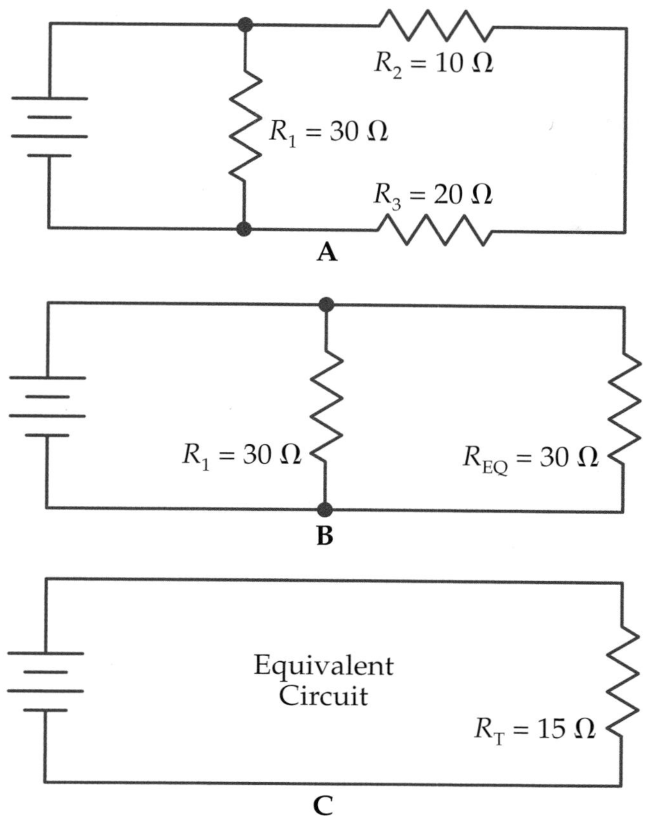

Figure 8-4. Series-parallel circuit with resistors of unequal value. A—First, solve the series element, R_2 and R_3. B—Equivalent resistance (R_{EQ}) is 30 Ω. C—Total resistance (R_T) is 15 Ω.

resistance (R_{EQ}) of two resistors in series, add the values of the individual resistors:

$$R_{EQ} = R_2 + R_3$$
$$= 10\ \Omega + 20\ \Omega$$
$$= 30\ \Omega$$

This gives us the equivalent of two 30-Ω resistances in parallel, shown in Figure 8-4B. To solve this equivalent circuit for total resistance, use the formula for parallel branches of equal value:

$$R_T = \frac{R}{N}$$
$$= \frac{30\ \Omega}{2}$$
$$= 15\ \Omega$$

Figure 8-4C shows the equivalent circuit, which has a total resistance of 15 Ω.

Circuit Analysis

Circuit analysis is the process of breaking down the full circuit network to determine the amount of current and voltage in each

Practical Application 8-1: Drawing Schematic Diagrams

Since circuits can be complex, there are basic rules of construction to follow to make circuits easier to read. The following are some rules which you will find useful at this stage in your study of electronics, along with a schematic drawing which incorporates these rules.

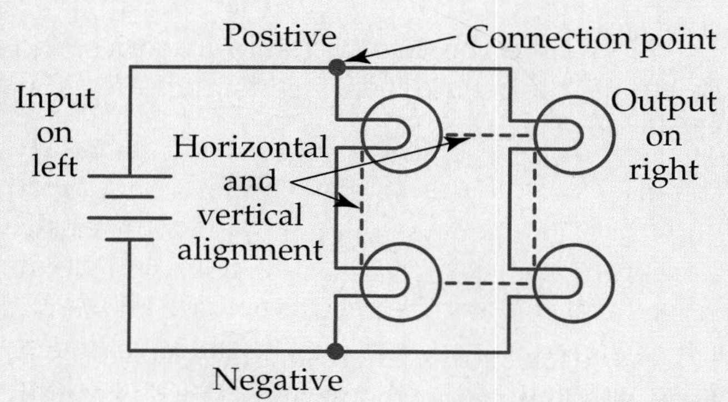

- Lay out the circuit so that the top part is positive and the bottom part is negative.
- Draw the input on the left and the output on the right of the circuit drawing. This means that the power supply is drawn on the left and the load is drawn on the right.
- Align components vertically and horizontally.
- Use a dot to indicate connection points.
- Leave ample space between components and wires. In other words, don't crowd components and wires together.

These drawing practices can save you a lot of time when interpreting schematic diagrams. This simple idea can be very valuable in accurately troubleshooting electrical circuits.

Project 8-1: TV Antenna Coupler

Three resistors can be wired in a special TV antenna coupler circuit so that you can hook up two television sets to one antenna. To build this project, carefully follow the construction procedure while studying the diagram. Few parts are needed:

No.	Item
3	910-Ω resistors
1	300-Ω TV feed line

This coupler eliminates your need for a second TV antenna. It also serves to isolate and cut down on interaction between the two operating TV sets.

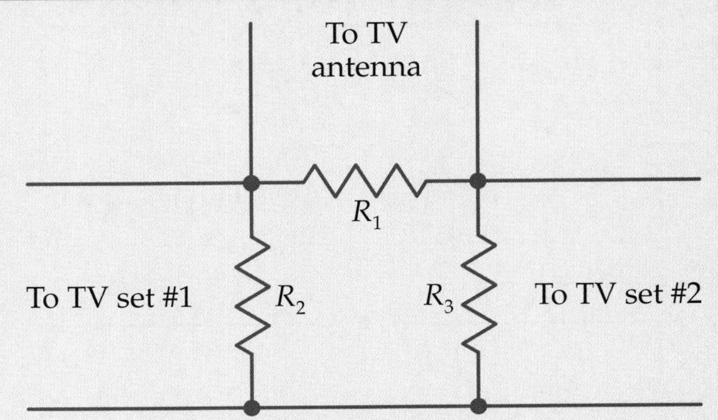

If you have cable TV or use a satellite dish, you can split the signal just as you did with the coupler in this project. Most electronics stores sell splitters that allow you to connect two, three, or more TVs to a single input signal. The splitters are marked for the incoming signal and the additional TVs. The splitter is small enough to fit in the palm of your hand.

of its paths. It includes finding the amount of current in each path and the voltage drop across each path and circuit component.

Circuit analysis also reveals open circuits and short circuits. An ***open circuit*** occurs when the path for current is broken. This can occur accidentally or through the use of a switch. A ***short circuit*** is one that has high current because it has practically no resistance. The current in a short circuit causes the fuse to blow or the circuit breaker to open.

As part of the circuit analysis, an open can be located with a meter. Track the circuit until you get a loss of the signal. That is your open. A blown fuse can be the result of a short circuit. Check the fuse with your meter. If it is blown, you need to track down the cause. Short circuits are very difficult to find because they may not show up until you reach some sequence in the circuit operation.

Now, look at **Figure 8-5.** It shows the circuit we will analyze and its equivalent circuits. In this analysis, we will determine the following:

- Total voltage (E_T).
- Resistance in each branch (R_{EQ}).
- Total resistance (R_T).
- Total current (I_T).
- Current in each branch (I_1 and I_2).

The series-parallel circuit shown in Figure 8-5A can be solved for total voltage (E_T), resistance in each branch (R_{EQ}), and total resistance (R_T) in just a few simple steps. Note that the power supplies are in series. This means that we can add the voltages of the two batteries and come up with the equivalent voltage shown in Figure 8-5B:

$$E_T = E_1 + E_2$$
$$= 12 + 12$$
$$= 24 \text{ V}$$

Now take the two resistors in series and calculate their equivalent resistance (R_{EQ}) as shown in Figure 8-5C:

$$R_{EQ} = R_1 + R_2$$
$$= 12\ \Omega + 12\ \Omega$$
$$= 24\ \Omega$$

Solve for equivalent resistance of two resistors of equal value in parallel. This, in turn,

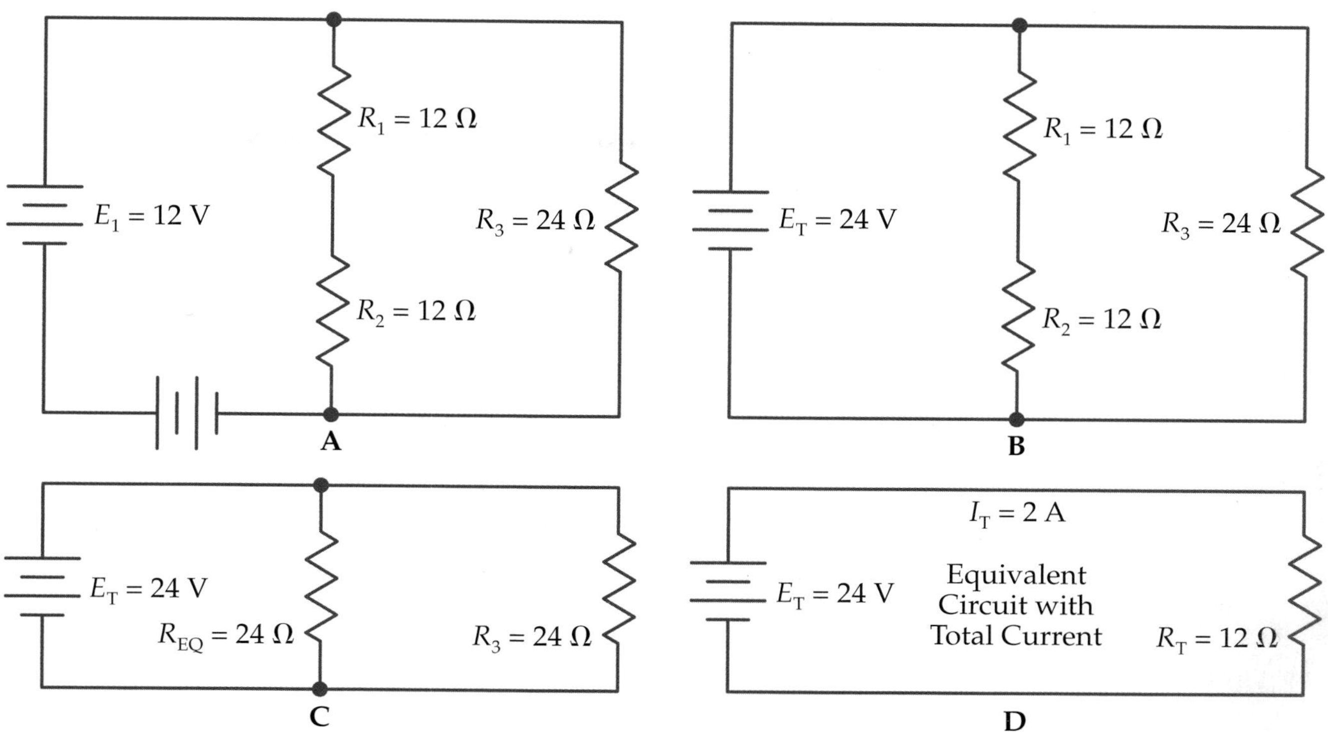

Figure 8-5. Solve this circuit for total voltage (E_T) and total resistance (R_T). A—Circuit has two power supplies and three resistors. B—Equivalent voltage (E_T) is 24 V. C—Equivalent resistance (R_{EQ}) is 24 Ω. D—Total resistance (R_T) is 12 Ω.

gives the total resistance (R_T) of the equivalent circuit shown in Figure 8-5D:

$$R_T = \frac{R}{N} = \frac{24\ \Omega}{2} = 12\ \Omega$$

In Figure 8-5D, the total voltage is 24 V and the total resistance is 12 Ω. Using these values in a formula based on Ohm's law, you can solve for total current (I_T):

$$I_T = \frac{E}{R_T} = \frac{24\ \text{A}}{12\ \Omega} = 2\ \text{A}$$

The total current is 2 A. This does not mean that 2 A flows in each branch of our circuit. It does mean that the battery must be able to supply two amps because that is how much the circuit will use when all the branches are combined.

Now let's find the current in each branch. **Figure 8-6** shows the circuit in Figure 8-5B. Notice that the voltage is combined into a single voltage source, but the parallel branches

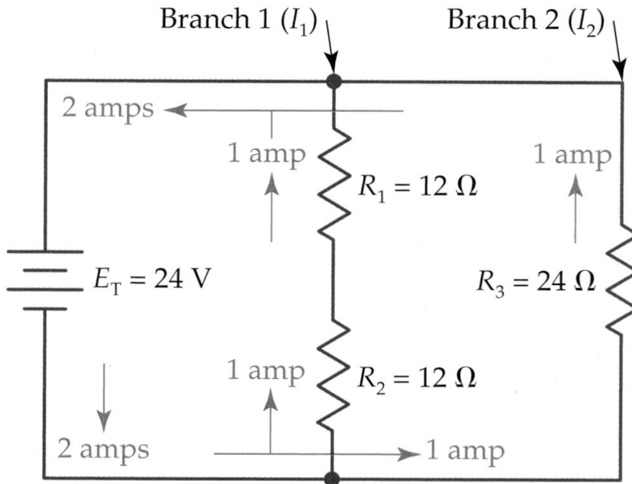

Figure 8-6. Analyze this circuit for the current output and the amount of current flowing through each element. Current output is 2 A, with 1 A flowing through each branch of the circuit.

are intact. To find the current in each branch, we need to divide the total resistance of each branch into the voltage applied across that branch. From our previous analysis of this circuit, we know that the total voltage applied across each branch is 24 V and the resistance in each branch is 24 Ω. Therefore,

$$I_1 = \frac{E_T}{R_{EQ}}$$

$$= \frac{24\text{ V}}{24\ \Omega}$$

$$= 1\text{ A}$$

Since R_{EQ} and R_3 are the same, the current in each branch is the same.

Example 8-3:

Find the total voltage (E_T), the total resistance (R_T), the total current (I_T), and the current through each branch of the parallel section of the circuit shown.

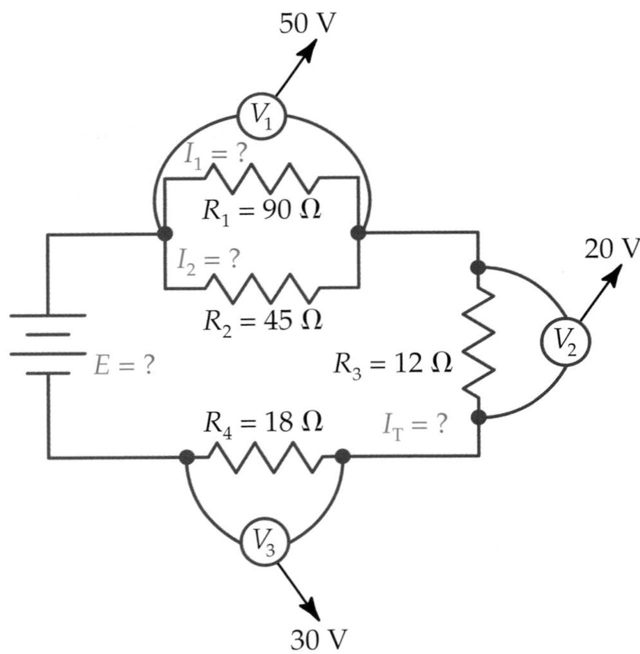

Start by finding the total voltage (E_T). Note the three voltmeter readings across the three portions of the circuit. Recall that the total voltage is equal to the sum of the voltage drops:

$$E_T = V_1 + V_2 + V_3$$

$$= 50\text{ V} + 20\text{ V} + 30\text{ V}$$

$$= 100\text{ V}$$

The total voltage for this circuit is 100 V. Use the formula for resistors in parallel to find the equivalent resistance (R_{EQ}) of the parallel part of the circuit:

$$\frac{1}{R_{EQ}} = \frac{1}{R_1} + \frac{1}{R_2}$$

$$= \frac{1}{90\ \Omega} + \frac{1}{45\ \Omega}$$

$$= \frac{1}{90\ \Omega} + \frac{2}{90\ \Omega}$$

$$\frac{1}{R_{EQ}} = \frac{3}{90\ \Omega}$$

Invert the equation to find the equivalent resistance (R_{EQ}):

$$R_{EQ} = \frac{90\ \Omega}{3}$$

$$= 30\ \Omega$$

The two resistors in parallel offer the equivalent resistance of 30 Ω. Use this value and the equation for resistors in series to find the total resistance (R_T) in the circuit.

$$R_T = R_{EQ} + R_3 + R_4$$

$$= 30\ \Omega + 12\ \Omega + 18\ \Omega$$

$$= 60\ \Omega$$

The total resistance for the circuit is 60 Ω. Now find the total current (I_T). Use Ohm's law and the calculated values for R_T and E_T to find I_T.

$$I_T = \frac{E_T}{R_T}$$

$$= \frac{100\text{ V}}{60\ \Omega}$$

$$= 1.67\text{ A}$$

The total current for the circuit is 1.67 A. Using Ohm's law, find the current in each branch of the parallel portion of the circuit. Start by finding the current (I_1) in the branch containing resistor R_1.

$$I_1 = \frac{E_{R1}}{R_1}$$

We know the voltage in this branch because it is the same in all branches of a parallel circuit. That means the voltage drop across these branches is 50 V.

$$I_1 = \frac{50\text{ V}}{90\ \Omega}$$
$$= 0.56\text{ A}$$

The current through this branch is 0.56 A. Now, do the same calculations for the branch containing resistor R_2.

$$I_2 = \frac{E_{R2}}{R_2}$$
$$= \frac{50\text{ V}}{45\ \Omega}$$
$$= 1.11\text{ A}$$

The current through this branch is 1.11 A. We can make sure this is correct by adding I_1 and I_2. If we get 1.67 A, which is the value we calculated for total current (I_T), we know we are correct.

$$I_1 + I_2 = 0.56\text{ A} + 1.11\text{ A}$$
$$= 1.67\text{ A}$$

We know we did the work correctly because I_1 and I_2 are equal to 1.67 A. Therefore, the total voltage (E_T) in this circuit is 100 V, the total resistance (R_T) is 60 Ω, the total current (I_T) is 1.67 A, and the currents through the branches (I_1 and I_2) are 0.56 A and 1.11 A.

Series-Parallel Capacitors

What if you have a circuit with capacitors in a series-parallel network? You can solve for total capacitance by the same methods used to handle series and parallel circuit problems individually. Look at **Figure 8-7.** Notice that the circuit in Figure 8-7A has three capacitors, each with a rating of 40 μF. To solve this circuit for total capacitance (C_T), first find the equivalent capacitance of the parallel branch:

$$C_{EQ} = C_2 + C_3$$

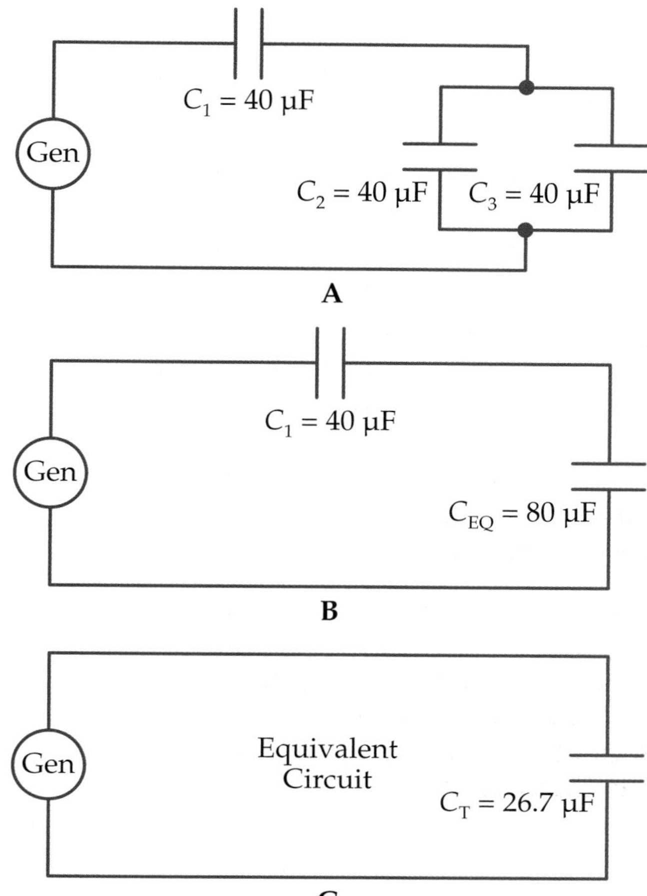

Figure 8-7. Find total capacitance (C_T) of this circuit. A—Three capacitors of equal value. B—Equivalent capacitance (C_{EQ}) of capacitors in parallel is 80 μF. C—Total capacitance (C_T) is 26.7 μF.

$$= 40\ \mu\text{F} + 40\ \mu\text{F}$$
$$= 80\ \mu\text{F}$$

The equivalent value of the two capacitors in parallel is 80 μF, as shown in Figure 8-7B.

Then, find the total capacitance (C_T) for capacitors in series:

$$\frac{1}{C_T} = \frac{1}{C_1} + \frac{1}{C_{EQ}}$$
$$= \frac{1}{40\ \mu\text{F}} + \frac{1}{80\ \mu\text{F}}$$
$$= \frac{2}{80\ \mu\text{F}} + \frac{1}{80\ \mu\text{F}}$$
$$= \frac{3}{80\ \mu\text{F}}$$

Invert both sides of the equation:

$$C_T = \frac{80\ \mu F}{3}$$

$$= 26.7\ \mu F$$

Total capacitance of this equivalent circuit is 26.7 µF, as shown in Figure 8-7C.

The total capacitance is smaller than the capacitance of any capacitor in series. It is also smaller than the equivalent capacitance of any parallel portion of the circuit. The total capacitance can be larger than the capacitance of an individual capacitor in parallel. Once you solve for total capacitance, check your answer. If it is larger than the individual values in series or the equivalent value of the capacitors in parallel, you have made a mistake.

You know how to find equivalent circuits for series-parallel circuits that contain either capacitors or resistors. Many circuits have both resistors and capacitors in them. Circuits of this type cannot be solved by means of the simple formulas we have used so far. Chapter 14 shows how to do this.

Example 8-4:

Find the total capacitance for the circuit shown.

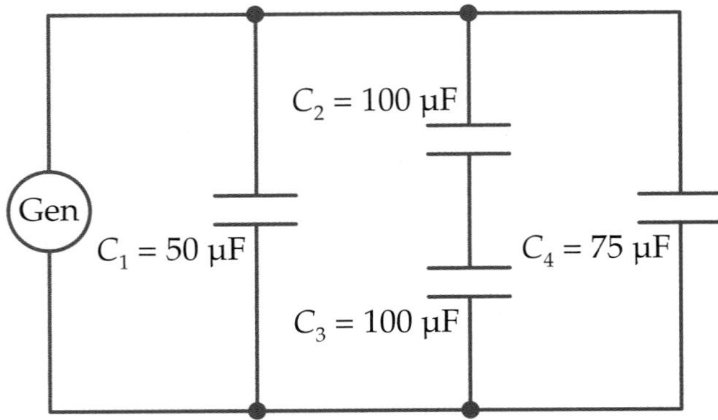

Start by finding the equivalent capacitance (C_{EQ}) for the middle branch that contains two capacitors, using the formula for capacitors in series:

$$\frac{1}{C_{EQ}} = \frac{1}{C_2} + \frac{1}{C_3}$$

$$= \frac{1}{100\ \mu F} + \frac{1}{100\ \mu F}$$

$$= \frac{2}{100\ \mu F}$$

$$= \frac{1}{50\ \mu F}$$

Invert both sides of the equation:

$$C_{EQ} = 50\ \mu F$$

The equivalent capacitance (C_{EQ}) is 50 µF. Use this value and the formula for capacitors in parallel to find the total capacitance (C_T) of the circuit:

$$C_T = C_1 + C_{EQ} + C_3$$

$$= 50\ \mu F + 50\ \mu F + 75\ \mu F$$

$$= 175\ \mu F$$

The total capacitance (C_T) for this circuit is 175 µF.

Ground

We have mentioned the term *ground* in previous chapters. A ***ground*** is an electrical connection between a circuit and the earth or a metallic object that takes the place of the earth. **Figure 8-8** shows two types of ground symbols that normally appear on drawings. These symbols are sometimes used interchangeably; although, there is a difference. The ***chassis ground*** is a grounding system in which the metal frame or box that supports a circuit board is used as a ground. For example,

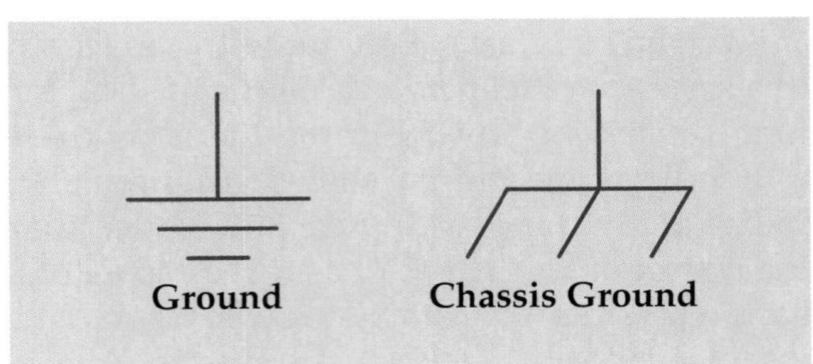

Figure 8-8. Two ground symbols used on circuit drawings.

your computer chassis is used to hold and connect circuit boards that provide graphics, sound, and video cards. You will also find a chassis on television monitors, fax machines, microwaves, telephones, and aircraft radios.

Figure 8-9 shows a diagram of a signaling system that can be used to turn a light on and off. Trace the ground circuit from the power supply through a wire to the load. Note that two ground symbols are used, indicating that the ground circuit is to act as a second wire in the system. When the switch is closed, the circuit is completed and the lamp lights. Radios and television sets also use this arrangement. However, in these applications, the chassis serves as the ground.

One-Wire System

Automobiles use a ***one-wire electrical system.*** This means that rather than having a wire return to the power supply, the frame of the vehicle serves as a path for ground. The one-wire system is easier to trace and simpler to service. **Figure 8-10** illustrates a series circuit and parallel circuit of this type.

All U.S. automakers ground the negative post of the battery. Some older vehicles may have a positive ground. Regardless of which system is used, all automotive electrical circuits are completed by using the car frame as the second wire or path.

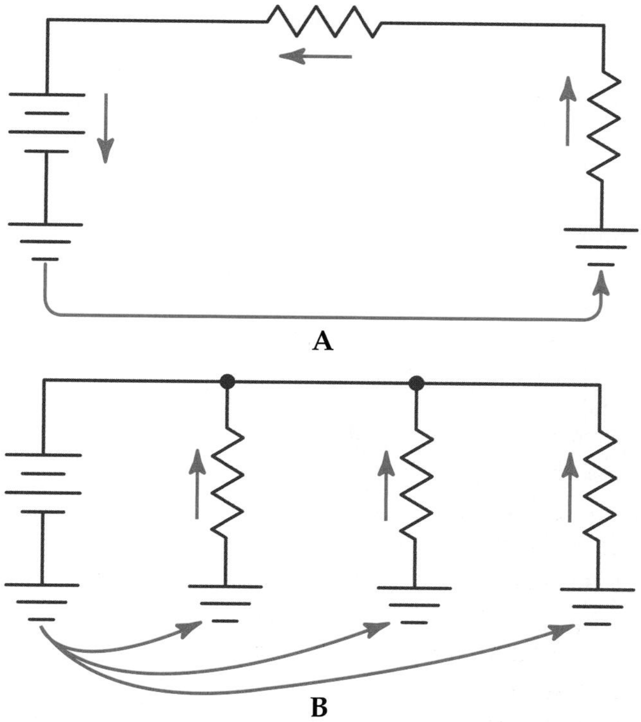

Figure 8-10. Simple one-wire circuits. A—One-wire series circuit. Note the path from ground connection to ground connection. B—One-wire parallel circuit. Note multiple paths to ground in this application.

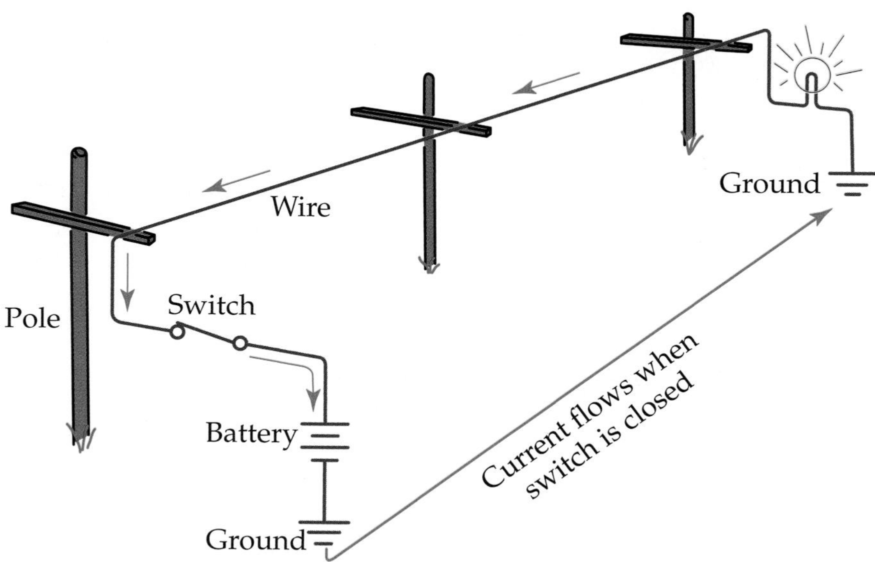

Figure 8-9. A one-wire signaling system using the ground circuit in place of a second wire in the system.

Figure 8-11 gives a simplified idea of some of the electrical circuits you can find on a car. The ground paths for the headlights, taillights, and alternator are shown. Notice that the negative terminal of the battery is connected to the frame. A circuit exists through the frame to each of the lights and the alternator, through the wire, and to the positive terminal of the battery.

The car frame and all the metal parts connected to it serve as the paths to the components. The wire on the component is the return path that completes the circuit. The ground path for the alternator is actually through its housing to the engine, then to the engine ground strap attached to the frame or to the negative post of the battery.

With that knowledge, electrical troubleshooting should start with the battery connection. Make sure that it is corrosion free. The cause of many electrical problems can be solved by cleaning the battery posts. This should be followed by checking for blown fuses. If the fuse is blown, you must try to find the reason that it has blown. After checking for loose electrical connections, proceed to testing with a 12-V continuity tester. You should recall from Chapter 6 that it consists of a light with a pointed probe and an alligator clip, which is connected to ground.

Note in **Figure 8-12** that you can make voltage checks by clamping the negative lead of your voltmeter or probe to a good ground point. The probe is then touched to the positive

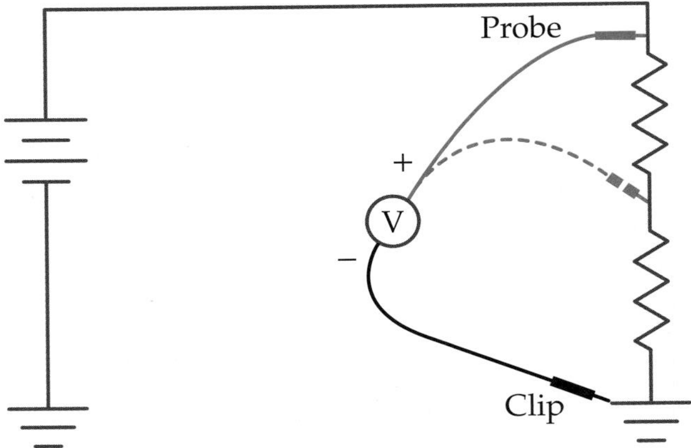

Figure 8-12. A one-wire system speeds voltmeter checks for continuity and voltage drops.

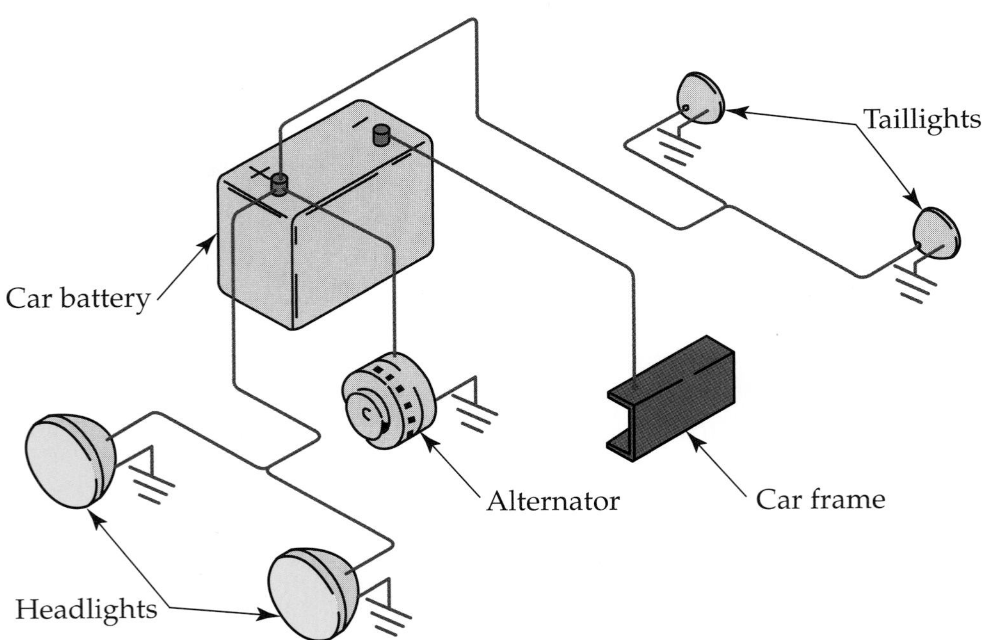

Figure 8-11. Automobiles use the frame for grounding the various electrical circuits.

side of the device in question. Continue to move the probe through the circuit until you locate the open or the loose connection.

> **Note**
>
> You can do the same series of tests using a voltmeter. Before you start any electrical test, check your meter to be sure that it is working. This sounds simple, but a lot of test results have to be done over because of meters with bad batteries or the failure to turn the meter on.

When troubleshooting a car's electrical system, you may notice extra ground straps in locations other than the connection to the negative pole of the battery. Why do you think there is a need for additional ground straps? What do you know about many of the new cars? If you said there are many parts now being made of plastic, you are on the right track. Plastic is an insulator. Therefore, none of the electrical components mounted on plastic have a source for current unless another ground strap is provided.

Polarity

Most one-wire system circuits are not as simple as previously shown. Many appear to have voltage changes, which do not seem to make any sense. To better understand these circuits, look at **Figure 8-13.** This figure shows different ways of drawing the same simple series circuit. Notice that Figure 8-13B is the equivalent to Figure 8-13A. The only difference in these circuits is that the resistors are closer to the power supply.

Figure 8-13C is similar to Figure 8-13B except that the wire has been removed from the right side of the circuit and has been replaced with ground symbols. This is similar to the auto circuit layout we examined earlier. Again, this eliminates the need for showing return wires on a drawing. It also keeps schematics simple and easier to read. Note that polarity is marked at both ends of each resistor. This makes it easier to understand some of the voltage readings that don't seem to make sense.

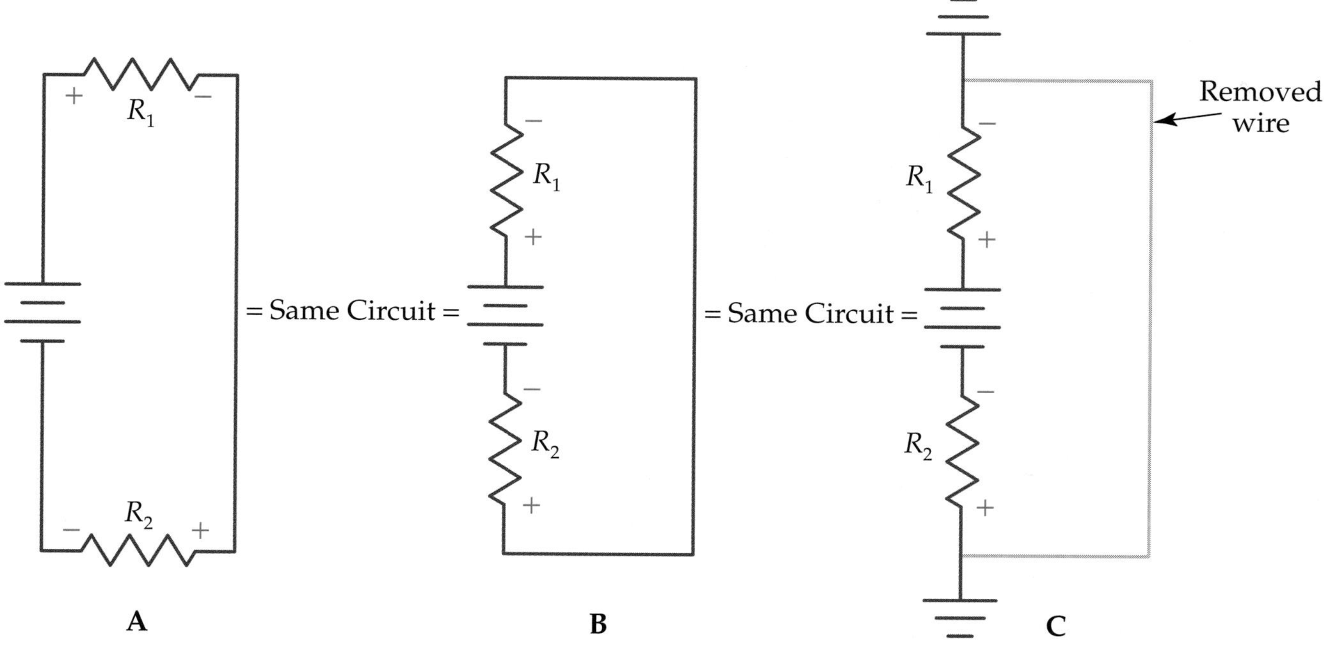

Figure 8-13. Different ways to draw the same series circuit. A—A simple series circuit. B—The same series circuit with resistors placed at different points. C—Wire removed from series circuit and replaced with ground symbols.

Resistors do not have polarity on their own, so when the polarity of a resistor is indicated on a drawing, it is assigned with respect to the polarity of the power supply. Look at **Figure 8-14.** Notice that Figure 8-14A shows that resistor R_1 has a positive side and a negative side. The side of R_1 that is connected to the positive post of the battery is marked plus (+), and the other side is marked minus (–). In the same way, the side of R_2 that is connected to the negative post of the battery has negative polarity. Therefore, the polarity of a resistor is always assigned with respect to the polarity of the power supply. The end of the resistor connected to the positive terminal of the power supply is positive, and the end of the resistor connected to the power supply's negative post is negative.

If you reconnect this circuit without the grounds, you will see the polarities shown in Figure 8-14B. Again, the two ends of each resistor are labeled with respect to the power supply. Although the lower end of R_1 is marked minus (–) and the upper end of R_2 is marked plus (+), there is no difference between these points. However, the polarity is really based on how each resistor is positioned in relation to the power supply. It will help if you keep this in mind as you take measurements across the various points of a circuit. This concept along with using a reference point will help prepare you for information on polarity of a sine wave in Chapter 11.

To illustrate how this form of polarity works, look at the simple series circuit in **Figure 8-15.** If you measure the voltage between point *A* and point *B*, you will get 15 V. Likewise, the voltage between *B* and *C* will be 15 V. The voltage across both resistors from *A* to *C* will read 30 V (voltage of the power supply).

Reference Point

To gain a better understanding of voltage drop, we could pick some reference point and call it zero. Everything above zero would be positive, while everything below zero would be negative.

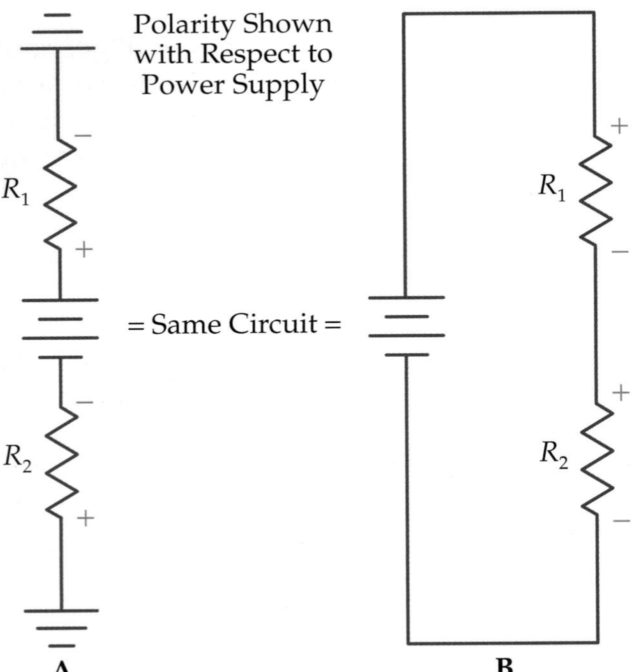

Figure 8-14. This series circuit has polarity indicated at the ends of the resistors. A—One-wire system grounded at two points. B—A complete circuit with return wire.

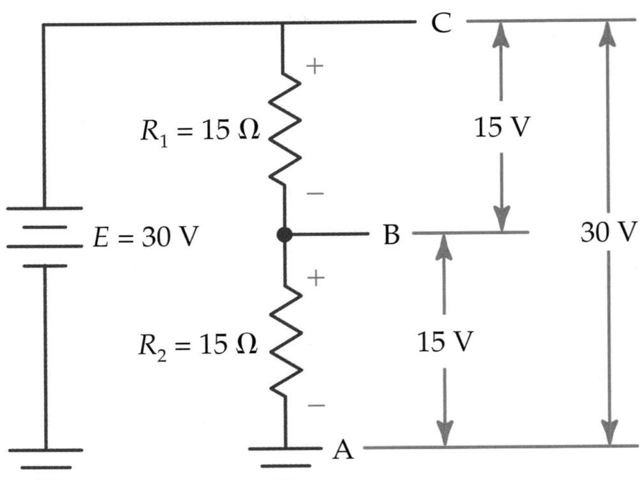

Figure 8-15. Values are given for voltage drop readings taken across each resistor and then across both resistors. Note the polarity of the resistors.

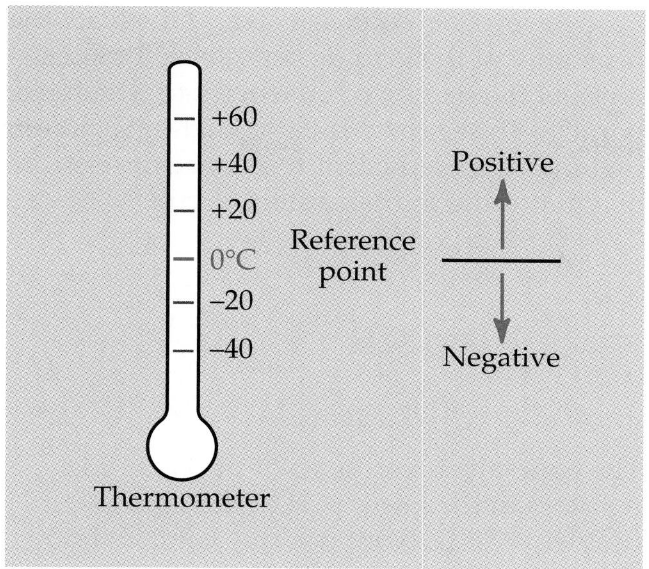

Figure 8-16. On a Celsius thermometer, the zero mark is considered the reference point. All readings above zero are positive. All readings below zero are negative.

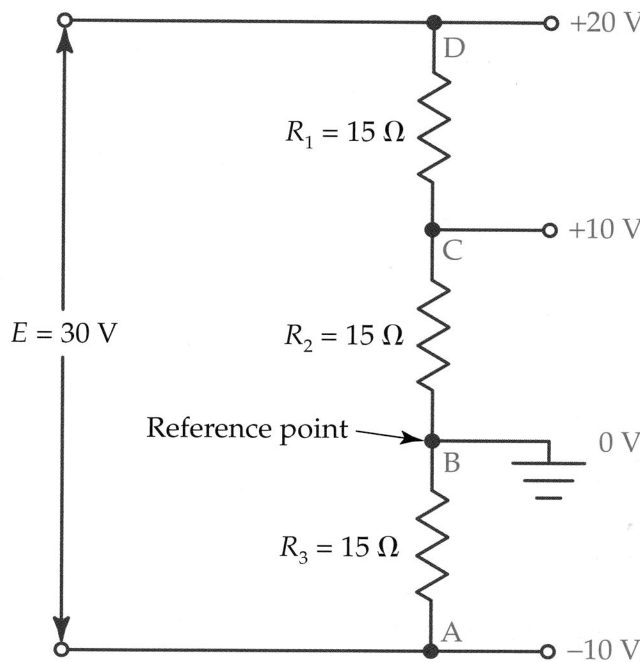

Figure 8-17. Voltage values are given for voltage drop readings taken across resistors from reference points *A* to *B* (–10 V), *B* to *C* (+10 V), and *B* to *D* (+20 V). Note that the circuit is grounded at reference point *B*.

This reference point can be compared with 0° on a Celsius thermometer. See **Figure 8-16.** Again, any reading above zero is positive; below zero is negative. You could get a reading of 20° above the reference point (20°C) or 20° below the reference point (–20°C). In this application, the reference point is where water freezes. With voltages, you can also establish a reference point.

In **Figure 8-17,** you have a power supply with a 30-V potential. By grounding the circuit at the reference point, you can make voltage measurements in relation to that point. In this situation, the ground is said to be a reference point from which voltage measurements are made.

If you measure between point *B* and point *A*, you will get a reading of –10 V. If you measure from *B* to *C*, you will get a reading of +10 V. Finally, measuring from *B* to *D*, you will get a reading of +20 V.

Note that from point *A* to point *D*, the potential difference is still 30 V (which matches the power supply). Since you are using the ground as a reference point, you will get both negative and positive voltages from the same power supply. This is the whole idea behind a voltage divider, which is covered in the following section.

Voltage Divider

If you want to have different voltages for different parts of a circuit, you can construct a ***voltage divider.*** A voltage divider is nothing more than a series of resistors. Due to voltage drops, these resistors allow different but fixed voltages at different parts of the circuit. A load can be connected in parallel to one or more of these resistors. The amount of voltage across that load will be different than the source voltage and can be changed by varying the value of the resistors in the voltage divider.

To design a good voltage divider, you must first look at the load resistance. Note in **Figure 8-18** that if you make a voltage divider with three 15-Ω resistors, you would get a 10-V drop across each one. The current (I) in this circuit can be calculated by finding the total resistance (R_T) and using Ohm's law:

$$R_T = R_1 + R_2 + R_3$$
$$= 15\ \Omega + 15\ \Omega + 15\ \Omega$$
$$= 45\ \Omega$$

$$I = \frac{E}{R}$$
$$= \frac{30\ \text{V}}{45\ \Omega}$$
$$= 0.67\ \text{A}$$

The current in this circuit is 0.67 A.

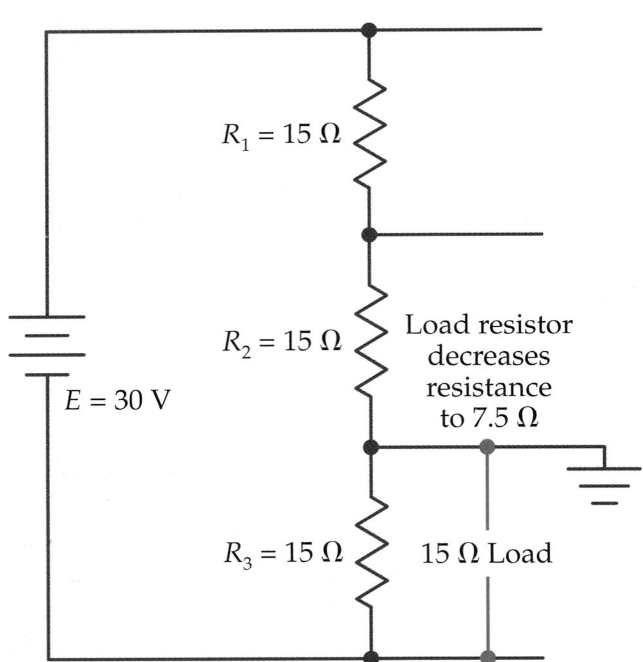

Figure 8-18. Three 15-Ω resistors and one 15-Ω load resistor serve as a voltage divider that provides different voltages to different parts of the circuit.

However, as soon as you add the load, the resistance of the load decreases the total resistance of the section of the circuit to which it is parallel. To see the effects of loading a circuit, first find the equivalent resistance of resistors of equal value in the parallel circuit:

$$R_{EQ} = \frac{R}{N}$$
$$= \frac{15\ \Omega}{2}$$
$$= 7.5\ \Omega$$

The equivalent resistance of the two 15-Ω resistors in the lower part of the voltage divider is 7.5 Ω. Now, use this value to find the total resistance and the new current in this circuit:

$$R_T = R_1 + R_2 + R_{EQ}$$
$$= 15\ \Omega + 15\ \Omega + 7.5\ \Omega$$
$$= 37.5\ \Omega$$

$$I = \frac{E}{R_T}$$
$$= \frac{30\ \text{V}}{37.5\ \Omega}$$
$$= 0.8\ \text{A}$$

As a result of this resistance change, the current in the circuit increases. In this example, it increases from 0.67 A to 0.8 A. With the addition of any load resistor, the circuit carries higher amperage.

When constructing a voltage divider circuit, it is important to watch both resistance and current values. Make sure the current does not become too high and overload the circuit as a result of the lowered resistance.

Example 8-5:

Find the total resistance (R_T) and current (I) of the circuit shown without and with the added load.

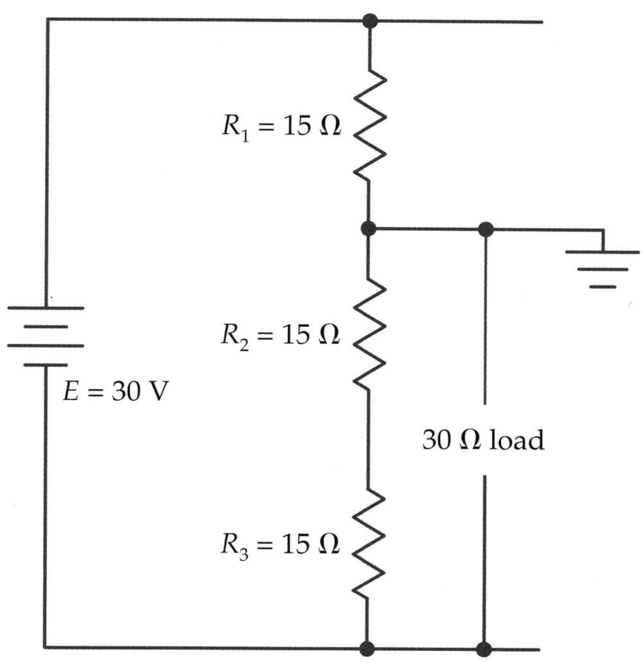

Start by finding the total resistance (R_T) without the load.

$$R_T = R_1 + R_2 + R_3$$
$$= 15\ \Omega + 15\ \Omega + 15\ \Omega$$
$$= 45\ \Omega$$

Then, use Ohm's law to find the current (I_T):

$$I_T = \frac{E}{R_T}$$
$$= \frac{30\text{ V}}{45\ \Omega}$$
$$= 0.667\text{ A}$$

Now, find the total resistance (R_T) with the load. The first step is to find the equivalent resistance ($R_{EQ(R2,R3)}$) of the two resistors in parallel with the load:

$$R_{EQ(R2,R3)} = R_2 + R_3$$
$$= 15\ \Omega + 15\ \Omega$$
$$= 30\ \Omega$$

Use this value to find R_{EQ}.

$$\frac{1}{R_{EQ}} = \frac{1}{R_{EQ(R2,R3)}} + \frac{1}{R_L}$$
$$= \frac{1}{30\ \Omega} + \frac{1}{30\ \Omega}$$
$$= \frac{2}{30\ \Omega}$$

Invert the equation:

$$R_{EQ} = \frac{30\ \Omega}{2}$$
$$= 15\ \Omega$$

Find the total resistance (R_T) for the circuit:

$$R_T = R_1 + R_{EQ}$$
$$= 15\ \Omega + 15\ \Omega$$
$$= 30\ \Omega$$

Now, use Ohm's law to find the current (I):

$$I = \frac{E}{R_T}$$
$$= \frac{30\text{ V}}{30\ \Omega}$$
$$= 1\text{ A}$$

The total resistance in this circuit is 30 Ω, and the current is 1 A. The circuit with a load draws more current (1 A) than without the load (0.667 A).

Wheatstone Bridge

A ***Wheatstone bridge*** is a highly accurate circuit used to measure resistance in series-parallel circuits. It is much more accurate than an ohmmeter. It works by comparing voltages across a bridge circuit. Look at **Figure 8-19.** Notice that the bridge circuit consists of four resistors: two of a known value (R_1 and R_3), a variable resistor (R_2), and one of an unknown value (R_x). The resistor with the unknown value is the resistor to be measured. A voltmeter is placed in the circuit so that it reads the difference in potential between point A and point B.

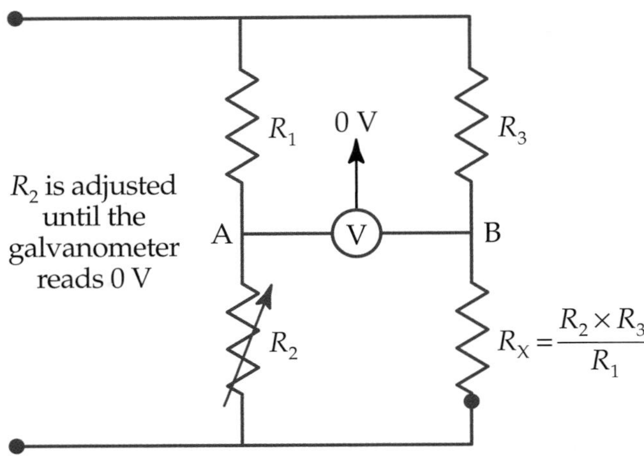

Figure 8-19. Typical Wheatstone bridge setup. Variable resistor R_2 can be adjusted to obtain the desired balance of voltages in the circuit. This, in turn, pinpoints the value of resistor R_X.

The variable resistor is adjusted until the voltmeter reads 0 V. When this occurs, the value of the unknown resistor (R_x) is equal to R_2 times R_3 divided by R_1:

$$R_x = \frac{R_2 \times R_3}{R_1} \tag{8-1}$$

The operation of a Wheatstone bridge can be compared to a scale in a science lab. The value of an unknown weight is found by moving the weights on the triple beam balance. When the weight of the object is found, the beam is balanced.

Math Focus 8-1: Ratios and Proportions

A *ratio* shows a relationship between two values. In electronics, ratios are used to show the relationship between such things as input and output voltage or the resistance v alues in a voltage divider circuit. Ratios can be written as a fraction, such as 2/3, or with a colon between the two values, such as 2:3.

A *proportion* is an equation that says two ratios are equal. To check if two ratios are equal, you can reduce them to their lowest terms:

$$\frac{2}{10} = \frac{1}{5}$$

$$\frac{2 \div 2}{10 \div 2} = \frac{1}{5}$$

$$\frac{1}{5} = \frac{1}{5}$$

Or, you can cross multiply:

$$\frac{2}{10} \times \frac{1}{5}$$

$$2 \times 5 = 10 \times 1$$

$$10 = 10$$

The cross multiply method can be used to find an unknown value in one of the ratios. For example, to find the value of x, you would do the following:

> **Note**
> In actual practice, a galvanometer is used in place of a voltmeter.

To better understand how a Wheatstone bridge operates, let us examine the circuit in **Figure 8-20.** We will first determine the value of the unknown resistor (R_x):

$$R_x = \frac{R_2 \times R_3}{R_1}$$

$$= \frac{32\ \Omega \times 9\ \Omega}{12\ \Omega}$$

$$= \frac{288\ \Omega}{12\ \Omega}$$

$$= 24\ \Omega$$

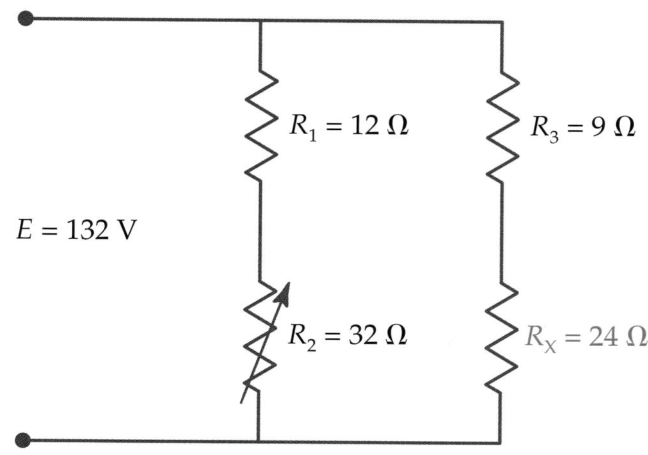

Figure 8-20. When this circuit is balanced, the value of resistor R_X is 24 Ω.

$$\frac{2}{x} \times \frac{1}{5}$$

$$2 \times 5 = 1x$$

$$\frac{2 \times 5}{1} = \frac{1x}{1}$$

$$10 = x$$

Remember the rules of algebraic equations you learned in Chapter 5? They are also used to find the unknown on one side of an equation. These rules are as follows:

- Values on both sides of the equal sign are equal.
- Values can be rearranged, as long as the equality of the formula is maintained.

Practice Problems

Find the unknown value for each of these proportions. Check your answers by either reducing the ratios to their lowest terms or by cross multiplying. Write your work and answers on a separate sheet of paper.

1. $\frac{16}{32} = \frac{8}{x}$
2. $\frac{45}{180} = \frac{x}{24}$
3. $\frac{75}{x} = \frac{35}{70}$
4. $\frac{x}{6} = \frac{36}{72}$
5. $\frac{10}{x} = \frac{3}{51}$
6. $\frac{24}{114} = \frac{x}{57}$
7. $\frac{60}{x} = \frac{12}{80}$
8. $\frac{x}{64} = \frac{21}{24}$

The value of the unknown resistor (R_x) is 24 Ω. Now, we will analyze the circuit to see how it is balanced. We will begin by determining the current in each branch of the bridge circuit. To do this, we need to determine the total resistance in each branch and then divide the voltage of each branch by these values.

$$R_{EQ1} = R_1 + R_2$$
$$= 12\ \Omega + 32\ \Omega$$
$$= 44\ \Omega$$

$$R_{EQ2} = R_3 + R_x$$
$$= 9\ \Omega + 24\ \Omega$$
$$= 33\ \Omega$$

$$I_1 = \frac{E}{R_{EQ1}}$$
$$= \frac{132\text{ V}}{44\ \Omega}$$
$$= 3\text{ A}$$

$$I_2 = \frac{E}{R_{EQ2}}$$
$$= \frac{132\text{ V}}{33\ \Omega}$$
$$= 4\text{ A}$$

Now that you have determined the current value in each branch, we will redraw the same circuit and add a voltmeter, **Figure 8-21.** The voltmeter will read the difference in potential across the resistors in both branches. Before taking this reading, let's find the voltage drop across each resistor:

$$V_{R1} = I \times R_1$$
$$= 3\text{ A} \times 12\ \Omega$$
$$= 36\text{ V}$$

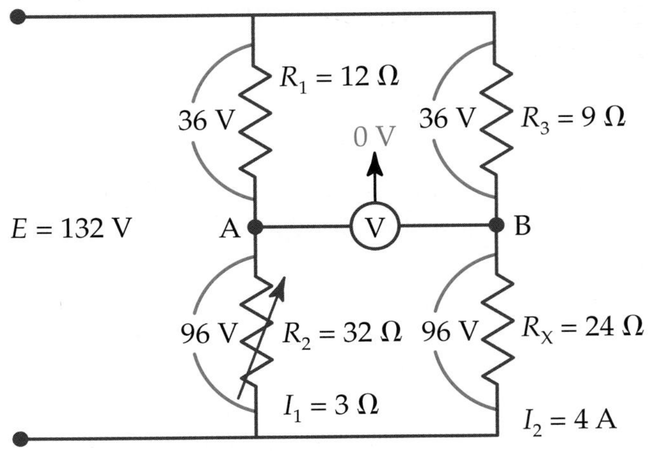

Figure 8-21. Wheatstone bridge principle: the voltmeter should read zero when the voltages are balanced on each side of the circuit.

$$V_{R2} = I \times R_2$$
$$= 3\text{ A} \times 32\ \Omega$$
$$= 96\text{ V}$$

$$V_{R3} = I \times R_3$$
$$= 4\text{ A} \times 9\ \Omega$$
$$= 36\text{ V}$$

$$V_{Rx} = I \times R_x$$
$$= 4\text{ A} \times 24\ \Omega$$
$$= 96\text{ V}$$

With 96 V at point *A* and point *B*, there is no potential difference between the circuit branches. Therefore, the voltmeter reads 0 V. The balance is represented by the ratio

$$\frac{V_{R1}}{V_{R2}} = \frac{V_{R3}}{V_{Rx}} \tag{8-2}$$

Using this ratio we can see that the circuit is balanced:

$$\frac{36 \text{ V}}{96 \text{ V}} = \frac{36 \text{ V}}{96\text{V}}$$

The principle behind a Wheatstone bridge is to get the voltmeter to read zero. A zero reading means there is no difference in potential between the elements of this resistor bridge. If the variable resistor (R_2) were adjusted to a value other than 32 Ω, the circuit would be imbalanced and a voltage would appear on the voltmeter. For example, **Figure 8-22** shows the same circuit in Figure 8-21. However, the variable resistor has been set to 36 Ω. This changes the amount of current flowing in this branch and the amount of voltage across R_1 and R_2.

$$R_{EQ1} = R_1 + R_2$$
$$= 12\ \Omega + 36\ \Omega$$
$$= 48\ \Omega$$

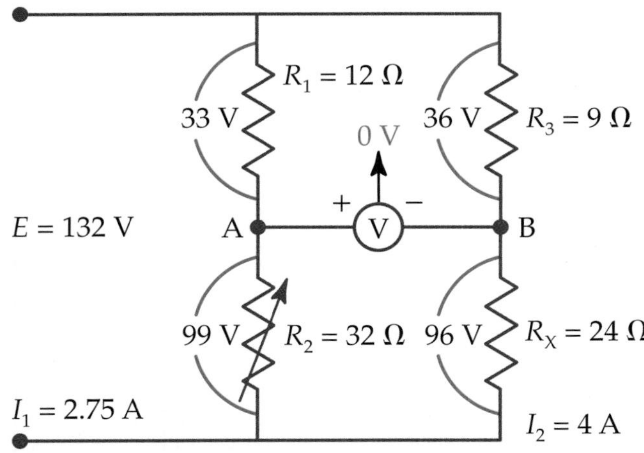

Figure 8-22. An imbalanced bridge circuit produces a difference in potential across point *A* and point *B*.

$$R_{EQ2} = R_3 + R_x$$
$$= 9\ \Omega + 24\ \Omega$$
$$= 33\ \Omega$$

$$I_1 = \frac{E}{R_{EQ1}}$$
$$= \frac{132 \text{ V}}{48\ \Omega}$$
$$= 2.75 \text{ A}$$

$$I_2 = \frac{E}{R_{EQ2}}$$
$$= \frac{132 \text{ V}}{33\ \Omega}$$
$$= 4 \text{ A}$$

$$V_{R1} = I \times R_1$$
$$= 2.75 \text{ A} \times 12\ \Omega$$
$$= 33 \text{ V}$$

$$V_{R2} = I \times R_2$$
$$= 2.75 \text{ A} \times 36\ \Omega$$
$$= 99 \text{ V}$$

With 99 V across R_2 and 96 V across R_x, there is a 3-V difference in potential between point *A* and point *B*. The circuit is, therefore, imbalanced.

Project 8-2: Climb the Ladder Game

Here is a game you can make using an electrical power supply, two pieces of copper tubing, and strips cut from old cans. The metal strips are wired so that it is difficult, yet possible, to move the copper plugs from hole to hole without breaking the circuit.

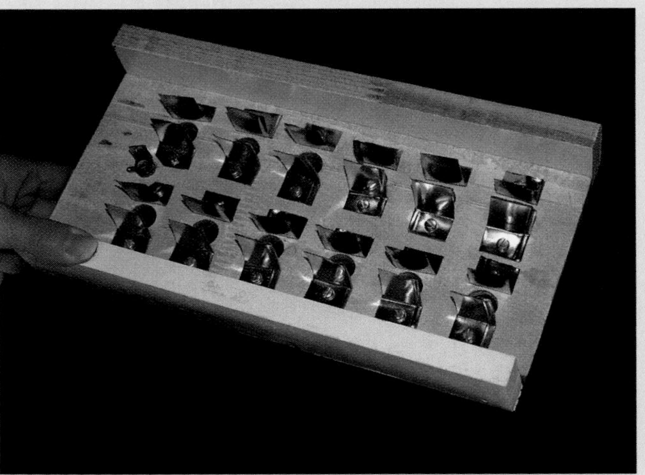

Find these items and follow the procedure to build the game.

No.	Item
1	1/3″ plywood base, 4″ × 10″
2	1/4″ plywood sides, 2″ × 10″
20	Small nails
2	Cans, cut into 20 1/4″ × 1 1/4″ strips
2	Lengths of copper tubing, 1/2″ × 3″
1	Flashlight bulb
1	6′ No. 18 wire
1	6-V battery

Procedure

1. Drill ten 1/2″ holes in the plywood base.
2. Drill one 1/4″ hole for the flashlight bulb.
3. Nail 20 metal strips to the underside of the base alongside the holes and bend them up as shown.
4. Squeeze one end of each piece of copper tubing to keep it from falling through the hole.
5. Make a small hole in the tail end of each strip so you can connect the wires.
6. Cut the wire to the lengths that span between two holes, and insert the wires into the holes in the strips. Wire the project as shown.

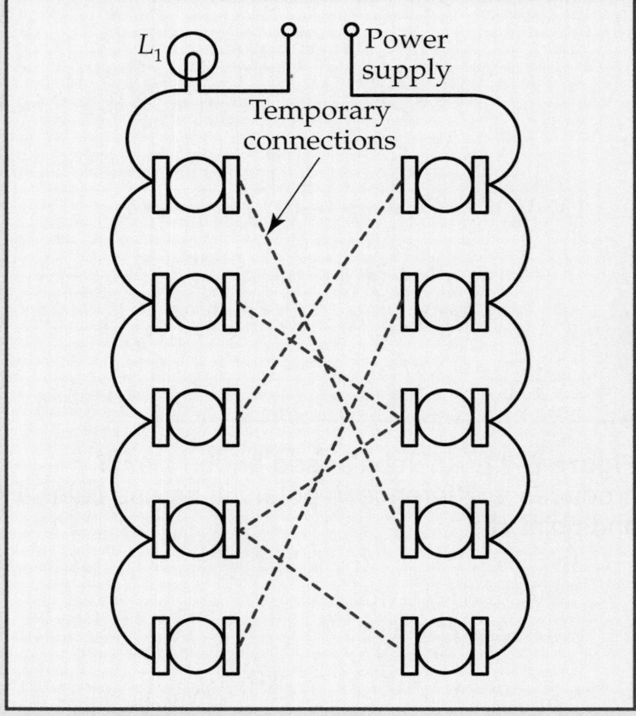

(Project by Randy Schafer)

Project 8-2: Climb the Ladder Game *(Continued)*

7. Hook up the battery and the bulb.
8. Test all the connections for continuity. Solder all wires except those between the two columns of holes (those depicted as dotted lines in the drawing). They will be moved to start a new game.
9. Nail plywood sides to the base.

You can modify the circuit to suit your needs. You could use four 1-1/2 cells connected in series, or a 120-V to 6-V step-down transformer as a power supply. The transformer power supply means you do not have to keep replacing the batteries. The step-down transformer is covered in Chapter 12.

To play the Climb the Ladder Game, allow the player to move only a single copper tube up one of the columns. If the light does not come on, the next player makes a move. The process repeats until both tubes reach the top. The process must take at least five moves before both tubes can be placed in the top two holes.

The player making the fewest incorrect moves is the winner.

You can change the order in which the wires are connected and start round two. A typical connection can be seen in the schematic. With the tubes at the starting point, the correct order of moves is as follows:

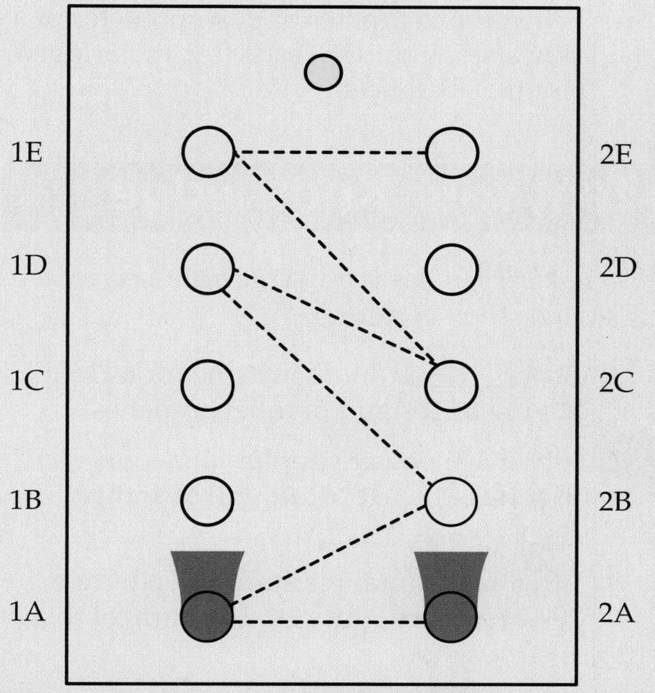

Summary

- A series-parallel circuit contains a combination of series and parallel circuits.
- The total resistance in a series-parallel circuit can be found by calculating the total resistance of the series and parallel sections of the circuit, then combining them using the equations for total resistance.
- Circuit analysis is the process of breaking down the full circuit network to determine the amount of current and voltage in each of its paths.
- The total capacitance in a series-parallel circuit can be found by calculating the total capacitance of the series and parallel sections of the circuit, then combining them using the equations for total capacitance.
- Grounding takes place when a circuit is completed by connecting an end of a wire to the earth or some metallic object that takes the place of the earth.
- A one-wire system is used to connect the electrical components on an automobile.
- The metal body of a car is used to complete the electrical connection in place of a return wire.

- A voltage divider provides different voltages for different parts of a circuit.
- A Wheatstone bridge is a highly accurate circuit used to measure resistance.
- A Wheatstone bridge uses a voltmeter or galvanometer in the middle of a four-resistor bridge.
- When the voltmeter or galvanometer of a Wheatstone bridge reads 0 V, the bridge circuit is balanced.

Test Your Knowledge

Do not write in this book. Write your answers on a separate sheet of paper.

1. A(n) ______ is another name for a circuit having series and parallel branches.
2. If two 3-V power supplies are connected in series, the net result will be either ______ V or ______ V.
3. What is the total resistance (R_T) of two 20-Ω resistors connected in parallel to a 40-Ω resistor?
4. *True or False?* The current in a series-parallel circuit is the same in all parts of the circuit.
5. What is the total capacitance of two 4 μF capacitors connected in parallel to two 4 μF capacitors?
6. What is a ground?
7. Explain how a one-wire system is configured in an automobile.
8. A load added to a voltage divider circuit (decreases, increases) the total resistance of the circuit.
9. A load added to a voltage divider circuit (decreases, increases) the total current of the circuit.
10. What is the purpose of a Wheatstone bridge?

Activities

1. Why do you see some cars with a black strap hanging down to the ground? Cars are not produced with these straps. Why not?
2. Modern electric drills are made with a 3-prong plug, one of which is a ground. Why do some people cut off the ground prong? Is that really the safe thing to do?
3. Is there a difference between a negative or positive ground found on different cars? Which system is better?
4. What will happen if you are not careful when using jumper cables on cars with weak batteries? Are European and U.S. cars "jumped" in the same way?
5. Metrics are used in most electrical measurements. Why did it take so long for other trades to convert to the metric system?
6. Why are there so many ways of drawing the same symbols?
7. Some people claim that they check for voltage by using their fingers. What could happen to those people?
8. Make a list of some of the places where grounds are used.

Multimeters

Learning Objectives

After studying this chapter, you will be able to do the following:

- Adjust the range selector switch on a multimeter to select the desired test.
- Connect meter probes in a circuit correctly and safely.
- Freeze a reading on a digital multimeter.
- Read the display of digital and analog multimeters.

Technical Terms

analog multimeter
autoranging multimeter
digital multimeter
multimeter
range selector switch

A ***multimeter*** is a meter that measures current, voltage, and resistance. It is a combination of an ohmmeter, ammeter, and voltmeter. A multimeter can be analog or digital. **Figure 9-1** shows both types. There is a wide range of multimeters from which you can choose. You can purchase them for prices ranging from 10 dollars to more than 2000 dollars. The less expensive ones have a smaller range, lower accuracy, and fewer accessories. Some of the more expensive ones can be used to read capacitance, inductance, temperature, and frequency.

Not all multimeters look alike. Therefore, the multimeters shown in this chapter may not look like the ones with which you are familiar, but they take the same kinds of measurements. This chapter focuses on their ability to read resistance, ac and dc voltage, and current.

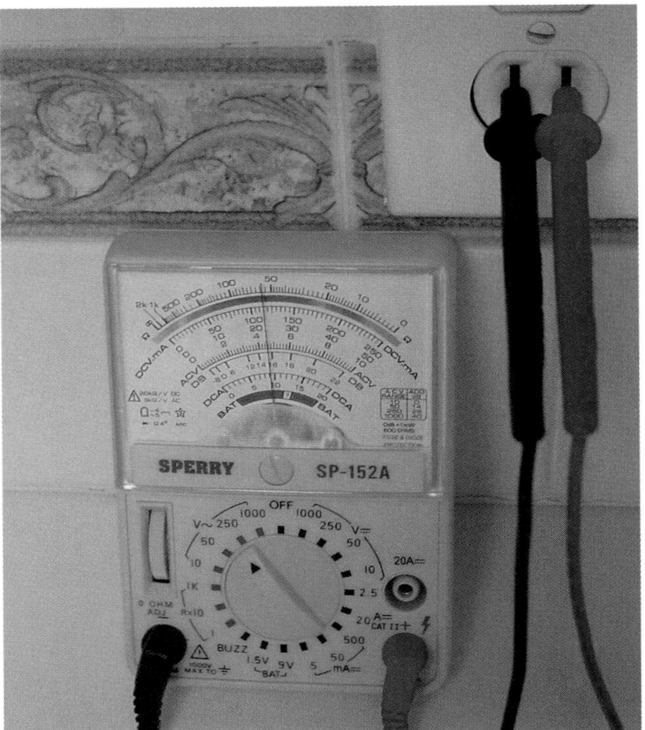

Analog

Digital

Figure 9-1. An analog and a digital multimeter.

Analog Multimeters

An ***analog multimeter*** has multiple graduated scales (ranges of electrical values) and a needle that points to a value on a scale. **Figure 9-2** shows a closeup of a typical display for this type of meter. Analog meters are typically used when reading a value that does not remain steady, such as when tweaking a Wheatstone bridge circuit or testing a capacitor.

This section does not cover the analog meter in depth. Instead, it covers its major features and provides some tips on its use. Check with your instructor for any in-depth information you may need.

Scale

An analog multimeter has various scales on its display. Look again at Figure 9-2. Notice that it has an ohms scale and several dc (direct current) and ac (alternating current) scales. The ohms scale is used when taking resistance readings. The dc scales are used when taking voltage and current readings in a dc circuit. The same scales are used when taking voltage

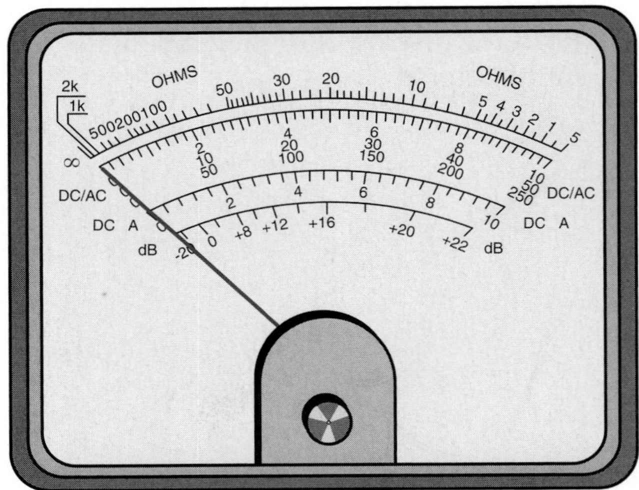

Figure 9-2. Analog multimeter display. Notice the various scales.

and current readings in an ac circuit. The scale read for these readings depends on the range selector switch setting, which we will discuss in the following section.

The most accurate readings can be obtained when the pointer is near the center of the scale. **Figure 9-3** shows the range where you can get the best precision. Try to avoid reading the meter at either end of the scale. Note in **Figure 9-4** that the high numbers at the left end of the top scale are very close together. This is the upper range of the ohms scale. The numbers are in the correct place, but it is very difficult to read with precision. If you tried to take a reading in this area, you could make a mistake of several hundred ohms and not even know it.

In addition to reading the correct scale on an analog meter, some tips on reading an analog meter scale are worth noting. In **Figure 9-5,** for example, a small mirror is located behind the pointer. To read this meter accurately, you must position yourself so that the pointer and its reflection in the mirror line up. If you are able to see the reflection, you are not in the right position to take an accurate reading. Move your head in either direction until you cannot see the reflection, and then take the reading. If the meter does not have a mirror behind the pointer, look straight at the pointer and meter face for best readings.

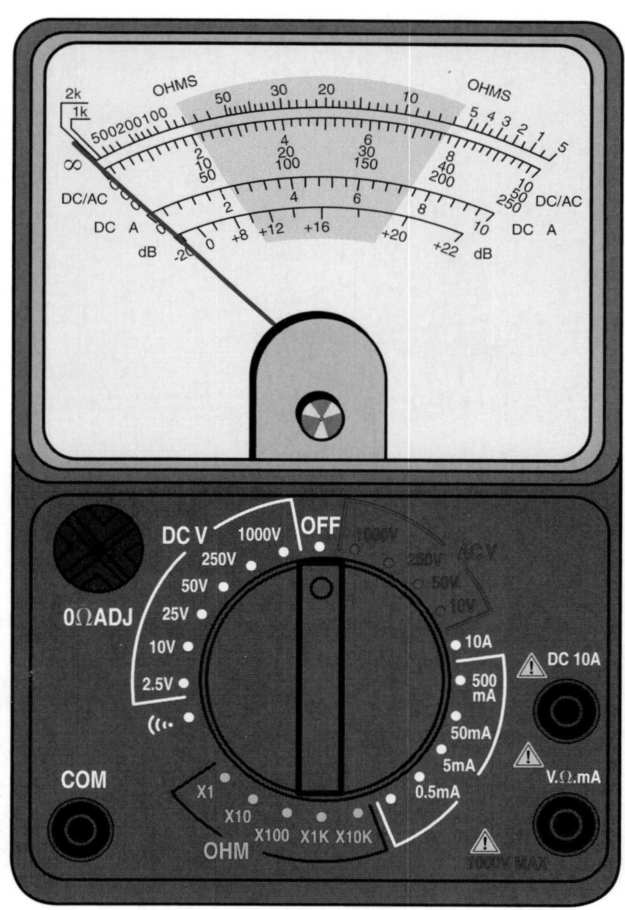

Figure 9-3. The portion of the meter's face in color indicates the area in which it is easiest to take accurate readings.

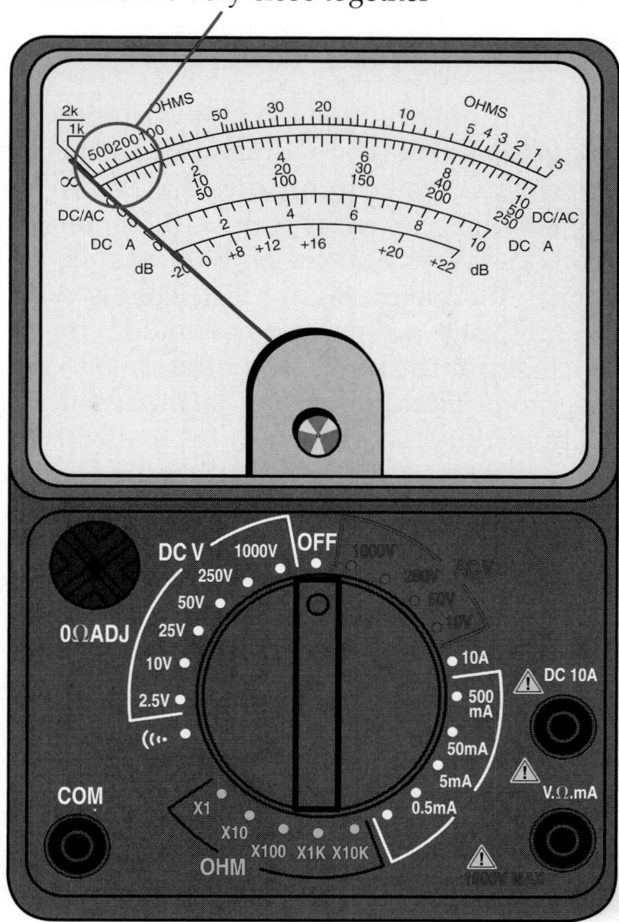

Figure 9-4. The circled area shows why you should avoid taking readings at the extreme ends of a meter scale.

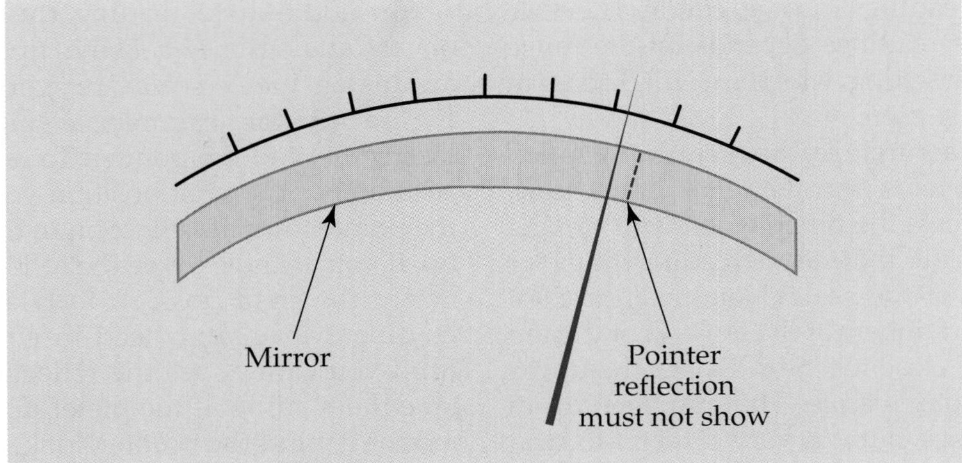

Figure 9-5. The pointer on the meter must be sighted at a right angle to the meter's face. When the reflection does not show, you have a right angle and can read the meter.

Range Selector Switch

When making voltage and current tests, it is possible that the reading under test is too high for the meter to handle. For example, a surge of high current will drive the pointer of an analog meter past the end of the scale and damage the meter, **Figure 9-6.** This is called "pegging the meter." Some analog meters have a peg at the right end of the scale to stop the pointer from going any farther, which is where pegging got its name.

To prevent this from happening, almost all meters have a ***range selector switch.*** When working with unknown voltage and current readings, you should start by adjusting the range selector to its highest value, **Figure 9-7.** Then, work down the range selector switch until you get a reading that can be measured accurately. Try to get the reading in the middle of the scale.

Caution

Remember, when reading unknown currents or voltages, always start with the range selector in the highest position. This helps prevent damage to your meter.

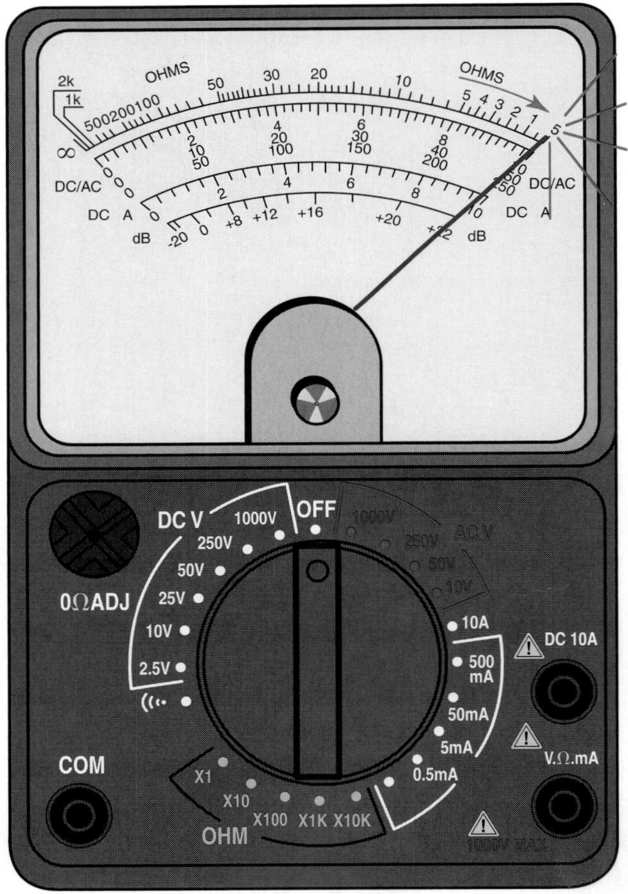

Figure 9-6. "Pegging the meter" occurs when the range selector is set too low and the pointer swings hard against the right end of the scale.

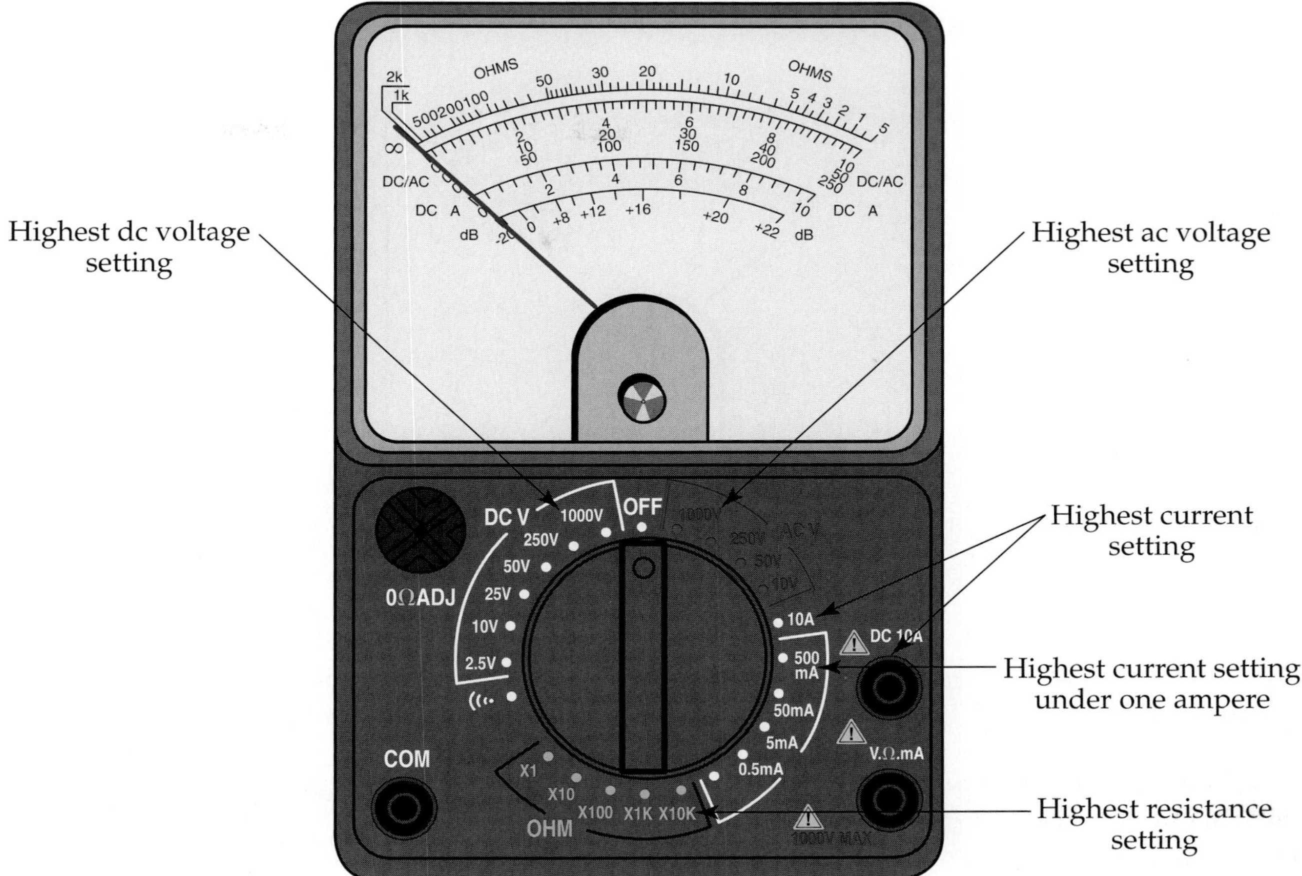

Figure 9-7. The range selector on this multimeter provides several ranges for testing current, voltage, and resistance. Always start with your meter set to the highest range when reading unknown values.

As previously mentioned, the range selector setting determines the scale to be read. **Figure 9-8** shows this relationship. Notice that all of the ohm ranges specified on the range selector are read on the same ohms scale. When the range selector switch is set to dc 2.5 V, dc 25 V, dc 250 V, or ac 250 V, the 0 to 250 DC/AC scale is read. All dc current, ac 50 V, and dc 50 V are read on the 0 to 50 DC/AC scale. The 10 V and 100 V dc and ac selector switch settings are read on the 0 to 10 DC/AC scale. Notice that for this meter there is a separate scale for reading dc current in the 1 A to 10 A range and a special input terminal for reading high current on the DC 10 A scale.

Ohmmeter Safety Precautions

Whenever you work with an ohmmeter, the power must be off in the circuit under test. All ohmmeters have dry cells for their power source. Ohmmeters and multimeters use these dry cells as the voltage supply needed to force current through the resistor being tested.

If you accidentally connect an ohmmeter to a live circuit, you may damage the meter. If you happen to leave the range selector in the ohms position and attempt to test voltage, the meter may explode because the voltage from the circuit's power source is charging the dry cells in the multimeter at a very high rate. This huge rush of electrons could cause the meter to blow apart.

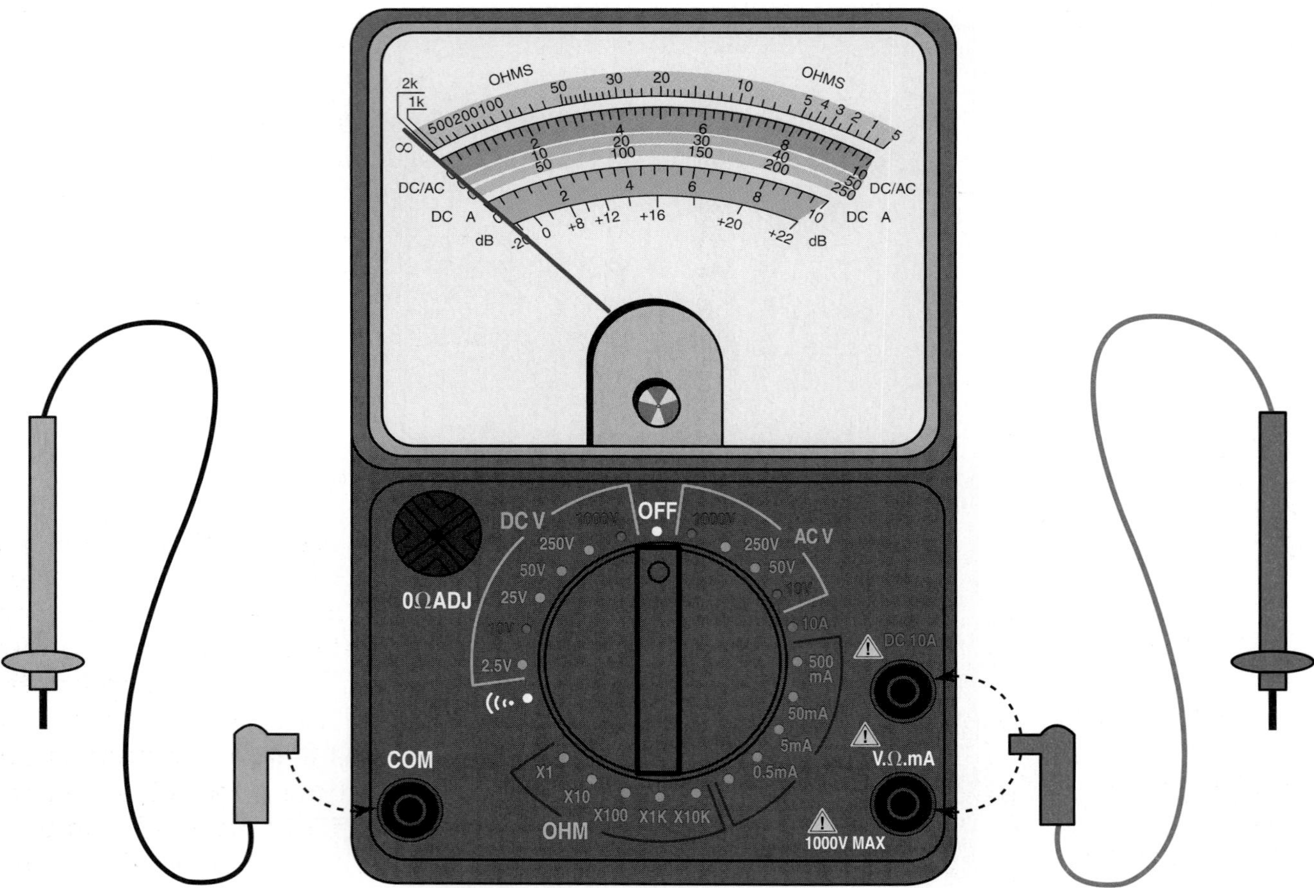

Figure 9-8. Meters with more than one scale generally have a range selector. Note the many positions on this meter's range selector and the corresponding scales. Also, notice the two input terminals for the red probe.

Most multimeters have a built-in overload device. If you attempt to take a voltage reading with the range selector switch in the ohms position, it will pop the reset button. If this happens, remove the probe from the circuit, set the range selector switch to the desired voltage range, and push the reset button.

Zeroing the Meter

When an analog multimeter is going to be used to measure resistance, the first thing to do is zero the meter, **Figure 9-9.** To do this, take the red probe (in the positive terminal) and touch it to the black probe (in the terminal marked "Common"). Note that when the probes touch, the pointer swings to the right, toward the zero on the top scale of the meter.

If the pointer does not stop exactly on zero, you will have to adjust the "zero ohms adjust" (marked "0ΩADJ" on our meter) dial until the pointer aligns with the zero mark. The zero ohms adjustment must be made each time you move the range selector switch to a new resistance position. This will ensure an accurate resistance reading every time. If you do not zero the ohmmeter, your reading will be off the same amount that your meter is out of adjustment.

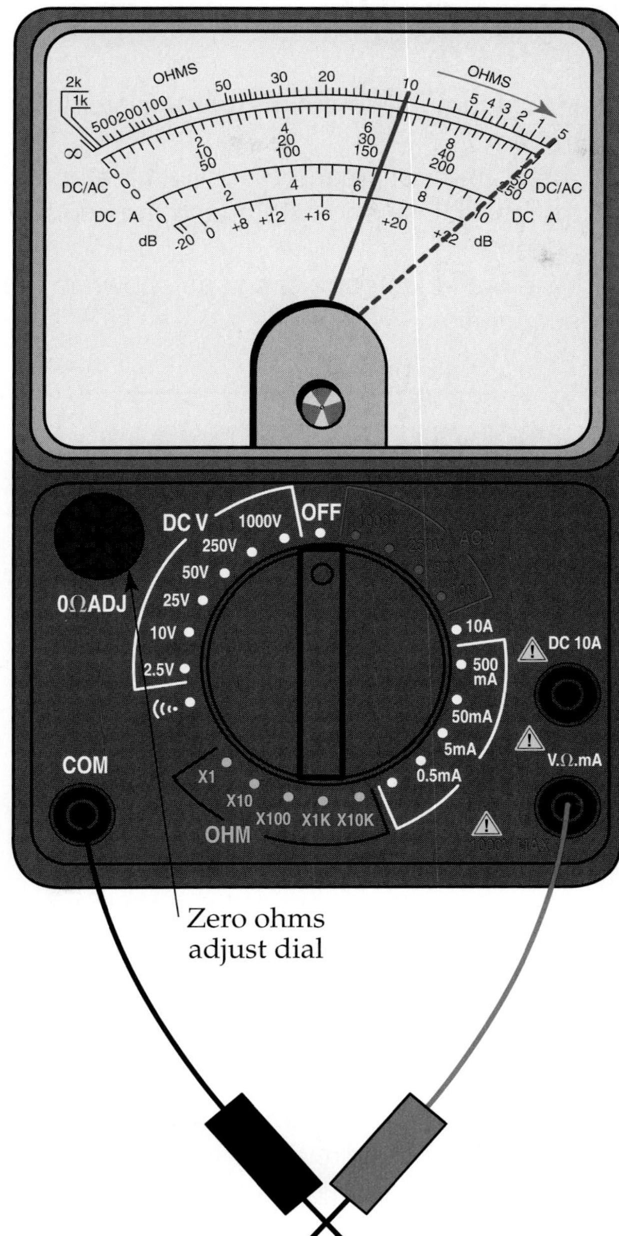

Figure 9-9. To zero your ohmmeter, touch the probes together and adjust the zero-ohms dial until the pointer aligns with zero.

Meter Protection and Maintenance

The best way to protect your meter is to disconnect the power source from the circuit before connecting an ohmmeter or multimeter. When you are done using a multimeter, put the range selector switch in the highest voltage position or in the off position if the meter has one. This takes the dry cells out of the circuit. This also lengthens the life of the dry cells.

As the dry cells inside the ohmmeter or multimeter begin to age, the "zero ohms pot" must be adjusted more often. When the dry cells inside the meter need to be replaced, the meter can no longer be zeroed and all readings are wrong.

Digital Multimeters

A ***digital multimeter,*** or ***DMM,*** spells out the specific value of the electrical unit under test. Most DMMs use segmented display numbers similar to the type found on many alarm clocks, ovens, and other appliances. One type of digital readout has groups of seven small segments that form a number on the meter face according to which segments light, **Figure 9-10.** In this way, all numbers from zero through nine can be shown by lighting the proper segments. For example, numbers two and six would light up as shown in **Figure 9-11.** The number eight lights up the entire seven segments. Can you figure out the rest of the numbers?

One advantage DMMs have over analog multimeters is they are accurate within 0.1 %.

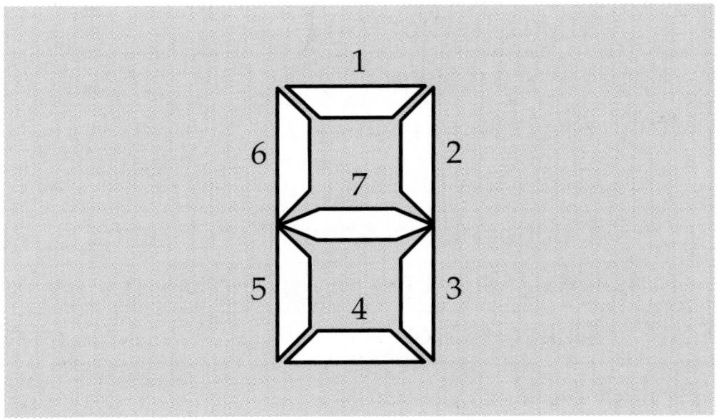

Figure 9-10. A seven-segment display.

Project 9-1: Lie Detector

A lie detector can use an analog meter to detect a change in resistance. One particular design is shown here.

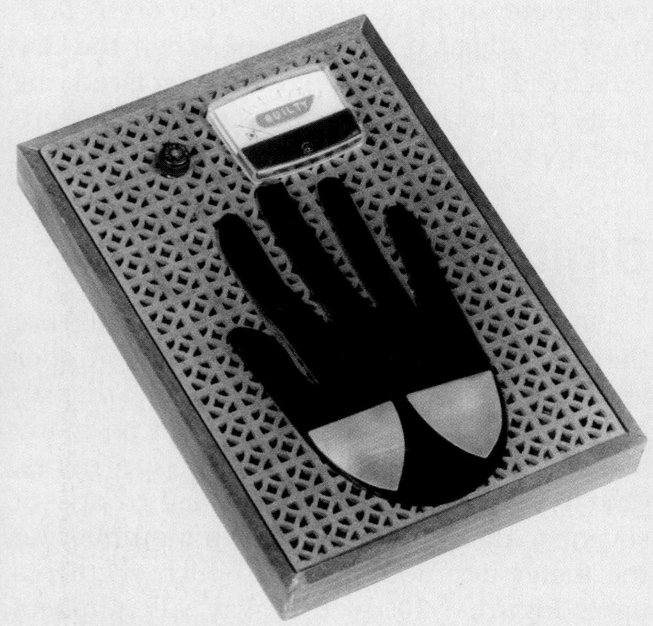

Gather the materials listed. Then assemble the materials according to the steps in the procedure. Be sure to mount the parts in compact fashion, and keep the connecting wires as short as possible.

No.	Item
1	15-V battery (B_1)
1	Milliammeter (M_1)
1	12-kΩ resistor (R_1)
1	1.5-kΩ resistor (R_2)
1	SK3005 transistor (Q_1)
1	SPST switch (S_1)
2	Copper plates, about 4″ × 4″ each
1	Box of your own design and wiring
1	Battery holder

(Project by Mike O'Berski)

Procedure

1. Mount the resistors and the transistor in the box.
2. Attach the switch and the milliammeter to the cover. Use epoxy to attach the copper plates.
3. Wire the lie detector circuit as shown in the schematic.

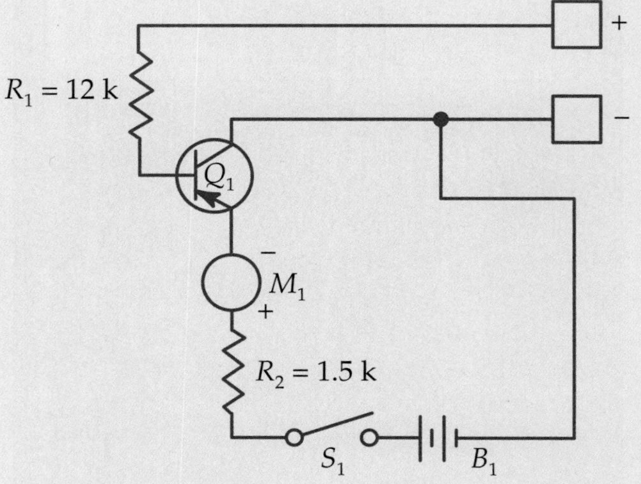

4. After all the other connections are made, install the battery between the switch and the transistor.
5. Install the cover on the box.
6. Turn off the switch and zero the meter with the mechanical adjustment screw.
7. Turn on the switch and use a jumper wire to complete the circuit across the copper plates. You should get a reading of at least 0.8 mA. If not, check the condition of the dry cell.

Now you have a simple ohmmeter with which to check body resistance. To use the lie detector, have someone put a hand on the copper plates. Turn on the switch and ask a question. Since many people's palms sweat when they are under stress, they may have moisture on their palms if they lie. This will reduce resistance, and the meter needle will deflect to the right. By this decrease in resistance, you know that this person has sweaty palms and could be telling a lie.

Figure 9-11. By lighting certain segments of a seven-segment display, numbers zero through nine can be displayed.

It is nearly impossible to read an analog meter with that degree of accuracy. Another advantage of DMMs is the display. You do not have to compare a selector switch setting to a line on the meter scale and interpret the value of the reading. For that reason, fewer mistakes are made in reading a value because the numbers are readily visible.

DMM Features

Many of the newer DMMs consist of a 6000-count display for a high-resolution digital readout, four push buttons (HOLD, MIN MAX, RANGE, and Hz), a range selector switch, and color-coded input terminals for attaching probes.

Examine the meter in **Figure 9-12.** Notice the four push buttons and the range selector switch settings. Also notice that, like the analog multimeter, the DMM has a special input terminal for 10 A readings. However, it does not have a special range selector setting for this purpose.

Some DMMs are capable of switching between ranges automatically. This type is known as an ***autoranging multimeter.*** They are read directly on the display screen without having to change selector switch settings. A DMM may have other functions besides ammeter, voltage, and ohms. These functions are beeper, capacitance, and diode. See **Figure 9-13** for typical range selector switch settings and their function.

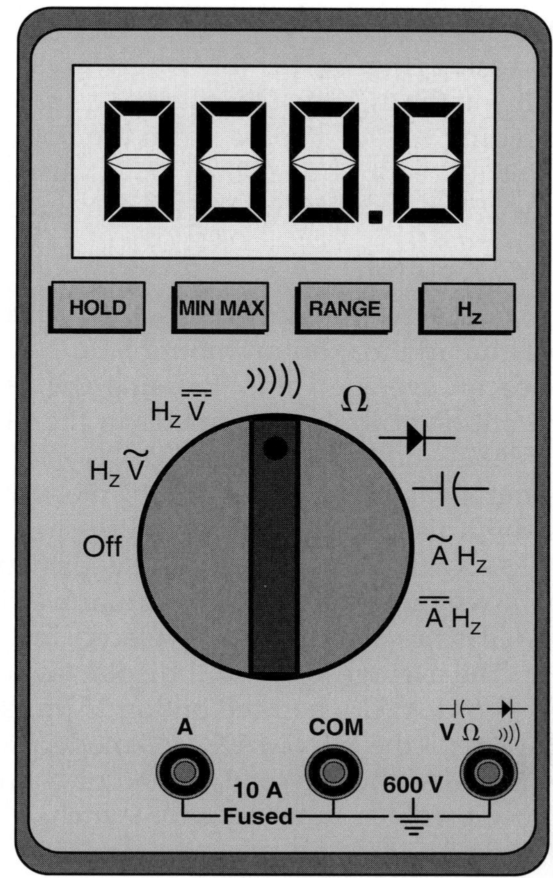

Figure 9-12. This DMM has four push buttons and many range selector switch settings. Can you explain what each push button and range selector switch setting is for?

Range Selector Switch Symbol	Function
$H_Z\tilde{V}$	AC volts and frequency
$H_Z\overline{V}$	DC volts and frequency
)))))	Beeper
Ω	Resistance
—▶|—	Diode test
—|(—	Capacitor test
$\tilde{A}\ H_Z$	AC amps and frequency
$\overline{A}\ H_Z$	DC amps and frequency

Figure 9-13. Range selector switch symbols and their description.

HOLD button

Pressing the HOLD button freezes the reading on the display. This allows you to lock the reading on the display if you are not in a position to read it. You can then move the meter so you can see and record the measurement.

MIN MAX button

The MIN MAX button is used to capture minimum and maximum input readings and provide an average value. Pressing the MIN MAX button will place the meter in the MIN MAX AVG mode. The word *MAX* and the maximum value recorded since the meter was placed in this mode will display. Pressing the MIN MAX button again will display the word *MIN*. It will also display the minimum value recorded since the meter was placed in this mode. The average value will display as well as the word *AVG* when the button is pressed again. To exit the MIN MAX AVG mode, press the MIN MAX button for 1 second or change the position of the range selector switch.

RANGE button

The RANGE button allows you to take control of the DMM by overriding the autoranging function and locking on to a specific range. When you push this button, it cycles you through the ranges from which you can make your choice. You can return to the autoranging function by holding down the RANGE button for two seconds. An area on the display shows the range you have selected or if the meter is set to autoranging.

Hz button

You may have noticed the yellow Hz button and the yellow "Hz" printed next to the voltage and current switch locations. Hz is short for Hertz, which is the frequency of the voltage and current. In the United States and several other countries, electricity is produced with a frequency of 60 Hertz. In Europe and several other locations, the frequency of their electricity is 50 Hertz. That is why people traveling from the U.S. to Europe with their shavers or other small appliances must purchase a converter. Otherwise, these appliances run the risk of being destroyed when they are plugged into a 50 Hz outlet. Many of these countries also use a different type of plug for their electrical outlets. More of an explanation is provided in Chapter 11.

If you push the yellow Hz button during a voltage or current test, the DMM will read the frequency of that test. In Chapter 8, we discussed a reference point on a thermometer. There is a similar reference point, sometimes called a *threshold level*, with an ac signal. The DMM can measure the number of times the ac signal crosses that point each second. This number appears on the display and is the frequency of the signal under test.

The DMM can read the voltage frequency in the range from 5 Hz to 50,000 Hz. It can read the current frequency in the range from 50 Hz to 5,000 Hz. If you push the yellow button again, it returns to the voltage or current under test.

Beeper Function

If you set the range selector switch to Beeper, you will hear an audio signal, which alerts you that there is continuity. This function can save you time because you do not need to look at your meter when you are testing for open wires.

> **Warning**
> If you test fuses for continuity using the Beeper function, make sure no power is applied to the fuse.

Capacitance Function

The DMM shown in **Figure 9-14** can be used to measure capacitance. It can read capacitance values from 1 μF to 9999 μF. Follow these steps to measure capacitance:

1. Connect the black probe to the COM terminal and the red probe to the terminal for capacitance, diode test, voltage, resistance, and continuity.

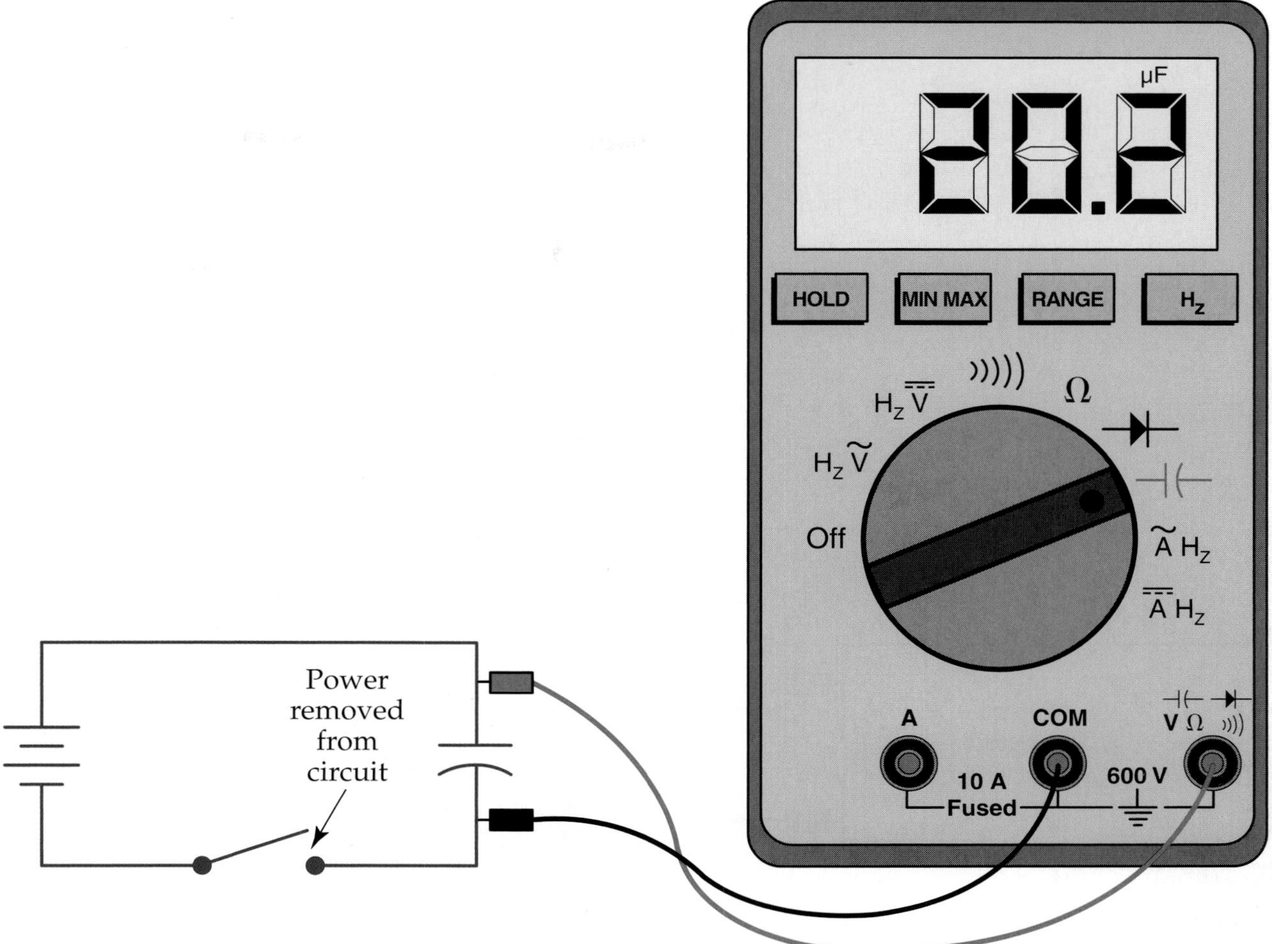

Figure 9-14. Power must be removed from the circuit before measuring capacitance.

2. Turn the range selector switch to the capacitor position.
3. Make sure the portion of the circuit or the capacitor being measured is not electrically "hot." If the circuit contains high-voltage capacitors, make sure they have been discharged before taking a capacitance measurement.
4. Touch the probes to the capacitor leads. If testing an electrolytic capacitor, be sure to touch the common probe to the negative capacitor lead and the positive probe to the positive capacitor lead.
5. Read the value of the capacitance on the display.

Diode Test Function

The diode test is used to check semiconductor devices such as diodes, transistors, and other items. Diodes are covered in Chapter 17.

Ammeter Function

An ammeter or multimeter is always connected in series with the power source and the load when measuring current flow through the load, **Figure 9-15.** The polarity of the meter must match the circuit's polarity when using it in a dc circuit. In a dc circuit, current flows in only one direction. Polarity is not important when hooking up your meter

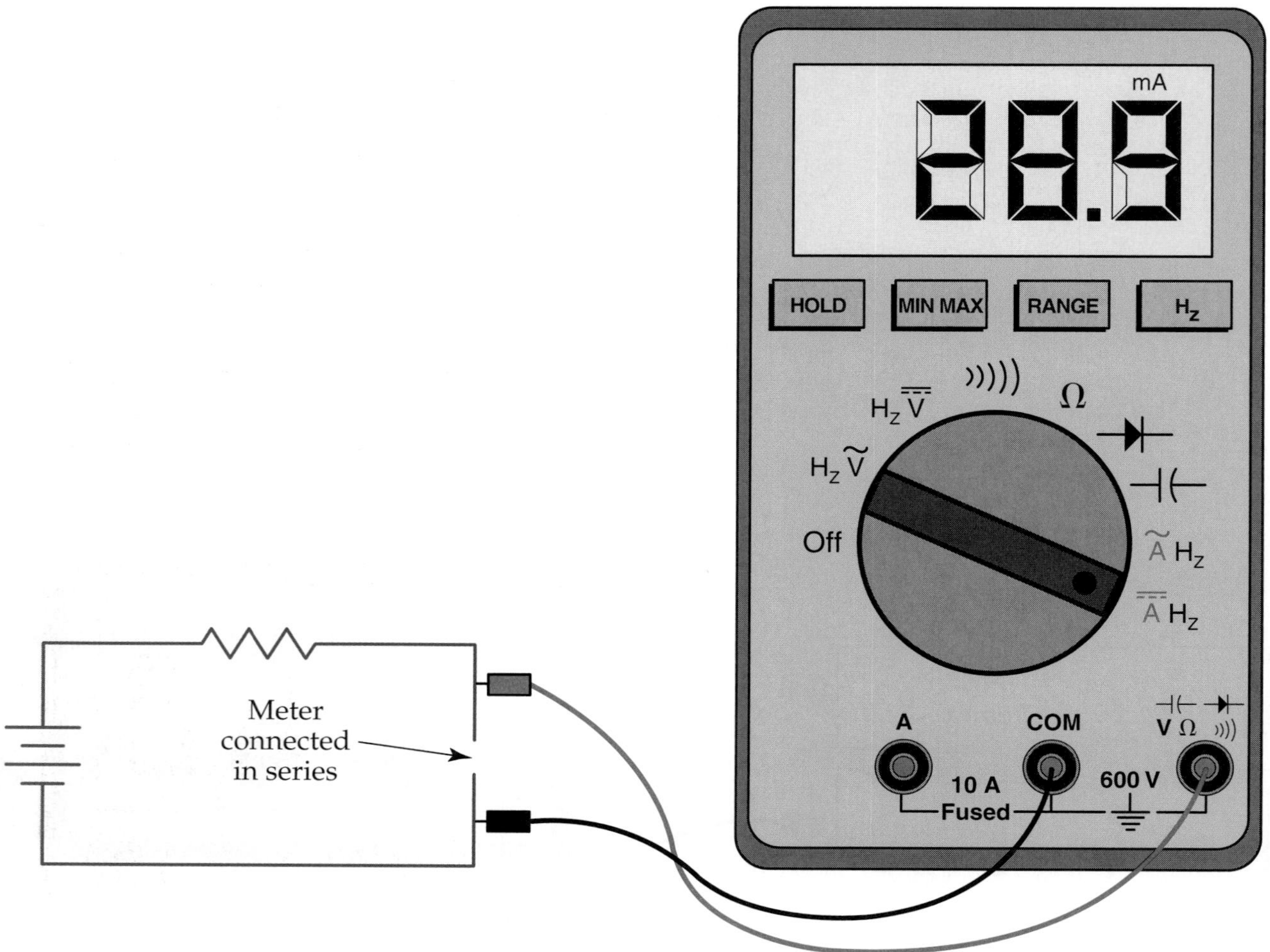

Figure 9-15. An ammeter or multimeter is always connected in series with the power source and load when measuring current.

to an ac circuit. In Chapter 11, you will learn why. To measure current, follow these steps:

1. Connect the black probe to the COM input terminal and the red probe to the input terminal for measuring current. If you place the red probe in the 10 A input terminal, you can measure from 0.001 A to 10 A. The meter can take a reading of a temporary overload up to 20 A.
2. Turn the range selector switch to the $\overline{\overline{A}}$ position when reading direct current and to the $\tilde{A}$ position when reading alternating current.
3. Make sure the circuit power is disconnected.
4. Open the circuit at some convenient point. Remember, when measuring current, the meter must be connected in series.
5. If measuring direct current, connect the black probe to the disconnected part of the circuit on the negative side of the power source. Connect the red probe to the disconnected part of the circuit that is on the positive side of the power source. If measuring alternating current, you do not have to worry about polarity.

Warning

Remember that current is flowing in this circuit during the test. Make sure you do not touch the metal on the probes, or you will be part of the circuit and may be shocked as a result.

6. Turn on the power to the circuit.
7. Read the meter.
8. Disconnect the power before removing the leads.

Note

If you change the position of the range selector switch to and from the amps position, some meters give you a warning like "LEAd." This warning is aimed at protecting the DMM from being damaged from unwanted overload conditions.

Voltmeter Function

A voltmeter or a multimeter is used to measure voltage values in an electrical circuit. It is placed in parallel to the voltage source or component, **Figure 9-16.** When measuring voltage in a dc circuit, polarity is an issue. The meter probes should be placed so that they match the polarity of the circuit. To measure voltage, follow these steps:

1. Connect the black probe to the COM input terminal and the red probe to the input terminal for measuring voltage.

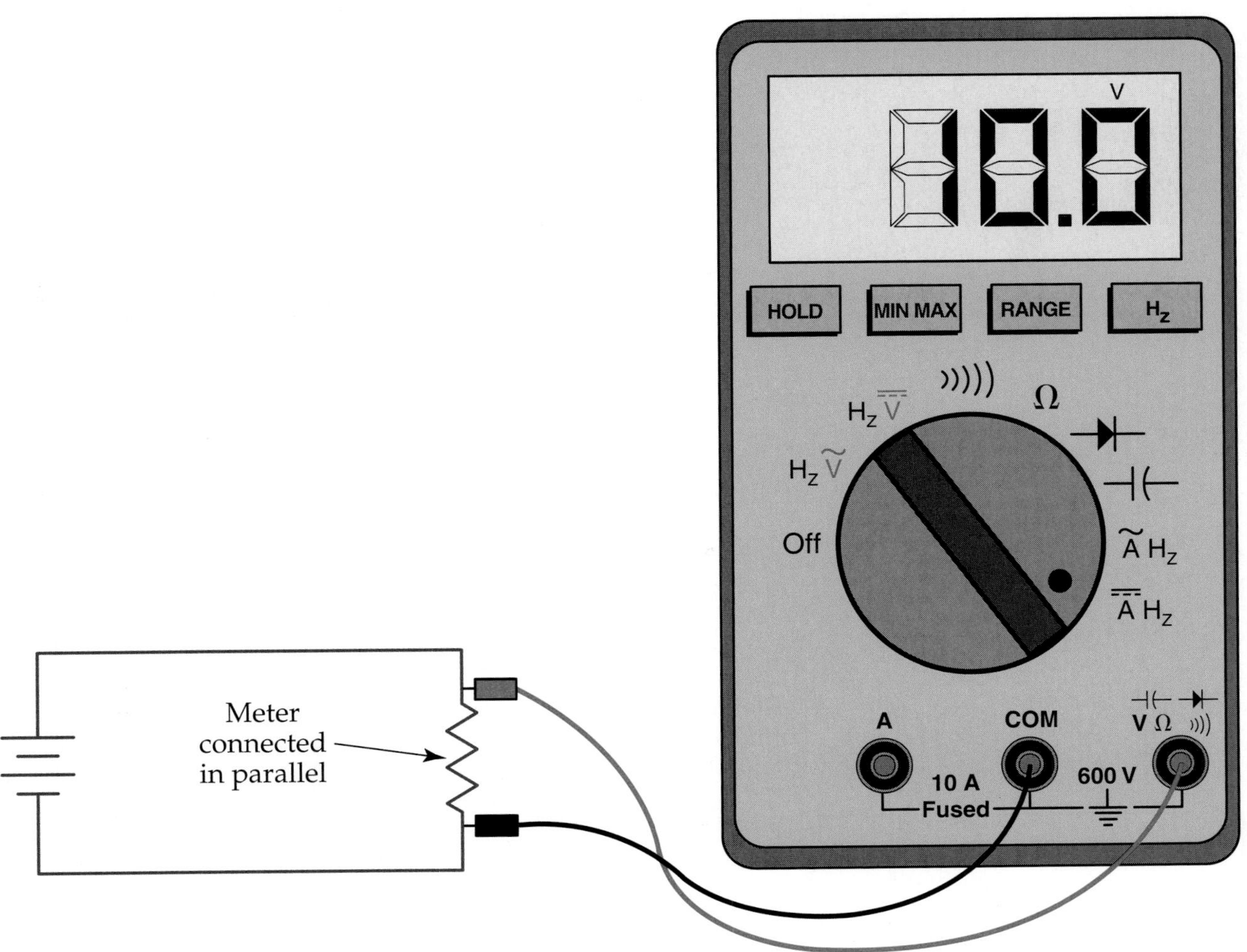

Figure 9-16. The probes of a voltmeter or a multimeter placed in parallel to the voltage source or component.

2. Turn the range selector switch to the $\overline{\overline{V}}$ position when reading dc voltage and to the $\tilde{V}$ position when reading ac voltage.
3. Connect the negative (black) probe to the negative side and the red probe to the positive side of the component or power source being measured.

Danger

Do not touch any meter leads or any part of an electrical circuit unless you are absolutely sure the power is off. It only takes about 0.1 A to kill you.

4. Read the value of the voltage on the display.

Ohmmeter Function

The ohmmeter or multimeter can be used to measure the resistance of a portion of a circuit or the value of resistors. Power must be removed from the circuit before placing the probes across the area to be tested, **Figure 9-17.**

A DMM typically has the ability to read resistance values from 0.1 Ω to 40 M Ω in six ranges or switch positions. The value can be read on the display. Use the following steps to measure resistance:

1. Connect the black probe to the COM input terminal and the red probe to the input terminal for measuring resistance.
2. Turn the range selector switch to the Ω position. The DMM is capable of reading resistance in six ranges. This will be done

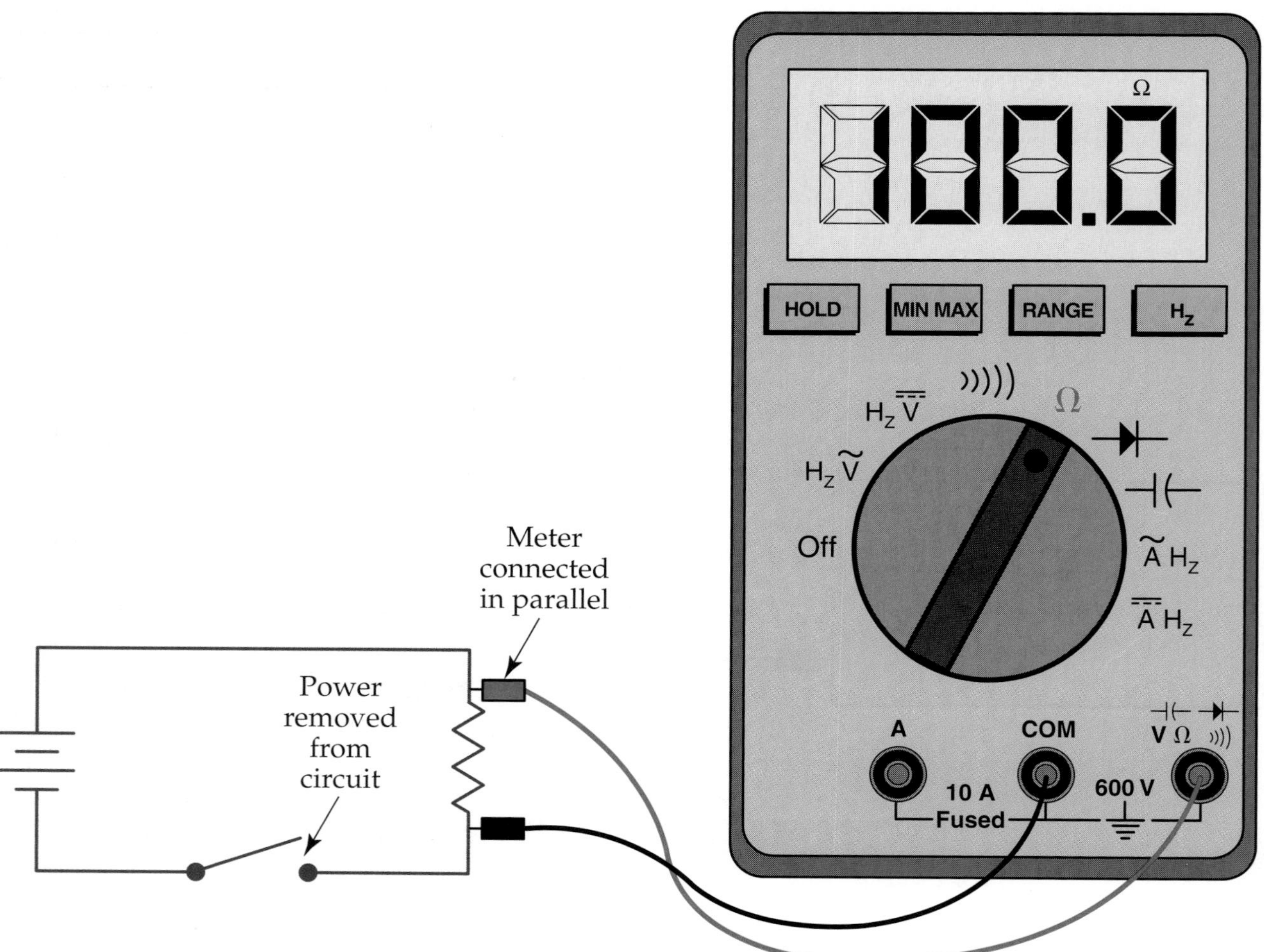

Figure 9-17. When measuring resistance, power is removed from the circuit and the test probes are placed parallel to the resistor or area of resistance.

automatically without having to change any switch positions.

3. Make sure the portion of the circuit or the resistor being measured is not electrically "hot." There should be no power on the item being tested.

Caution

If the circuit contains high-voltage capacitors, make sure they have been discharged before taking a resistance measurement. A capacitor may contain "hidden power" to a portion of the circuit being tested.

4. Place the meter probes across the resistance to be measured.
5. Read the value of the resistance on the display.

Accuracy of Readings

There is an accuracy tolerance for every test you make with a meter. This is similar to the tolerance you calculated with the bands on a resistor. The accuracy of each test, whether it is the measurement of current, voltage, or resistance will vary. For example, the accu-

Project 9-2: Voltage-Polarity Checker

At times when you want to check whether a circuit has ac or dc voltage and a meter is not handy, you can use the voltage-polarity checker. This device can also check the polarity of dc circuits.

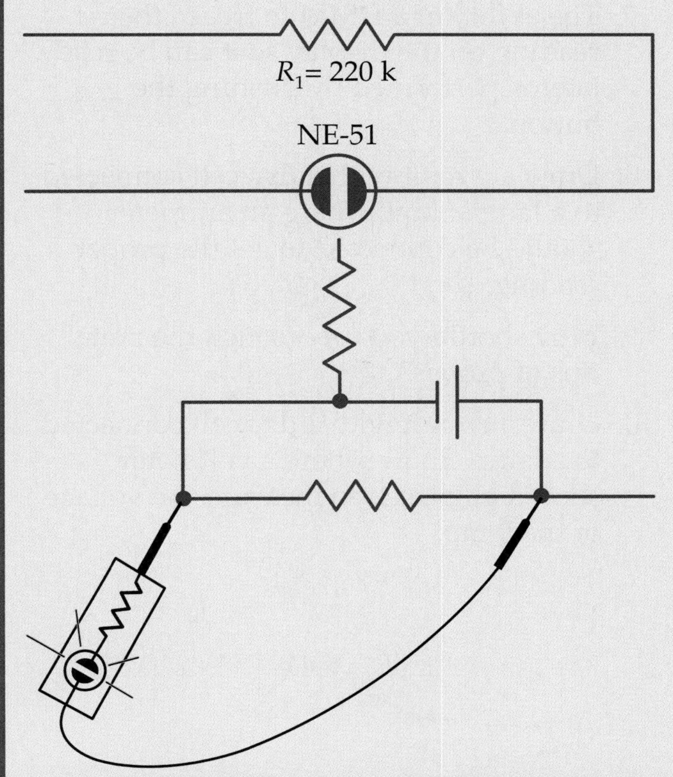

You can check circuits in the range of 110 V to 480 V. Higher voltages will cause arcing in the neon lamp. Gather the following parts:

No.	Item
1	NE-51 neon lamp
1	220-kΩ, 1-W resistor
2	Probes
3	Two 12″ and one 3″ long, AWG 18 insulated wire

Procedure

1. Attach a probe to each of two pieces of wire.
2. Connect the other end of one wire to the resistor and the end of the second wire to the neon lamp.
3. Connect a wire between the resistor and the lamp.

To check voltage, touch the probes across the circuit to be tested. The NE-51 lamp will light when voltage is present. If both electrodes in the lamp glow, the voltage is ac. If only one electrode glows, the voltage is dc. The lit electrode is connected to the probe touching a negative point in the circuit.

racy of measuring resistance is ± (0.9% of the reading + 1). If you have readout of 30 Ω, your accuracy would be:

$$\pm (0.9\% \text{ of } 30\ \Omega + 1)$$

$$\pm (0.270\ \Omega + 1)$$

$$\pm 1.27\ \Omega$$

Summary

- A multimeter is used to measure current, voltage, and resistance.
- An analog multimeter has several graduated scales and a pointer.
- A range selector switch allows you to adjust the range of values the meter is measuring.
- It is easiest to take an accurate reading in the middle of an analog meter scale, where the numbers on the face are farthest apart.
- If an analog meter has a mirror behind the pointer, the meter should be read when the pointer and its reflection are lined up so the user cannot see any reflection.
- When using an analog ohmmeter or multimeter, always zero the meter before taking a measurement.
- A digital multimeter, or DMM, uses a segmented display to show the actual value measured.
- Ammeters are always connected in series with the circuit being measured, while voltmeters and ohmmeters are connected in parallel.
- The circuit's power supply must be off or disconnected and high-voltage capacitors have to be discharged prior to measuring resistance and capacitance.

Test Your Knowledge

Do not write in this book. Write your answers on a separate sheet of paper.

1. Where can the most accurate readings be obtained on an analog meter?
2. A(n) _____ is used on a meter to change between voltage, current, and resistance tests.
3. When using an analog meter to measure unknown currents or voltages, always start with the range selector switch set at the (lowest, highest) position.
4. *True or False?* When an analog meter has a mirror behind the pointer, the correct reading is taken when the pointer lines up with its reflection.
5. What are some advantages of a digital meter?
6. What is the name of the feature that allows a DMM to switch between ranges automatically?
7. The ability of a DMM to freeze the reading on the display so it can be read later is performed by pushing the _____ button.
8. Draw a circuit with a dry cell connected to a lamp. Show where an ammeter should be connected to get the proper reading.
9. Why should you *never* touch the metal tips of probes?
10. Draw a circuit with a dry cell connected to a lamp. Show where a voltmeter should be placed to measure the voltage of the lamp.

11. *True or False?* Before you disconnect meter test leads from a circuit, you should make sure the power source is turned off to prevent electrical shock.
12. It only takes about _____ A to kill you.

Activities

1. Why are there so many different types of meters?
2. Compare the cost of digital meters with analog meters. How much have they changed in the last five years? Why?
3. How many different types of lighted segments can be found on digital meters? Why do they differ?
4. What is so different about the inner workings of a DMM when compared to an analog multimeter? Do DMMs contain any type of computer chip? How do they handle the high electrical loads?
5. List the costs of different brands of meters that read the same current values. Why do the costs vary?
6. In some situations, electricians working in industry do not turn off the electricity before making measurements. Why not?
7. Does the utility company turn off the electrical power when you have a power line down because of a storm?
8. Telephone company technicians can tell within a couple of miles where a broken wire is located. How can they do this without leaving their office?
9. Why does the telephone in your home have such small wire?
10. Some meters made by the same company must have their dry cells replaced more often than others. Why?
11. The dry cells in some modern meters will not run down if you accidentally leave the meter set in the ohms range for a long time. Why not?
12. How does a digital wristwatch get its power to operate?
13. Ask a number of electricians which meter is the best. Why do they disagree?

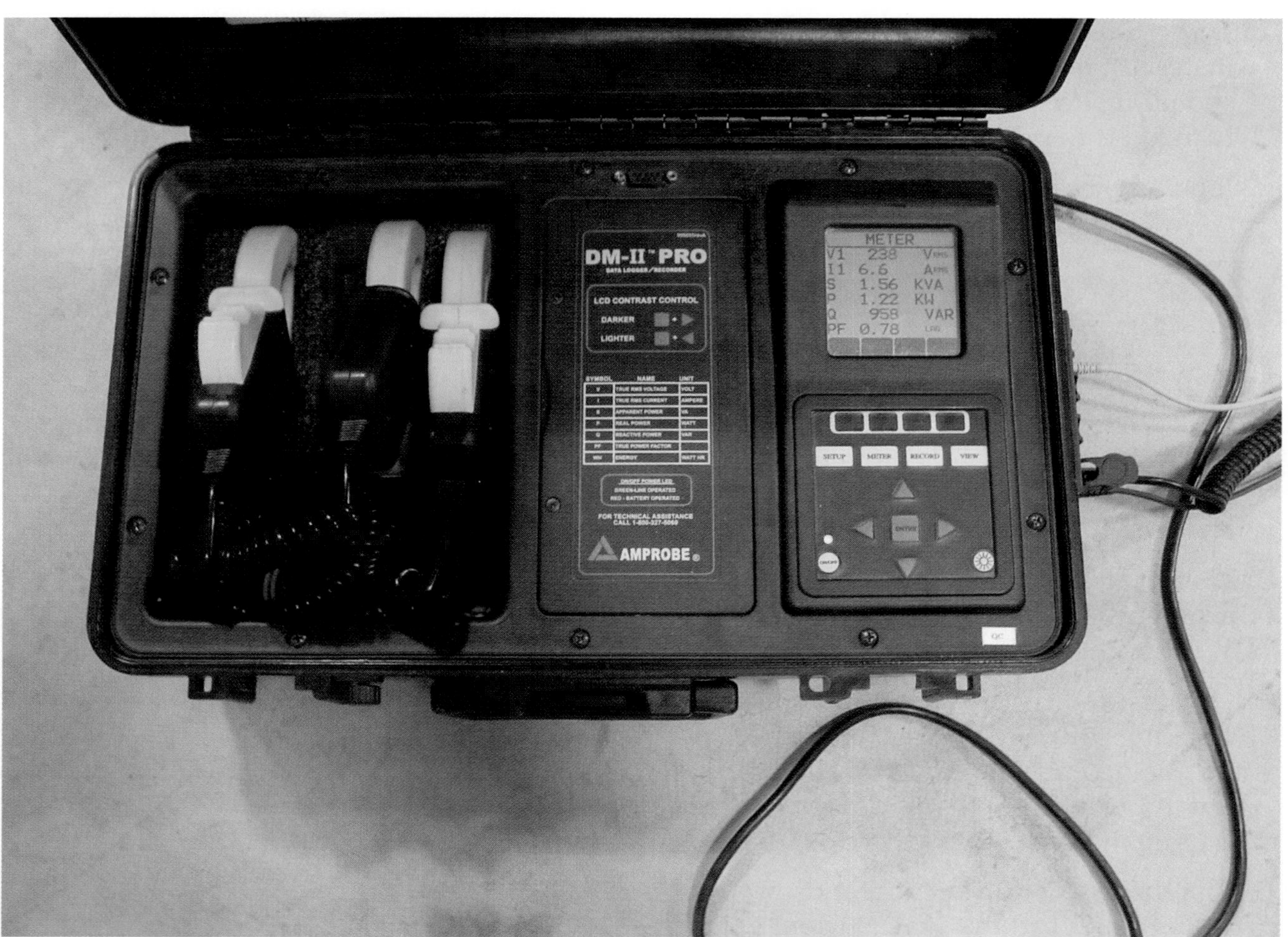

This meter by Amprobe is able to record data, such as voltage and current. A technician can view the data on the meter's LCD display or download the data to a PC and view it on the PC's monitor.

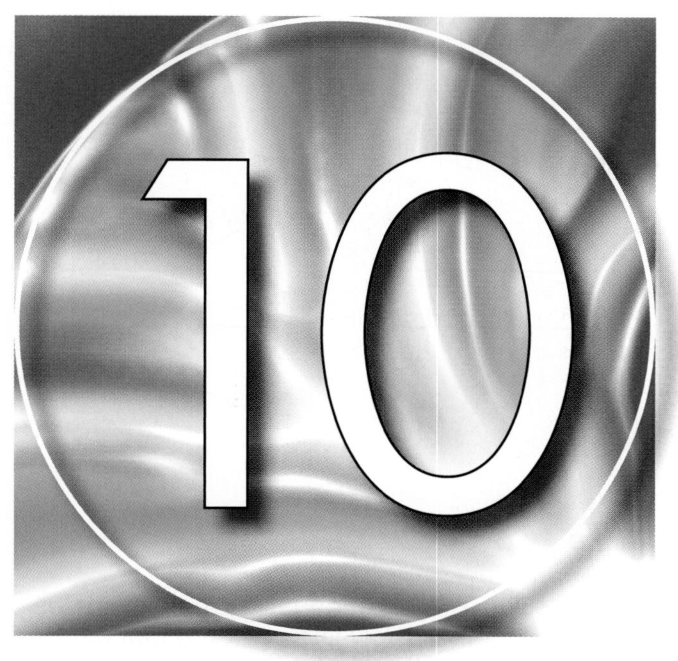

Magnetism

Learning Objectives

After studying this chapter, you will be able to do the following:

- State the two basic principles of magnetism.
- Explain the pattern of magnetic lines of force.
- Demonstrate how to create electricity with magnetism.
- Describe how to increase the amount of current flow created through magnetism.
- Prove that a magnetic field is created when current flows through a conductor.
- Use the left-hand rule to identify the north pole of a magnetic field around a coil.

Technical Terms

ferrous metals
left-hand rule
lines of force
magnetic north
nonferrous metals

Magnetism has many uses in the areas of electricity and electronics. As you have learned in Chapter 2, magnetism can be used to produce electricity. Some other uses of magnetism are turning motors, opening and closing switches, increasing and decreasing voltage levels, and transmitting and receiving radio signals.

In this chapter, you will learn about the principles of magnetism and how magnetism is used to create electricity. This chapter also provides you with information that will prepare you for upcoming chapters.

Basic Principles of Magnetism

In the 1800s, an Englishman named Michael Faraday first performed the experiment with iron filings and discovered lines of force. He also experimented with the interaction between two magnets. From this work, he was able to state two basic principles of magnetism:

- Like poles repel.
- Unlike poles attract.

When two poles are the same—for example, two north poles or two south poles—they will repel one another, **Figure 10-1.** When two poles are opposite—a north pole and a south pole—they will attract one another, **Figure 10-2.**

> **Note**
> Always remember that like poles repel and unlike poles attract.

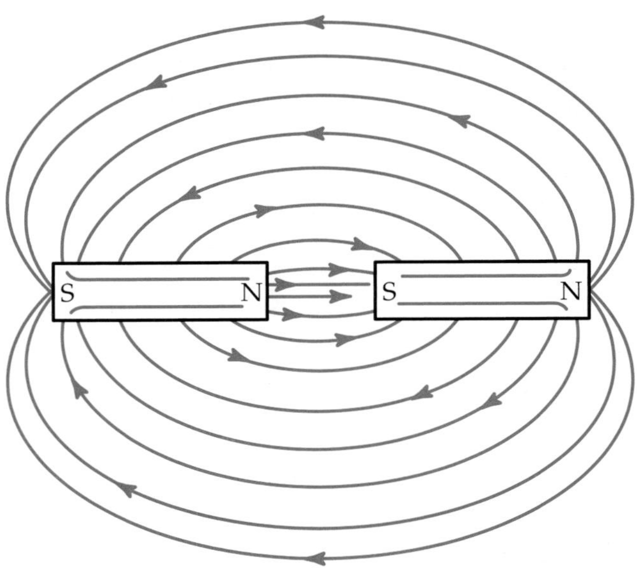

Figure 10-2. Two bar magnets with unlike poles facing each other show lines of force that merge into a pattern of attraction. Note that the lines of force always go from north to south.

A magnetic field surrounds the earth. This is what attracts the needle of a compass to help the user find north. The needle of a compass is

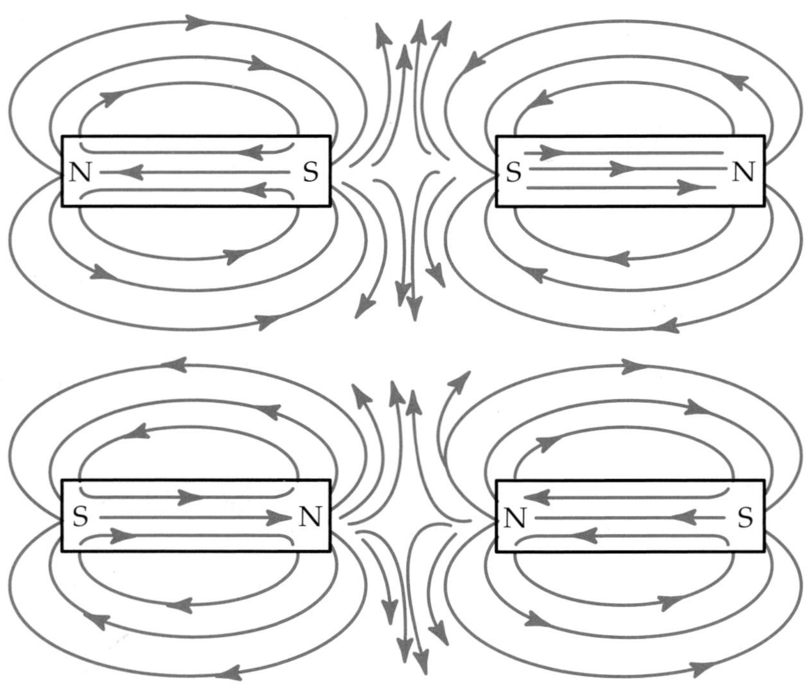

Figure 10-1. The opposing lines of force of these two bar magnets show that like poles repel one another.

a magnet, and its north pole seeks the Earth's northern pole, **Figure 10-3.** The north pole of any magnet that is allowed to freely turn will do the same. However, you have just learned that unlike poles attract. This means that the north pole of the compass is being attracted by a south magnetic pole. Hence, the Earth's north pole has south magnetic polarity. To lessen the confusion, the Earth's pole that attracts the north pole of a compass needle is called ***magnetic north,*** **Figure 10-4.**

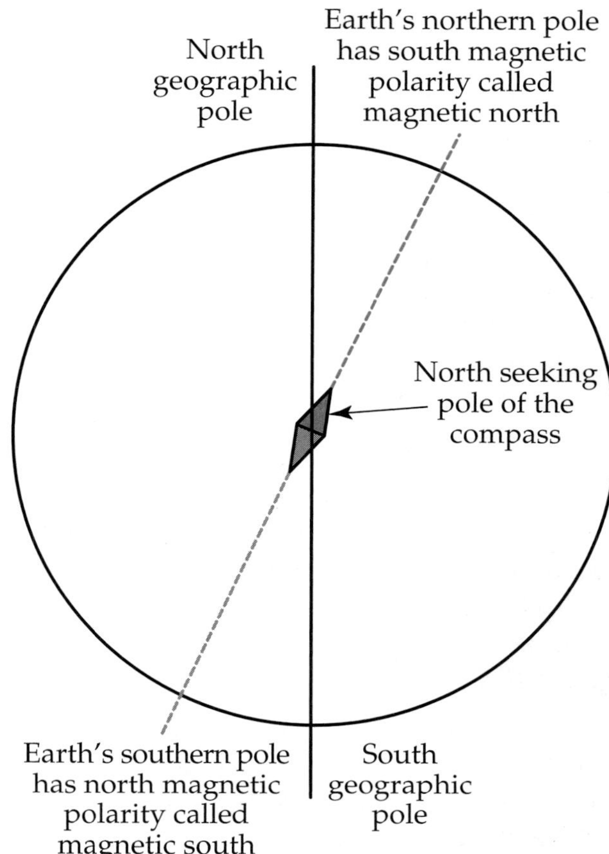

Figure 10-4. The north pole of the compass needle points at the magnetic pole with southern polarity, which is called "magnetic north" because it is close to the geographic north pole.

Magnetic Materials

All material is either magnetic or nonmagnetic. Most ***ferrous metals*** (those that contain iron) are magnetic, while most ***nonferrous metals*** (those that do not contain iron) are not magnetic. Copper, brass, and aluminum are examples of nonferrous, nonmagnetic metals. An easy way to find out whether or not something is magnetic is to place a magnet near it. If it is attracted to the magnet, then that material is magnetic.

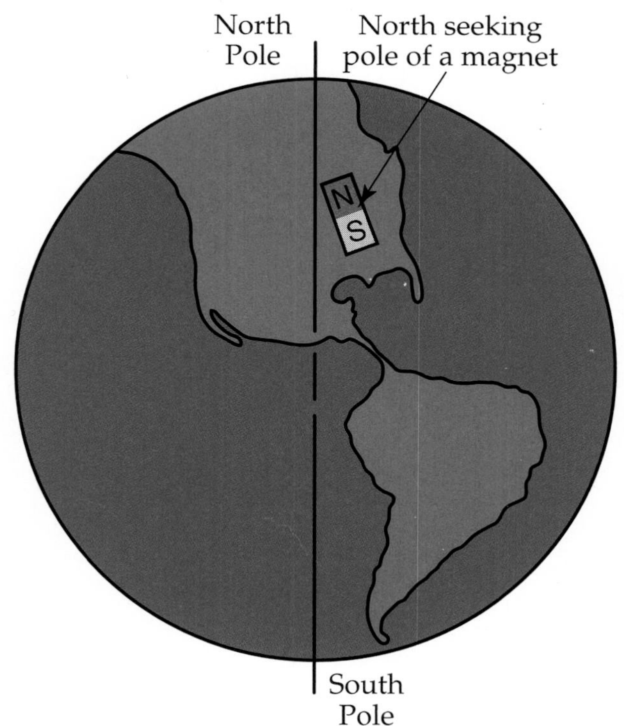

Figure 10-3. The north seeking pole of a magnet or the pointer of a compass always points in the general direction of the north geographic pole.

If you look at a magnet attracting a piece of steel, you can see the movement of the steel toward the magnet. However, you cannot see the magnetic field because it is invisible. To observe a magnetic field, place a piece of paper over a bar magnet. File a steel rod or bar to obtain some fine steel particles. Sprinkle the filings on the paper over the magnet, and gently tap the paper. You will see the filings align into a pattern of attraction between the north and south poles of the magnet. The pattern will look like that shown around the magnet in **Figure 10-5.** Do not use particles that are too large or too heavy because they will not move into place.

Magnetic Lines of Force

The lines shown in Figure 10-5 are called ***lines of force*** or ***lines of flux.*** Lines of force around a horseshoe magnet look like those in **Figure 10-6.**

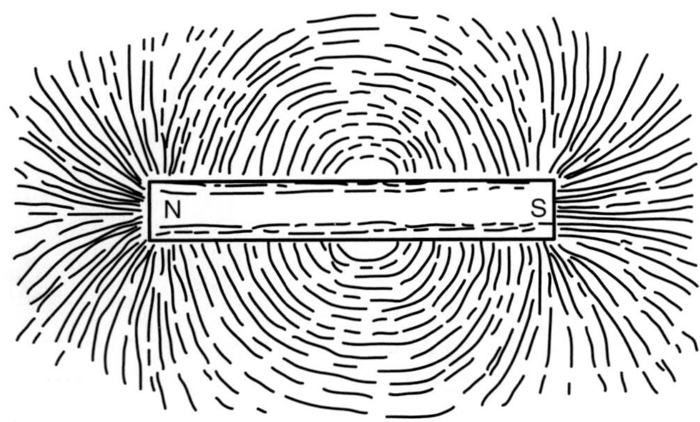

Figure 10-5. Magnetic lines of force around a bar magnet.

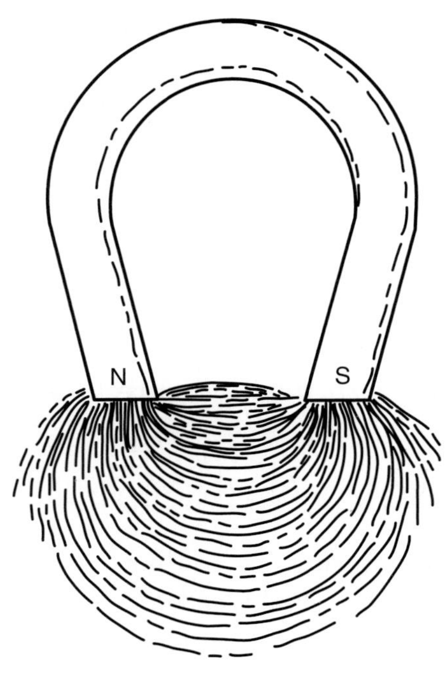

Figure 10-6. Magnetic lines of force around a horseshoe magnet.

Figure 10-7 reveals that the lines of force start from one end of the magnet and are attracted to the other end. The lines of force

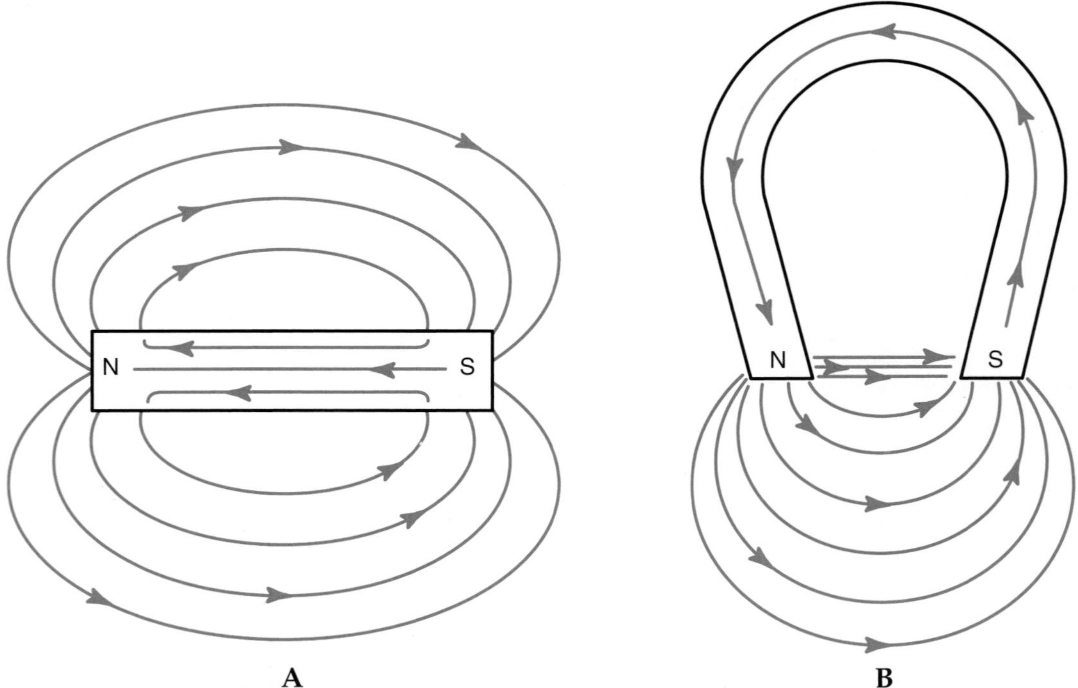

Figure 10-7. Direction of lines of force. A—Bar magnet. B—Horseshoe magnet.

are always drawn as leaving the magnet from the north pole and entering the magnet at the south pole. The direction of the lines of force is arbitrary because nothing is really moving. However, assigning them a direction can help with understanding the behavior of magnets and magnetic fields. The areas where the lines of force are closest together are where the magnetic field is strongest. The areas where the lines are farther apart are where the field is weakest.

Earlier in this chapter, we sprinkled iron filings on paper placed over a magnet to show the lines of force. Paper is an insulator, which means that it does not allow electrons to flow through it easily. Glass, air, and wood are also insulators.

Although insulators block the flow of electrons, they do not block the flow of the magnetic lines of force. This is why you are able to see the lines of force formed on the paper. You can use plastic, glass, or other insulators over the magnet and still perform the experiment with iron filings. **Figure 10-8** shows how magnetic lines of force can pass through most nonmagnetic materials. However, there are ways of bending the lines of force so that they can take another path through the magnetic material.

Bending lines of force is a very important factor when dealing with magnetism. Motors, generators, and transformers use this concept to increase their speed, strength, or their power. These features will be demonstrated in later chapters.

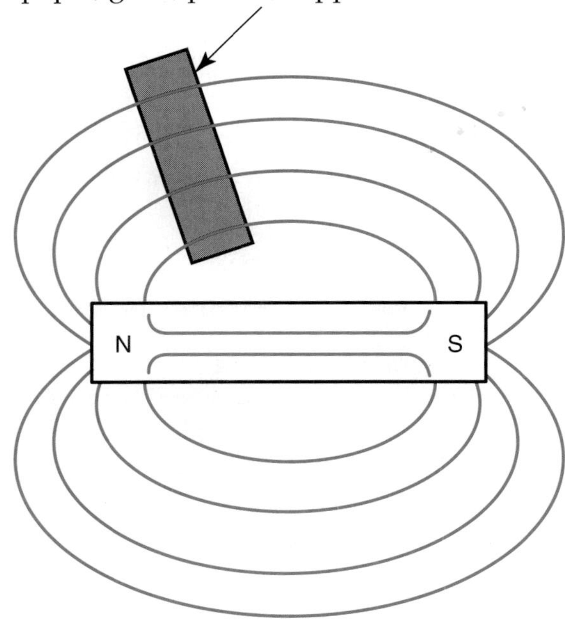

Figure 10-8. Magnetic lines of force are unaffected by insulators and most nonmagnetic materials.

Generating Electricity with Magnetism

As you learned in Chapter 2, magnetism is the most common source of electricity. Most of the electricity we use daily is generated by magnetism. To demonstrate how electricity is generated through magnetism, perform the following experiment using a galvanometer. First, take a coil of wire and attach the two ends to the terminals of the galvanometer, **Figure 10-9.** Next, take a bar magnet and quickly insert it into the coil while watching the meter pointer move. Quickly remove the

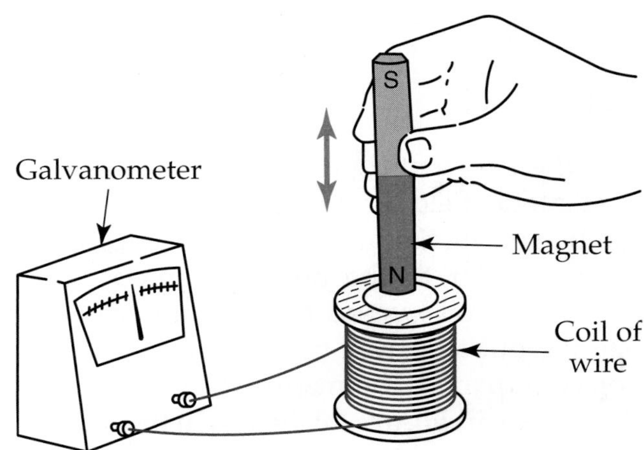

Figure 10-9. Moving a magnet into a coil of wire causes the needle of a galvanometer to move in one direction. Taking the magnet out of the coil causes the needle of the galvanometer to move in the other direction.

magnet and note that pointer movement is in the opposite direction.

The meter pointer moves because the magnetic field passes through the insulation on the wire and affects the conductive core of the wire. The magnetic field moves the electrons in the wire, which creates a current in the wire. The current causes the pointer of the meter to move. This would also work if the magnet remains still, and the coil was moved back and forth around it.

When the electrons in a wire are forced to move more, the current in the wire increases. This action is seen in the meter readings. The key to creating this current is relative movement between the magnetic field and the conductor. This means that current is created by moving either the magnet through the coil or the coil over the magnet. Later, we will see how all these ideas are used to build motors and generators to do specific jobs.

The size of the movement of the electrons (amount of current flow) in the wire can be controlled by the size of the magnet, the number of turns in the coil, and how quickly the magnet is moved in and out of the coil. As the movement of electrons in the wire increases, the current in the wire increases. This can be seen in the meter readings.

Creating Magnetism with Current

When current flows through a conductor, a magnetic field is created. You can prove this by setting up an experiment with metal filings similar to the one performed earlier in this chapter. Instead of using a permanent magnet, put a current-carrying wire through a piece of paper. The magnetic field will look like the one in **Figure 10-10.** It will be made up of circles around the wire. The only problem with this magnetic field is that it is too weak to do much good. Therefore, we must find a way to make the field stronger. One of the easiest ways to do this is to wrap the wire into the shape of a coil.

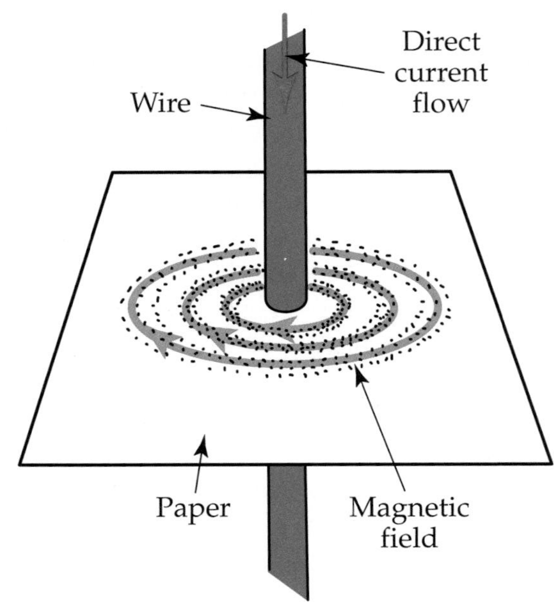

Figure 10-10. Metal filings show the magnetic field around a wire.

To make the coil, wind the wire around some insulating material, **Figure 10-11.** Remember, insulation cannot stop the magnetic lines of force. The magnetic field will cut through the insulator. This coil makes the magnaetic field stronger. Unlike the field created by the single wire, the field formed by the coil has a north pole and a south pole.

If you took an ordinary magnet and placed it in a magnetic field, the poles always line up in a certain way. However, with the coil magnet,

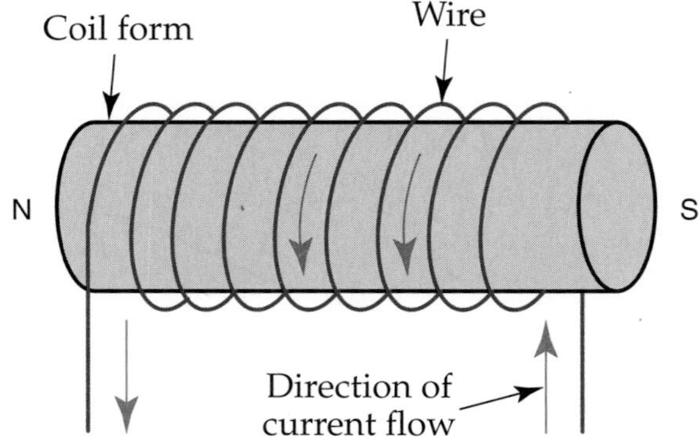

Figure 10-11. Wire wrapped on a coil form provides the means for making a magnetic field experiment.

the poles can switch ends. Then, if you pass a direct current through the wire in one direction, the north pole will be established at one particular end of the magnet. However, if the direction of the current is reversed, the north pole will be at the other end.

To identify the north pole of the magnetic field around a coil, use the ***left-hand rule.*** Simply, curl the fingers of your left hand in the direction the current flows through the coil. Your thumb will point to the north pole of the magnetic field, **Figure 10-12.** Remember, current flows from the negative terminal of a power source to the positive terminal.

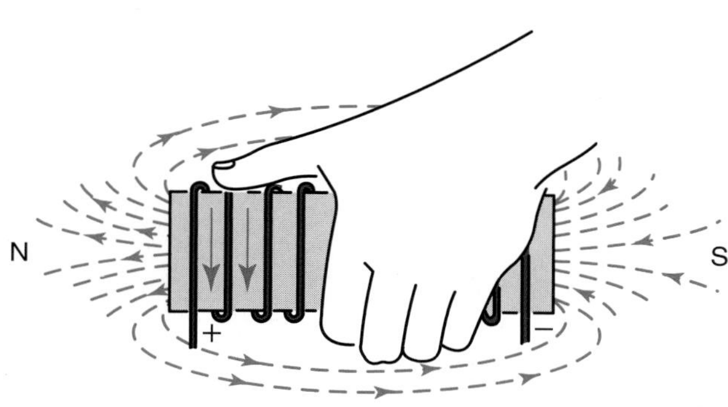

Figure 10-12. The left-hand rule for coil polarity determines which end of the coil is the north pole.

Summary

- Like magnetic poles repel and opposite magnetic poles attract.
- The Earth has a north and south magnetic pole.
- The north pole of the Earth is called the *magnetic north* because it is has a south magnetic polarity.
- Material is either magnetic or nonmagnetic.
- Ferrous material contains iron and is magnetic.
- Nonferrous material does not contain iron and is not magnetic.
- Magnetic lines of force leave a magnet from the north pole and enter the magnet at the south pole.
- Insulators do not block the flow of the magnetic lines of force.
- Current can be created by moving either a magnet through a coil or a coil over a magnet.
- The amount of current flow induced by magnetism can be controlled by the size of the magnet, the number of turns in the coil, and how quickly the magnet is moved in and out of the coil.
- A magnetic field is created around a conductor when current flows through it.
- The left-hand rule is used to identify the north pole of the magnetic field around a coil.

Test Your Knowledge

Do not write in this book. Write your answers on a separate sheet of paper.

1. The basic laws of magnetism are that ______ poles repel and ______ poles attract.
2. Magnetic lines of force come out of the ______ pole and go into the ______ pole of a permanent magnet.
3. Current is created by moving either a(n) ______ through a(n) ______ or a(n) ______ over a(n) ______.
4. Describe three ways to control the amount of current flow created through magnetism.
5. Explain how to use the left-hand rule.

Activities

1. Why is a keeper used on a magnet?
2. Where are the poles on a ring magnet?
3. Why does striking a magnet with a sharp blow demagnetize it?
4. Why does a magnet grow weaker with age?
5. Do the transformers on telephone poles produce a magnetic field? Why do power company workers (linemen) place protective covers on power lines near where they are working? Does this have anything to do with magnetic fields?

Alternating Current

Learning Objectives

After studying this chapter, you will be able to do the following:

- Describe how alternating current is produced.
- Determine the frequency, period, and amplitude of a sine wave.
- Explain what is meant by the RMS value of a sine wave.
- Calculate the RMS value of a current or voltage.
- Demonstrate how to view a sine wave on an oscilloscope.

Technical Terms

alternating current (ac)
alternator
amplitude
armature coil
audio frequency (AF) waves
brushes
channel
commutator
cycle
dc generator
eddy currents
field coils
frequency
generator
graticule
hertz (Hz)
in phase
Lissajous figures
oscilloscope
peak voltage (V_p)
peak-to-peak voltage (V_{pp})
period
radio frequency (RF) waves
root mean square (RMS)
rotor
signal generator
sine wave
slip rings
square wave
stator
three-phase system

When you studied Ohm's law in Chapter 5, you learned about electrical current that flowed in only one direction. The electrons left the negative post of the battery, flowed through the circuit, and returned to the positive post of the battery. This is called *direct current (dc)*.

This chapter covers ***alternating current (ac).*** Alternating current is any current that reverses its direction of flow at regular intervals. The symbol most commonly used to indicate an ac power supply is given in **Figure 11-1.** The wavy line in the center of the circle represents a sine wave. A ***sine wave*** is the waveform of a single frequency of alternating current.

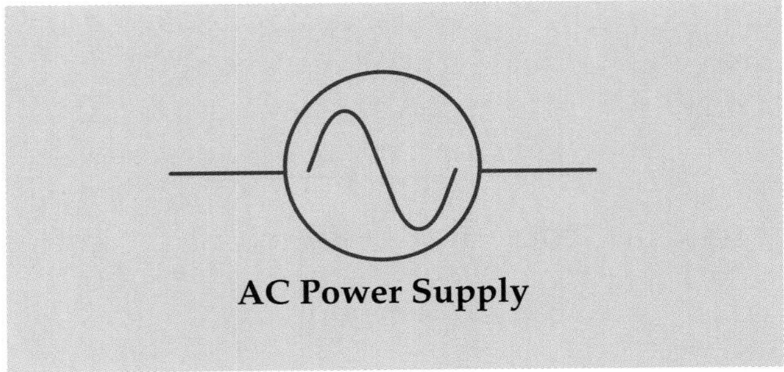

Figure 11-1. This symbol for the ac power supply is used when drawing schematics.

How AC Is Produced

Imagine that you have a pair of magnets like the ones shown in **Figure 11-2.** The dotted lines indicate that there is a magnetic field between the poles of the two magnets. Now, quickly pass a piece of wire through the field so that the wire cuts through the lines of force, **Figure 11-3.** This induces or creates current in the wire. This is called *induction* and is explained in more detail in Chapter 12. To understand the example shown in Figure 11-3, there are three important points to remember:

- Voltage is induced in the wire only when the wire cuts across the lines of force (up and down), not when it is moving parallel (left and right) to the lines of force.
- As the speed of the wire cutting through the lines of force increases, the voltage induced in the wire increases.
- When the motion of the wire is reversed, the polarity of the induced voltage is reversed. In other words, the current changes direction.

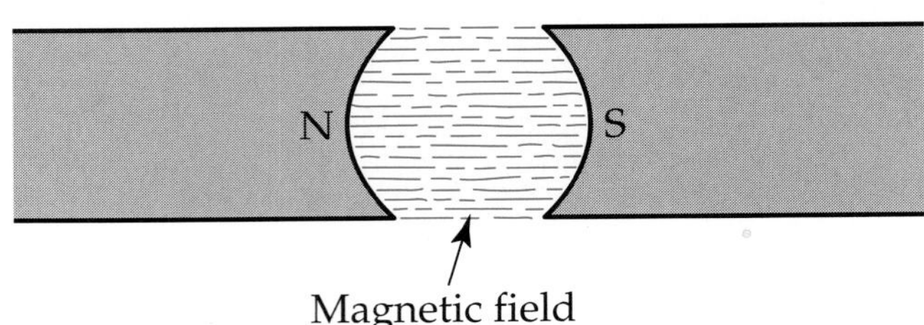

Figure 11-2. Lines of force between north and south poles of magnets show the extent of the magnetic field.

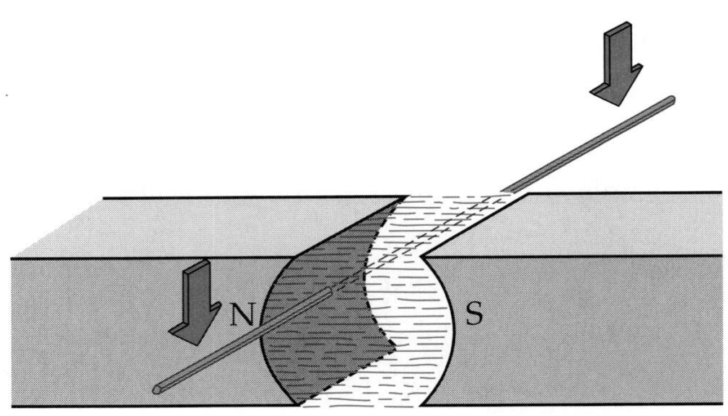

Figure 11-3. A wire passing through a magnetic field cuts the lines of force, which induces electric current in the wire.

Keeping these points in mind, look at the coil of wire in the magnetic field in **Figure 11-4.** If you cut across the coil at the dotted line, you will have the end view shown in **Figure 11-5.** The cut ends of the wire are labeled *A* and *B* so that you can see what happens to each side of the coil as it rotates inside the magnetic field.

To help you understand how the coil rotates, imagine a wagon or bicycle turned upside-down so it cannot move. Mark two spots on the wheel, and call one *A* and one *B*, **Figure 11-6.** If you were to watch those spots as you spin the wheel, tracing the path of *A* and *B* for one complete revolution, you would see that each point returns to its original starting point. Likewise, when the coil of wire cuts through the lines of force of the magnetic field in **Figure 11-7,** you can trace *A* and *B* throughout the complete revolution. At the end of a single revolution, both *A* and *B* return to their original starting points. In electrical terms, one complete revolution of the coil is called a ***cycle.***

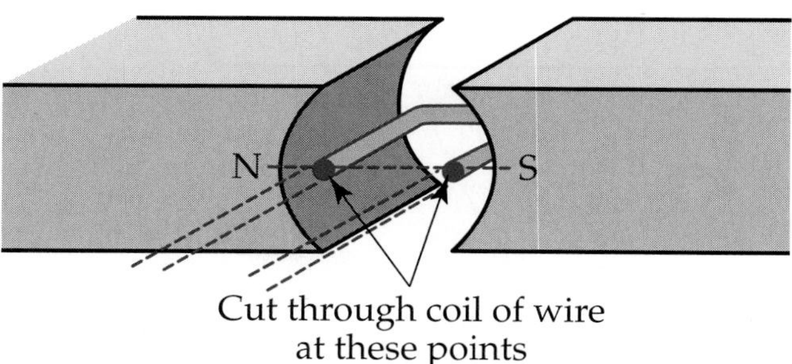

Figure 11-4. A coil of wire is placed inside the magnetic field.

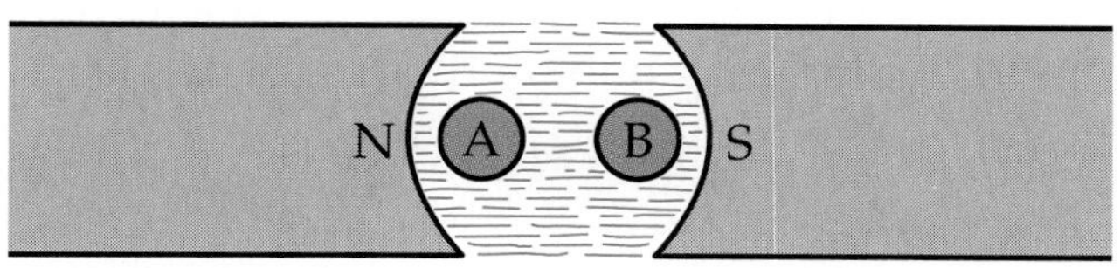

Figure 11-5. End view of the coil from Figure 11-4. The cut ends of the coil are marked *A* and *B.*

Sine Waves

Figure 11-7 shows a sine wave. Sine waves are often used to describe the cycle of an electrical signal. They can provide much information about a given signal. To experiment with the coil and magnet, consider point *1* on the sine wave shown. At this point, the coil of wire moves parallel to the lines of force in the magnetic field. It does not cut the lines of force. Therefore, no voltage is induced in the coil.

Imagine the coil rotating around its axis. At point *2,* the coil has turned clockwise 90°. *A* and *B* move at right angles to the lines of force. This induces a maximum amount of voltage in the coil and is, in effect, the highest point the sine wave will reach.

At point *3,* like point *1,* no voltage is induced. Note that as *A* and *B* turn clockwise, they start to cut the lines of force in the opposite direction. This causes the voltage induced in the coil to reverse its polarity.

At point *4,* the voltage reaches another peak equal to the peak at point *2.* However, at this point, the sine wave reaches its negative peak voltage. Finally, at point *5,* the coil has returned to its starting point. No lines of force are cut by the coil, so the sine wave returns to zero on the reference line.

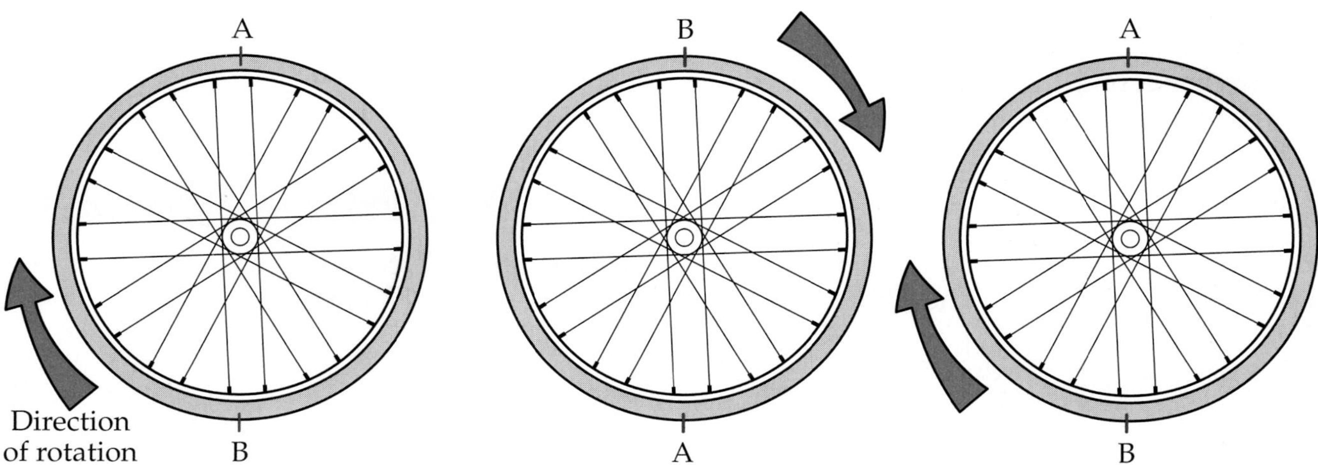

Figure 11-6. A rotating bicycle wheel illustrates the path followed by marks *A* and *B.* Marks *A* and *B* in Figure 11-5 follow a similar path.

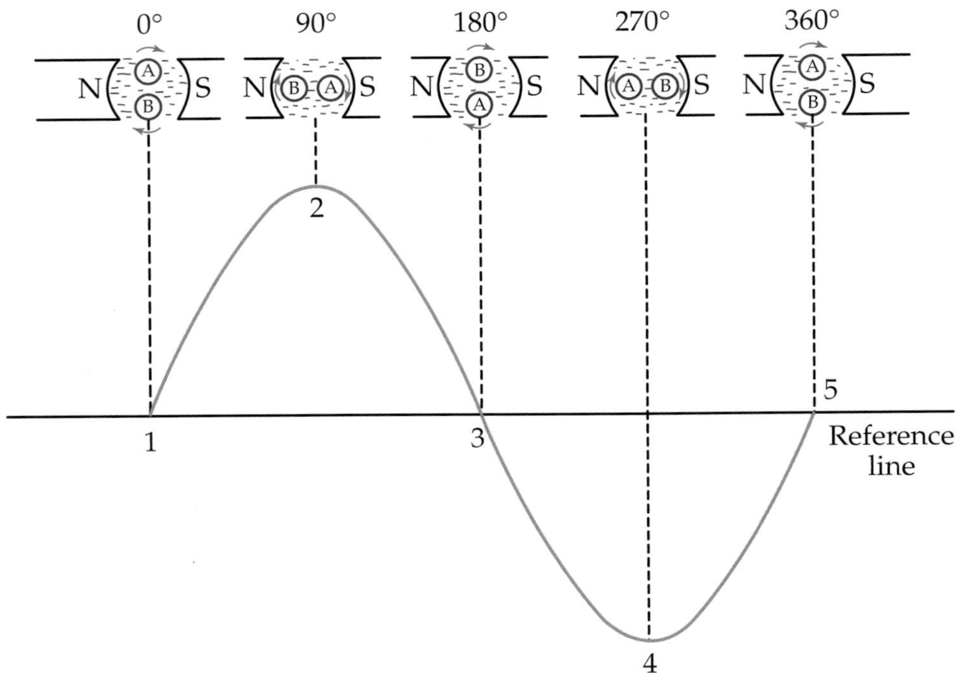

Figure 11-7. One sine wave is generated through one complete rotation of the coil inside the magnetic field. The coil cuts through the maximum number of lines of force in positions *2* and *4*.

Sine Wave Characteristics

A sine wave, like the one in **Figure 11-8,** has many characteristics that explain a lot about the circuit being tested. These characteristics are amplitude, frequency, and period.

The ***amplitude*** of a sine wave is its height above or below the reference line (zero) at any given time. ***Frequency*** is the number of cycles per second. Imagine if the coil shown in Figure 11-7 had rotated at a speed of five revolutions per second. It would have produced five complete sine waves in that second like that in Figure 11-8. The frequency of this sine wave is five cycles per second, or five hertz (5 Hz). One ***hertz (Hz)*** is equal to one cycle per second. If the coil rotates faster, the frequency increases. For example, current in the United States is 60 Hz, which means that sixty complete revolutions, or cycles, are produced in one second.

A ***period*** is the time needed to complete one cycle. Since five complete cycles were completed in one second in Figure 11-8, the period of the sine wave is one-fifth of a second. The time it takes to complete one cycle can be found by dividing one by the frequency (f):

$$\text{Period} = \frac{1}{f} \qquad (11\text{-}1)$$

Using this formula to determine the period of the sine wave in Figure 11-8, you would do the following:

$$\text{Period} = \frac{1}{f}$$
$$= \frac{1}{5\text{ Hz}}$$
$$= \frac{1}{5\text{s}} \text{ or } 0.2\text{s}$$

Example 11-1:

The current in U.S. homes has a frequency of 60 Hz. What is the period?

$$\text{Period} = \frac{1}{f}$$
$$= \frac{1}{60\text{ Hz}}$$
$$= \frac{1}{60s} \text{ or } 0.017s$$

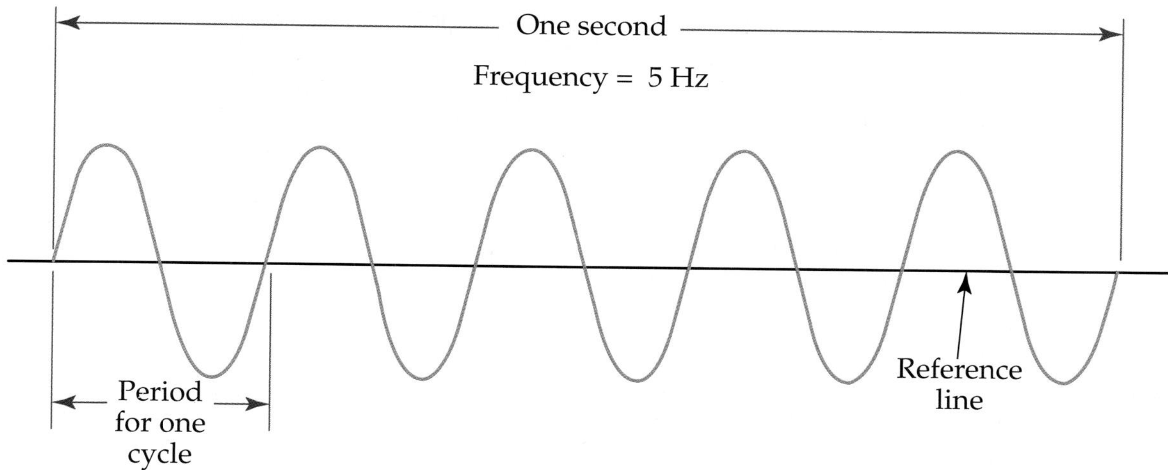

Figure 11-8. If five sine waves are produced in one second, they have a frequency of 5 Hz and a period of one-fifth of a second.

This period, one-sixtieth of a second, is very short. This example shows that as frequency increases, the period decreases.

Peak Voltage

As you have previously observed, a sine wave will reach its maximum amount of voltage in the positive direction at one point in the cycle of a sine wave. This is called ***peak voltage*** ***(V_p)*****, Figure 11-9.** At another point, it will reach its maximum amount of voltage in the negative direction. This is also called the *peak voltage*. However, it is in the negative direction.

Notice in Figure 11-9, that peak voltage is measured from a reference line, which is at 0 V. Therefore, we always have two peaks on a sine wave: a negative peak and a positive peak. The voltage difference between the negative peak (V_{-P}) and positive peak (V_{+P}) is commonly called the ***peak-to-peak voltage***

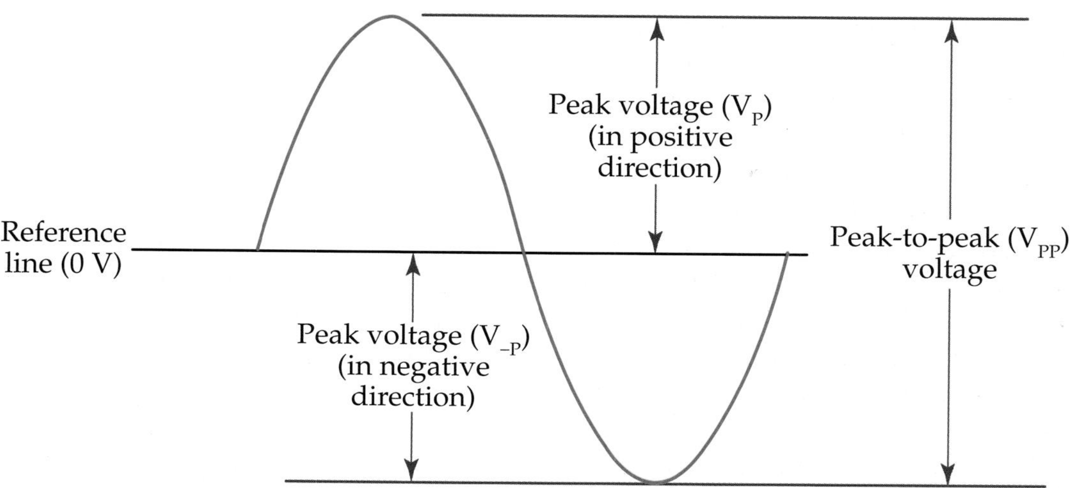

Figure 11-9. Peak voltage is the maximum amplitude. Peak-to-peak voltage is the sum of the two peak voltages in a single cycle.

(V_{pp}) value of a sine wave. It is expressed by the formula

$$V_{pp} = V_{+p} + |V_{-p}| \quad (11\text{-}2)$$

The lines (called bars) around V_{-p} mean that you should use the absolute value of V_{-p}. The absolute value of a number is the distance of that number from zero. This means that absolute values are always positive. For example, the absolute value of –3 is 3 (–3 is 3 units away from zero). The absolute value of 3 is also 3 (3 is 3 units away from zero).

In **Figure 11-10,** the ac sine wave has a peak voltage of 100 V. The voltage reaches 100 V twice during each cycle. The voltage reaches 0 volts three times during each cycle. The peak-to-peak voltage can be found by using formula 11-2:

$$\begin{aligned} V_{pp} &= V_{+p} + |V_{-p}| \\ &= 100\text{ V} + |-100\text{ V}| \\ &= 100\text{ V} + 100\text{ V} \\ &= 200\text{ V} \end{aligned}$$

The peak-to-peak voltage is 200 V.

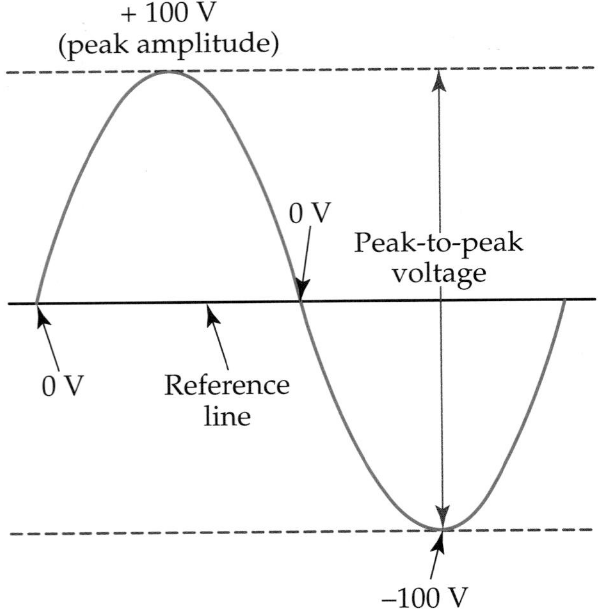

Figure 11-10. In this example of a sine wave, the peak amplitude is 100 V above and below the reference line. The peak-to-peak voltage, therefore, is 200 V.

Example 11-2:

Find the peak-to-peak voltage of the following sine wave.

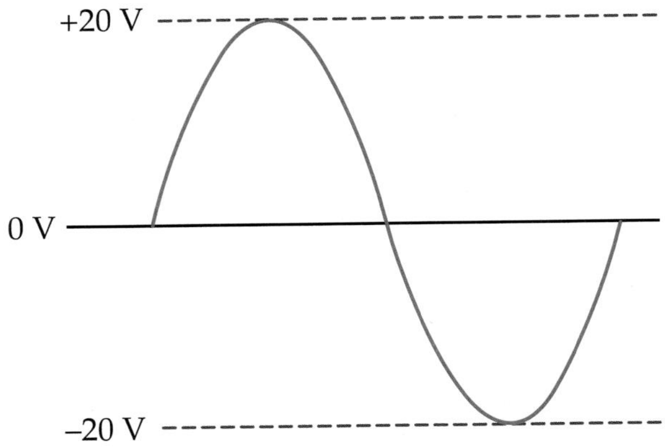

The positive peak voltage is 20 V and the negative peak voltage is –20 V. Use the formula for calculating peak-to-peak voltage:

$$\begin{aligned} V_{pp} &= V_{+p} + |V_{-p}| \\ &= 20\text{ V} + |-20\text{ V}| \\ &= 20\text{ V} + 20\text{ V} \\ &= 40\text{ V} \end{aligned}$$

The peak-to-peak voltage is 40 V.

Peak voltage, peak-to-peak voltage, and current are very important in the study of electronic devices such as diodes, rectifiers, and transistors. These devices can be overloaded very easily. If we do not know the peak voltage, a diode could be destroyed in just a fraction of a second.

RMS Value

In Chapter 5, you read that a certain amount of the heat is developed in a resistor when the circuit is completed. For comparison, **Figure 11-11** shows similar ac and dc circuits. One has a 100-V ac power supply connected to a 50-Ω resistor. The other circuit has a 100-V dc power supply connected to an identical 50-Ω resistor. Which voltage would you expect to produce more heat? Remember,

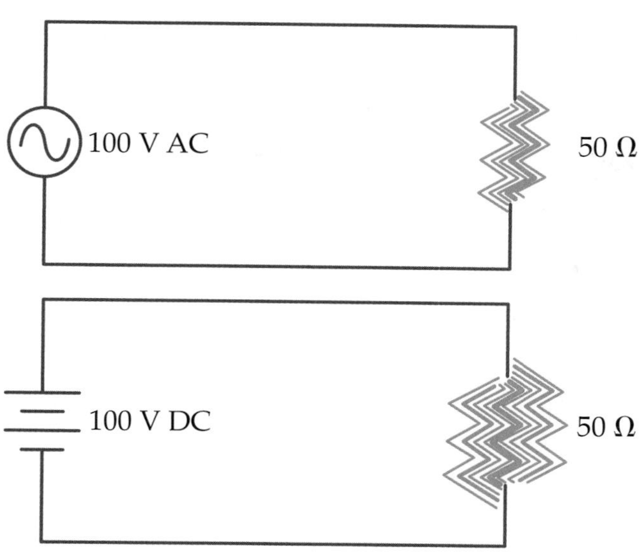

Figure 11-11. In a test of heat values, identical resistors in similar ac and dc circuits are compared.

ac voltage reaches 100 V only twice in each cycle, while dc voltage remains a steady 100 V. The answer is dc. DC voltage produces more heat in the resistor. See **Figure 11-12.**

After many experiments, scientists have come up with a practical way of comparing the heating values of ac and dc voltages. They found that if you multiplied the peak value of an ac sine wave by the number 0.707, you would get a ***root mean square (RMS)*** value (effective value) for an ac wave.

Therefore, RMS values provide a way to compare the heating values of ac and dc voltages. If the peak voltage of an ac sine wave is multiplied by 0.707, the product is the dc voltage that produces the same amount of heat. The formula is

$$\text{RMS} = V_p \times 0.707 \qquad (11\text{-}3)$$

We can find the RMS value of an ac circuit with a 100 V peak voltage using this formula:

$$\begin{aligned} \text{RMS} &= V_p \times 0.707 \\ &= 100 \text{ V} \times 0.707 \\ &= 70.7 \text{ V} \end{aligned}$$

The RMS value of 100 V is therefore 70.7 V, **Figure 11-13.** The RMS value of 70.7 V means that the 100-V peak ac voltage will have as much heating value as a dc voltage of 70.7 V.

> **Note**
> Sine waves can be used to represent voltage and current. Therefore, the RMS formula works for either of these electrical values.

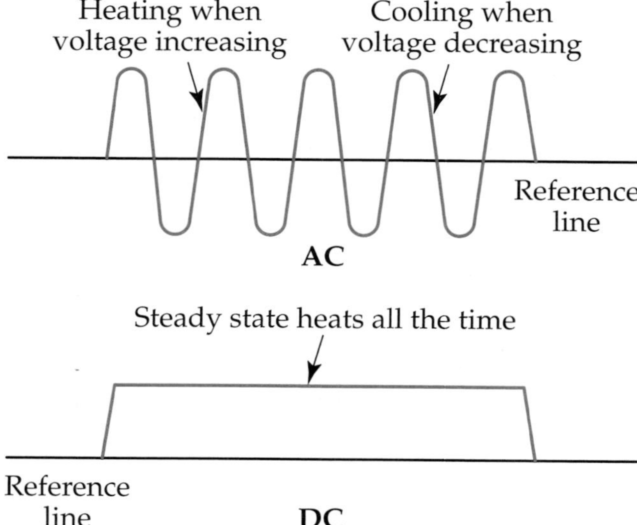

Figure 11-12. More heat is created by dc voltage. Note that ac voltage heats on the rise of a sine wave only.

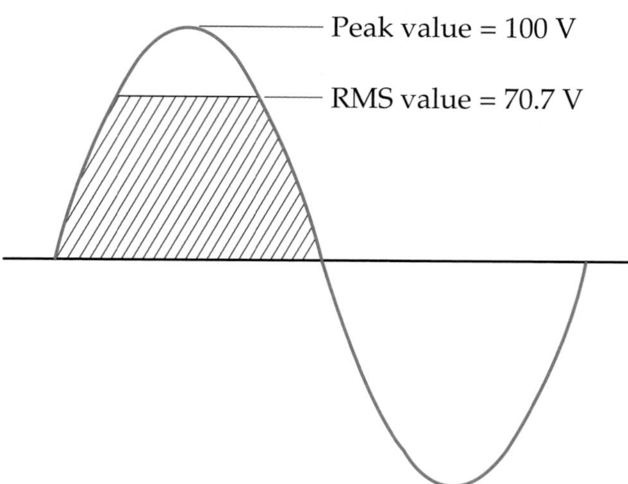

Figure 11-13. The RMS value of an ac sine wave is 70.7% of its peak value.

Example 11-3:

Suppose an ac power supply provides electricity with a peak voltage of 90 V. Find the RMS value for this voltage using the RMS formula:

$$\begin{aligned} RMS &= V_p \times 0.707 \\ &= 90\ V \times 0.707 \\ &= 63.63\ V \end{aligned}$$

A dc power supply that provides 63.63 V produces the same amount of heat as this ac power supply with a peak voltage of 90 V.

RMS value is so important that whenever people in electrical and electronic work speak of an ac voltage or current, they usually are talking about the RMS value. An example of this is the 110-V power supply found in your home. This number actually refers to the RMS value. Peak voltage is higher. This can be proved mathematically by rearranging the RMS formula:

$$RMS = V_p \times 0.707$$

$$\frac{RMS}{0.707} = \frac{V_p \times 0.707}{0.707}$$

$$\frac{RMS}{0.707} = V_p$$

This is the same as

$$RMS \times \frac{1}{0.707} = V_p$$

Therefore, since 1 divided by 0.707 equals 1.414, we have the following formula:

$$V_p = RMS \times 1.414 \qquad (11\text{-}4)$$

Now, let us examine the 110-V RMS value in our example. We will convert it to peak voltage using this formula:

$$\begin{aligned} V_p &= RMS \times 1.414 \\ &= 110\ V \times 1.414 \\ &= 155.54\ V \end{aligned}$$

The peak voltage value of 110-V RMS is 155.54 V. As you can see, peak voltage is higher than the RMS value. You now have a simple way to convert from peak voltage to RMS and from RMS to peak voltage.

Note

Some people are confused about whether their house voltage is 110 V, 115 V, or 120 V. Actually, they are all the same number. This is possible because of the tolerance allowed on the voltage delivered to your home. The power company has to deliver 240 V, which is divided into two-120 V legs in your home's electrical box. However, there is a ±5% tolerance allowed by the power companies. This results in +/–12 volts, which falls between 100 V and 120 V.

Square Waves

A ***square wave*** is another type of wave that is commonly used in the field of electricity and electronics. A square wave is easy to produce with a circuit containing a battery, a switch, and a resistor, **Figure 11-14.** When the switch is open, the voltage across the resistor is zero. When the switch is closed, the voltage across the resistor is the voltage provided by the battery. It stays at the voltage of the battery until the switch is opened. A square wave is created with this process.

If the switch is closed or opened every second, it produces a square wave with a frequency of 1 Hz. Look at **Figure 11-15.** As you can see, sine waves and square waves have three things in common: frequency, amplitude, and period. Since the voltage in the dc circuit in Figure 11-14A never changes direction, the wave never goes below zero. Zero is the lowest point, so the peak-to-peak voltage is also 10 V. The period of this wave is 1 second.

If the square wave was produced by an ac source, the square wave would look like that in **Figure 11-16.** Notice that the reference line

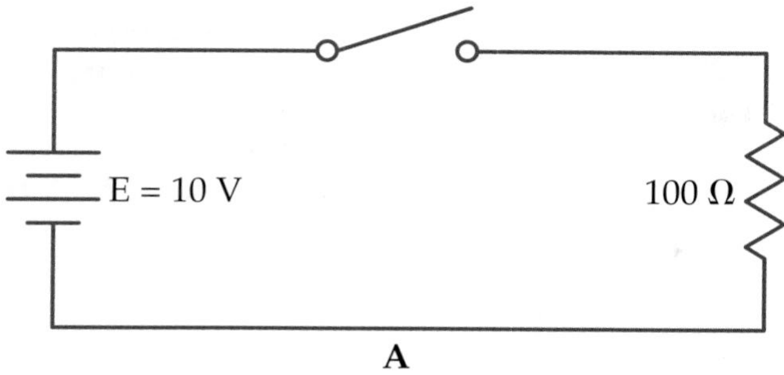

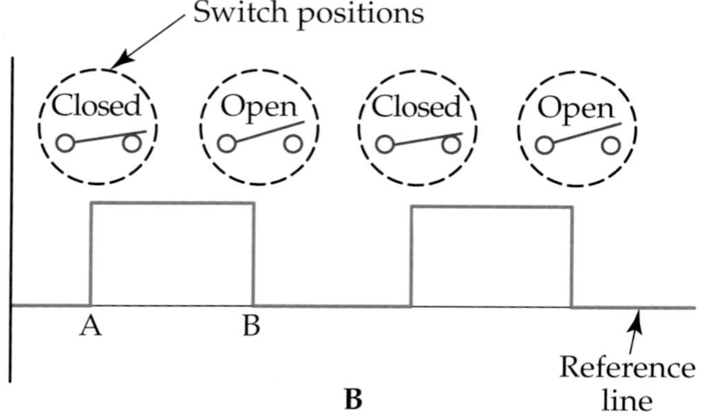

Figure 11-14. Producing a square wave. A—This simple circuit produces a square wave. B—A square wave can be produced by turning the switch on and off at regular intervals.

has been moved halfway up the wave. The peak amplitude of the wave is 5 V. The voltage changes direction and travels 5 V below the reference line. The peak-to-peak voltage would still be 10 V. The period of this wave is 1 second.

Sine Waves and Sound

If you were to connect a speaker in an ac circuit with a frequency of 400 Hz and an RMS voltage of 5 V as shown in **Figure 11-17,** you would hear a sound. The sound is called a *tone.* Increasing the voltage of the wave makes the tone louder. Increasing the frequency of the wave makes the pitch higher.

Most people can hear sounds as low as 20 Hz and as high as 20,000 Hz. Although, as people age or suffer hearing damage, this range tends to become narrower. Frequencies between 20 Hz and 20,000 Hz are called ***audio frequency (AF) waves.***

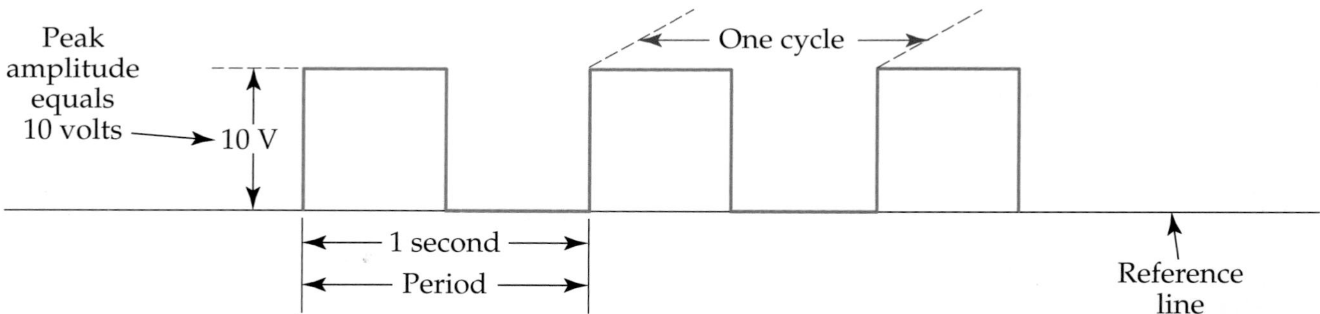

Figure 11-15. By opening or closing the switch once every second, we can produce a square wave with a frequency of 1 Hz.

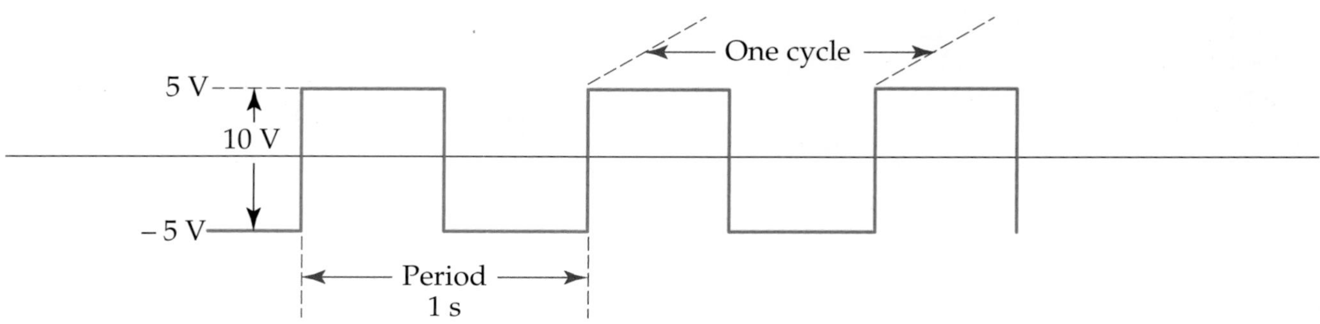

Figure 11-16. A square wave produced by an ac source. Notice that the waveform extends above and below the reference line.

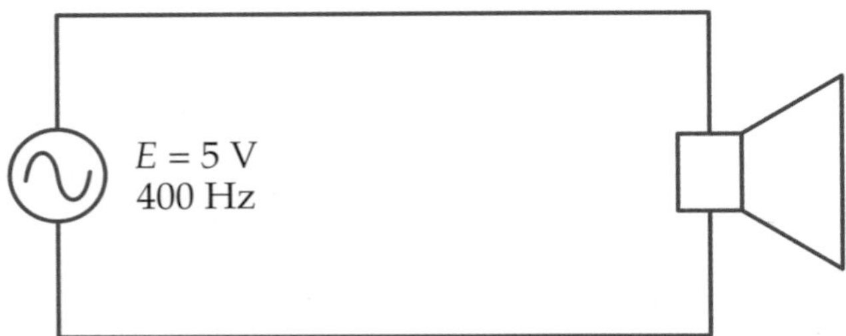

Figure 11-17. We can produce sound by putting a loud speaker in the circuit.

Signal Generators

Most ac waves you will deal with in electronics can be produced by a machine called a ***signal generator,* Figure 11-18.** These machines are used to produce ac voltages with a frequency as low as 20 Hz or as high as 200,000,000 Hz, **Figure 11-19.**

Signal generators can have many control knobs, but three common control knobs are found on all of them, **Figure 11-20.** The *frequency control knob* allows the operator to choose the number of waves, Figure 11-20A. The *amplitude control knob* allows the operator to adjust the height of the wave produced, Figure 11-20B. The *function control knob* allows the operator to choose either a sine wave, sawtooth wave, or a square wave, Figure 11-20C.

Figure 11-18. Signal generators produce ac voltages with a broad range of frequencies.

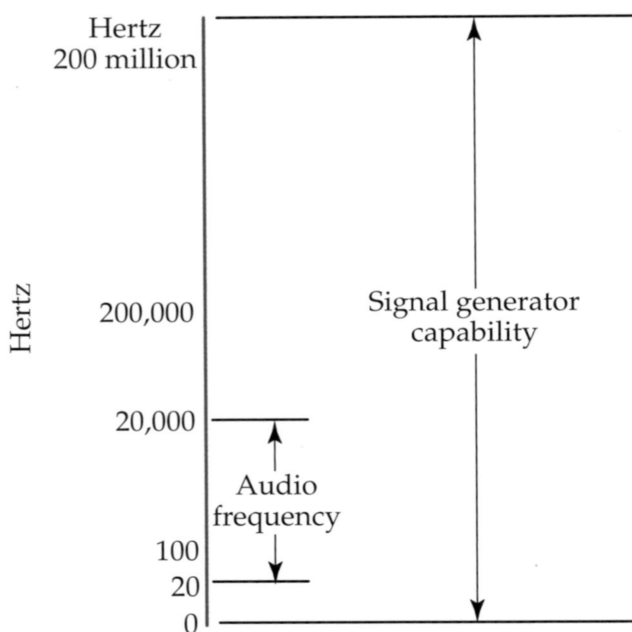

Figure 11-19. A signal generator can produce sounds far beyond the range of human hearing.

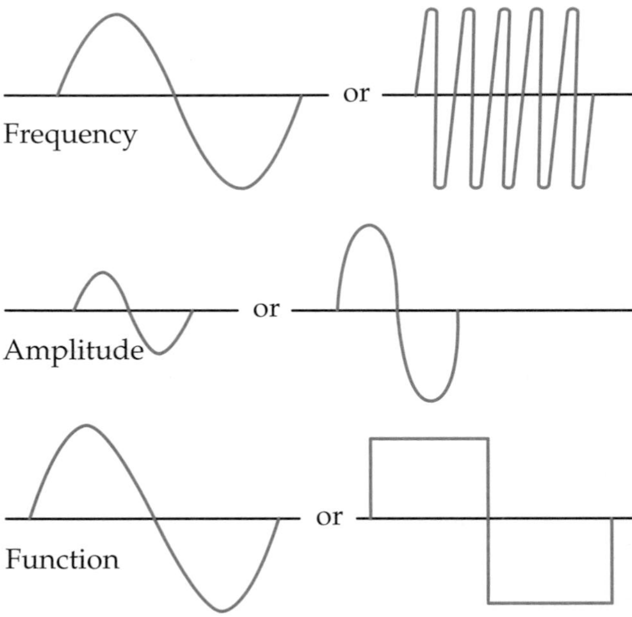

Figure 11-20. All signal generators allow the user to adjust these three characteristics of a sine wave.

> **Note**
> There are many different types of signal generators. Because of this, you may have noticed several control knobs or buttons on your generator that are not mentioned here. If you are unsure about operating any of the controls, ask your instructor for an explanation.

Signal generators produce frequencies higher than 20,000 Hz, which humans cannot hear. **Figure 11-21** shows the frequency range for radio, television, telephone, and astronomy equipment. Radio stations transmit voices and music through the air by means of waves that a radio can pick up and change into frequencies that people can hear, **Figure 11-22.** Waves at these high frequencies are called ***radio frequency (RF) waves.*** These RF waves are all around us and are a very important part of electronics.

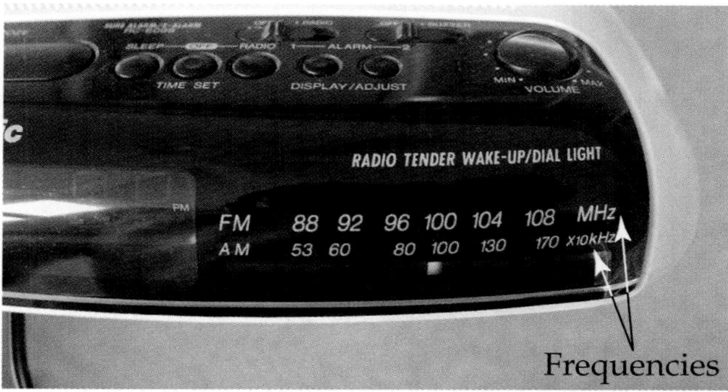

Figure 11-22. This AM/FM radio dial covers an AM band from 530 kHz to 1700 kHz.

Description	Abbrev.	Frequencies	Examples of Use
Extremely high frequency	EHF	30 GHz–300 GHz	• Microwave
Superhigh frequency	SHF	3 GHz–30 GHz	• Microwave
Ultrahigh frequency	UHF	300 MHz–3 GHz	• Cellular phones and towers • Pagers • *Television stations 14–69
Very high frequency	VHF	30 MHz–300 MHz	• FM Radio • Pagers • *Television stations 2–13 • Weather Radio
High frequency	HF	3 MHz–30 MHz	• CB and Shortwave Radio • Radio Astronomy
Medium frequency	MF	300 kHz–3 MHz	• AM Radio • CB and Shortwave Radio • Radio navigation and maritime/aeronautical mobile
Low frequency	LF	30 kHz–300 kHz	• Radio navigation and maritime/aeronautical mobile
Very low frequency	VLF	3 kHz–30 kHz	• Audio range for humans • Radio navigation and maritime/aeronautical mobile
Voice frequency	VF	300 Hz–3 kHz	• Audio range for humans • Radio navigation and maritime/aeronautical mobile
Extremely low frequency	ELF	Below 300 Hz	• Audio range for humans • Military stations for submarine communication

*These frequencies are to be allocated to other uses when conversion to digital TV has been completed.

Figure 11-21. Telephone, television, and astronomy equipment transmit and receive in a wide range of frequencies well above that of human hearing.

Oscilloscopes

The ***oscilloscope*** is a testing instrument designed to give a technician information about a wave, such as its waveform, frequency, period, and amplitude. Most people just call an oscilloscope a "scope."

There are many types of scopes. A common type used by those beginning the study of electronics is shown in **Figure 11-23.** Some scopes are handheld, **Figure 11-24,** and are used by technicians who do testing in the field instead of at a bench.

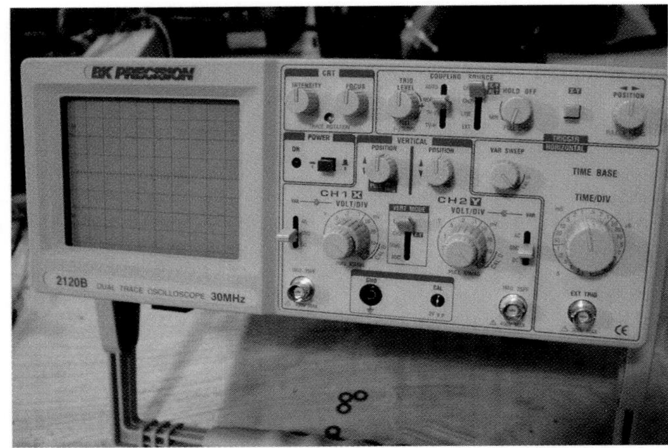

Figure 11-23. Typical oscilloscope used in the classroom.

Each company makes its scope a little different from the others. Most scopes, however, have the same basic parts and will read the same basic values in a circuit. The only way to really learn how to set them up and take readings is through practice. Too often people see the scope and are afraid to use it. Some companies have scopes in their electrical shop just collecting dust.

Once you learn how to read a scope, you will have one of the best tools available for checking circuits and correcting faults. A scope can save many hours of troubleshooting when used by people who know how to read them.

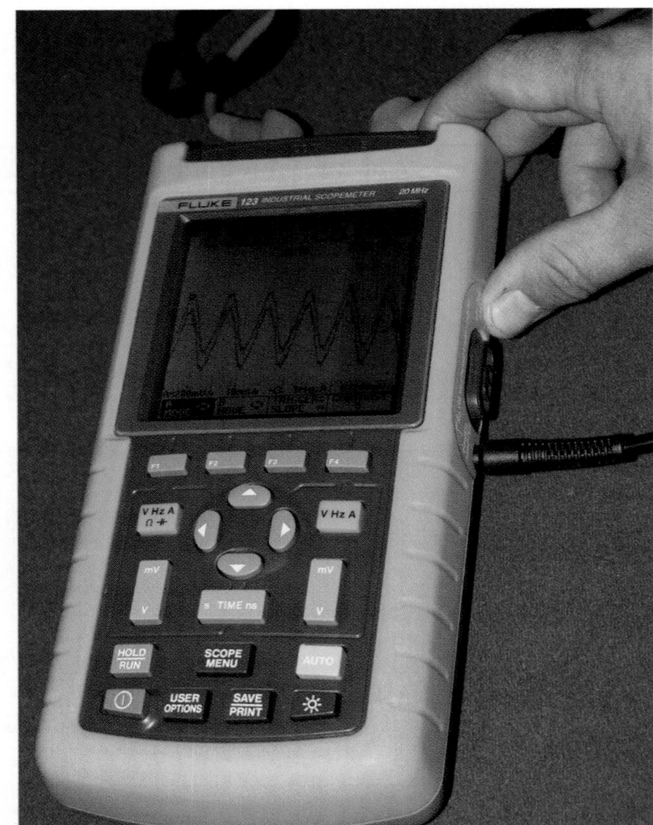

Figure 11-24. Handheld oscilloscope.

Graticule

The scope screen has a ***graticule,*** or ***grid,*** which helps in taking measurements. A graticule includes a horizontal axis and vertical axis. Time is shown on the horizontal axis and voltage is shown on the vertical axis, **Figure 11-25.** The horizontal axis is called the *X-axis*. The vertical axis is called the *Y-axis*.

Probes

There are several types of probes. The one that you will use depends on the type of circuit to be tested. Commonly used probes include the following:

- Low-capacitance probe.
- RF probe.
- Demodulator probe.
- High-impedance probe.

Figure 11-26 shows two types of probe tips. The tip with the hook can be attached to the circuit, allowing the technician to free his

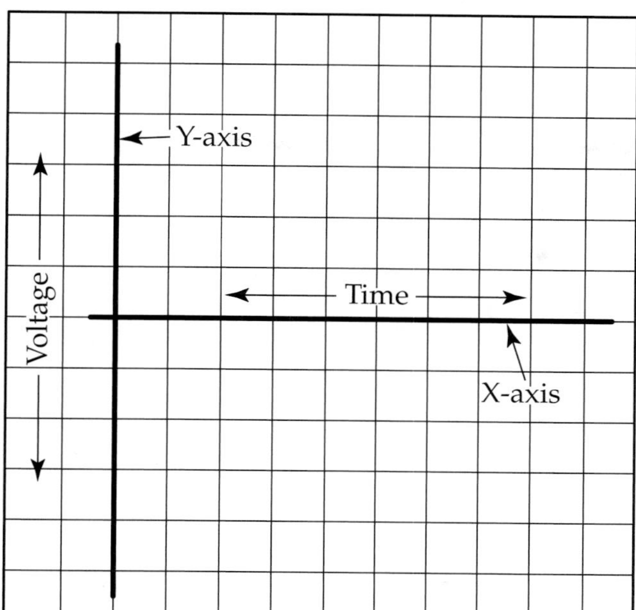

Figure 11-25. The horizontal axis on a scope screen shows the time. The vertical axis shows the voltage.

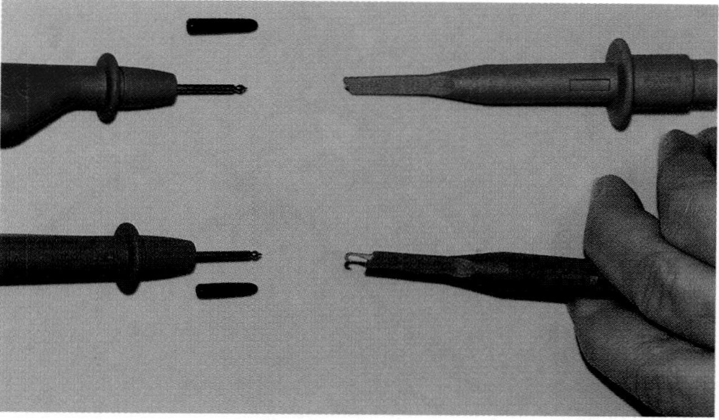

Figure 11-26. Two kinds of probe tips.

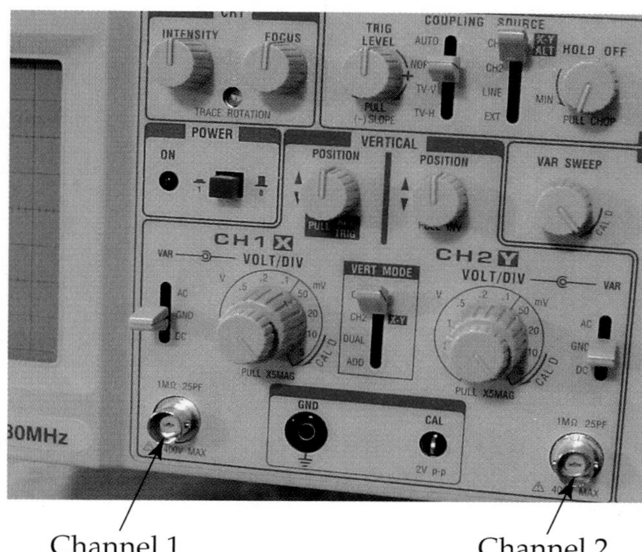

Figure 11-27. The BNC sockets on this scope are used for plugging in probes. They provide a channel through which a signal can be received and displayed.

or her hands for making scope adjustments. There are times, however, when a technician will use the straight-tipped probe because he or she is only looking for voltages or certain wave shapes. Then, the technician will just touch the probe to the circuit and go on to the next test point.

The jacks or BNC sockets on the scope are used for plugging in probes, **Figure 11-27.** Each jack provides a ***channel.*** By plugging a probe into a channel and connecting the tip of the probe to a circuit, a technician is able to see the signal under test. Scopes have two or four channels so that a technician can compare two different signals at one time. For example, you could examine voltage and current.

Dials

There are a number of dials on the front of a scope. Each is used to make certain types of measurements. The dials with which you should be familiar are the Volts/Div, Sec/Div, Vertical Position, and Horizontal Position dials. These will allow you to make simple measurements of amplitude and frequency.

The Volts/Div dial sets the measurement of the vertical axis. For example, if the Volts/Div dial were set to 5 mV (millivolt), each major division would be equal to 5 mV, **Figure 11-28.** Each small division would therefore be equal to 1/5 of a millivolt, or 1 mV. The Sec/Div dial sets the measurement of the horizontal axis. If the Sec/Div dial were set to 1 ms (millisecond), each major division would be equal to 1 ms, and each small division would be equal to 0.2 ms.

The Vertical Position and Horizontal Position dials position the waveform vertically

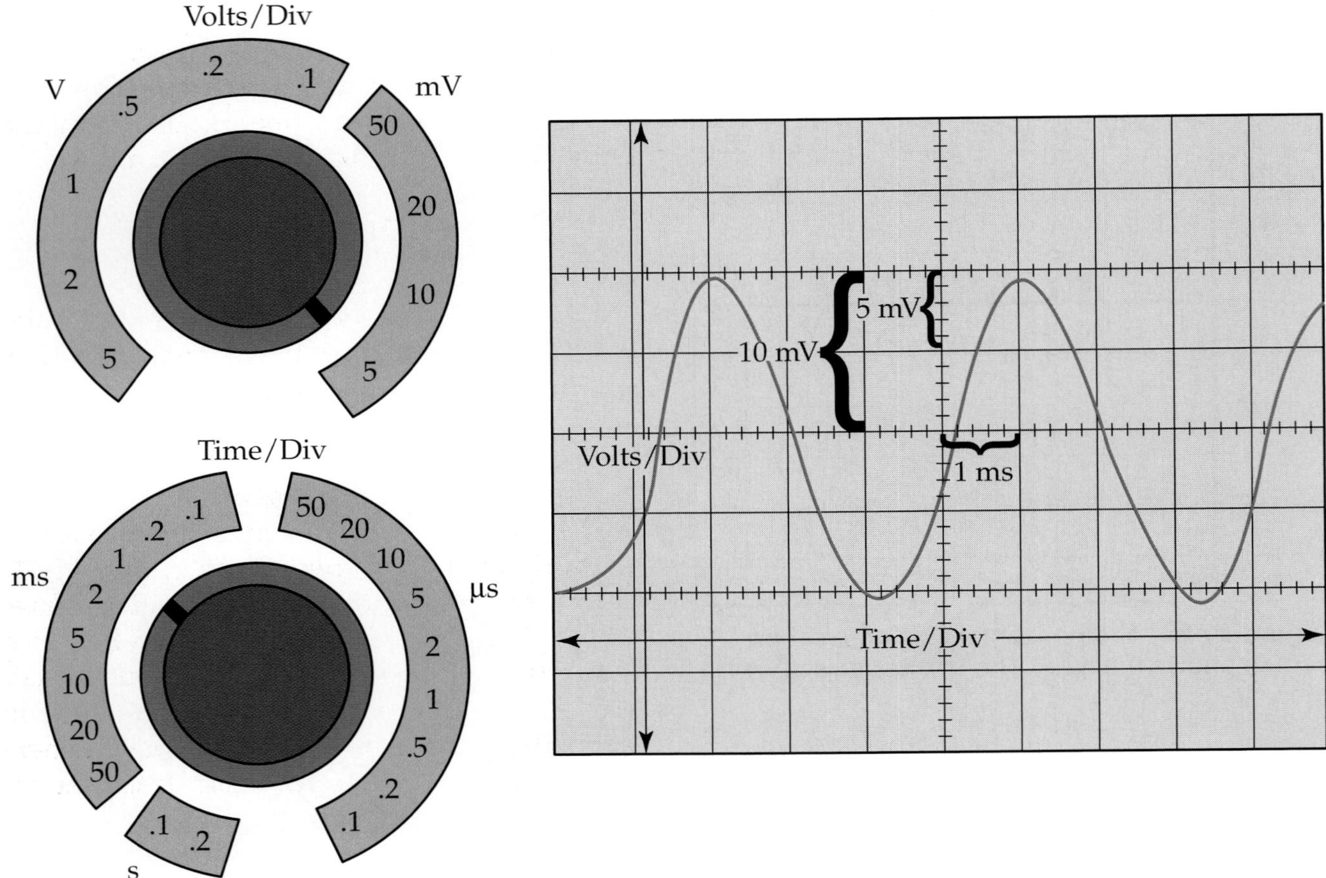

Figure 11-28. The Volt/Div dial sets the measurement for the vertical axis. The Time/Div sets the measurement for the horizontal axis.

or horizontally. A technician can therefore position the waveform so that it better aligns with the major divisions of the graticule. This makes it easier for the technician to make a reading.

Measuring Unknown Frequencies

The number of things you can measure using a scope is almost endless. For example, a scope can be used for the measurement of unknown frequencies. To do this, you will need to use a sine wave generator in addition to the scope.

The sine wave generator is an instrument marked off with frequencies that you can set. This check is made by using the known frequency of the sine wave generator to find the value of the unknown frequency. It is often used when working on television receivers.

Lissajous Figures

To check for an unknown frequency, you must know about ***Lissajous figures.*** These are a group of figures of various shapes named in honor of the French scientist Lissajous. He discovered that these figures could be developed in several ways. However, we are interested in using them to measure frequencies.

We can demonstrate how Lissajous figures help us find unknown frequencies. First, hook up the scope to the sine wave generator and to the unknown frequency as shown in **Figure 11-29.** Then, adjust the sine wave generator until a figure appears on the scope.

Suppose the figure on the scope is a circle. Next, draw a right angle and place it by the figure on the scope, **Figure 11-30.** Place it so that the left side of the figure touches the vertical line. Next, set it so that the horizontal line touches the bottom of the figure. Actually, you do not need to draw the right angle because of the graticule on the scope. Finally, count the places where the figure touches each line.

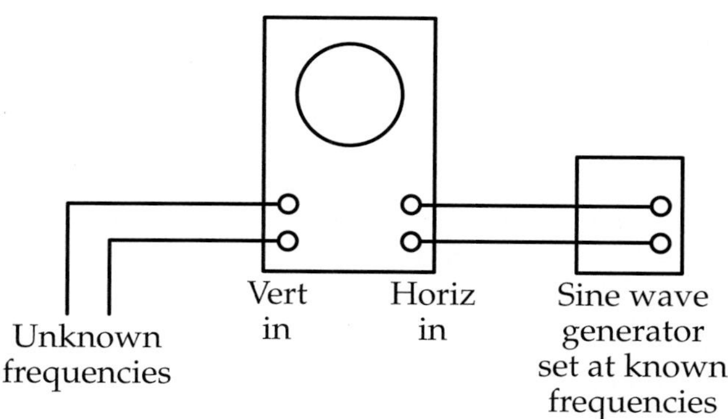

Figure 11-29. Using the scope and sine wave generator to find unknown frequencies. Lissajous figures help explain results.

Ratios for Lissajous Figures

Our circle in Figure 11-30 touches the horizontal line in one place, and it touches the vertical line in one place. Therefore, it has a one to one ratio (1:1). This means that the unknown frequency is equal to the frequency set on the sine wave generator.

Suppose that our Lissajous figure touched the horizontal line in two places, **Figure 11-31.** Now, we have a two to one ratio (2:1). Therefore, the vertical input to the scope would be twice the frequency of the horizontal input signal. This means that our unknown frequency is twice the frequency of the sine wave generator. If the generator (f_{Gen}) had been set on

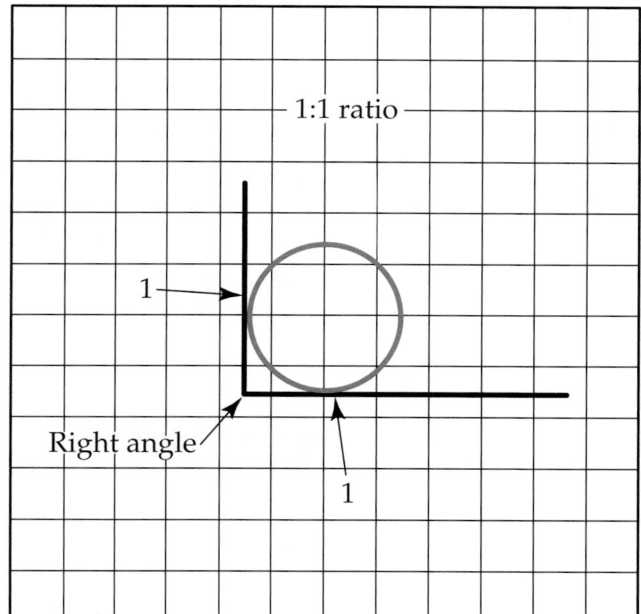

Figure 11-30. This circle is a Lissajous figure which tells you that the two frequencies being measured are equal. The figure touches each line in one place.

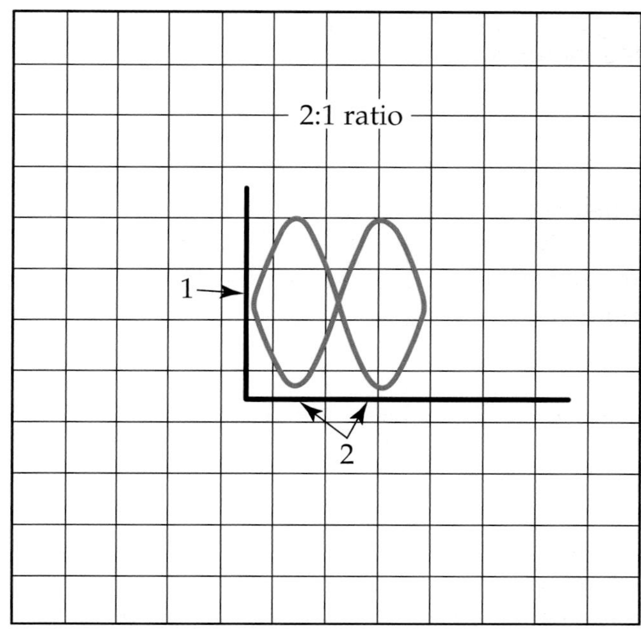

Figure 11-31. A two to one ratio Lissajous figure touches the horizontal reference line in two places and the vertical reference line in one place.

300 Hz, the unknown frequency ($f_{Unknown}$) would be 600 Hz:

$$f_{Unknown} \times \frac{\text{Points touching}}{\text{vertical line}} = f_{Gen} \times \frac{\text{Points touching}}{\text{horizontal line}}$$

$$f_{Unknown} \times 1 = 300 \text{ Hz} \times 2$$

$$f_{Unknown} = 600 \text{ Hz}$$

If we had a three to one ratio (3:1), **Figure 11-32,** the unknown frequency would be three times as large as the known frequency. If our sine wave generator (f_{Gen}) had been set at 400 Hz, our unknown frequency ($f_{Unknown}$) would be equal to 1200 Hz:

$$f_{Unknown} \times \frac{\text{Points touching}}{\text{vertical line}} = f_{Gen} \times \frac{\text{Points touching}}{\text{horizontal line}}$$

$$f_{Unknown} \times 1 = 400 \text{ Hz} \times 3$$

$$f_{Unknown} = 1200 \text{ Hz}$$

Not all of the Lissajous figures end up with high ratios on the horizontal axis. Some will have the greater number touching the vertical axis, **Figure 11-33.** In this case, the ratio is one to three (1:3). From this, it is easy to figure out what the unknown frequency would be.

Getting Readable Ratios

In many cases, you will get figures that touch the horizontal axis and vertical axis several times and end up with ratios of three to five (3:5), four to six (4:6), or some other combination. What happens when you have trouble counting the number of points where the figures touch the horizontal and vertical lines? You should adjust the sine wave generator to different frequencies until you can get a ratio that is easy to read.

In some cases, the figures will tend to turn slowly. When this happens, the sine wave is not adjusted so that the ratio reads in whole numbers. The Lissajous figure may turn in either direction. It still means the same thing. Again, the best solution is to change the known frequency until you get a figure that is easier to read. If this is not possible, you will have to try another method to check for unknown frequencies.

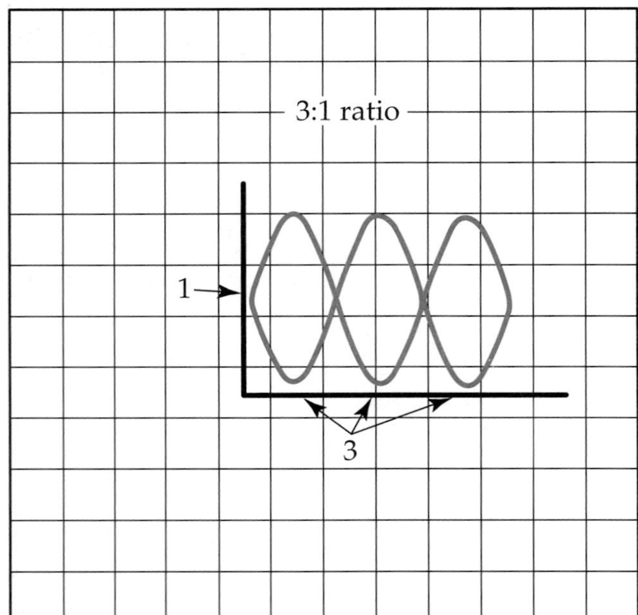

Figure 11-32. In a three to one ratio Lissajous figure, the figure touches the horizontal reference line in three places and vertical reference line in one place.

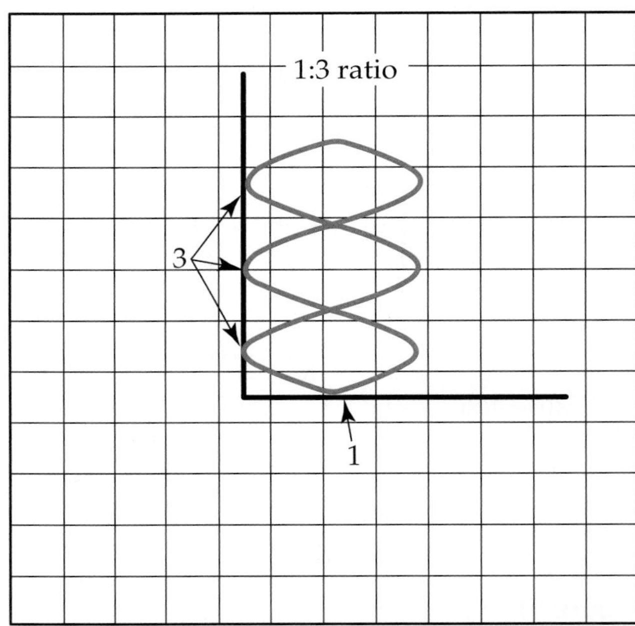

Figure 11-33. In a one to three ratio Lissajous figure, the figure touches the horizontal reference line in one place and the vertical reference line in three places.

Practical Application 11-1: Reading Voltage on an Oscilloscope

The graticule of an oscilloscope screen with its horizontal and vertical axes is used to help you take measurements. This practical application will show you how to figure out the voltage measurements displayed on the scope.

When reading the voltage of a wave, you must use the vertical axis of the graticule. The major divisions are based on the VOLTS/DIV setting. For example, say the VOLTS/DIV dial is set to 5 V as shown in the following illustration:

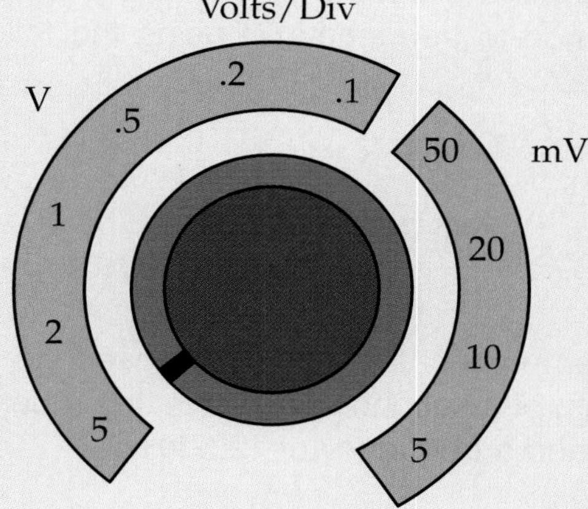

Each major division is, therefore, equal to 5 V. To find the peak voltage of a wave, count the number of major divisions above the horizontal reference line to the positive peak amplitude level.

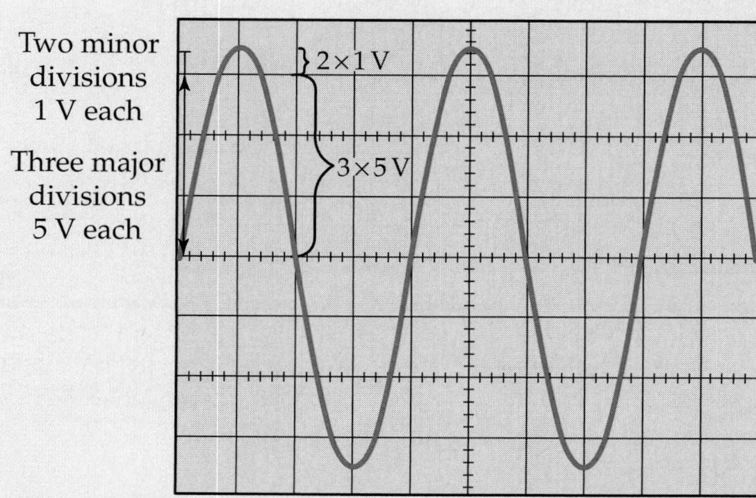

Multiply that number by the major division voltage value.

$$V_p = \text{major divisions} \times \text{VOLTS/DIV setting}$$

For the sine wave shown, this would be

$$\begin{aligned} V_p &= \text{major divisions} \times \text{VOLTS/DIV setting} \\ &= 3 \times 5\text{ V} \\ &= 15\text{ V} \end{aligned}$$

When the wave extends above the last major division, count the number of minor divisions to its peak. For the sine wave shown, this is two. Each minor division is equal to 1 V because there are five minor divisions within a major division.

Minor division voltage = VOLTS/DIV setting/of minor divisions

$$\begin{aligned} &= 5\text{ V}/5 \\ &= 1\text{ V} \end{aligned}$$

Since the wave extends above the last major division by two minor divisions, add 2 V to the total voltage. That means that the total peak voltage (V_p) is equal to:

$$\begin{aligned} V_p &= 2\text{ V} + 15\text{ V} \\ &= 17\text{ V} \end{aligned}$$

The peak voltage for this wave is 17 V. To calculate the peak-to-peak voltage, use the formula you learned earlier in this chapter.

$$\begin{aligned} V_{pp} &= V_{+p} + |V_{-p}| \\ &= 17\text{ V} + |17\text{ V}_{-p}| \\ &= 34\text{ V} \end{aligned}$$

Therefore, the peak voltage (V_p) of this wave is 17 V and the peak-to-peak voltage (V_{pp}) is 34 V.

Practical Application 11-2: Reading Frequency on an Oscilloscope

This practical application will teach you how to figure out the period and frequency measurement displayed on the oscilloscope screen. When reading the period and frequency of a wave, you must use the horizontal axis of the graticule. The measurement of the major divisions is based on the TIME/DIV setting. For example, set the TIME/DIV dial to 20 μs as shown in the following illustration:

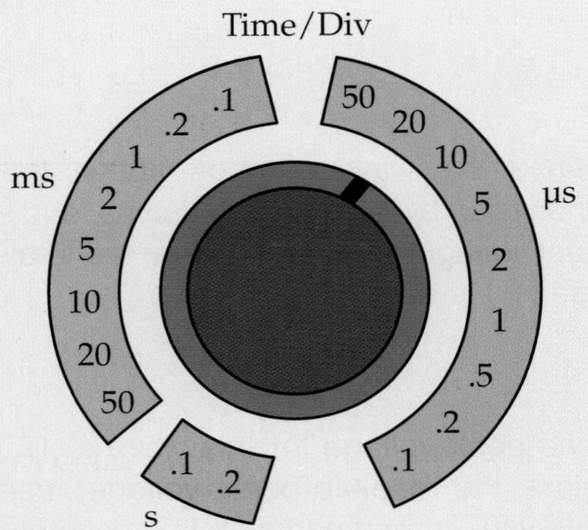

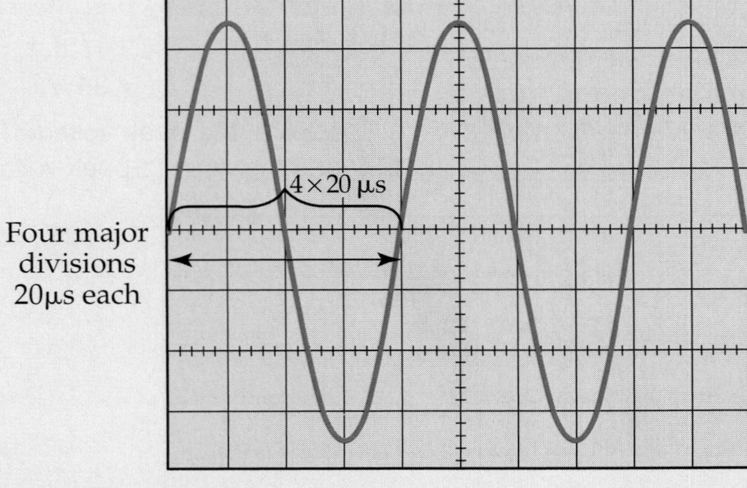

Each major division is, therefore, equal to 20 μs. To find the period of a wave, count the number of major divisions along the horizontal axis for a complete cycle. Multiply this number by the TIME/DIV setting.

Period = Number of major divisions × TIME/DIV setting

For example, for one cycle of the wave shown, there are four major divisions. The TIME/DIV dial is set to 20 μs.

To determine the period of this wave, multiply 4 times 20 μs.

$$\text{Period} = 4 \times 20\ \mu s$$
$$= 80\ \mu s$$

To find the frequency of the wave, convert the period to frequency (f) using the following formula:

$$f = \frac{1}{\text{Period}}$$
$$= \frac{1}{80\ \mu s}$$
$$= 12{,}500, \text{ or } 12.5\ \text{kHz}$$

Remember that one microsecond is a millionth of a second. This wave has a period of 80 μs and a frequency of 12.5 kHz.

Phase Relationship

To understand how voltage and current relate to one another, we will examine their phase relationship. Look at the two waves, voltage and current, in **Figure 11-34.** Notice that both waves reach their maximum amplitude at the same time, and both reach zero at the same time. When this happens, we say that voltage and current are ***in phase.*** In circuits that have resistors as the only load, the voltage and current will be in phase with one another.

Most circuits contain resistors, capacitors, inductors, and other components. The next chapter introduces you to inductors. When resistors, capacitors, and inductors are combined in various circuits, these circuits have an out-of-phase relationships between voltage and current, **Figure 11-35.** Notice that the current is leading the voltage in this example.

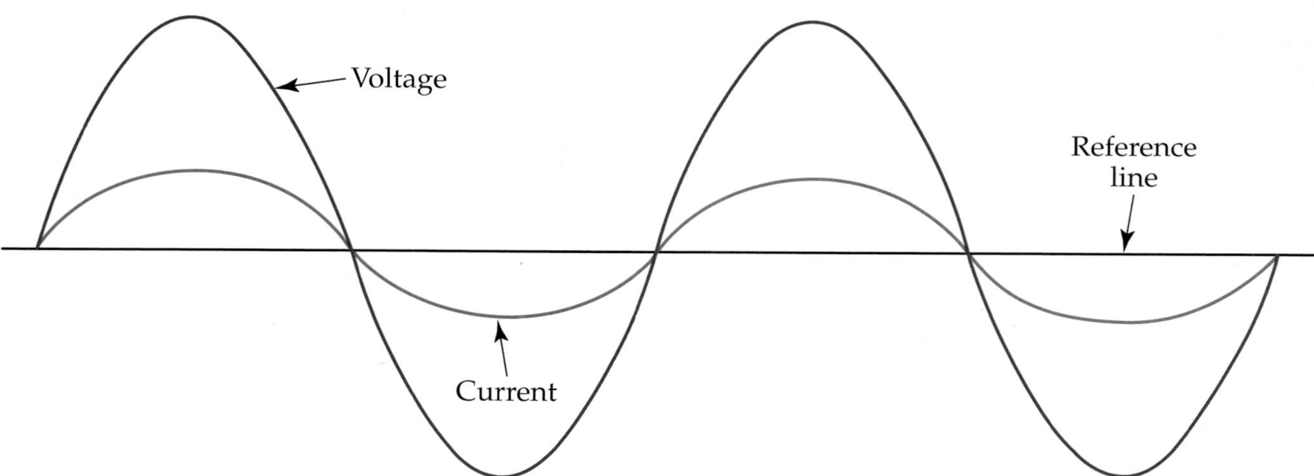

Figure 11-34. These voltage and current waves are in phase. Note that they peak at the same time and reach zero at the same time.

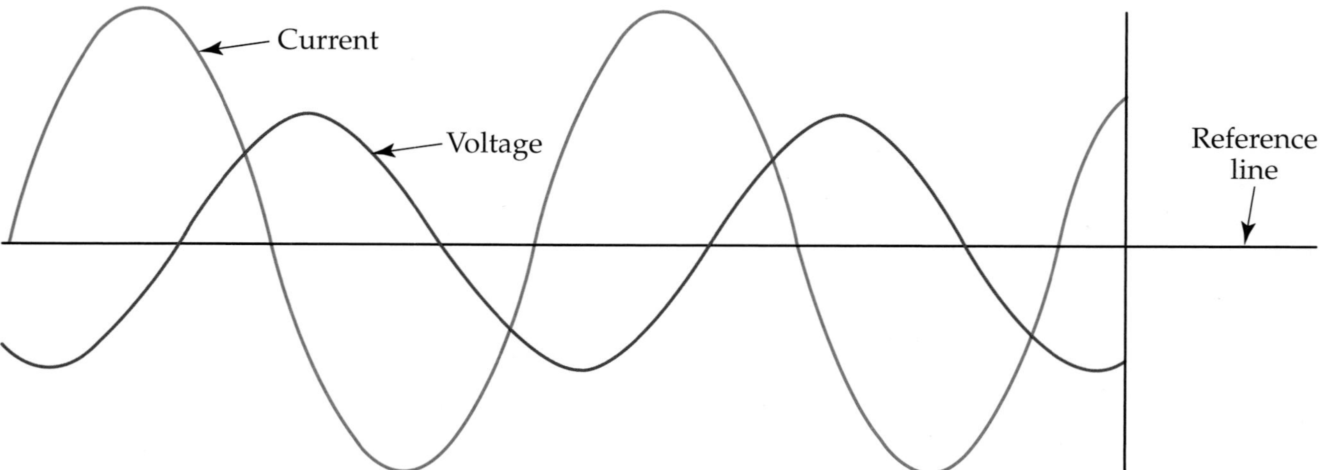

Figure 11-35. These voltage and current waves are out of phase. Note how the peaks vary and that the waves do not meet at the same point on the reference line.

Generators

A ***generator*** is a rotating machine that converts mechanical energy into electrical energy. This mechanical energy could be supplied by a gasoline engine, waterfall, coal, or nuclear power or in other ways described in Chapter 2. The schematic symbol for a generator is shown in **Figure 11-36.**

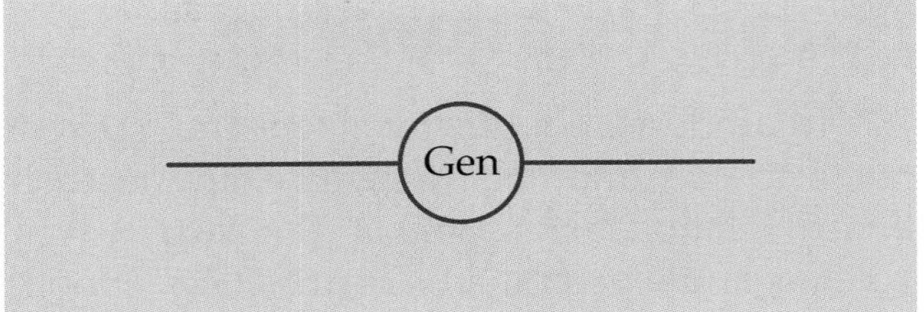

Figure 11-36. The schematic symbol for a generator.

You should recall from Chapter 10 that there are two ways to generate electricity through magnetism: passing a magnet by a wire coil or passing a coil by a magnet. The generator uses the same principles to generate electricity. In a generator, a coil of wire can be rotated inside a stationary magnetic field, or a magnetic field can be rotated inside a stationary coil of wire. The stationary part of a generator is the ***stator.*** The rotating part of the generator is known as the ***rotor.*** The magnetic field can be created with permanent magnets or electromagnets. **Figure 11-37** shows the stator and rotor of a disassembled generator.

Figure 11-37. Disassembled generator showing the stator (part that is stationary) and rotor (part that rotates).

DC Generator

Figure 11-38 shows a simple dc generator. A ***dc generator*** is a device that uses rotary motion to create direct current. The stator in this generator uses permanent magnets. Some stators consist of copper windings called ***field coils.*** An electromagnetic field is created in the field coil by feeding it a direct current. Field coils enable a stronger electrical current to be produced.

The rotor in this simple dc generator consists of an ***armature coil*** and ***commutator.*** The commutator is made of two semicircular metal bars. Each segment of the commutator is connected to one end of the armature coil and is insulated from the other to prevent arcing.

The rotor is turned by some means of mechanical power, causing the armature coil to rotate within the magnetic field. When the armature coil cuts through the magnetic lines of force, a current is induced. The current flows through the armature coil to the commutator. The two segments of the commutator pass the induced current to the ***brushes.*** The brushes are made of a graphite material, so they are

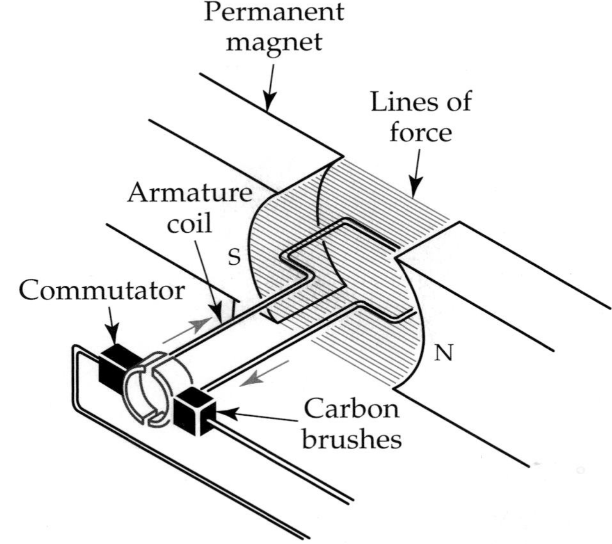

Figure 11-38. The commutator in this simple generator consists of two pieces of metal. Carbon brushes contact the commutator in order to carry the flow of the current that is generated in the armature.

good conductors of electricity. Brushes are attached with springs to the stator and are, therefore, stationary. The springs keep them in position so they are kept in contact with the turning commutator.

DC Generator Output

The output from a dc generator is not pure dc. A pure dc waveform is a steady straight line above the reference line. The output from a dc generator is pulsing dc, **Figure 11-39.** **Figure 11-40** shows what a generator with a four-piece commutator would produce. Notice the two overlapping sets of pulsing dc. If you look at the peaks in **Figure 11-41,** you can see that this gets closer to the straight line you think of as being dc. By using a capacitor, **Figure 11-42,** we can filter the dc to take out most of the ripple. This is covered further in Chapter 16.

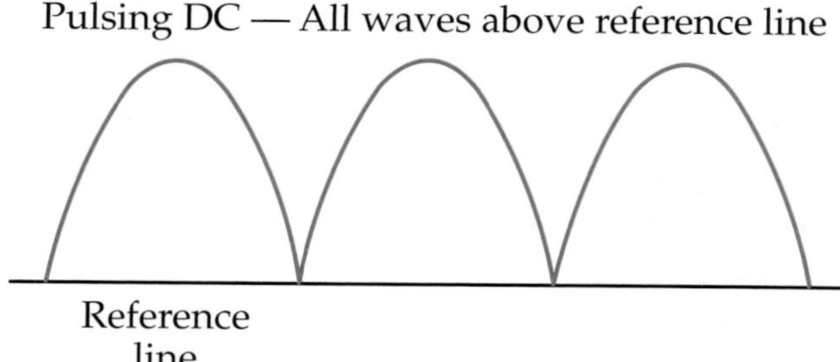

Figure 11-39. A dc generator produces a pulsing waveform with all waves above the reference line.

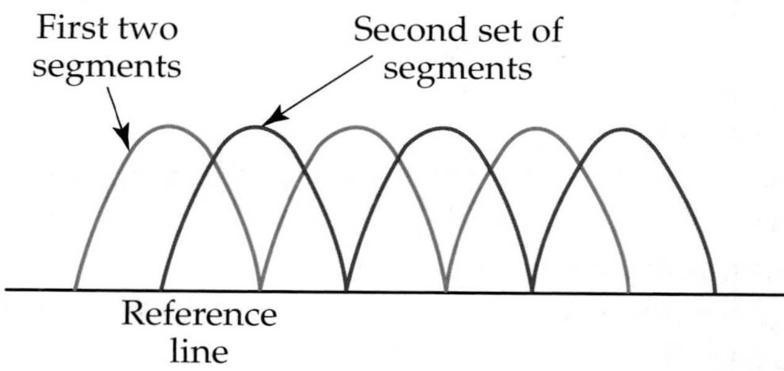

Figure 11-40. A four-piece commutator produces overlapping sets of pulsing dc. This is called a *ripple pulse*.

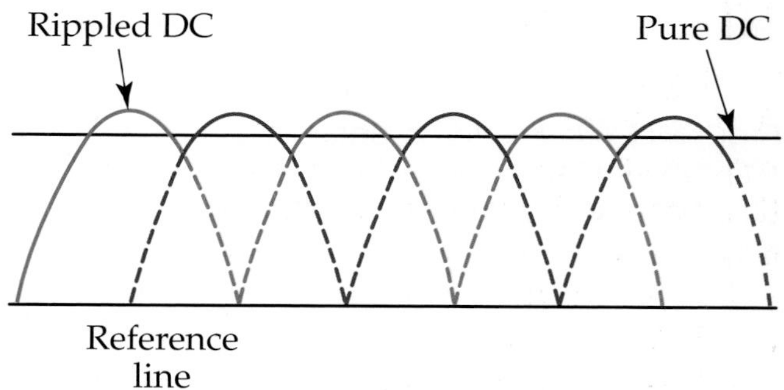

Figure 11-41. With rippled dc, the resulting waveform more closely follows the pure dc wave.

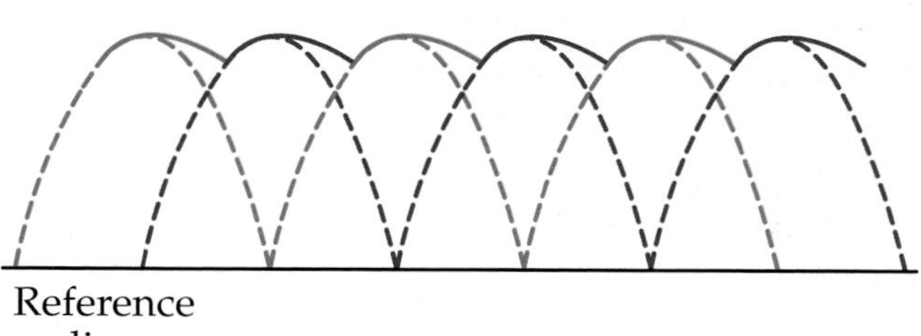

Figure 11-42. Capacitors can be added to a dc generator circuit to smooth the ripples from rippled dc.

AC Generator/Alternator

The ac generator is called an ***alternator,*** especially when it is used in an automotive charging system. The ac generator uses ***slip rings,*** **Figure 11-43,** in place of the commutator found on dc generators. The slip rings are designed to let the coil turn without twisting or breaking any wires. Each slip ring is soldered to one end of the armature coil. As the coil turns, the slip rings slide past the fixed brushes.

During the first half of the rotation, the coil cuts across the field's north pole. This produces current with the upper half of the sine wave. On the second half of the rotation, the reverse is true. This produces the lower half of the sine wave.

Three-Phase Alternator

Many alternators use a ***three-phase system,*** **Figure 11-44.** Three phase means that a sine wave starts every 120° instead of the normal 360°. In this manner, the valleys left in a normal sine wave are filled so they produce less of a ripple.

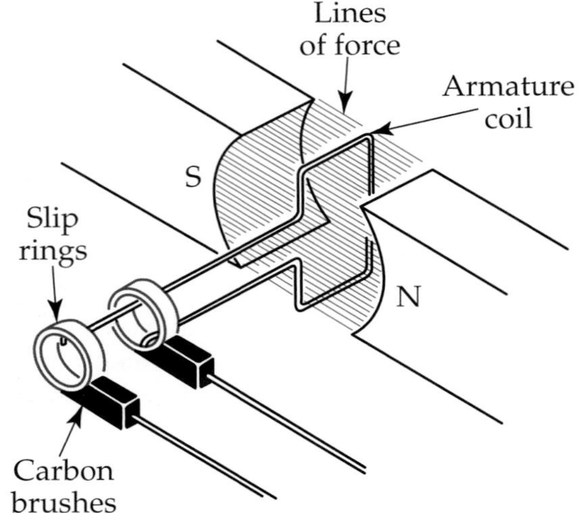

Figure 11-43. This simple ac generator has two slip rings and two brushes. Note that each end of the coil wire end leads to a separate slip ring.

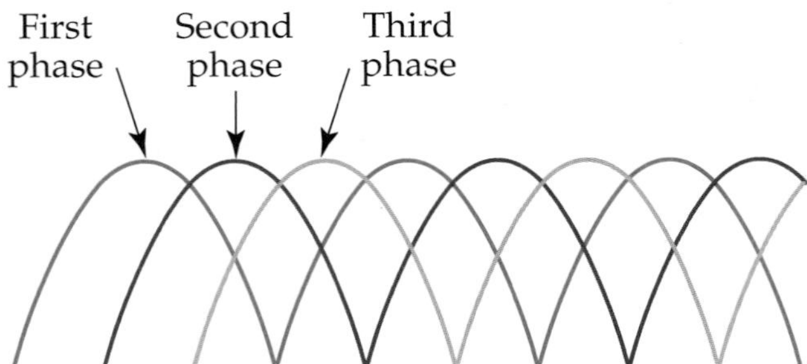

Figure 11-44. The three-phase system used in alternators starts a new sine wave every 120° (instead of every 360°) to level off dips in the sine waves.

Three-phase alternators are commonly used in automobiles to charge the battery and operate electrical devices. To do this, a rectifier or diode assembly is used. Rectifiers and diodes are discussed in Chapter 17. For now, just think of a rectifier, or diode, as a one-way street. It limits electrical flow to one direction only. In an automotive alternator, the diodes change the ac to dc.

The alternator is driven by mechanical energy produced by the automobile engine. It generates current through all three phases of its windings. Because of this, the alternator is able to charge the battery at low engine speed while the dc generator cannot.

Eddy Currents

Eddy currents are found in generators, transformers, and many other devices that involve a moving magnetic field that cuts through a conductor. When you row a boat, you will see swirls (eddies) in the water coming off the tips of the oars. When a generator armature cuts through the magnetic field, current is induced in the armature windings. Current also circulates within the metal in the form of eddies. These are known as ***eddy currents.*** The eddy currents in generators and transformers produce unwanted heat as they push their way through the metal.

If you leave your oars in the water and push them in the opposite direction, you break up the swirls in the water. Since you cannot do that with a generator, the armature is made of several pieces which are laminated together, **Figure 11-45.** The eddies are broken up because they cannot move through the layers.

Eddy currents are not always a bad thing. They can be used to an advantage. Eddy currents are used in braking systems on some hybrid cars. By creating a magnetic field around a spinning piece of metal, it is possible to create an eddy current brake. Eddy currents can stop roller coasters and stamping presses that produce car bodies, without the use of a friction brake. This type of braking system also turns the kinetic energy into electricity that is used to recharge batteries. This system is known as *regenerative braking.*

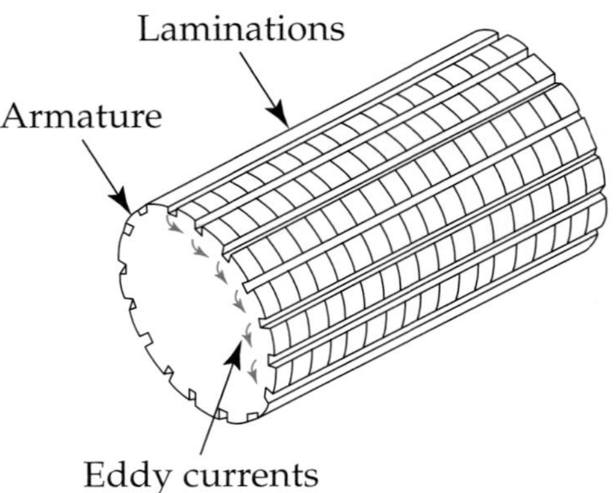

Figure 11-45. The metal core of a generator armature is laminated to reduce the flow of eddy currents.

Project 11-1: Multiple Outlet Extension Cord

How many times have you found yourself in need of another electrical outlet? After you hook up an extension cord, you run into the problem of trying to run two power tools. The extension cord outlet box in this project has four receptacles that can be turned on and off with a switch. It also has indicator lamps to tell you whether or not the power is on. If you should overload either side, a fuse protects it.

To build the project, obtain the following materials:

No.	Item
1	Outlet box
2	NE-2 lamps with dropping resistors
2	Fuse assemblies (with 3 A–4 A fuses)
2	DPST switches
2	110-V duplex receptacle with ground
1	Male plug (cap) with ground
1	18/3 extension cord
1	Grommet or cord strain relief

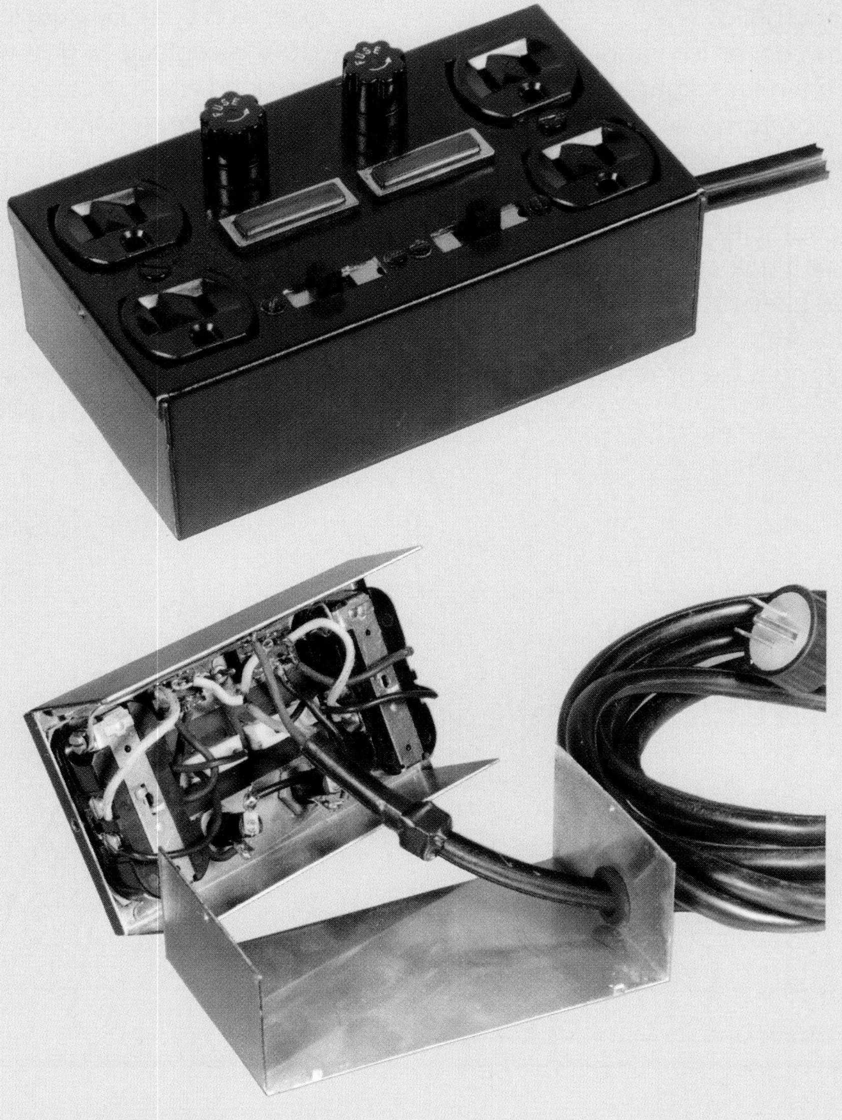

(Project by Mike O'Berski)

Project 11-1: Multiple Outlet Extension Cord *(Continued)*

Check all the parts and decide on the size of the box. You can either build one or buy it from a local electrical supply store. Follow the procedure to build this project.

Procedure

1. Using the following schematic as a reference lay out the location of the fuses, lamp, resistors, switches, and receptacles. Be sure to allow enough room to make it easy to wire and build.
2. Cut holes in the cover for various parts. Some boxes have knockouts (partially cut holes ready to be removed). If holes must be drilled, be sure to wear safety glasses to protect your eyes from metal chips.
3. Mount the receptacles in the cover and wire them. Use black covered wire to attach the brass electrical fitting on the receptacle. Attach the white wire to the silver fitting. Use green wire for grounds.
4. Mount the resistors and indicator lights, splicing them into line. Keep all connections well insulated, especially if you use a metal box. Be sure to use grommets, seals, and connectors to protect the wire insulation from being cut.
5. Mount the fuses and switches and wire them in position.
6. When the wiring is completed, close either switch. The indicator lamp should come on. If not, a fuse may be blown.
7. Wire an extension cord of the desired length to the box as shown. Use a wireman's knot inside the box, so wire cannot be pulled out during use.
8. Mount the male plug on the other end of the extension cord. Be sure to use a grounded 110-V receptacle to give you the protection you need when using portable power tools.

Plug the extension cord into the 110-V receptacle and unwind as much of the cord as you need to bring the multiple outlet box to the job. You may wish to make a stand on wheels for the multiple outlet box, so you can roll the outlet box and stand to each job. In addition, you can drill a 1/2″ hole at each corner of the stand and install short wooden dowels, which you can wrap the cord around for storage.

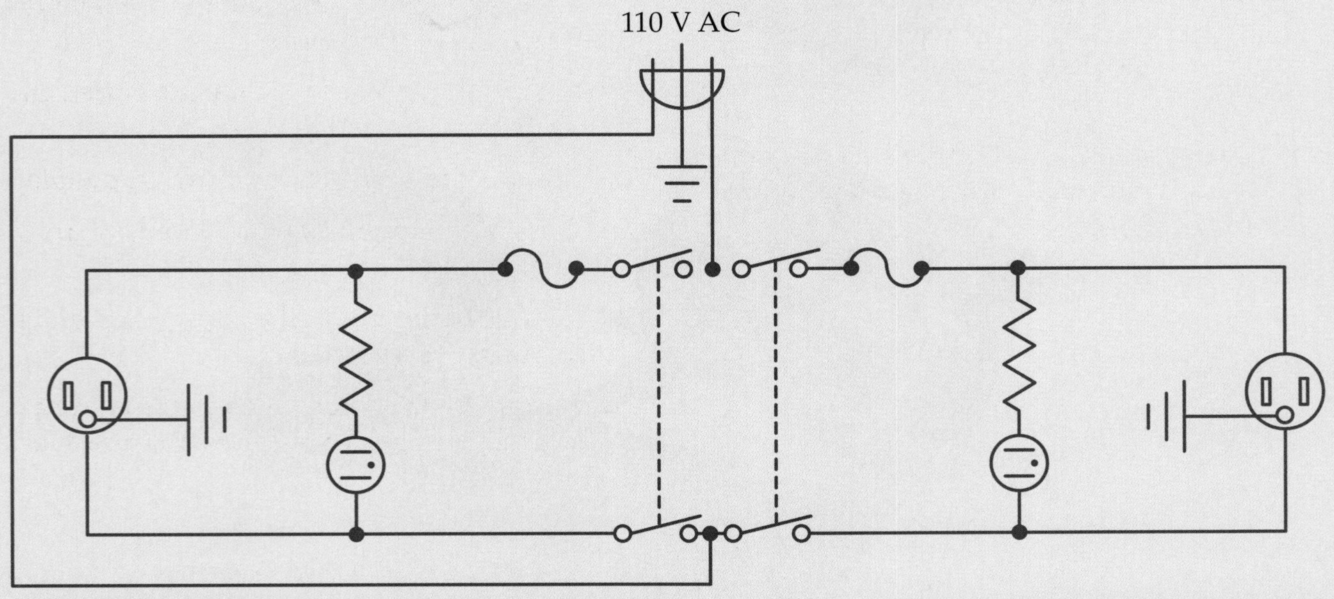

Summary

- An alternating current (ac) reverses its polarity at regular intervals.
- Passing coils of wire through a magnetic field produces alternating current by induction.
- The frequency of the ac produced in the United States is 60 Hertz (Hz).
- Frequency, period, amplitude, peak voltage, and peak-to-peak voltage of a circuit can be determined by examining a sine wave on an oscilloscope.
- To determine the dc heat equivalent value for an ac wave, you can use this formula:

$$\text{RMS} = V_p \times 0.707$$

- RMS stands for root mean square.
- Square waves are produced by a signal generator or with a circuit that switches on and off at regular intervals.
- Square waves, like sine waves, have a frequency, period, and amplitude.
- Current and voltage in a resistor-based circuit are in phase when the sine waves they produced reach their highest value at the same time and cross zero at the same time.
- A generator is a rotating machine that converts mechanical energy into electrical energy.

Test Your Knowledge

Do not write in this book. Write your answers on a separate sheet of paper.

1. Voltage is induced in a wire only when the wire (cuts across, moves parallel to) the lines of force.
2. As the speed of the wire cutting through the lines of force increases, the voltage induced in the wire (decreases, increases).
3. The ______ of a sine wave is its height above or below the reference line (zero) at any given time.
4. The number of cycles per second is called ______.
5. Convert the following:
   1600 kHz equals ______ Hz.
   200 MHz equals ______ kHz.
   20,000 Hz equals ______ kHz.
6. What is the frequency of the current in your home?
7. What is the period of a 300 Hz sine wave?
8. If an ac signal had a 100-V peak, would it give you more or less heat than 100 volts RMS? Show your work.
9. Is the voltage in your home measured in RMS or peak voltage?
10. What is the audio range for most people?
11. Time is shown on the ______ axis of an oscilloscope's graticule.
12. Voltage is shown on the ______ axis of an oscilloscope's graticule.
13. *True or False?* A signal generator is used to produce sine waves and square waves.

Activities

1. Why does the power company transport its electricity in ac instead of dc?
2. Recently, a few companies have switched to dc as a means of transporting power for long distances. Where are these companies? Why are they doing it?
3. Make a list of some of the different frequencies used by radio stations in your city.
4. Make a list of the different frequencies used by television stations around your area of the country. Why are these different than radio stations?
5. The National Electric Code governs most electrical construction. What is it? Why do building inspectors follow it? How do you tie the wireman's knot used in Project 11-1?
6. Besides the sine wave, what are some of the other types of waves found in electrical work?
7. What happens to your hearing as you get older? Why does that affect the range of frequencies that you can hear?
8. Why can a dog hear higher frequencies than humans?
9. Microwave ovens are used to cook food in just a few minutes instead of the long periods of time previously needed. How do they work? Would the waves from these ovens hurt you if they were turned on without the door being in place?
10. Where are most of the major power companies located in your area of the country?
11. Why are so many people unwilling to have nuclear power plants located near their homes?
12. Where is the biggest generating plant in your area located? How much electricity does it generate in a single day? How long would a city of 30,000 people be able to operate on that much energy?
13. How much electricity does your family use in a single year?
14. Make a list of the five countries that consume the most electricity.

Electromagnetic Induction

Learning Objectives

After studying this chapter, you will be able to do the following:

- Demonstrate how to induce a voltage using a coil and a magnet.
- Explain four things that affect the amount of inductance in a coil.
- Draw the symbols for a fixed and adjustable inductor.
- Calculate the power and current on the secondary of a transformer.

Technical Terms

closed-form transformer
counter-electromotive force
henry (H)
induced emf
inductance
inductor
magnetic coupling
mutual inductance
primary
secondary
shell-form transformer
step-down transformer
step-up transformer
tap
total inductance
transformers
volt-ampere

In Chapter 10, you learned that a magnetic field is created around a conductor when current flows through it. You also learned that moving a magnetic field across a conductor or moving a conductor across a magnetic field creates current in the conductor. Both of these occur in a circuit that contains a coil, or an *inductor*. This chapter takes a closer look at these principles of magnetism and how they work together to create *inductance*. It also looks at the characteristics of inductors and transformers.

Induced Voltage

In Chapter 10, you performed a simple experiment with a coil of wire, a galvanometer, and a magnet. When you pushed the magnet into the coil and quickly removed it, the galvanometer detected a slight current, **Figure 12-1.** The current was present only when the magnet was moving through the coil. Current can only be created in a conductor when there is a change in the magnetic field around it. The changing magnetic field induces voltage in the coil, which creates current in the coil. This voltage is called ***induced emf.***

There is always an opposition to change in electricity. As current starts to flow, a force opposes the flow. The opposing force is greatly reduced once the flow of current reaches its maximum. When the flow of current is shut off, a force opposes the flow until the current reaches zero. You have seen examples of this outside of the electrical field, such as the following:

- Lower gas mileage getting a car going from 0 miles per hour to 40 miles per hour.
- Time to bring a pressure washer or saw blade up to speed.
- Effort required to pedal a bike from a standing start.

Some people say these examples are the result of friction or some other function. In electricity, there is a ***counter-electromotive force*** (also called cemf or counter emf) which opposes the applied voltage. This results in a reduction or delay in any increase (at start-up) or decrease (at shutdown) in the flow of current.

Look at the dc circuit in **Figure 12-2.** This circuit contains a dc power source, switch, and coil. When the switch is closed, circuit current begins to increase. The current creates a magnetic field around the coil, Figure 12-2A. As the current increases, the magnetic field expands. Because it is changing, the magnetic field induces voltage in the coil that opposes the polarity of the source voltage.

Once the circuit current reaches its maximum and remains constant, the magnetic field stops expanding and also remains constant, Figure 12-2B. There is no longer any induced emf in the coil because a constant magnetic field does not induce voltage in a coil.

Suppose the dc power source in this circuit is replaced by an ac power source, **Figure 12-3.** The current in an ac circuit is always changing. This means there is always a changing magnetic field around the coil and, therefore, always induced emf in the coil. As alternating current changes its direction, the polarity of the coil reverses so that it opposes the polarity of the ac power source.

Note in **Figure 12-4** that the changing field set up by each turn of the coil passes through other turns of the coil. This is known

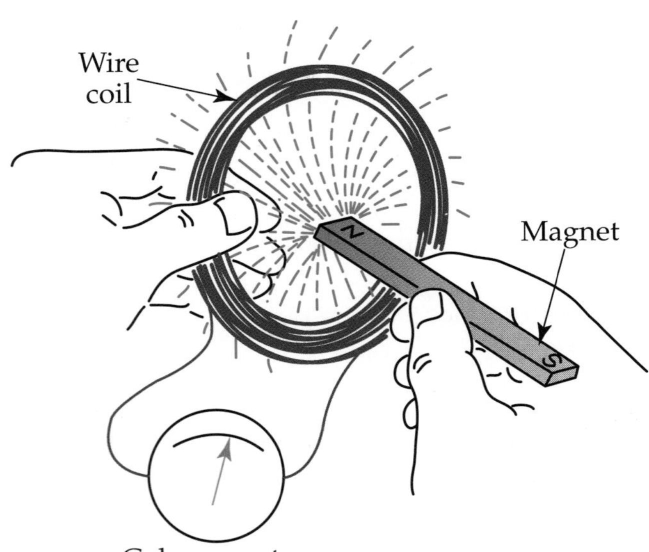

Figure 12-1. Recall that the changing magnetic field created by the sudden movement of a magnet inside a coil of wire will create current in the wire.

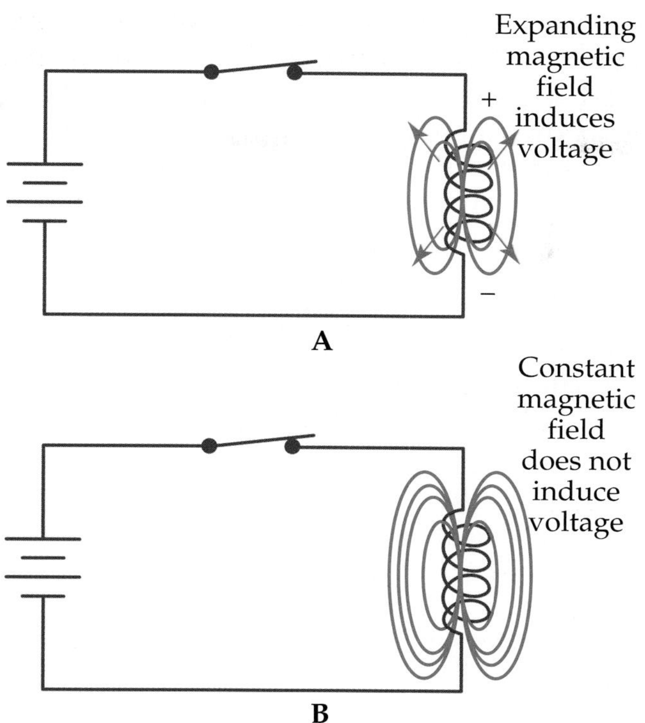

Figure 12-2. A dc circuit with a switch and a coil. A—Voltage is induced in the coil when the switch is first closed. This induced voltage opposes the polarity of the source voltage. B—No voltage is induced in the coil once the current reaches its maximum and becomes constant.

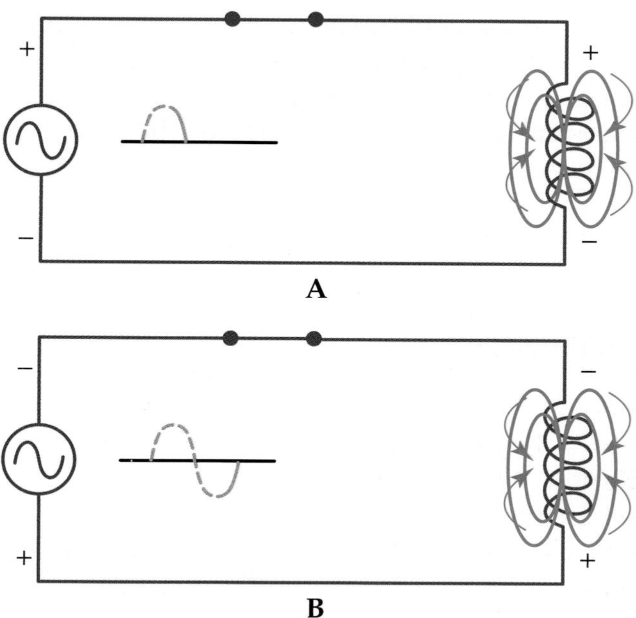

Figure 12-3. An ac circuit with a switch and a coil. When the field collapses during the positive and negative cycles of alternating current, the polarity of the coil reverses so that it opposes the polarity of the ac power source. A—Collapsing field during positive cycle. B—Collapsing field during negative cycle.

as ***magnetic coupling.*** The changing magnetic field in one turn of the coil produces an emf in the other turns of the coil through which the field passes.

Inductance

The induced voltage produced in a coil by a changing magnetic field affects the flow of circuit current through the coil. Because the induced voltage opposes the voltage of the power source in the circuit, the induced voltage opposes change in the current flow of the circuit. This opposition to a change in current flow is called ***inductance.***

Let us look again at what happens in a circuit consisting of a dc power source, a switch, and a coil. This time, we will look at how induced voltage and the current it produces affects the current from the circuit's power source. To distinguish this current from the induced current, we will refer to it as *circuit current.*

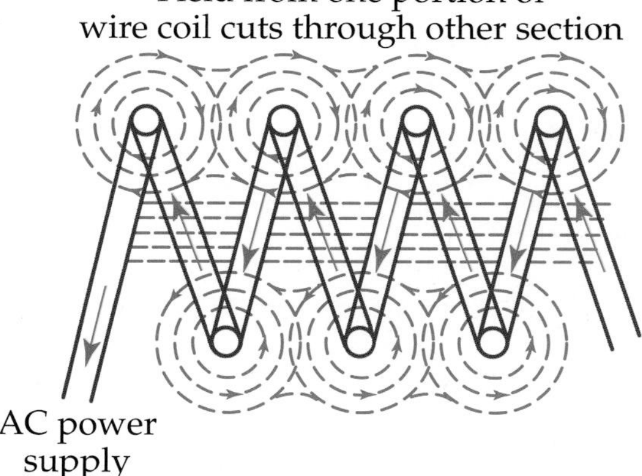

Figure 12-4. The changing magnetic field created by alternating current in each turn of the coil cuts through the other coils, creating a magnetic coupling.

In **Figure 12-5,** the switch 1 (S_1) has just been closed. The circuit current is increasing, so the magnetic field around the coil is expanding. This induces emf in the coil. The induced emf opposes the voltage of the power source, so it opposes the change in circuit current. This means that it slows the increase of circuit current. In other words, a circuit with a coil takes longer to reach its maximum circuit current than the same circuit without a coil.

When current in a dc circuit reaches its maximum and becomes constant, the magnetic field stops expanding and becomes constant, too. Since there is no change in the magnetic field, there is no induced voltage and no opposing current. Circuit current is allowed to flow freely through the circuit. The magnetic field surrounding the coil is constant.

When the power source is removed from the circuit (S_1 opened and S_2 closed), **Figure 12-6,** the circuit current starts to decrease toward

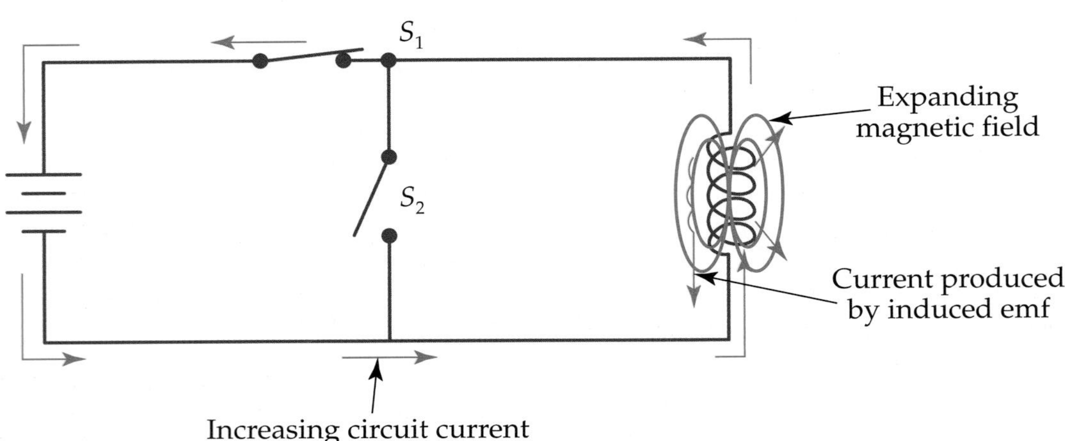

Figure 12-5. The induced emf produces current that opposes the circuit current, slowing down the circuit current's increase.

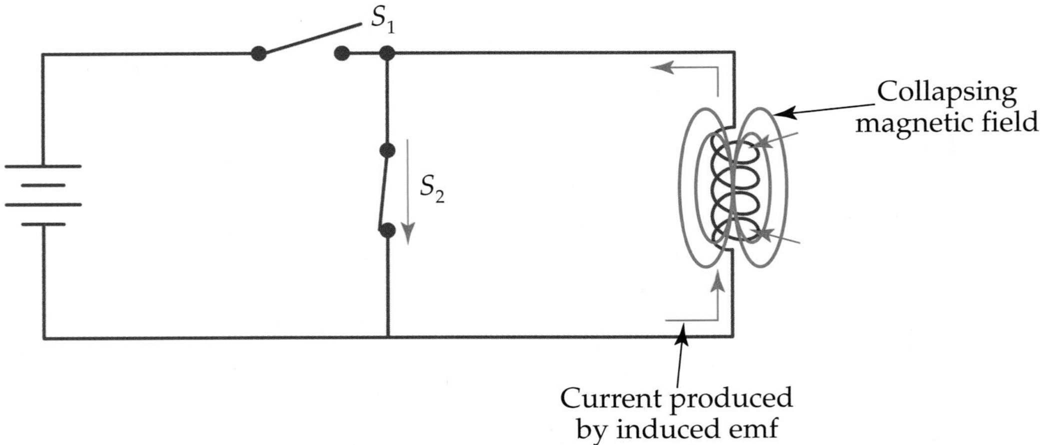

Figure 12-6. When the magnetic field is decreasing, the current produced by the induced emf flows with the circuit current. This slows the circuit current's decrease to zero.

zero. This decrease in circuit current causes the magnetic field around the coil to collapse. The collapsing magnetic field again induces a voltage in the coil. The induced emf produces a current that flows until the magnetic field has completely collapsed. In other words, the current of a circuit with a coil in it takes longer to drop to zero after the switch has been opened than the same circuit without a coil. When the magnetic field either expands or collapses, it induces a current in a coil.

I
Switch closed
time
Switch opened
Circuit without Coil
A

I
Switch closed
time
Switch opened
Circuit with Coil
B

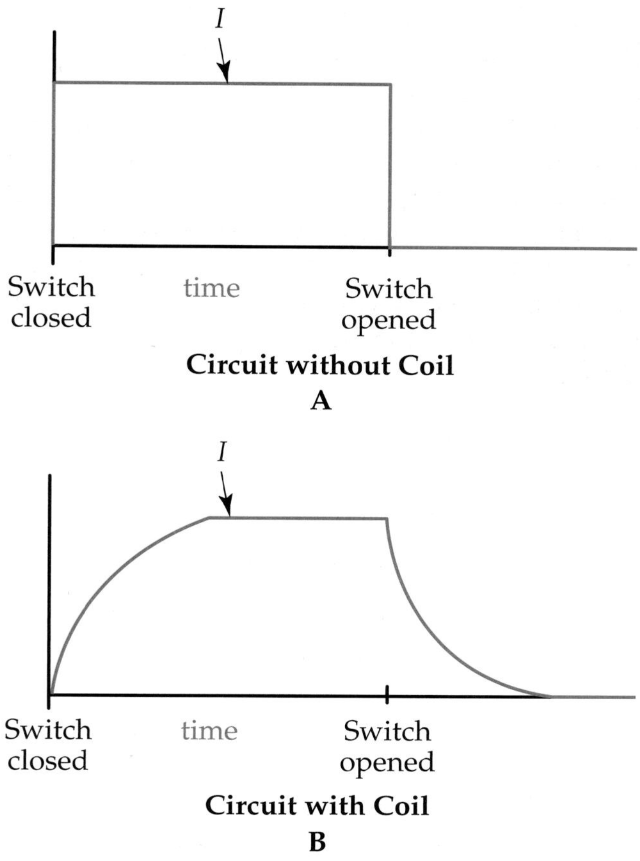

Figure 12-7. Graphs of current waveforms in a dc circuit. A—Current waveform in a dc circuit without a coil. B—Current waveform in a dc circuit with a coil.

Figure 12-7 shows the waveforms of a dc circuit without a coil and one with a coil. Notice that when the circuit does not have a coil, the wave takes the shape of a square wave when the switches are closed and opened, Figure 12-7A. When the dc circuit has a coil in it, the wave does not immediately reach its peak. Instead, the climb to its peak amplitude is slower. The wave is a curve on a scope before and after it levels out. The induced emf tends to oppose the change in current, Figure 12-7B.

Most of the time, a coil has little or no effect on a dc circuit. It only affects a dc circuit when the switch is opened or closed because those are the only times the circuit current is changing. However, a coil has a large effect on an ac circuit because the circuit current in an ac circuit is always changing.

Inductors

Any part of a circuit that has the property of inductance is called an ***inductor.*** The most common form of an inductor is a coil of wire. There are several symbols for inductors. The symbol for a basic inductor and the symbols for more complex inductors are shown in **Figure 12-8.** A circuit board containing inductors is shown in **Figure 12-9.**

Occasionally, an inductor is called a *choke.* **Figure 12-10** shows a radio frequency (RF) choke for an automobile. An RF choke is used to reduce circuit current or to stop ac from getting into a dc circuit. The core of a choke can be either air or adjustable ferrite (soft iron).

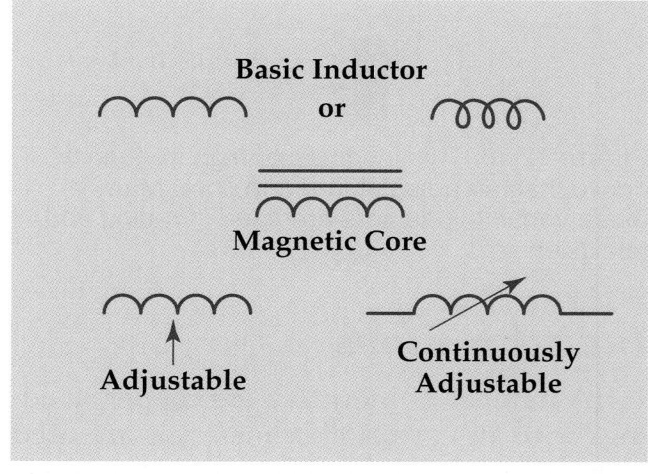

Figure 12-8. These are the symbols commonly used to indicate the use of different types of inductors in electrical circuits.

Figure 12-9. Inductors are labeled with the letter *L* on a circuit board and come in a wide variety of shapes, sizes, and colors, like the blue and green ones shown here.

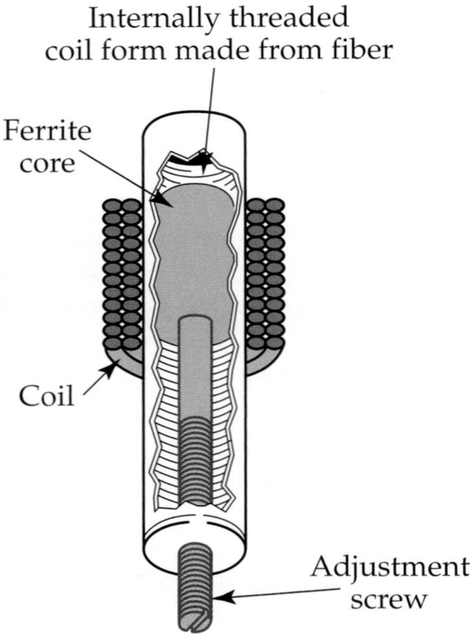

Figure 12-10. This inductor, or high frequency choke, has an adjustable ferrite core. Many of these variable inductors are used in radios and television sets.

Inductor Cores

A number of inductors are copper windings with air cores. Other materials are used inside the coil to reduce their size, make them more efficient, and cut their overall cost. Some of the newer inductors are built with powder cores. Some of these cores contain an organic binder that has a tendency to breakdown with high temperature applications. Although iron tends to heat up in some circuits, ferrite (iron) and steel cores are still being used in audio frequency circuits. Selecting the right core for the right job requires a fair amount of knowledge and study.

Unit of Inductance

The unit of inductance is the ***henry (H).*** An inductor has an inductance of one henry when one volt is induced while the current is changing at the rate of one ampere per second. The majority of inductors falls into the range of millihenrys (mH) to the smaller range of microhenrys (μH).

Inductor Applications

Inductors are combined with other components for use in many specific types of circuits. For example, inductors are combined with capacitors to form a wide variety of filter circuits. Adjustable ferrite core inductors are used in radios to tune a circuit by passing or rejecting the specific frequencies.

Inductors have uses in signal processing, switching power supplies, metal detectors, dc to dc converters, analog circuits, and adjustable speed drives. Part of their job is to limit voltage spikes, smooth out ac ripple, and reduce surges.

There is one use for inductors that almost everyone overlooks, even though it is very large and is encountered almost every day. In fact, it is so large, you have to drive your car through it. The next time you pull up to a signal light, you might notice something that looks like lines cut in the road in the shape of a big box. This is an inductor circuit. A wire is placed into a cut in the concrete and covered with a rubber material to seal it in place. When you drive your car into the "box," you are placing a large piece of steel inside the inductor. This creates a magnetic field which triggers a timer that changes the light signal.

You will also notice that a human being walking across the street passing through the coil does not trigger a light change. Most motorcycles are too small to trigger a light change. There are some motorcycles that have magnets mounted on them in an effort to create a large enough magnetic field to activate the coil and timer.

Factors of Inductance

Look at **Figure 12-11.** Note that there are several things that affect the amount of inductance in a coil:

- Number of turns of wire in the coil.
- Type of core.
- Cross-sectional area of the core.
- Spacing between the turns of wire.

The more turns of wire on a coil, the more lines of flux there will be to act on other turns. That means the inductance will increase as the number of turns increases. If you had only one turn of wire in the coil, the inductance would be very small and could be ignored. With a large number of turns, a small amount of inductance from each coil of wire adds up, creating a large impact on the circuit.

If you use a core material such as soft iron, you will get more inductance than if you use hardened steel. A magnetic field can be created in the core depending on the material from which the core is made. The magnetic field of the coils combines with the magnetic field of the core and creates a greater counter emf than the coil would have created alone. Inductors with magnetic cores have greater inductance than those with air cores or insulating cores.

Soft iron cores have the ability to increase the flux density much better than hardened steel cores. Steel is produced by using alloys while it is in the liquid state. After it cools to a solid state, it is hardened by drawing it through dies to reduce the size. High internal stress and hardness caused by drawing it through dies reduces the magnetic properties of the steel.

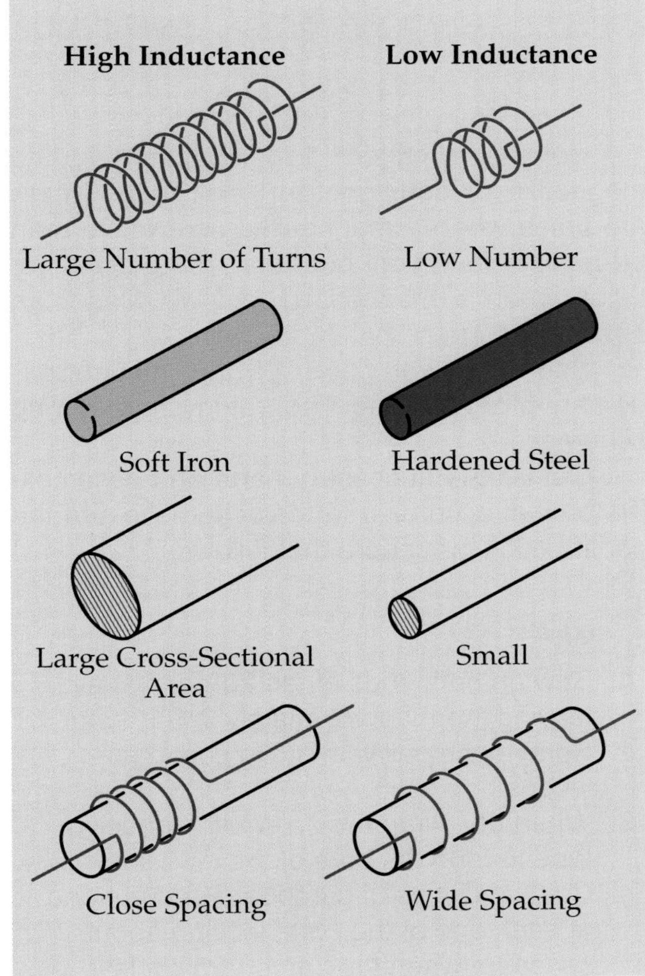

Figure 12-11. The amount of inductance in a coil of wire is affected by the physical makeup of the coil.

If the core of the inductor has a large cross-sectional area, it will create more flux lines than a small core. An increase in flux lines means an increase in flux density and an increase in inductance.

Spacing the turns of wire in the coil affects the amount of inductance in the circuit. If the turns are widely spaced, the weaker outer edge of the magnetic field around a turn of wire would be all that cut through other turns, or the lines of flux from one turn may not cut through any other turns at all. If either of these situations occurred, either less voltage or no voltage would be induced.

Project 12-1: Magnetizer-Demagnetizer

Here is a project that will allow you to magnetize or demagnetize small tools. There are times when a magnetized screwdriver can be a big help in tight quarters. There are other times when a magnetized tool will hinder an assembling operation.

See the two views of the magnetizer-demagnetizer pictured. Then, study the following schematic.

Gather the necessary parts and assemble the project according to the schematic and the construction procedure that follows.

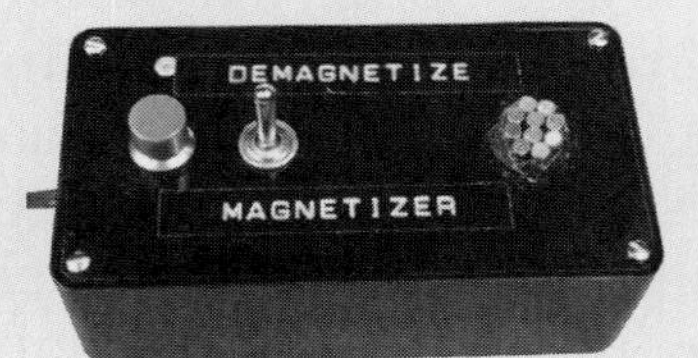

(Project by Robert Wong)

No.	Item
1	Handmade inductor
1	Electrolytic capacitor, 20 μF, 200 V
1	Silicon diode, 400 mA or higher, 300 V
1	Wirewound resistor, 200 Ω, 5 W
1	Momentary contact push-button switch
1	DPDT toggle switch
1	Lamp cord, 110 V
1	Plastic box and cover, 6″ × 3″ × 2″
	Epoxy, tape, solder, and soldering iron

Procedure

1. Saw off both ends of ten 20-penny nails, leaving them approximately 2 1/4″ long.
2. Spray each nail with a thin coating of lacquer.
3. When dry, wrap the nails with adhesive tape to form the core of the inductor.
4. Cut two circles of cardboard and put one at each end of the core, leaving about 1″ of space between the cardboard circles.
5. Wrap layers of No. 36 magnet wire around the core, totaling about 4500 turns.
6. Before cutting the wire, scrape the enamel to bare a spot for attaching an ohmmeter lead. Bare the starting end of the coil wire and connect the other ohmmeter lead to it. The resistance should be between 425 Ω and 450 Ω. If the resistance is lower, add turns of wire to the coil and recheck the resistance value.
7. Cut a hole in the cover just large enough for one end of the core to protrude about 1/4″ through the cover.

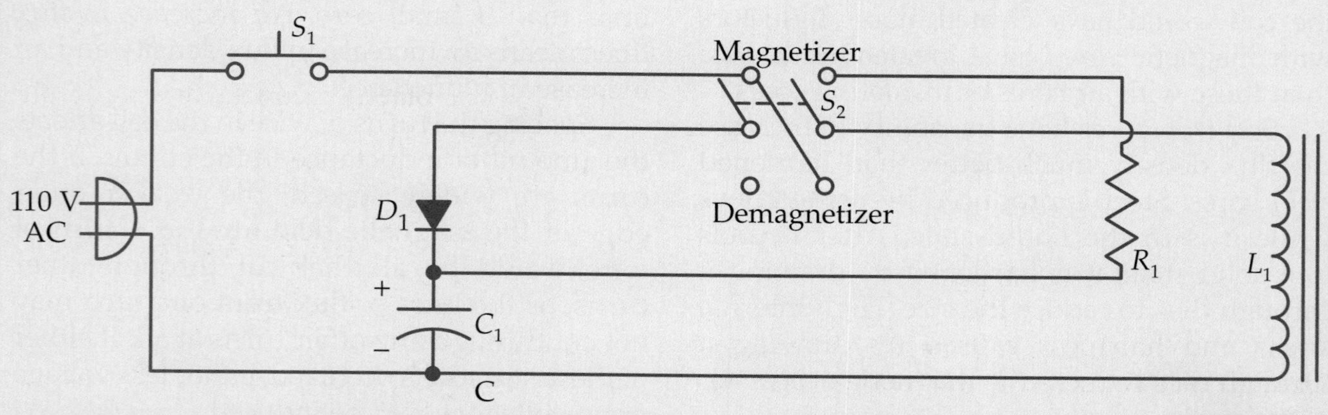

Project 12-1: Magnetizer-Demagnetizer *(Continued)*

8. Tape the finished coil, leaving both wire ends outside the tape. Scrape the enamel from wire ends and solder two hookup wires to them. Tape the soldered joints to the coil.
9. Epoxy the core to the cover so that it cannot move. Allow 24 hours for drying.
10. Install and wire the other parts to the underside of the cover. Observe the correct polarity when connecting the silicon diode (D_1) and the electrolytic capacitor (C_1).

To use the magnetizer-demagnetizer, plug the unit into a 110-V ac outlet. Place the toggle switch in the magnetizer position. While depressing the push button, pass the small tool across the core a few times to magnetize the tool. To demagnetize a tool, put the toggle switch in the demagnetize position. While depressing the push button, pass the tool across the core several times. The tool will no longer be magnetic.

Inductors in Series

When an inductor is shown in a circuit drawing, it is labeled with a capital *L*. Inductors are found in series, parallel, and series-parallel circuits. In **Figure 12-12,** three inductors (L_1, L_2, L_3) are used in series with an ac power supply.

When analyzing a series circuit for ***total inductance,*** we use the following formula:

$$L_T = L_1 + L_2 + L_3 + \cdots \qquad (12\text{-}1)$$

For the circuit shown in Figure 12-12, the total inductance for this circuit is 9 mH.

$$\begin{aligned} L_T &= L_1 + L_2 + L_3 \\ &= 2\text{ mH} + 3\text{ mH} + 4\text{ mH} \\ &= 9\text{ mH} \end{aligned}$$

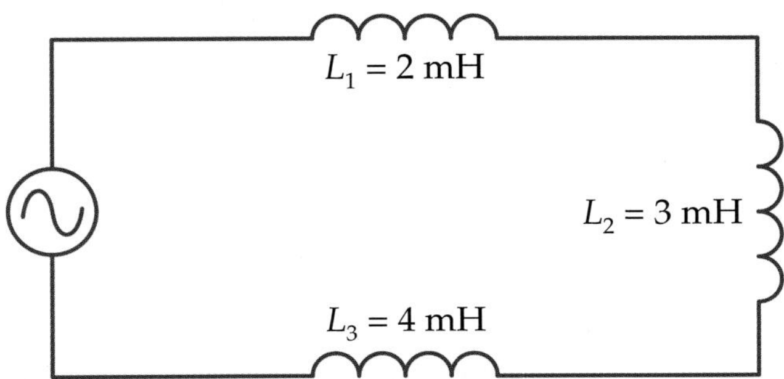

Figure 12-12. Inductors of three different millihenry values are used in this ac-powered series circuit.

Example 12-1:

Suppose that there are two inductors in series with the following values: $L_1 = 7\ \mu\text{H}$ and $L_2 = 4\ \mu\text{H}$. What is the total inductance (L_T) in the circuit? Start with the formula for total inductance in series.

$$\begin{aligned} L_T &= L_1 + L_2 \\ &= 7\ \mu\text{H} + 4\ \mu\text{H} \\ &= 11\ \mu\text{H} \end{aligned}$$

The total inductance for this circuit is 11 μH.

Inductors in Parallel

If the same three inductors were found in a parallel circuit, **Figure 12-13,** we would use this formula:

$$\frac{1}{L_T} = \frac{1}{L_1} + \frac{1}{L_2} + \frac{1}{L_3} + \cdots \qquad (12\text{-}2)$$

Solving for total inductance, we would find:

$$\begin{aligned} \frac{1}{L_T} &= \frac{1}{L_1} + \frac{1}{L_2} + \frac{1}{L_3} \\ &= \frac{1}{2\text{ mH}} + \frac{1}{3\text{ mH}} + \frac{1}{4\text{ mH}} \\ &= \frac{6}{12\text{ mH}} + \frac{4}{12\text{ mH}} + \frac{3}{12\text{ mH}} \\ &= \frac{13}{12\text{ mH}} \end{aligned}$$

Math Focus 12-1: Square Roots

In upcoming chapters and in the following section, you will need to calculate the square root of numbers. A square root is related to the base of a number that is *squared*, or that has an exponent of 2, such as 4^2. You should recall from Chapter 3 that an exponent is a number that tells how many times to multiply another number by itself. When a number is squared, it is only multiplied by itself one time. Therefore,

$$4^2 = 4 \times 4$$

You know that 4^2 is equal to 16. The square root of 16 then is a number that when multiplied by itself gives that value.

$$4^2 = 4 \times 4 = 16$$

A formula that says you need to find the square root of 16 is written as follows:

$$\sqrt{16}$$

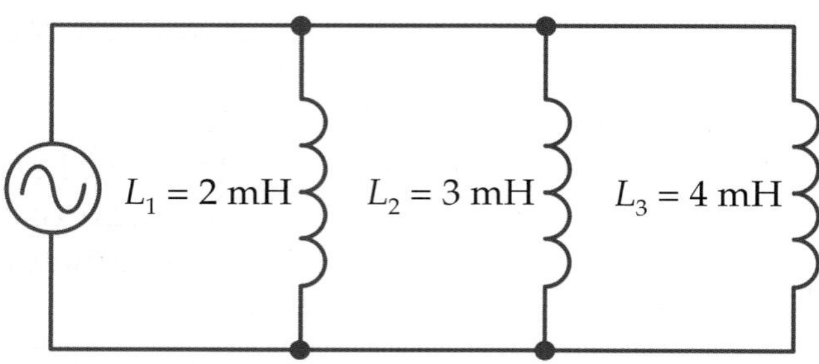

Figure 12-13. Three inductors are wired in parallel with the ac power supply.

Invert both sides of the equation:

$$L_T = \frac{12}{13 \text{ mH}}$$
$$= 0.923 \text{ mH}$$

This circuit has a total inductance of 0.923 mH, or 923 μH.

Example 12-2:

Suppose a parallel circuit has two branches, and each branch contains an inductor. The first inductor (L_1) has an inductance of 130 μH, and the second inductor (L_2) has an inductance of 260 μH. What is the total inductance (L_T) for this circuit?

$$\frac{1}{L_T} = \frac{1}{L_1} + \frac{1}{L_2}$$
$$= \frac{1}{130 \text{ μH}} + \frac{1}{260 \text{ μH}}$$
$$= \frac{2}{260 \text{ μH}} + \frac{1}{260 \text{ μH}}$$
$$= \frac{3}{260 \text{ μH}}$$

Invert both sides of the equation:

$$L_T = \frac{260}{3 \text{ μH}}$$
$$= 86.7 \text{ μH}$$

The total inductance for this circuit is 86.7 μH.

The same concept used to solve total resistance in series-parallel circuits can be used to analyze inductor values. Consider

This means that you must find the square root of 16. In the electronics formulas that you will see, you will have to first do other computations before finding the square root of the number. For example, you may have to multiply some numbers first before finding the square root:

$\sqrt{2 \times 8}$

$\sqrt{16}$

4

Practice Problems

Find the square root of the following perfect squares.

1. $\sqrt{100}$
2. $\sqrt{25}$
3. $\sqrt{144}$
4. $\sqrt{64}$
5. $\sqrt{81}$
6. $\sqrt{3 \times 3}$
7. $\sqrt{2 \times 3 \times 6}$
8. $\sqrt{2 \times 6 \times 12}$

the circuit shown in **Figure 12-14.** You recall that an inductor has no effect on a dc circuit, except when the switch is closing or opening.

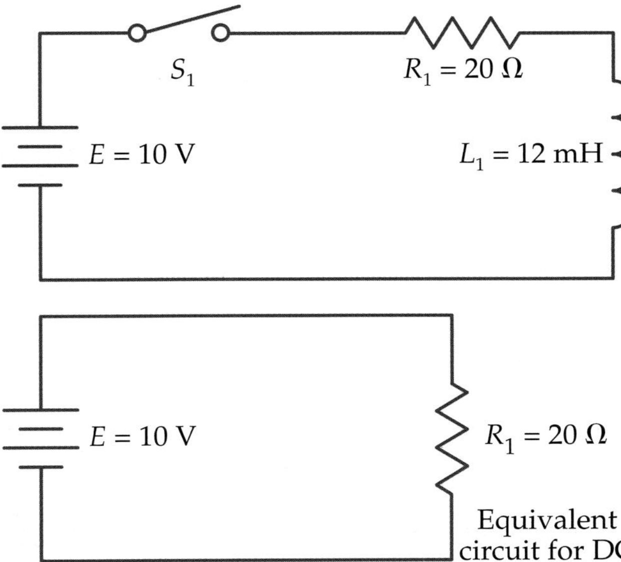

Figure 12-14. Since an inductor has no effect on the constant current flow of a dc power supply, it can be left out when solving circuit problems.

Therefore, we can analyze this circuit as if the inductor is not there.

Once the switch is closed, the magnetic field expands and remains constant. At that point, the current levels off and you can use Ohm's law to solve for *I*, *E*, and *R*. The inductor will have no effect on the circuit.

Note in the equivalent circuit shown in Figure 12-14 that only voltage (*E*) and resistance (*R*) are given. Based on these values, you can solve this circuit for current (*I*) by using Ohm's law:

$$I = \frac{E}{R}$$

$$I = \frac{10\ \text{V}}{20\ \Omega}$$

$$I = 0.5\ \text{A or } 500\ \text{mA}$$

More information is required before you can analyze these circuits with inductor and resistor combinations in ac circuits. Chapter 14 explains how to solve these more difficult circuits.

Mutual Inductance

One use of inductance can be demonstrated by placing two coils of wire close to each other, **Figure 12-15.** If a changing current (ac) is connected to coil *1*, a changing magnetic field is produced around it. Some of the lines of force of the magnetic field around coil *1* cut through the windings of coil *2*. The changing magnetic field induces voltage in the windings of coil *2*. This induced voltage causes a current to flow in coil *2*. A current flows in the second coil even though no wire joins the two coils.

When two coils of wire are close enough to be linked by a magnetic field, they are said to have ***mutual inductance.*** The amount of mutual inductance in the two coils depends on three things:

- Inductance of coil *1*.
- Inductance of coil *2*.
- Number of lines of flux set up by coil *1* that cut across coil *2*.

The cutting action of the lines of flux across the coils depends on other issues. The distance between the coils and the voltage applied to the coils are among the most important. These will be explained more fully in the next section dealing with transformers. The formula for mutual inductance is

$$M = K\sqrt{L_1 L_2} \quad (12\text{-}3)$$

where

M = Mutual inductance.

K = Coefficient of coupling (percent of coupling).

L_1 = Inductance of coil *1*.

L_2 = Inductance of coil *2*.

Suppose we have two coils placed near each other. One has an inductance of 12.5 H, and the other has an inductance of 2 H. Alternating current is passed through coil *1*, creating lines of force. When this occurs, 50% of these lines of force cut across coil *2*. We can use the formula to find the mutual inductance between the two coils.

$$M = K\sqrt{L_1 L_2}$$

We know that L_1 = 12.5 H and L_2 = 2 H. We also know that K = 0.5 or 50%. Substituting these values into the mutual induction formula, we have

$$M = 0.5\sqrt{12.5\text{ H} \times 2\text{ H}}$$
$$= 0.5\sqrt{25\text{ H}}$$
$$= 0.5 \times 5\text{ H}$$
$$= 2.5\text{ H}$$

These two coils have a mutual inductance of 2.5 H.

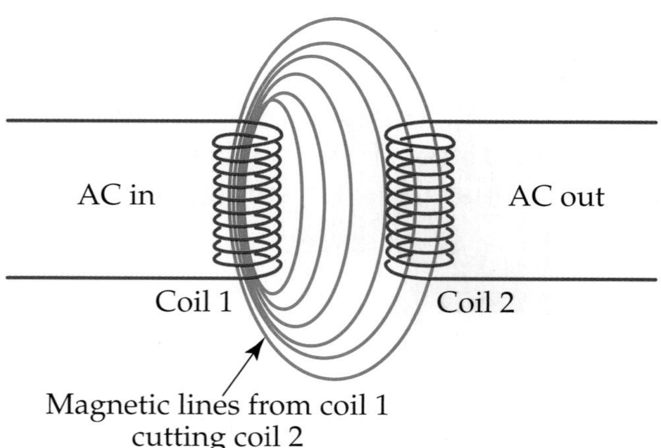

Figure 12-15. In this demonstration of mutual inductance, ac flow in coil *1* induces voltage across coil *2*.

Example 12-3:

Suppose two coils of wire have an inductance of 20 mH and 5 mH, and their coefficient of coupling is 75%. What is the mutual inductance of the two coils? Begin with the formula for mutual inductance:

$$M = K\sqrt{L_1 L_2}$$

We know the values of L_1, L_2, and K. Substitute these values into the formula.

$$M = 0.75\sqrt{20\text{ mH} \times 5\text{ mH}}$$
$$= 0.75\sqrt{100\text{ mH}}$$
$$= 0.75 \times 10\text{ mH}$$
$$= 7.5\text{ mH}$$

The mutual inductance of these two coils is 7.5 mH.

Transformers

The most common use for mutual inductance is in ***transformers.*** A transformer's job is to transfer electrical energy from one circuit to another by means of induction. It usually consists of a pair of coils wound around the same iron core or shell. The coils are wound on the same core to ensure that the magnetic coupling between the two coils is as high as possible. To help you get an understanding of how this is done, a simplified version of a transformer is shown in **Figure 12-16.** Actual construction is shown on the following pages.

The transformer winding connected to the energy source is called the ***primary.*** The winding with the induced energy is called the ***secondary.*** A transformer and its symbols are shown in **Figure 12-17.** When shown on printed circuit boards, the letter *T* is used.

There are numerous ways to construct transformers depending on the size and output performance required. There is a conflict on some of their names, but the types and uses will be discussed as follows.

One type of transformer is called a ***closed-form transformer,*** **Figure 12-18.** In a closed-form transformer, the core is a closed ring with the primary wrapped around one side

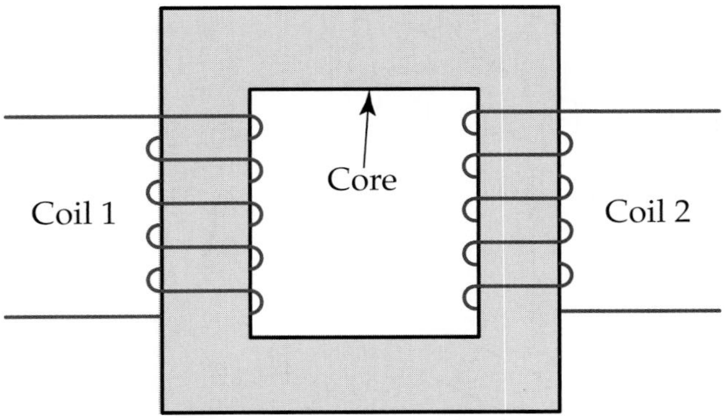

Figure 12-16. The basic idea for transformer design is a pair of wire coils wound around the same core.

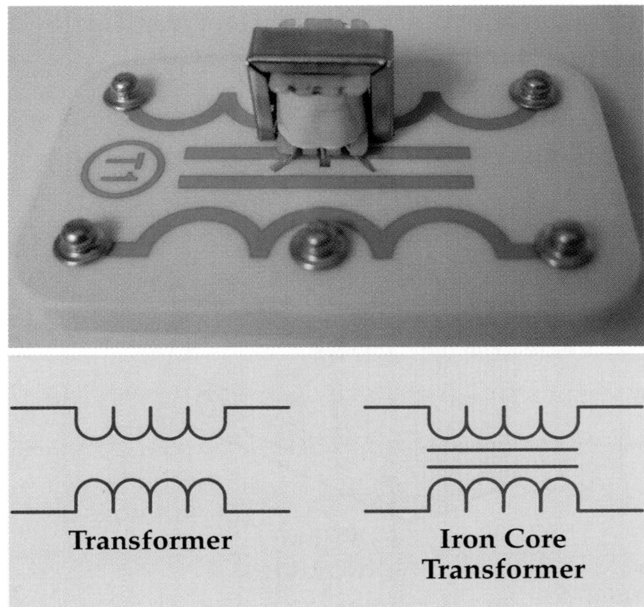

Figure 12-17. A transformer and two schematic symbols.

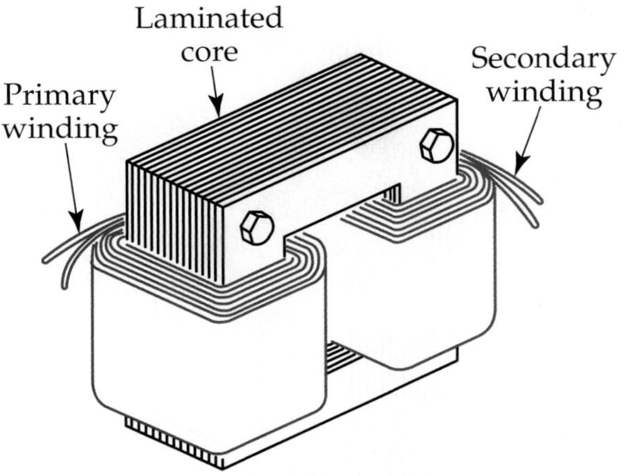

Closed-Form Transformer

Figure 12-18. This type of transformer is called *closed form.*

of it and the secondary wrapped around the other side. Some manufacturers insert layers of insulation between the windings to reduce the possibility of a shout circuit.

Another popular kind of transformer is the ***shell-form transformer,*** **Figure 12-19.** It is constructed by wrapping the primary around a cardboard core and then wrapping the secondary over the primary. Insulating paper separates each layer of windings from the other, producing a transformer with a high mutual inductance.

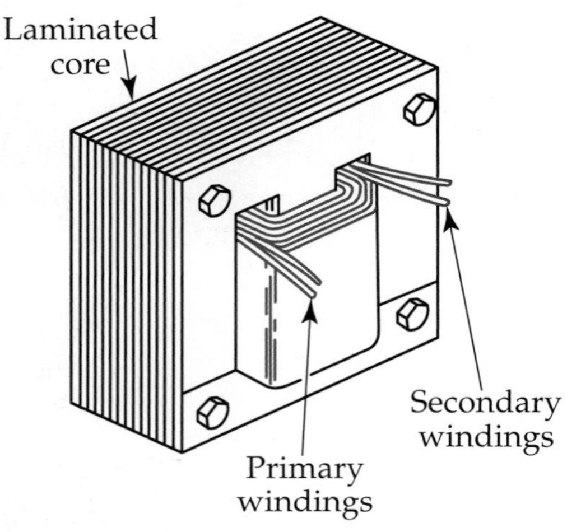

Figure 12-19. This type of transformer is called *shell form.*

The E-shaped pieces of the laminated core are inserted in the center of and around the outsides of the windings. The direction of the E-shaped and I-shaped pieces are alternated, **Figure 12-20.** Bolts that are insulated from all parts of the layers pull the stack together. Insulating the bolts and constantly reversing the E-shaped and I-shaped pieces further reduces eddy currents in the laminations.

Laminated Cores

Transformer cores are made of thin strips of metal that are laminated and stacked together. If a solid core were used in a transformer, or if the thin strips of metal in a laminated core were stacked one on top of the other without being laminated, a lot of heat would be produced. When heat is produced, energy is lost, so this heat must be avoided.

Heat is created in the transformer when eddy currents flow in the core because of induced voltage, **Figure 12-21.** As the voltage created by the magnetic field increases, the amount of heat produced by the eddy currents in the transformer increases. Thus, this means the greater the heat, the greater the energy loss. This energy loss is called *eddy current loss.*

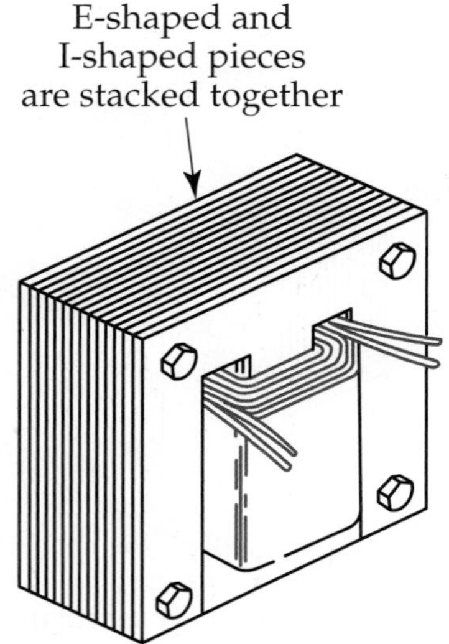

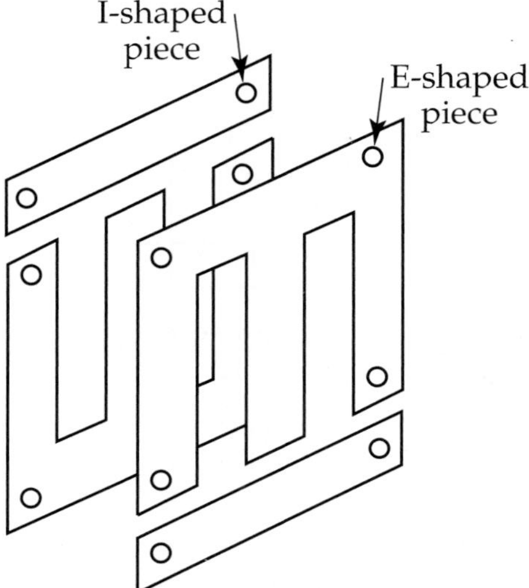

Figure 12–20. The core of a shell-form transformer consists of E-shaped and I-shaped pieces stacked together.

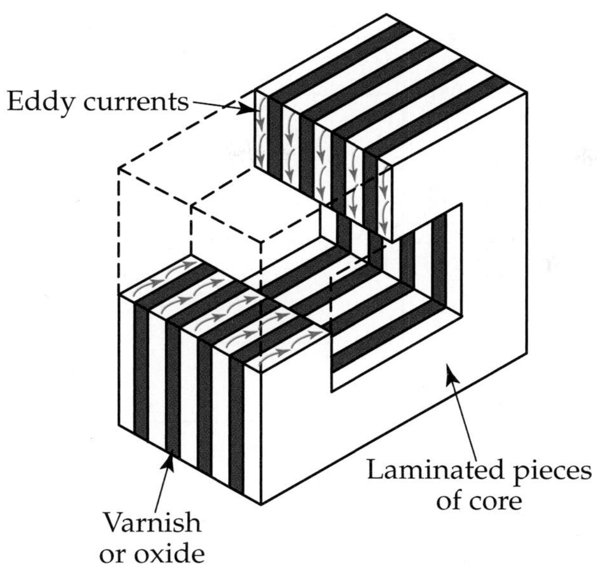

Figure 12-21. Alternating current flow in the transformer core breaks down into tiny loops or eddy currents.

Figure 12-22. Transformer windings made of copper wire. (Power Partners, Inc.)

Eddy currents cannot be eliminated. However, there are several things that can be done to reduce them. Reducing the thickness of the steel laminations that make up the core will reduce eddy currents. Thick laminations allow for larger loops and larger eddy currents. Another way to reduce eddy currents is to coat each metal strip with varnish or oxide, similar to capacitor construction. This insulation separates each piece in the stacked core.

Primary and secondary windings are made of copper wire, **Figure 12-22,** or aluminum strip, **Figure 12-23,** depending on the size of the transformer. The larger transformers, usually found on poles in local neighborhoods, use the aluminum strip. The wire is coated with an insulating material such as varnish. Notice that the layers of the windings must also be separated by an insulating material which is a thermally coated paper. The insulating material on the wire and between its layers is necessary to prevent short circuits in and between the windings.

The windings and cores are bolted or banded together. The larger transformers are put into cans, filled with cooling oil, and prepared for shipping. The cooling oil reduces heat after the transformer is put into operation. All these steps are necessary to keep the core material as close together as possible to reduce eddy currents. The insulation prevents short circuits. Reducing heat,

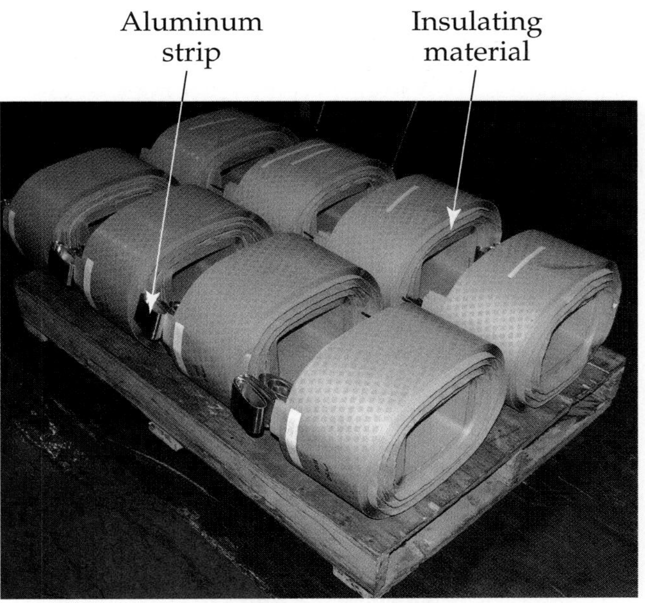

Figure 12-23. Transformer windings made of aluminum strips. (Power Partners, Inc.)

reducing eddy currents, and eliminating short circuits are the only ways to produce an efficient transformer.

Step-Up Transformers

Transformers are widely used in electronics because they can change the source voltage to the voltage required for a circuit. The transformer shown in **Figure 12-24** has 100 turns of wire in the primary winding and 600 turns in the secondary winding. Remember that the number of turns of wire in the coil of an inductor affects the amount of inductance. If we connect a 110-V ac source to the primary windings, 660 volts of ac are induced in the secondary. This is an example of a ***step-up transformer*** because the output voltage in the secondary winding is greater than the voltage in the primary winding. The relationship between voltage and the number of turns can be expressed by the following equation:

$$\frac{N_P}{N_S} = \frac{E_P}{E_S} \qquad (12\text{-}4)$$

where N is the number of turns and E is the voltage. Using the previous example, here is how we find the voltage in the secondary windings (E_S):

$$\frac{N_P}{N_S} = \frac{E_P}{E_S}$$

$$\frac{100}{600} = \frac{110 \text{ V}}{E_S}$$

Cross multiply by multiplying 100 by E_S and 600 by 110 V.

$$100 \times E_S = 600 \times 110 \text{ V}$$

$$100E_S = 66{,}000 \text{ V}$$

Now divide both sides of the equation by 100:

$$\frac{100E_S}{100} = \frac{66{,}000 \text{ V}}{100}$$

$$E_S = 660 \text{ V}$$

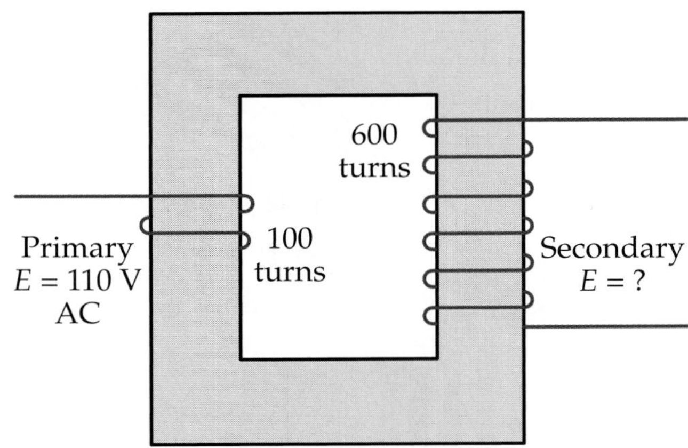

Figure 12-24. A simplified step-up transformer. The output voltage (E) is greater than the voltage in the primary since there are more turns in the secondary.

With 110 V in the primary, this transformer provides 660 V in the secondary. You can see by this increase why this is called a step-up transformer.

Example 12-4:

Suppose there are 150 turns in the primary winding and 750 in the secondary winding of a transformer. If the primary is carrying 80 V, what is the voltage of the secondary windings (E_S)?

$$\frac{N_P}{N_S} = \frac{E_P}{E_S}$$

$$\frac{150}{750} = \frac{80 \text{ V}}{E_S}$$

Cross multiply by multiplying 150 by E_S and 750 V by 80.

$$150 \times E_S = 750 \times 80 \text{ V}$$

$$150E_S = 60{,}000 \text{ V}$$

Divide both sides of the equation by 150:

$$\frac{150E_S}{150} = \frac{60{,}000 \text{ V}}{150}$$

$$E_S = 400 \text{ V}$$

The output of this transformer is 400 V.

Step-Down Transformers

Next, we will look at a ***step-down transformer,*** **Figure 12-25.** In a step-down transformer, the voltage in the secondary is less than the voltage in the primary. Suppose we apply 100 V to a primary winding of 80 turns and the secondary winding has 10 turns. We can use the same formula to find the output:

$$\frac{N_P}{N_S} = \frac{E_P}{E_S}$$

$$\frac{80}{10} = \frac{100\text{ V}}{E_S}$$

Cross multiply by multiplying 80 by E_S and 10 by 100 V.

$$80 \times E_S = 10 \times 100\text{ V}$$

$$80E_S = 1000\text{ V}$$

Divide both sides by 80:

$$\frac{80E_S}{80} = \frac{1000\text{ V}}{80}$$

$$E_S = 12.5\text{ V}$$

The secondary voltage is 12.5 V.

Example 12-5:

Suppose you have a transformer with 1000 turns on the primary windings and 250 turns on the secondary windings. The primary's voltage is 600 V. What is the voltage in the secondary winding (E_S)?

$$\frac{N_P}{N_S} = \frac{E_P}{E_S}$$

$$\frac{1000}{250} = \frac{600\text{ V}}{E_S}$$

Cross multiply by multiplying 1000 by E_S and 250 by 600 V.

$$1000 \times E_S = 250 \times 600\text{ V}$$

$$1000\ E_S = 150{,}000\text{ V}$$

Divide both sides of the equation by 1000.

$$E_S = \frac{150{,}000\text{ V}}{1000}$$

$$E_S = 150\text{ V}$$

The secondary voltage is 150 V.

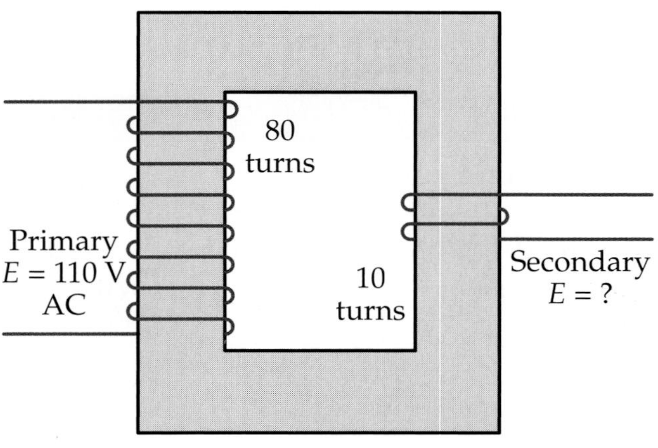

Figure 12-25. A simplified step-down transformer. The output voltage (E) is less than the voltage in the primary since there are less turns in the secondary winding.

The majority of the transformers that you will encounter in electronics will be step-down transformers. Circuitry with solid-state devices requires low voltages. Phone circuits use about 40 V, and battery chargers used for home tools use less than 20 V. Step-down transformers adjust the voltage to the proper amount for these devices.

Tapped Transformers

If you want to use a transformer that can cover a full range of voltages, you must use a tapped transformer. A ***tap*** is a lead that is connected or soldered to the wire in the secondary. It allows the turns ratio between the primary and secondary to be changed so that a voltage lower than the full voltage of the secondary can be used.

Look at the transformer in **Figure 12-26.** It is designed to provide voltages that range from 3 V to 12 V. The voltages increase in increments of 1.5 V. To achieve these increments, the transformer has taps that change the ratio of winding between the primary and secondary.

To understand how taps work, let us examine the turns ratio and output of the

Figure 12-26. Even small step-down transformers are used to deliver a range of voltages. Here is one with taps and an output from 3 V to 12 V.

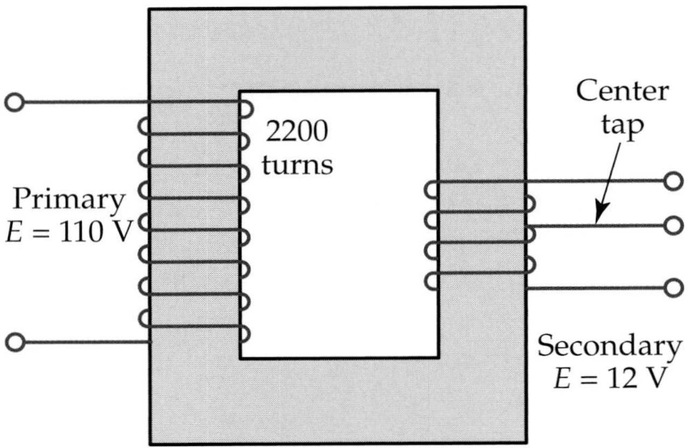

Figure 12-27. This transformer has a center tap that divides the number of turns in half.

tapped transformer in **Figure 12-27.** It has a 110-V primary consisting of 2200 turns of wire and a 12-V secondary with a center tap. The center tap divides the number of turns in half. First, let us calculate how many turns of wire are on the secondary. Substituting the variables of the formula with the known information, you would have

$$\frac{E_P}{E_S} = \frac{N_P}{N_S}$$

$$\frac{110\text{ V}}{12\text{ V}} = \frac{2200}{N_S}$$

$$110\text{ V} \times N_S = 12\text{ V} \times 2200$$

$$110\text{ V } N_S = 26{,}400\text{ V}$$

Divide both sides by 110 V.

$$\frac{110\text{ V } N_S}{110\text{ V}} = \frac{26{,}400\text{ V}}{110\text{ V}}$$

$$N_S = 240$$

The secondary of the transformer has 240 turns of wire.

Now, let us determine how many turns would be made before the center tap is soldered in place. The easy way to find this answer is to divide the number of turns on the secondary by two. The center tap would be attached after winding number 120. To determine the voltage at the center tap, divide the secondary voltage by two. The voltage at the center tap for this transformer is, therefore, equal to 12 V divided by 2. The center tap would be labeled 6 V. You could use the ratio formula to get the same answer.

$$\frac{N_P}{N_S} = \frac{E_P}{E_S}$$

$$\frac{2200}{120} = \frac{110\text{ V}}{E_S}$$

$$2200 \times E_S = 120 \times 110\text{ V}$$

Divide both sides by 2200 V.

$$\frac{2200\ E_S}{2200} = \frac{13{,}200}{2200}$$

$$E_S = 6\text{ V}$$

Example 12-6:

For the transformer in Figure 12-27, determine the number of turns that would need to be made before a tap would be soldered in place to achieve 3 V. Use the following formula:

$$\frac{E_P}{E_S} = \frac{N_P}{N_S}$$

Substitute the known values:

$$\frac{110\text{ V}}{3\text{ V}} = \frac{2200}{N_S}$$

Cross multiply:

$$110 \text{ V} \times N_S = 3 \text{ V} \times 2200$$

Divide both sides by 110 V:

$$\frac{110 \text{ V } N_S}{110 \text{ V}} = \frac{6600}{110 \text{ V}}$$

$$N_S = 60$$

The tap would be attached after winding number 60.

Transformer Phase Relationships

When the primary and secondary voltages are in-phase with one another, you have "like-wound transformers." To signify this, two dots are placed above the windings (both primary and secondary) of the schematic transformer symbol to show they are in-phase. If the dots were on opposite ends of the primary and secondary windings, you have "unlike-wound transformers." See **Figure 12-28.** This means that the primary and secondary are out of phase with one another. This relationship would show up on a dual trace scope.

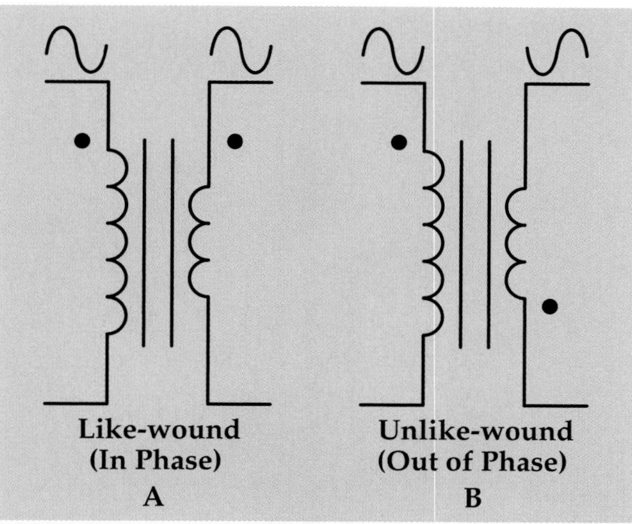

Figure 12-28. Two dots are placed above the windings of a transformer symbol to show they are in phase. When the dots are placed at opposite ends of the primary and secondary windings, the primary and secondary windings are out of phase.

Voltage-Current Relationships

Transformers are used in electrical circuits because they are very efficient. Almost all the power applied to the primary winding is transferred to the secondary winding, **Figure 12-29.** Recall from Chapter 5 that power is equal to the voltage times current ($P = EI$). We can take this formula and use it in transformer circuit analysis. If we say that transformers are 100% efficient and have no losses, then

$$P_P = P_S \qquad (12\text{-}5)$$

where P_P is the primary power and P_S is secondary power. Since $P = EI$, this equation is equal to

$$E_P I_P = E_S I_S \qquad (12\text{-}6)$$

If the power on the primary winding (P_P) of a transformer is 600 W and the voltage of the secondary (E_S) is 150 V, what is the current of the secondary? Use the following formula:

$$P_P = P_S$$

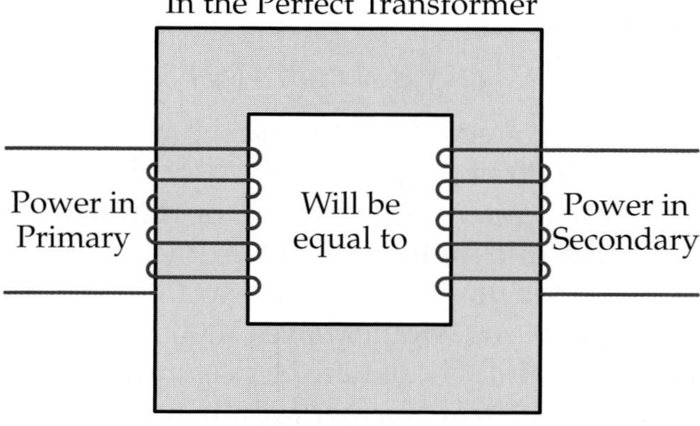

Figure 12-29. Since transformers are very efficient, you can assume that power applied to the primary equals power produced by the secondary.

Since $P_S = E_S I_S$, you can replace P_S with $E_S I_S$:

$$P_P = E_S I_S$$

You know P_P and E_S, so you can solve for I_S:

$$600 \text{ W} = 150 \text{ V} \times I_S$$

$$\frac{600 \text{ W}}{150 \text{ V}} = \frac{150 \text{ V} \times I_S}{150 \text{ V}}$$

$$4 \text{ A} = I_S$$

The current in the secondary windings is 4 A.

Example 12-7:

Suppose the power in the primary windings of a transformer (P_P) is 850 W and the current in the secondary (I_S) is 5 A. What is the voltage in the secondary windings (E_S)? Use the power formula:

$$P_P = P_S$$

Since $P_S = E_S I_S$, you can replace P_S with $E_S I_S$:

$$P_P = E_S I_S$$

You know P_P and I_S, so you can solve for E_S:

$$850 \text{ W} = E_S \times 5 \text{ A}$$

$$\frac{850 \text{ W}}{5 \text{ A}} = \frac{E_S \times 5 \text{ A}}{5 \text{ A}}$$

$$170 \text{ V} = E_S$$

The voltage in the secondary windings (E_S) is 170 V.

High-Voltage Transformers

High-voltage transformers have ratings of over 24,000 volt-amperes (VA) of electricity. The ***volt-ampere*** is the result of multiplying the volts and amps on the primary of the transformer. If the primary was 12,000 volts at 2 amps, you would have 24,000 VA. This particular unit is called a 24 kVA transformer. This type of transformer would not be used in school power supplies or around your home. High-voltage transformers are found in industries where large amounts of power are needed.

High-voltage transformers are also found in substations, **Figure 12-30.** The transformers on power poles, **Figure 12-31,** are used to step down the voltage from 1200 V to 240 V and 120 V for use in businesses and homes.

Figure 12-30. The transformers in this substation supply the power needs of a community.

Figure 12-31. Three power transformers are mounted on a pole. The transformer windings are placed inside a protective container filled with cooling oil.

Practical Application 12-1: Using a Current Clamp

One major problem with measuring current is opening the circuit to insert a meter. In some cases, this can be very dangerous. There is another solution to measuring circuit current that involves the principles of induction and transformer action. A *current clamp*, also known as a *current probe* can be placed over the wire. All you have to do is squeeze the clamp, place it over the wire and release.

The magnetic field from the wire will cut through the clamp. This induces a current that travels through the leads, is measured, and then is displayed on a digital readout. You can measure current in the range of 0.1 A to 300 A with this current clamp.

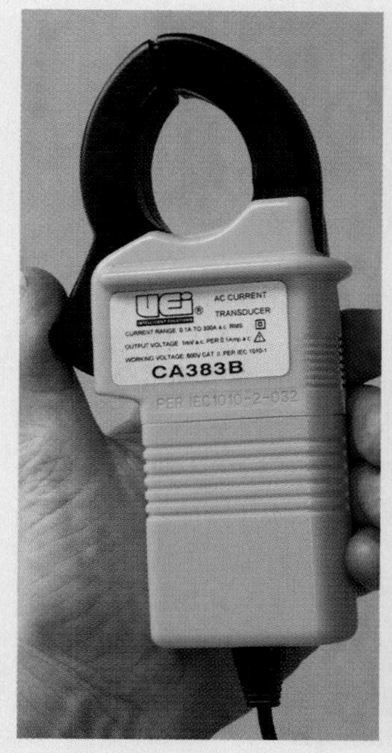

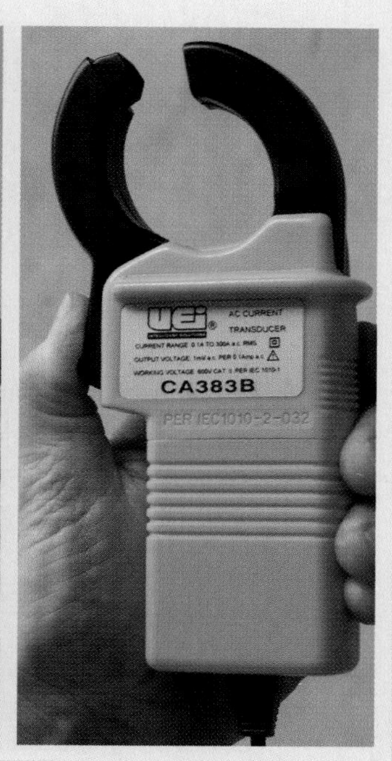

Project 12-2: Repulsion Coil

You can demonstrate induction and transformer action by building the repulsion coil shown.

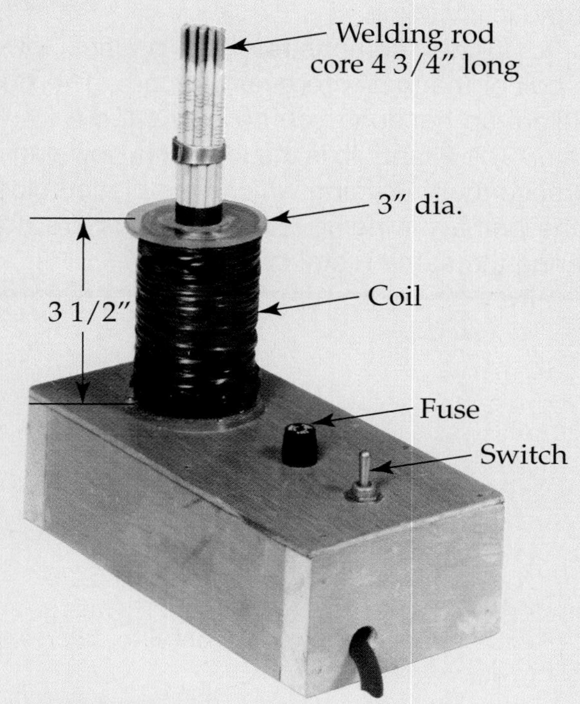

(Project by Perry Jossefides)

Gather the parts listed in the table. Then, assemble the project by following procedure and using the schematic.

Procedure

1. Cut the plastic or fiber ends, and glue them to tubing to make a coil form 3″ in diameter and 3 1/2″ long.
2. Wind 2 1/2 lb. of No. 22 magnet wire on the coil form. Wrap the wire in even layers.
3. Cut welding rods (coat hangers, 16–18 gauge steel wire, or a large bolt) to 4 3/4″ lengths.
4. Tape the rods together or use epoxy at both ends to hold the individual rods together and form the core of the repulsion coil.
5. Make a base to fit your needs.
6. Mount the coil on the base with wood screws.
7. Attach the white wire, black wire, switch, and fuse holder to the coil. See the schematic.
8. Attach the lamp cord to the fuse holder.

Project 12-2: Repulsion Coil *(Continued)*

9. Wrap 30′ of 18–22 gauge magnet wire around a 2″ pipe. Tape the wire together and remove the pipe.
10. Solder the lightbulb to the coil.

When the project is completed, plug the lamp cord into a 110-V electrical outlet. Drop the ring of aluminum tubing over the core welding rods and turn on the switch for the project. The ring will shoot up in the air in a surprising demonstration of inductance. The energized coil induces a current in the aluminum ring. A magnetic field is produced by the induced current. This magnetic field interacts with an opposing field in the core which repels the ring with such a force, it flies into the air. If you were able to balance a heavy ring that somewhat matches the opposing force, you would have taken the first step toward magnetic levitation.

To further demonstrate the project, lower the coil of magnet wire over the core. The bulb will light up because you have created a transformer. The flashbulb is the load. It is now part of the secondary winding which was placed close to the primary winding. Both windings are now sharing the same metal core.

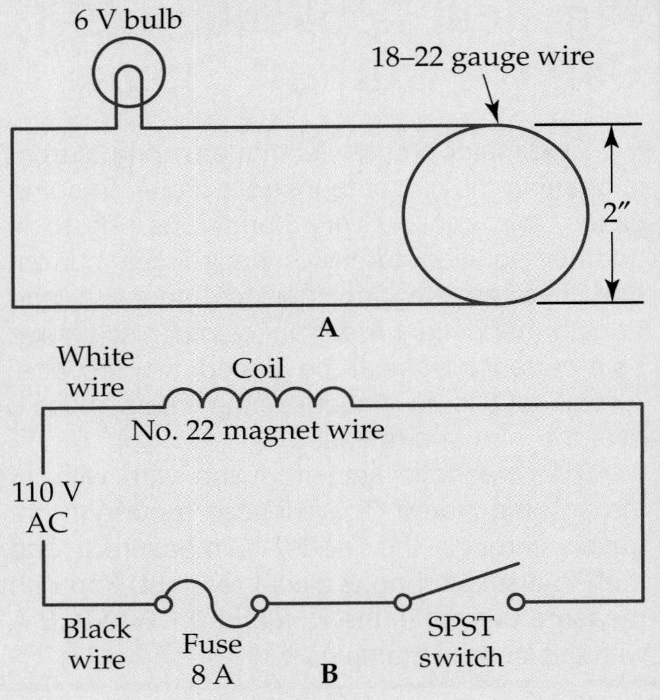

No.	Item
1	Plastic or fiber tubing, 3 1/2″ long with a 3/4″ to 1″ inside diameter
2	Plastic or fiber ends for tubing, 3″ diameter
20	Welding rods, cut to 3/4″ lengths
1	Toggle switch, 115-V, 15-A
1	Fuse, 8 A
1	Fuse holder
1	Male plug and lamp cord
2	Hookup wires, 4′ long
1	Magnet wire, 2 1/2-lb. No. 22
1	Lightbulb, 6 V
1	Magnet wire, 18–22 gauge, 30′ long
1	Aluminum tubing, 1 1/2″ diameter
	Wood for base

Summary

- Voltage can be induced in a coil of wire either by passing a magnet through the coil or by holding the magnet stationary and moving the coil over it.
- Voltage can be induced in a wire by changing the magnetic field.
- An inductor is a device that resists a change in current.
- Inductance is a measurement of how much an inductor opposes current change.
- Inductance is affected by the number of turns of wire in a coil, the type of core, the cross-sectional area of the core, and the length of coil compared to the number of turns of wire.
- When two coils of wire are close enough to be linked by a magnetic field, they are said to have mutual inductance.
- The formula for mutual inductance is $M = K\sqrt{L_1L_2}$.
- Transformers use the principle of mutual inductance to increase or decrease the amount of voltage and current in a circuit.
- The ratio between the numbers of turns equals the ratio between the voltages in the primary and s econdary windings of a transformer:

$$\frac{N_P}{N_S} = \frac{E_P}{E_S}$$

Test Your Knowledge

Do not write in this book. Write your answers on a separate sheet of paper.

1. Voltage created by a changing magnetic field cutting through a coil is called ______.
2. The opposition to a change in current flow is called ______.
3. An inductor will have no effect on ______ circuits.
4. The part of a circuit that opposes a change in current is called a(n) ______.
5. Draw the symbols for a fixed and variable inductor.
6. The unit of inductance is the ______.
7. Name three things that affect the inductance of a coil.
8. If we had three inductors connected in series, each having a value of 4 mH, what would the total inductance be? Show your work.
9. Name the two windings found on a transformer.
10. Is ac or dc produced in the secondary winding of a transformer?
11. Laminations on the core of a transformer help cut down on the ______ loss.
12. The two types of transformers that get their name from how they affect the voltage are ______ and ______.
13. A transformer connected to a 120-V ac source has 400 turns on the primary winding. If we want to obtain 6 V from the second winding, how many turns of wire do we need for the secondary? Show your work.

Activities

1. Visit a plant that makes transformers or inductors. How do they wind the coils? Why do these components cost more than resistors?
2. Make a list of the places you find transformers around your house.
3. Some neighborhoods have a device that looks like a large gray can attached to their telephone poles. These are transformers. What voltages do they produce?
4. Why do industrial plants operate so much of their equipment at 460 V instead of 120 V?
5. On what voltage do the large electromagnets found in junkyards operate?
6. Why do power companies clear the area under all the high tension transmission lines?
7. How is it possible to run power lines underground without fear of electrocuting people walking over them?
8. The coil on a car is another type of inductor. What would it look like if you cut it into two pieces?
9. Can you make a radio without an inductor?
10. What kind of wire is needed to make an inductor or transformer? What would happen if the wire were not coated with insulation?
11. How are transformers encapsulated?
12. If the core of a transformer came apart, what would it sound like? What would happen to the transformer?

Motors

Learning Objectives

After studying this chapter, you will be able to do the following:

- Draw the symbol for a motor.
- Explain how a motor converts electrical energy into mechanical energy.
- Contrast induction motors with synchronous motors.
- Calculate the number of poles required for a given rpm of a synchronous motor.

Technical Terms

field magnets
induction motor
mechanical energy
motor
slip
squirrel cage induction motor
synchronous motor
torque

In the 1830s, Michael Faraday experimented with the interaction of magnets. From his experiments he was able to state two basic principles of magnetism: Like poles repel and unlike poles attract. These basic principles of magnetism are the basis of operation for the electric motor. This chapter covers these principles, along with motor types and their applications.

What Is a Motor?

A ***motor*** is a rotating machine that converts electrical energy into mechanical energy. **Figure 13-1** gives the symbol used to represent a motor in electrical drawings.

Like generators, motors are made up of coils (windings), brushes, an armature, and a commutator, **Figure 13-2.** However, a motor uses these parts to create mechanical energy, not electrical energy.

Mechanical energy can be defined as the ability to do work. For example, a motor turns the agitator in a washing machine, the drum in a clothes dryer and the blades of a fan. Motors drive much of the machinery found in industry. Today, electric motors provide power for producing most of the mechanical energy used in the U.S. and are used to make everyday tasks easier, **Figure 13-3.**

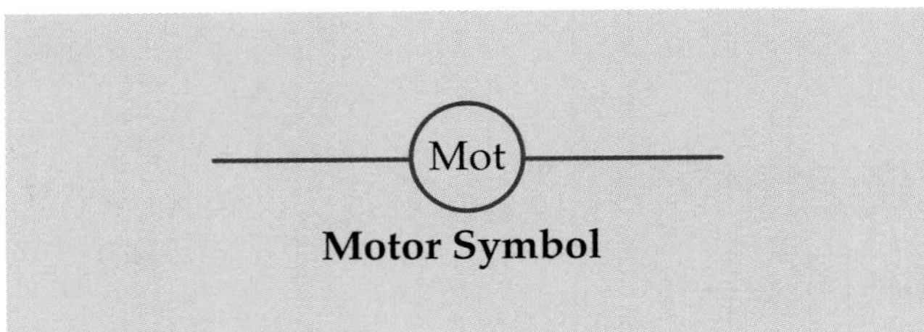

Figure 13-1. This symbol is used to designate motors in electrical diagrams.

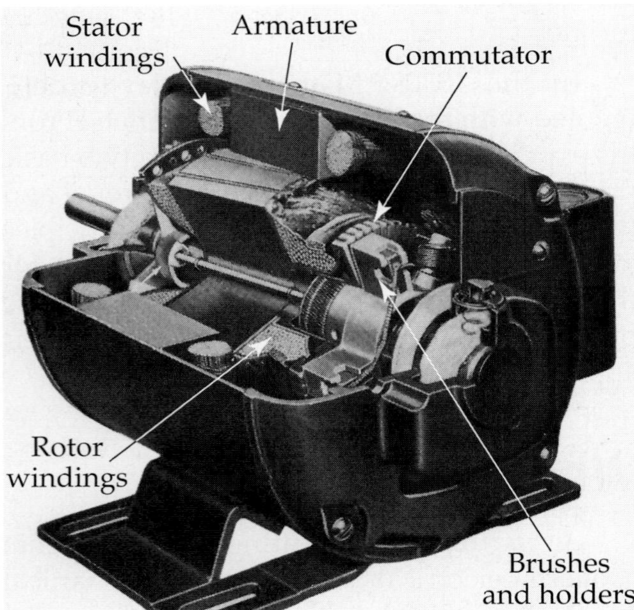

Figure 13-2. Cutaway view shows major internal parts of a heavy-duty electric motor.

Figure 13-3. DC motors play an important part in the movement of material and equipment. Batteries allow them to move without being attached to an electrical cord. This dc cart moves large numbers of shopping carts from the parking lot to the store.

Basic Motor Operation

A motor and a generator work in a similar manner. The basic difference is that electrical energy (current) is fed into the motor to produce mechanical energy—turning the armature. The magnetic principles of repulsion and attraction are used to turn the armature. Let us look at how this is so.

There are two ways you can make a magnet turn, which are shown in this simple experiment using two magnets. Lay down the first magnet so that it is free to move around a pivot point, **Figure 13-4.** Then bring the north pole of the second magnet close to the north pole of the first magnet. Since like poles repel, the first magnet will turn away from the second. This also happens if two south poles are used in the experiment.

Another way of making a magnet turn is by setting it up as before. However, instead of using like poles, bring the north pole of the second magnet close to the south pole of the first magnet. Since unlike poles attract, the first magnet is attracted to and turns toward the second magnet, **Figure 13-5.**

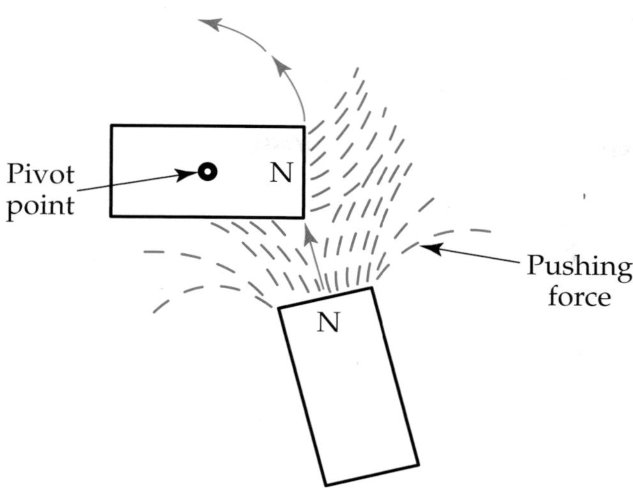

Figure 13-4. Like poles of two magnets repel each other. If one magnet is pivoted at the center, the pushing force from the other magnet can cause it to turn.

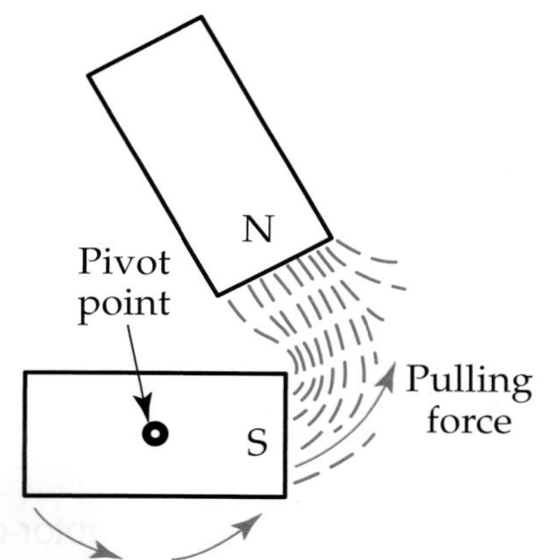

Figure 13-5. Unlike poles of two magnets attract each other. If one magnet is pivoted at the center, the pulling force from the other magnet can cause it to turn.

A coil can be substituted for the first magnet and a pair of magnets for the second, as shown in **Figure 13-6.** A coil of wire is wrapped around a pivoting metal bar, and a current is passed through the coil. This creates a magnetic field around the coil.

When the coil and its magnetic field are placed between the opposite poles of a pair of magnets, the coil will turn. This turning is the result of either attraction or repulsion,

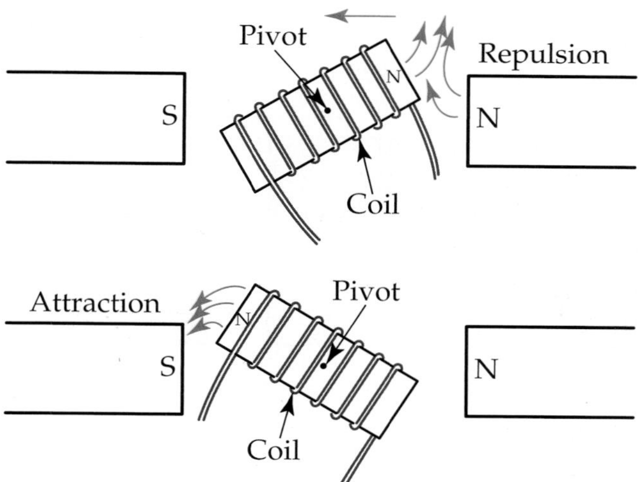

Figure 13-6. To strengthen the magnetic field and improve the turning action of the pivoting magnet, coils of wire are wrapped around the metal bar to form an electromagnet.

depending on the polarity of the magnets. As you can see, the laws of attraction and repulsion are the basis of electric motor operation.

Classifying Motors

There are hundreds of different kinds of motors and thousands of different uses for motors. Therefore, there are many different ways of classifying motors. One way to classify them is as dc or ac, **Figure 13-7.** (Keep in mind that the motor types listed in Figure 13-7 are not complete.)

> **Note**
>
> Single-phase ac power is available from home outlets. Polyphase means more than one phase.

Another way to classify motors is by size. Some manufacturers classify them according to the following sizes:

- Mini—small fractional size motors of less than 1 horsepower.
- Sizes less than 37 1/2 horsepower.
- Sizes above 37 1/2 horsepower.

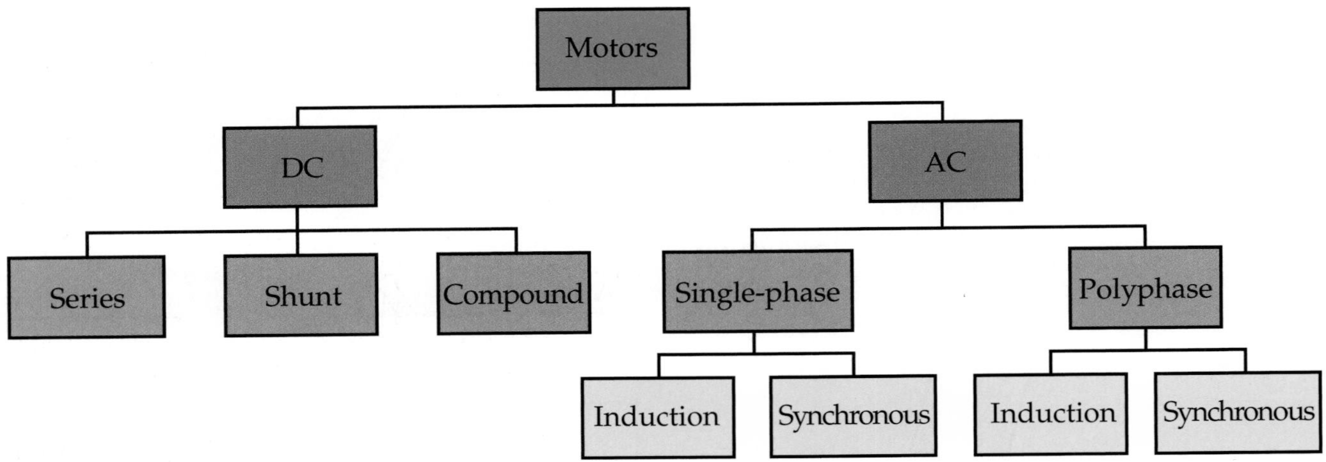

Figure 13-7. There are many ways to classify motors. In this example, motors are classified as either dc or ac.

You might ask why we must classify motors. Since there are hundreds of different kinds of electric motors, you must pick the one way to classify motors which best suits your needs. Motors are courses unto themselves in the electrical and electronics fields. In this chapter, we cover only three of the most commonly used motors: dc, induction, and synchronous.

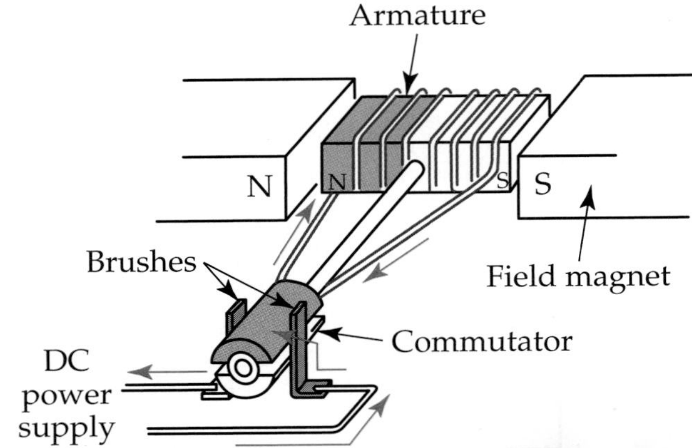

Figure 13-8. A simple dc electric motor can be formed by connecting the leads of the coil in the armature to the commutator. Fixed brushes carry dc electricity to and from the commutator.

DC Motors

To relate the attraction/repulsion experiment directly to electric motor operation, we will start with a simple dc motor, **Figure 13-8.** When a current runs through this circuit, it goes into the commutator through one brush, to the coil, back to the commutator, and out the second brush. When this happens, the magnetic field of the coil opposes the magnetic field of the pair of magnets, called ***field magnets.*** The repulsion of like poles makes the armature turn, **Figure 13-9.** As the armature turns, its north pole is attracted to the south pole of the field magnet. Just as the two unlike poles are about to line up, the commutator reverses the polarity of the armature, **Figure 13-10.** Again, the two like poles repel each other, and the armature continues to spin.

The dc motor in the previous figures has been simplified to make it easier to trace the current and see how the magnetic fields make the motor spin. However, instead of using a single coil, most motors have several sets of coils in the armature. More coils enable the armature to start more smoothly and to produce more torque. ***Torque*** is a measure of the force that creates a spin or rotation on the axis of a motor. The unit of torque in the English system is a foot-pound (ft-lb). In the international system of units (SI system), it is

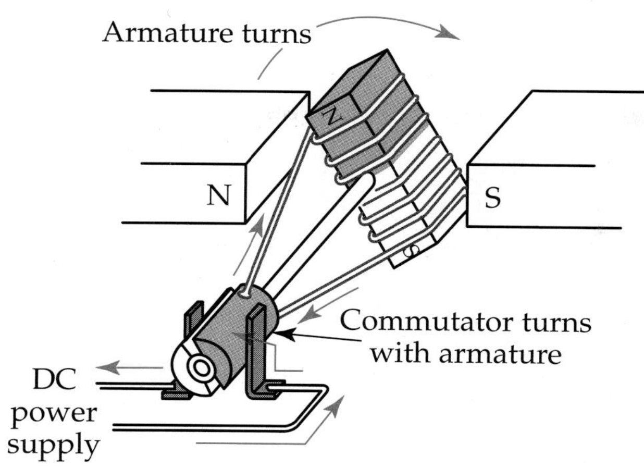

Figure 13-9. Like poles repel each other, causing the armature to start to turn. Once the armature completes a quarter turn, the unlike poles attract each other and the armature continues to turn.

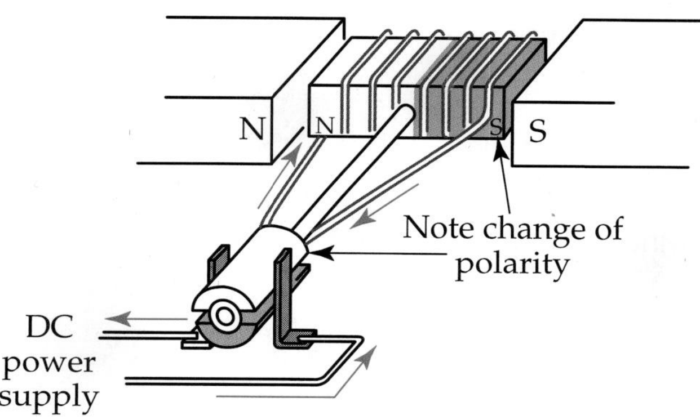

Figure 13-10. After one-half turn of the armature, its poles are reversed by the commutator and the repelling/attracting process is repeated. Note how the split commutator changes the polarity of the armature.

the Newton-meter (N•m). It is important to have the appropriate torque for the load on the motor.

Induction Motors

A second major type of electric motor is the ***induction motor.*** It does not use a commutator or brushes. Instead, it is constructed with a rotor that spins inside a stator.

The *stator* consists of a winding or windings on a laminated steel core. The *rotor* is made of laminated steel and has copper bars that fit into slots on the steel core. Two end shields are fitted with bearings to support the rotor shaft. This particular construction is called a ***squirrel cage induction motor,* Figure 13-11.**

A magnetic field is created in the stator by passing alternating current through its windings. Since the poles of the stator change polarity due to the changing direction of the alternating current, the magnetic field produced in the stator appears to revolve around the rotor.

This magnetic field also induces a current in the copper bars of the rotor. The rotor, in turn, becomes an electromagnet, and its magnetic field is attracted to the field of the stator. This attraction between the rotor and stator created by induction is what causes the motor to spin.

You can see in **Figure 13-12,** that the north pole created in the stator winding attracts the south pole of the rotor winding. As the magnetic field revolves around the stator windings, it exerts a magnetic pull on the rotor, causing it to turn. Note in **Figure 13-13** that the rotor has turned, while the stator has not. Eventually, the rotor will be pulled in a full circle, and it will begin its revolution again.

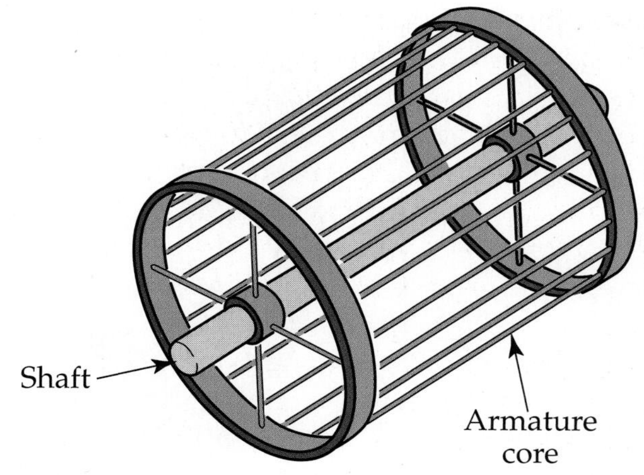

Figure 13-11. A squirrel cage rotor serves as the core for an induction motor. Coils of wire are wrapped on the core to form the moving part of the motor.

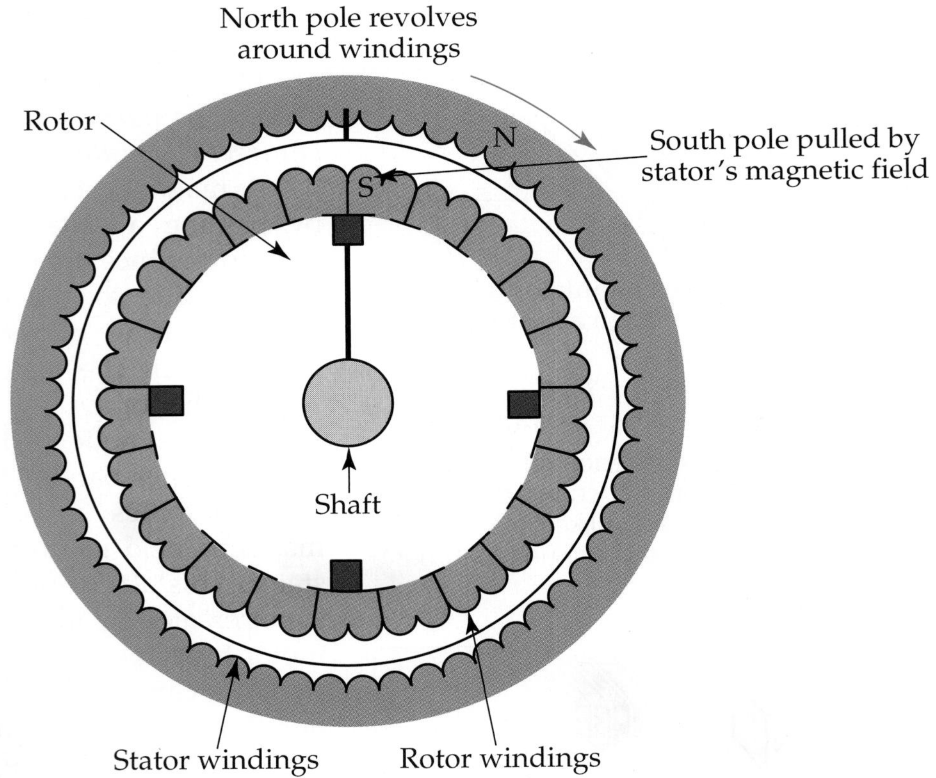

Figure 13-12. In an induction motor, the magnetic field around the stator windings creates a north pole, which, in turn, attracts the south pole of the rotor windings, causing the rotor to turn.

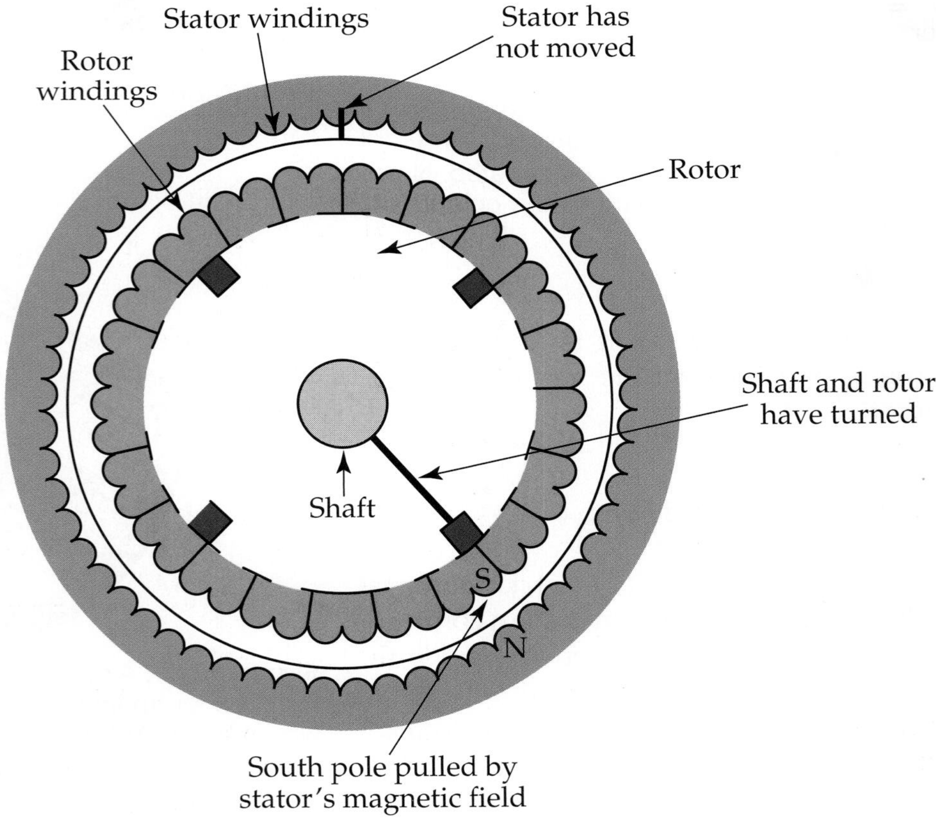

Figure 13-13. The magnetic field revolving around the stator windings moves just ahead of the rotating rotor.

Since the rotor can turn on its bearings, it starts to rotate just behind the magnetic field of the stator. These magnetic fields follow one another around the squirrel cage, with the rotor always behind the magnetic field of the stator. If the rotor matched the rotating field of the stator, the conductors in the rotor would be standing still with respect to the rotating field. If this happens, no voltage would be induced in the squirrel cage and there would be no current and no magnetic poles. There would, therefore, be no attraction between the rotor and the rotating field in the stator.

For proper operation of an induction motor, the rotor must revolve slightly behind the rotating field in the stator so the magnetic field can cut through the rotor conductors. This difference is called ***slip.*** Without slip, the rotor will not turn. For most purposes, the squirrel cage induction motor is considered to be a constant speed device. If the load on the rotor is increased, such as when making it turn a heavy weight, there would be a greater slip and the motor would run more slowly.

There are a number of different types of induction motors. Some are designed for low torque and are used in electric fans. However, those used in washing machines, appliances, and clothes dryers, need a higher starting torque.

Synchronous Motors

A third type of motor is the ***synchronous motor.*** You now know from induction motors that as you pass a current through the stator, you set up a magnetic field. The rotor, in turn, is pulled along by this magnetic field with some slip. The synchronous motor operates in a similar manner, **Figure 13-14.** However, the rotor spins in exact synchronization with the rotating magnetic field of the stator. The rotating magnetic field of a synchronous motor, therefore, "locks onto" the rotor, and the magnetic field and rotor turn at the same speed. The motor's speed then becomes predictable.

Figure 13-14. Arrows show how a synchronous motor works. The rotating magnetic field locks onto the rotor and turns it at a natural speed.

The synchronous motor gets its name from a term used to describe the natural speed of the rotating magnetic field of the stator. In the U.S., the natural speed is tied to the frequency of applied ac power. The power companies regulate this frequency, keeping it at 60 Hz.

A synchronous motor only operates at one speed (revolutions per minute). The speed is determined by two things: frequency of the sine wave coming into the stator and number of magnetic poles used in the rotor. You will learn how to calculate the speed in a later paragraph.

By building synchronous motors to operate at a specific constant speed, they are useful for many applications. For example, synchronous motors can be found in various timing devices, such as electric clocks. Since power companies regulate frequency at 60 Hz, these clocks keep very accurate time. The speed of large horsepower synchronous motors are so stable, they are used to drive dc generators. Synchronous motors are also used as timer motors in pumps, fans, mixers, and refrigerators.

Each motor runs only at the speed for which it is built. To get a synchronous motor to operate at another speed, the manufacturer must produce one with a different number of magnetic poles.

The speed of a motor is measured in revolutions per minute (rpm). The formula for calculating the speed of a synchronous motor is

$$\text{rpm} = \frac{f \times 120}{P} \tag{13-1}$$

where f is the frequency and P is the number of poles in the motor.

Since the frequency of applied ac power in the U.S. is 60 Hz, all we have to do to change the rpm of a synchronous motor is change the number of poles. Suppose that a motor has two poles. We can find its speed using this formula.

$$\begin{aligned}\text{rpm} &= \frac{f \times 120}{P}\\ &= \frac{60 \times 120}{2}\\ &= \frac{7200}{2}\\ &= 3600\end{aligned}$$

This means that a two-pole synchronous motor will run at a constant speed of 3600 rpm. Synchronous motors do not have variable speeds. If a certain speed is necessary, the number of poles needed in the motor can be calculated, also using this formula. To get a motor that will run at 1200 rpm, calculate how many poles would have to be built into the motor by using the same formula.

$$\text{rpm} = \frac{f \times 120}{P}$$

Multiply both sides by P and divide both sides by rpm:

$$\frac{P \times \text{rpm}}{\text{rpm}} = \frac{f \times 120}{P} \times \frac{P}{\text{rpm}}$$

Anything divided by itself is 1, so the "rpm/rpm" on the left side of the equation becomes 1, and the "P/P" on the right side of the equation becomes 1.

$$P = \frac{f \times 120}{\text{rpm}}$$

Insert the known values:

$$\begin{aligned}P &= \frac{60 \times 120}{1200}\\ &= \frac{7200}{1200}\\ &= 6 \text{ poles}\end{aligned}$$

Therefore, to get 1200 rpm from a synchronous motor, six poles would have to be built into it. Synchronous motors always have an even number of poles, ranging from 2 to 90 poles. A 90-pole motor of this type would run at 80 rpm.

Example 13-1:

Suppose you have a 36-pole synchronous motor. At what speed does it run?

Start with the formula for finding rpm:

$$\begin{aligned}\text{rpm} &= \frac{f \times 120}{P}\\ &= \frac{60 \times 120}{36}\\ &= \frac{7200}{36}\\ &= 200\end{aligned}$$

This motor runs at 200 rpm.

Example 13-2:

Suppose you need a motor that runs at 900 rpm. How many poles should the motor have?

Start with the modified formula that has P alone on the left side of the equation.

$$P = \frac{f \times 120}{\text{rpm}}$$

Insert the known values.

$$\begin{aligned}P &= \frac{60 \times 120}{900}\\ &= \frac{7200}{900}\\ &= 8 \text{ poles}\end{aligned}$$

A motor with 8 poles is needed.

Motor Maintenance and Troubleshooting

One of the major causes of electric motor failure is bearing wear. Since motors must spin freely, it is important that the bearings have proper lubrication. Some bearings need just two or three drops of oil. If you use too much oil, it will attract dirt and dust that can keep the bearings from turning freely. Prelubricated bearings do not need any oil at all. These bearings usually have the right amount and kind of lubricant sealed inside.

Another major cause of electric motor failure is worn brushes. Brushes are usually made of a carbon material. Because of this, the brushes tend to wear out and must be replaced. If the brushes are not making proper contact with the commutator, the motor will have a weak starting torque. Worn brushes, a weak brush spring, or brushes sticking in their holders can also cause bad contact.

Brushes usually are easy to remove. Many can be taken out by using a screwdriver to loosen the brush holder, **Figure 13-15.** Most brushes are spring loaded to keep them in contact with the commutator. Be careful not to lose these springs when you remove the brushes. Insert the replacement brushes and tighten the cover on the brush holder. Some hand tools have to be taken apart to gain access to the brush holder.

Warning

If you replace brushes in an electric motor, avoid electric shock by unplugging the power cord before you start to work. Brushes carry the current that makes motors spin.

A dirty commutator can cause poor electrical contact. The commutator is made from copper. It should be cleaned with very fine sandpaper, around 1200 grit, or you may have shorting problems later. When a commutator needs to be cleaned, it has a dull, blackened look. However, as it is cleaned, it should regain a shiny appearance.

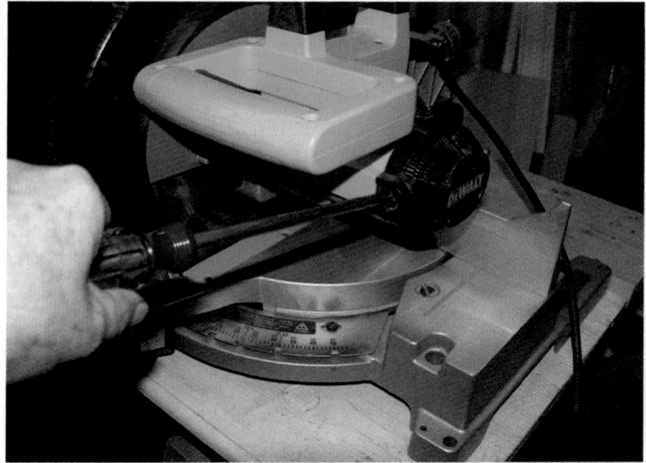

Figure 13-15. Brushes can be removed by loosening the brush holder with a screwdriver.

A number of different troubles can cause a motor to fail to start. A voltmeter should be used first to see if there is voltage at the motor terminals. If there is no voltage at the terminals, a part other than the motor is the problem. Other voltage checks can be made without first taking the motor off the machine. Some motors have to be lifted by means of a crane because of their size. So, doing these preliminary checks with the motor in place can save a lot of time and work.

When replacing a motor, replace it with a motor of the same specifications, such as size, horsepower, and voltage. This information is listed on the nameplate attached to the motor cover, **Figure 13-16.**

Figure 13-16. The nameplate on a motor contains information about the characteristics of the motor, such as its rated voltage and frequency.

Motor Selection

To find the right motor for an application, first consider what the motor is required to do. For example, is the motor going to turn the drum of a clothes dryer or a heavy conveyor belt? Then, determine the torque (twisting force) and speed needed for the load. Start narrowing down the available types of motors to one that meets the specific needs. Begin with the most general classification: ac or dc. From there, move to more specific levels of classification until the type of motor needed is clear.

The following examples are provided to show just some of the requirements or specifications you will need to know to select a suitable motor. This is *not* an all-inclusive list. These examples are given to show you that motor selection can be a difficult task for designers, engineers, and other users.

- Acceleration time.
- Bearing type and load specifications.
- Breakaway torque.
- Continuous load demand.
- Horsepower.
- Operating cost.
- Overload demands.
- Speed.
- Voltage and current.

To understand the importance of these features, consider the second item on the list. Is the motor going to be in a dusty environment or one subject to water or rain? Can the bearings be open or must they be sealed? Will the bearings be subject to quick starts and rapid stops with a heavy load? Will the bearing be used in a location that has temperatures at 40° below zero, or will they be used in the desert. All of these questions are important in the selection of the proper motor.

Summary

- A motor is a rotating machine that converts electrical energy into mechanical energy.
- The main parts of a motor include brushes, a commutator, an armature, and windings.
- The laws of attraction and repulsion are the basis for the operation of electric motors.
- Motors can be classified in many ways because there are so many different types of motors.
- More coils in the armature enable the armature to start more smoothly and to produce more torque.
- Torque is a measure of the force that creates a spin, or rotation, on the axis of a motor.
- Induction motors have a current-carrying stator with a rotating magnetic field, which attracts the magnetized rotor and makes the rotor spin.
- The rotor of a synchronous motor spins at a speed that matches the rotating speed of the stator.
- The speed of a synchronous motor can be calculated through the following formula:

$$\text{rpm} = \frac{\text{frequency} \times 120}{P}$$

- The number of poles in a synchronous motor that runs at a given rpm can be calculated using the following formula:

$$P = \frac{\text{frequency} \times 120}{\text{rpm}}$$

Test Your Knowledge

Do not write in this book. Write your answers on a separate sheet of paper.

1. Draw the symbol for a motor.
2. A motor converts ______ energy into ______ energy.
3. Which magnetic principles are used to turn the armature of a motor?
4. What happens when current flows through the armature windings of a dc motor?
5. What is the purpose of having several sets of coils in the armature of a motor?
6. A(n) ______ motor does not have brushes or a commutator.
7. The rotor of a(n) ______ motor spins in exact synchronization with the rotating magnetic field of the rotating magnetic field of the stator.
8. Why do motors have a varied number of poles?
9. Find the rpm of a motor that has four poles and is operated on 60 Hz. Show your work.
10. Name one of the major causes of motor failure.

Activities

1. Get an old motor and take it apart. Label the various parts.
2. Why are brushes for motors made from carbon (the same material used for resistors)?
3. Make a list of places in your home where you use motors. Be careful. Motors may be hidden in places you might not think about.
4. Research to find the specific sizes of the largest and smallest motors made. Where are they used?
5. Generators on cars have been replaced by alternators. What is the big advantage of an alternator?
6. Many motors have sealed bearings. How are they made? What makes them so different from regular bearings?
7. Many electrical devices are laminated. What holds them together? List some of the assemblies that are laminated.
8. The alternator on an automobile engine uses a three-phase system. Get the electrical diagram of such a system and see if you can figure out how it works. Where else do you find three-phase electrical systems?

In this chocolate fountain, an electric motor is used to drive six pumps, which circulate two tons of chocolate.

Reactance and Impedance

Learning Objectives

After studying this chapter, you will be able to do the following:

- Draw graphs to illustrate the differences between leading and lagging voltage and current in inductive and capacitive circuits.
- Describe the principle of vector addition.
- Calculate the value for inductive reactance, capacitive reactance, and impedance in a circuit.
- Explain the concept of transformer loading and its effect on the secondary.
- Calculate the time required to charge a capacitor in a circuit using the RC time constant formula.
- Use the correct formula to find the voltage drops in inductive and capacitive circuits.

Technical Terms

capacitive reactance
impedance
inductive reactance
open secondary
phase angle
RC time constant
right angle
right triangle
RL circuit
vector

In this chapter, we will look at the behavior of inductors in an ac circuit. From Chapter 11, you know that an inductor opposes a change in current. In a dc circuit, like that in **Figure 14-1,** once that current reaches its maximum, the inductor no longer affects the current. Therefore, the bulb will glow with full brightness.

Suppose the 12-V battery in Figure 14-1 is replaced with a 12-V ac source, **Figure 14-2.** Since the RMS value of the ac wave is 12 V, you would expect the bulb to glow as brightly as in the dc circuit. However, when the switch is closed, the light is not as bright. This means that the inductor resists the alternating current.

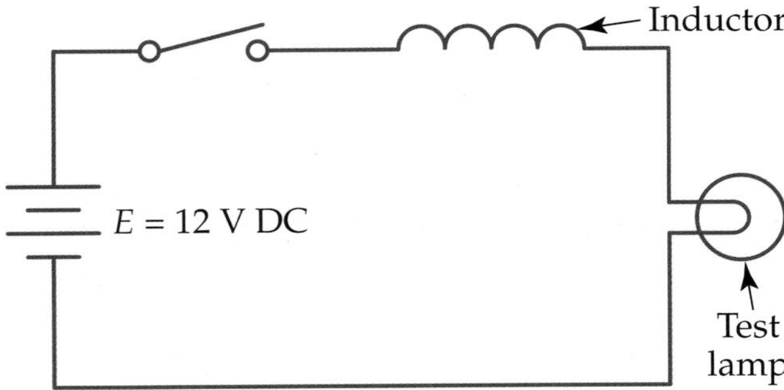

Figure 14-1. An inductor opposes a change in direct current flow. When the switch is closed, the lamp will be dimly lit until current flow in the circuit reaches its maximum value.

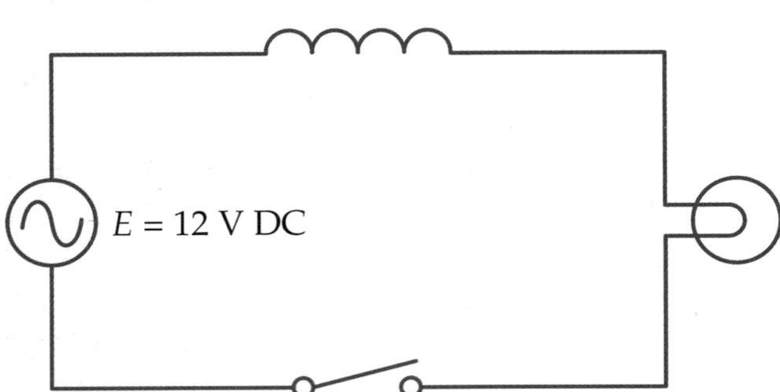

Figure 14-2. An inductor placed in an alternating current circuit offers inductive reactance to current flow. Adding the inductor dims the lamp.

Inductive Reactance

The opposition offered by inductors to any change in current is called ***inductive reactance.*** The amount of reactance that an inductor presents to an alternating current depends on two factors:

- Value of the inductor.
- Frequency of the alternating current.

The symbol for inductive reactance is X_L. Its value can be found by working out the following formula:

$$X_L = 2\pi fL \tag{14-1}$$

where π is the constant 3.14, *f* is the frequency of the applied voltage in hertz, and *L* is the inductance of the coil in henrys. The resulting value for inductive reactance is given in ohms.

Suppose we have the circuit shown in **Figure 14-3.** It has a 200-V ac source with a frequency of 318 Hz applied to an inductor with value of 0.5 H. This formula can be used to find the inductive reactance of the circuit:

$$\begin{aligned} X_L &= 2\pi fL \\ &= 2 \times 3.14 \times 318 \text{ Hz} \times 0.5 \text{ H} \\ &= 999\ \Omega \end{aligned}$$

This circuit has an inductive reactance of 999 Ω.

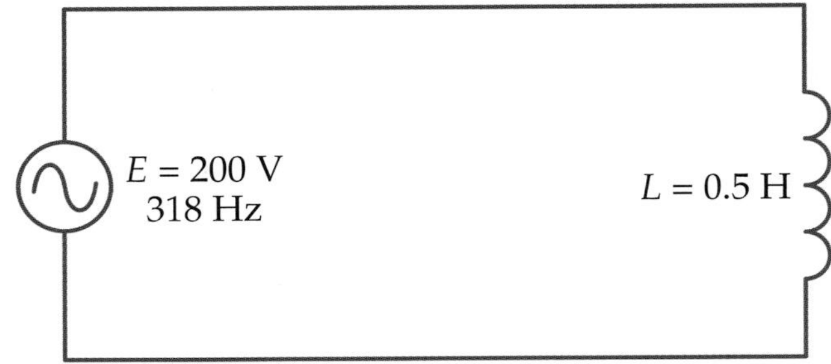

Figure 14-3. The inductive reactance (X_L) of this ac circuit is 999 Ω, found by using the formula $X_L = 2\pi fL$.

Example 14-1:

What is the inductive reactance of the circuit shown?

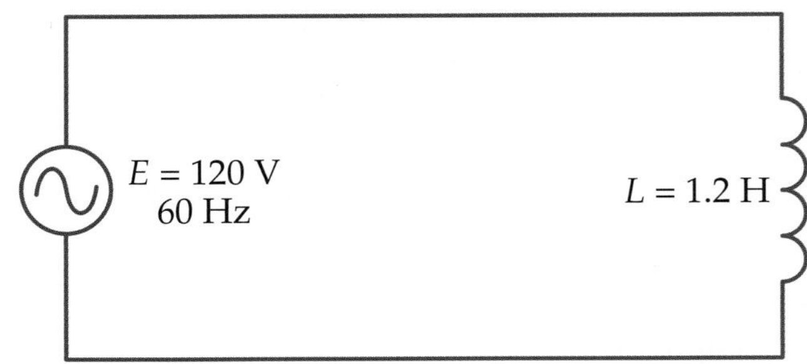

This circuit has a 120-V ac source with a frequency of 60 Hz. The inductance of the inductor is 1.2 H. Use the formula to find the inductive reactance:

$$\begin{aligned} X_L &= 2\pi fL \\ &= 2 \times 3.14 \times 60 \text{ Hz} \times 1.2 \text{ H} \\ &= 452\ \Omega \end{aligned}$$

This circuit has an inductive reactance of 452 Ω.

Example 14-2:

What is the inductive reactance of the circuit shown?

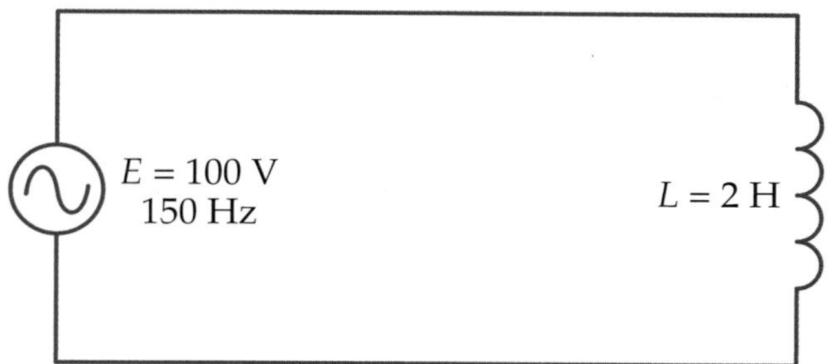

This circuit has a 100-V ac source with a frequency of 150 Hz. The inductance of the inductor is 2 H. Use the formula to find the inductive reactance:

$$X_L = 2\pi fL$$
$$= 2 \times 3.14 \times 150 \text{ Hz} \times 2 \text{ H}$$
$$= 1884\ \Omega$$

This circuit has an inductive reactance of 1884 Ω.

The value of X_L is measured in ohms because reactance affects ac the same way that resistance affects dc. In fact, referring to the circuit in Figure 14-3, we can use the voltage of the ac source to compute the amount of current flowing in the circuit. Use Ohm's law, replacing the resistance (R) with inductive reactance (X_L).

$$I = \frac{E}{X_L}$$
$$= \frac{200 \text{ V}}{999\ \Omega}$$
$$= 0.2 \text{ A}, \textit{ or } 200 \text{ mA}$$

Since this circuit contains only an inductor, the reactance of the inductor behaves just like a resistor.

Phase between Voltage and Current

If you look at one cycle of alternating current on a sine wave, **Figure 14-4,** you will see that the cycle is divided into 360°. This is the same number of degrees that the generator had to turn to produce a full cycle of ac.

When studying the sine wave, note that the current flow is changing at the fastest rate at 0°, 180°, and 360°. Also note that at exactly 90° and 270°, current flow changes direction.

In Chapter 12, we established that an inductor produces a counter electromotive force (counter emf) that depends on how fast the current is changing. The counter emf is highest when the current is changing the fastest. With this in mind, where would you expect counter emf to be highest? If you said 0°, 180°, or 360°, you are right. At these particular times in each cycle, the current is changing at its fastest rate, and the voltage across the inductor will be at a maximum.

Based on the sine wave, where would you expect the counter emf to be at a minimum? The answer is at 90° and 270°. At these points, the current stops changing for a moment and the counter EMF of the coil drops to zero.

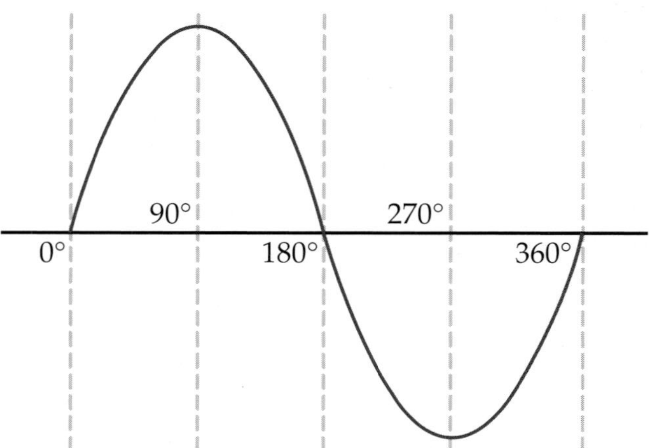

Figure 14-4. One cycle of alternating current in the form of a sine wave can be divided into 360 electrical degrees.

Graphing Current and Voltage

If you make one graph showing both current flow and EMF in the coil, **Figure 14-5,** you can see that the current and voltage are both sine waves. It is normal for current and voltage waves to have different amplitudes, but at this time we are only concerned with the difference between the phases of the waves. Note that the zero points of the two waves differ by 90°. Therefore, we can say that the voltage across the inductor is 90° out of phase with the current through it.

The voltage wave crosses the reference line (x-axis) at 90°, while the current wave crosses the line at 180°. Since current crosses the line 90° after voltage, we say that current *lags* voltage by 90° and that voltage *leads* current by 90°. In this circuit, when the voltage across the inductor is at a maximum, current through the inductor drops to zero. Current lags voltage in all inductors.

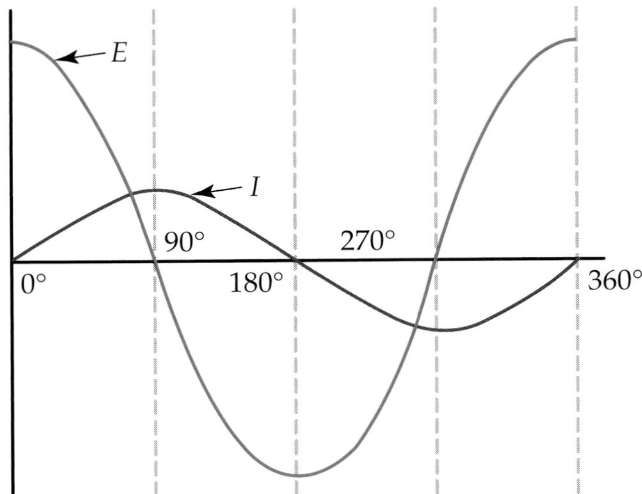

Figure 14-5. This graph compares out-of-phase sine waves of voltage and current in an inductive circuit.

AC Behavior of Series RL Circuits

The next step in the study of inductive reactance is to see what happens when we connect an inductor and a resistor in series. This type of circuit is called an ***RL circuit*** (*R* for resistance, *L* for inductance). Recall from Chapter 6 that when two elements are in series, they have the same amount of current passing through them.

Look at the series RL circuit in **Figure 14-6.** A voltmeter used to measure the voltage across the inductor and the voltage across the resistor should get a reading of 7.5 V across the inductor and 9.4 V across the resistor.

According to what you have learned about voltage drops in series, these readings should add up to the applied voltage, which is 12 V. You can see that 7.5 V plus 9.4 V does not add up to 12 V. A new method is needed for adding components in series in an RL circuit.

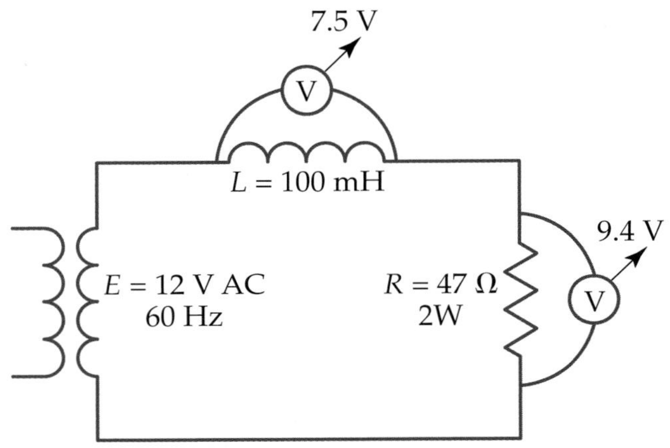

Figure 14-6. A series RL circuit contains both an inductor and a resistor. Note the voltage drop readings across these components.

Consider that an inductor has a maximum voltage across it when the current through it drops to zero. A resistor, on the other hand, has zero volts across it when the current drops to zero. In order to add quantities that behave like this, we must use ***vectors.***

Vectors

Vectors are drawn as arrows and are used to represent the magnitude and direction of a quantity, **Figure 14-7.** The length of the arrow

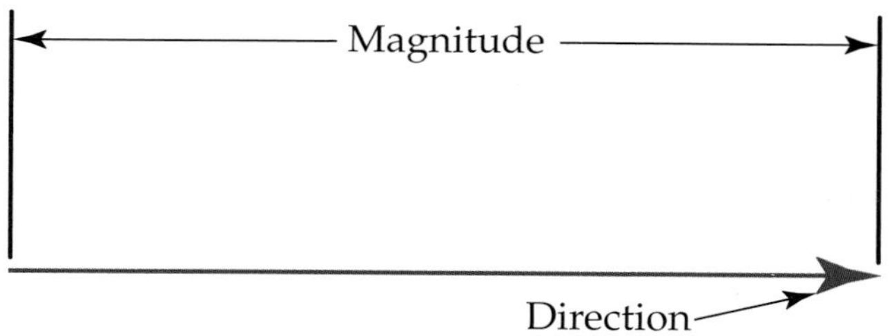

Figure 14-7. A vector generally is drawn as an arrow. A vector represents direction and magnitude of a given value.

represents the magnitude or size of the vector, while the arrowhead shows the direction of the vector. Vectors can be used to show the value of many different quantities, such as force, movement, voltage, and current.

For example, suppose you walked straight north for three miles, then turned right and walked straight east for four miles, **Figure 14-8.** If we let vector *A* represent your northward travel and vector *B* represent your eastward travel, you can get a good picture of what has occurred. Note that *B* is longer than *A* because *B* has a larger magnitude.

The little square drawn at the right turn means that the angle between vector *A* and vector *B* is 90°. An angle that is 90° is called a ***right angle.*** There are two questions to answer:

- How far have I walked?
- How far am I from my starting point?

To answer the first question is very simple. By simply adding the distances, you find that you have walked a total of seven miles.

The second question requires you to look at the magnitude and direction of your walk. In other words, you must look at each part of the walk as vectors and use vector addition. To do this, join the tail of vector *B* to the head of vector *A*. You turned 90°, so when the vectors are joined they will look like Figure 14-8. The distance you have walked from the starting point is shown by vector *C* in **Figure 14-9.** The figure formed is called a ***right triangle*** because it has a 90° angle in it. In a right triangle, the two shorter sides are called legs, and the longest side is called the hypotenuse.

The lengths of the sides of right triangles have a special relationship. If you take the lengths of the two legs of the triangle, square them, and add these totals, you will get the square of the length of the hypotenuse. This means that in Figure 14-9 vector *C* can be found by using the following formula, which is called the Pythagorean Theorem:

$$C^2 = A^2 + B^2 \tag{14-2}$$

where *A* is the length of side *A*, *B* is the length of side *B*, and *C* is the length of side *C*. This

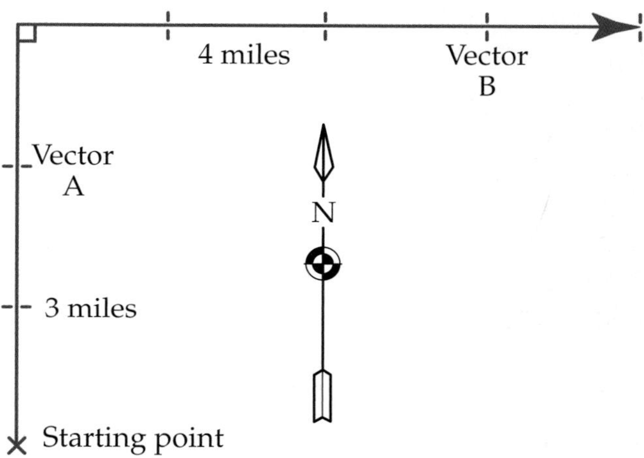

Figure 14-8. To illustrate the use of vectors, the route of a seven-mile walk is charted with vectors *A* and *B*.

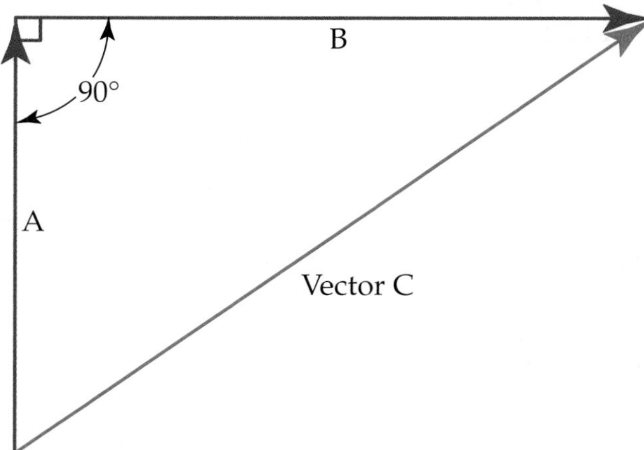

Figure 14-9. Using vectors *A* and *B* from Figure 14-8, vector *C* is drawn to complete the right triangle. The formula for solving the length of vector *C* is $C^2 = A^2 + B^2$.

formula is only true for right triangles. To use the formula for the example in Figure 7-8, insert the known values:

$$C^2 = 3^2 + 4^2$$
$$= 9 + 16$$
$$= 25$$
$$\sqrt{C^2} = \sqrt{25}$$
$$C = 5 \text{ mi}$$

The length of vector *C* is the actual distance from the starting point. Note that even though you walked a total of seven miles, you ended up only five miles from your starting point. Using vector addition, we can solve the problem of putting an inductor and a resistor in the same series circuit.

RL Voltages

Take another look at the series RL circuit shown in Figure 14-6. Remember that the values of the voltage drops across the inductor and the resistor did not add up to the source voltage. Also recall the following points:

- The 7.5 V across the inductor leads the current through the inductor by 90°.
- The 9.4 V across the resistor is in phase with the current through the resistor.
- The same current flows through both the inductor and the resistor.

If you keep these three points in mind, you can see that the voltage across the inductor is 90° out of phase with the voltage across the resistor. **Figure 14-10** shows these voltages drawn as vectors with a 90° angle between them since they are out of phase by 90°. If you want to add these voltages, you have to use the Pythagorean Theorem, substituting the lengths of vectors V_L and V_R for *A* and *B* and the symbol E_T^2 (total or applied voltage) for *C*.

$$E_T^2 = V_L^2 + V_R^2$$
$$= 7.5 \text{ V}^2 + 9.4 \text{ V}^2$$
$$= 56.25 \text{ V} + 88.36 \text{ V}$$
$$= 144.61 \text{ V}$$
$$\sqrt{E_T^2} = \sqrt{144.61 \text{ V}}$$
$$E_T = 12.0 \text{ V}$$

When these two voltages are added as vectors, we get an answer that equals the applied voltage.

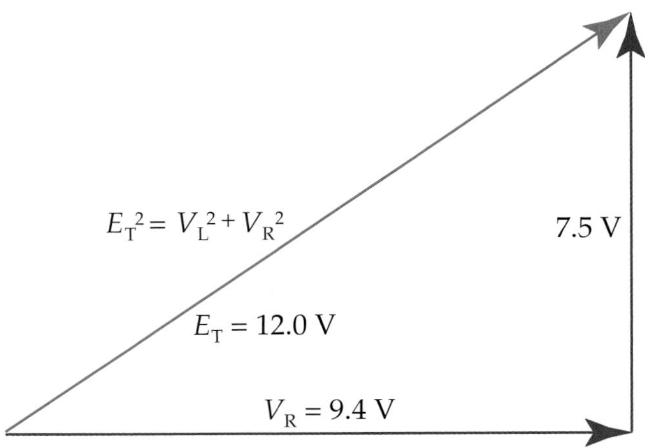

Figure 14-10. Substituting voltage values for vectors V_R and V_L and E^2 for C^2, use the formula shown to solve for the total voltage drop in the circuit.

Example 14-3:

Suppose you have a circuit with an inductor and a resistor in series. The voltage drop across the inductor is 7 V and the voltage drop across the resistor is 6 V. What is the applied voltage of this circuit?

Use the Pythagorean Theorem:

$$E_T^2 = V_L^2 + V_R^2$$
$$= 7 \text{ V}^2 + 6 \text{ V}^2$$
$$= 49 \text{ V} + 36 \text{ V}$$
$$= 85 \text{ V}$$
$$\sqrt{E_T^2} = \sqrt{85 \text{ V}}$$
$$E_T = 9.2 \text{ V}$$

The applied voltage is 9.2 V.

Example 14-4:

Suppose you have a circuit with an inductor and a resistor in series. The voltage drop across the inductor is 15 V and the voltage drop across the resistor is 20 V. What is the applied voltage of this circuit?

Use the Pythagorean Theorem:

$$E_T^{\,2} = V_L^{\,2} + V_R^{\,2}$$

$$= 15\text{ V}^2 + 20\text{ V}^2$$

$$= 225\text{ V} + 400\text{ V}$$

$$= 625\text{ V}$$

$$\sqrt{E_T^{\,2}} = \sqrt{625\text{ V}}$$

$$E_T = 25\text{ V}$$

The applied voltage is 25 V.

Impedance

The combination of reactance and resistance in a circuit is the impedance of the circuit. ***Impedance*** is the total opposition that an electrical circuit offers to an alternating current at a given frequency. It is represented by the symbol Z.

You can find the impedance of an RL circuit by adding the reactance (X_L) and resistance (R) like vectors using the Pythagorean Theorem. The letter Z is used to represent impedance in formulas and in circuit analysis. The reason the Pythagorean Theorem works is because resistance and reactance are 90° out of phase with one another.

$$Z^2 = X_L^{\,2} + R^2 \qquad (14\text{-}3)$$

Look at the series RL circuit shown in **Figure 14-11** as an example. The inductor has an value of 9 Ω; the resistor has a value of 12 Ω. The impedance of the circuit can be found using this formula:

$$Z^2 = X_L^{\,2} + R^2$$

$$= 9\ \Omega^2 + 12\ \Omega^2$$

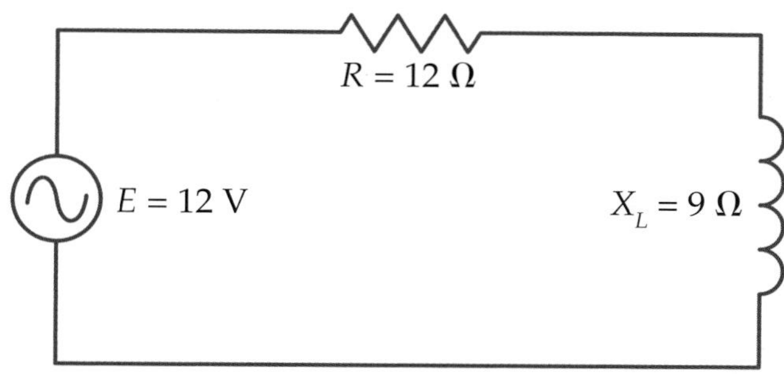

Figure 14-11. The resistor (*R*) and the inductor (*L*) in this series RL circuit offer opposition to the flow of alternating current.

$$= 81\ \Omega + 144\ \Omega$$

$$= +225\ \Omega$$

$$\sqrt{Z^2} = \sqrt{225\ \Omega}$$

$$Z = 15\ \Omega$$

The impedance is 15 Ω. This means there are 15 Ω of opposition to the ac flowing through the circuit.

Now that you know the total impedance of the circuit, you can use this value to find total current. To find total current when working with impedance values, use the following formula:

$$I = \frac{E}{Z} \qquad (14\text{-}4)$$

You know that the applied voltage (E) is 12 V and the impedance (Z) is 15 Ω. Therefore,

$$I = \frac{E}{Z}$$

$$= \frac{12\text{ V}}{15\ \Omega}$$

$$= 0.8\text{ A, } or \text{ 800 mA}$$

Total current is 800 mA. If you were to take this value and multiply it times the reactance (X_L) and resistance (R), you would have the voltages in **Figure 14-12.** Use formula 14-2 to check if your current calculation is right. If it is right, the vector addition of V_L and V_R will equal the applied voltage of 12 V.

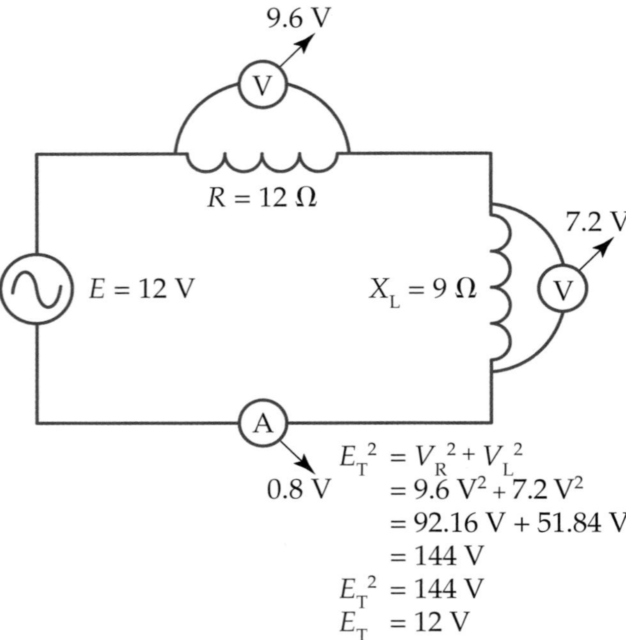

Figure 14-12. To find the voltage drops across the resistor and inductor, circuit current (I) is multiplied times the reactance (X_L) and resistance (R). The Pythagorean Theorem is used to verify these calculations.

Example 14-5:

Find the impedance and current of the circuit shown.

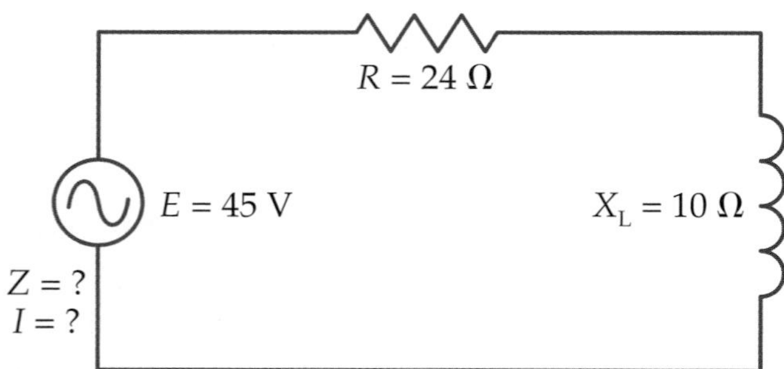

First, find the impedance.

$$Z^2 = X_L^2 + R^2$$
$$= 10\ \Omega^2 + 24\ \Omega^2$$
$$= 100\ \Omega + 576\ \Omega$$
$$= 676\ \Omega$$
$$\sqrt{Z^2} = \sqrt{676\ \Omega}$$
$$Z = 26\ \Omega$$

The impedance of this circuit is 26 Ω. Use this impedance value and the applied voltage to find the value of circuit current.

$$I = \frac{E}{Z}$$
$$= \frac{45\ V}{26\ \Omega}$$
$$= 1.73\ A$$

The circuit current is 1.73 A.

Phase Angle

The angle by which the voltage sine wave leads or lags the current sine wave in a circuit is known as its ***phase angle.*** Usually a phase angle is included with the impedance value. If we draw the two kinds of opposition, **Figure 14-13,** we will see that the inductive reactance and the resistance are 90° apart. The phase angle is given the symbol θ, which is the Greek letter theta. The phase angle is the same as the angle between the applied voltage and the resulting current. In other words, the phase angle tells how far out of phase the applied voltage and current are.

Impedance is a vector, and the phase angle tells its direction. However, you do not need to concern yourself with the phase angle. This text deals only with the size, or magnitude, of the impedance.

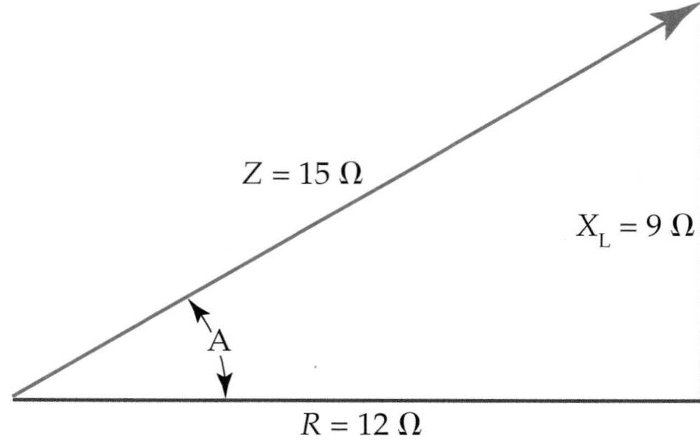

Figure 14-13. Total impedance is found by forming vectors X_L and R, and using the formula $Z^2 = X_L^2 + R^2$. The phase angle is labeled θ.

Transformer Loading

You should recall that we studied transformer operation in some detail in Chapter 12. We found that transformers transfer electrical energy from one circuit to another by electromagnetic induction. When a transformer has no load on the secondary, it acts differently than when the secondary is loaded. This is a problem of phasing.

Unloaded Transformer

When the transformer's secondary does not have a load, it is said to have an ***open secondary.*** This means that the primary of the transformer acts like an inductor, and the primary current lags the applied voltage by 90°, **Figure 14-14.** Counter emf, which is caused by the magnetic field created when current is flowing in the primary, lags current by 90°. You can see that the applied voltage and the counter emf have the opposite polarity. Most transformers are built so that the counter emf will be high in the primary when the secondary is open. This is done in an effort to keep the primary current as low as possible when there is no secondary current.

Keeping the primary current low when there is no load on the secondary keeps the primary windings and the transformer as cool as possible. As the heat of the windings increases, the efficiency of the transformer decreases. Therefore, in the long run, the transformer will not be able to deliver as much current as needed. Brownouts are the result of the inability to deliver electricity during peak demand days, which are usually the hottest days of the summer.

Even though no current flows in the secondary, the expanding and collapsing magnetic field in the primary still cuts through the turns of the secondary winding. This action induces a voltage in the secondary. The induced voltage, however, lags the current in the primary by 90°. Therefore, the secondary voltage will be a total of 180° out of phase with the primary voltage.

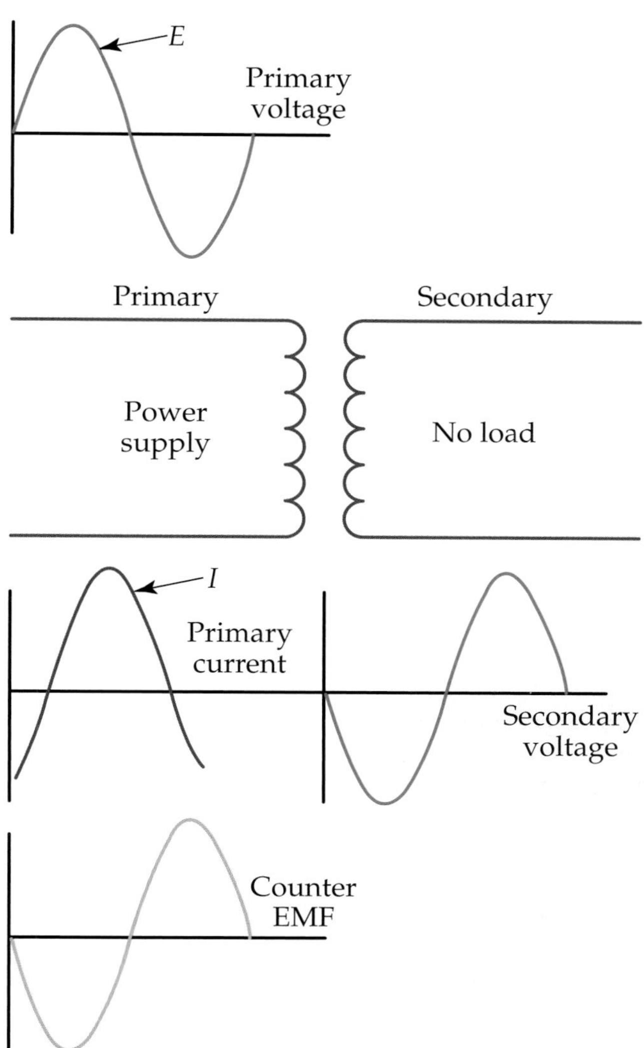

Figure 14-14. The primary current of a transformer with no load on the secondary lags the applied voltage by 90°.

You should recall from Chapter 12 that the way the primary and secondary coils of transformers are wound determines if the induced secondary voltage is in phase or out of phase with the primary voltage. In **Figure 14-15,** dots are used to show you whether the terminals are in phase or out of phase.

Loaded Transformer

When the transformer's secondary has a load, primary current automatically increases. If the secondary is overloaded or the secondary becomes shorted, the current in the primary

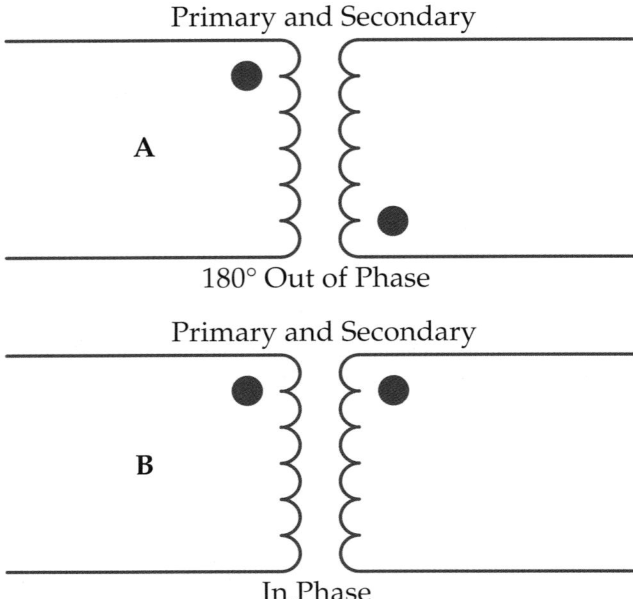

Figure 14-15. Dots indicate the phasing of the terminals on a transformer. A—Primary and secondary windings are out of phase. B—Secondary is wound to bring windings in phase.

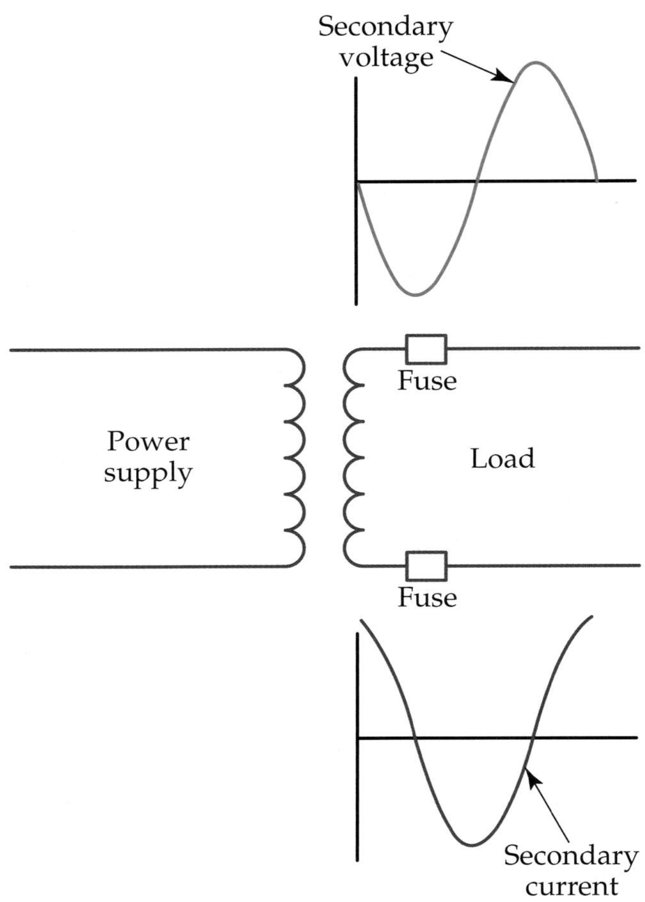

Figure 14-16. When a transformer's secondary has a load, the current in the secondary winding lags the secondary voltage by 90°.

will increase greatly. Therefore, the secondary winding must be fused to protect the primary winding, **Figure 14-16.** If it is not fused and the current in the primary surges, the transformer may be destroyed.

Capacitive Reactance and Impedance

We have studied two of the three main components in electrical circuits as they relate to impedance: the resistor and the inductor. The third main component is the capacitor. You should recall from Chapter 4 that a capacitor is a device that stores electrons until they are needed by the circuit. In this section, you will learn how a capacitor responds in resistive dc and ac circuits.

Capacitors in DC Circuits

In Chapter 4, we also found that if a capacitor is placed in a dc circuit, no current will flow once the capacitor becomes charged. Taking certain measurements on the circuit shown in **Figure 14-17** verifies this fact. This test circuit has a 12-V dc battery connected in series with a 210-μF capacitor and a 4700-Ω resistor. A voltmeter is connected across the capacitor and an ammeter is placed in series.

As the capacitor charges, there is a buildup of electrons on its plates. The larger the capacitor, the more electrons it can store. Voltage builds up slowly, taking about five seconds to reach battery voltage in the circuit. When the capacitor is fully charged, the ammeter shows zero because current has stopped flowing. The voltmeter shows the full source voltage of 12 V. Therefore, we can say that a capacitor in a dc circuit blocks current flow.

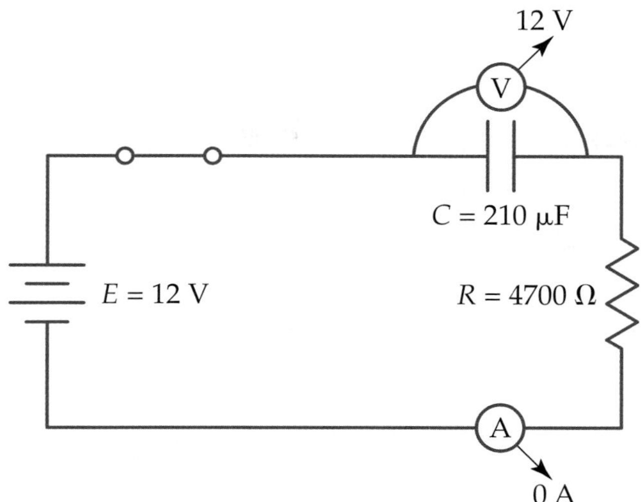

Figure 14-17. The effect of a capacitor can be demonstrated with this circuit. Voltmeter and ammeter tests show that once the capacitor is fully charged, no current flows.

RC Time Constant

The time required for a capacitor to become fully charged depends on the resistance (R) in the circuit and the capacitance (C) of the capacitor. If you multiply the resistance by the capacitance, you will get a number called the ***RC time constant*** of that circuit. The RC time constant is represented by the Greek letter tau (τ)

$$\tau = R \times C \tag{14-5}$$

where R is the resistance measured in ohms and C is the capacitance measured in farads. The total time in seconds (T) that it takes to charge a capacitor can be found by multiplying the RC time constant by five:

$$\mathrm{T} = 5\tau \tag{14-6}$$

Suppose that a 5000-μF capacitor and a 2000-Ω resistor are connected in series with a 150-V battery, as shown in **Figure 14-18.** How long will it take the capacitor to charge to circuit voltage? Start by converting microfarads to farads. A farad is equal to one million microfarads.

$$5000\ \mu\mathrm{F} = \frac{5000}{1{,}000{,}000\ \mathrm{F}} = 0.005\ \mathrm{F}$$

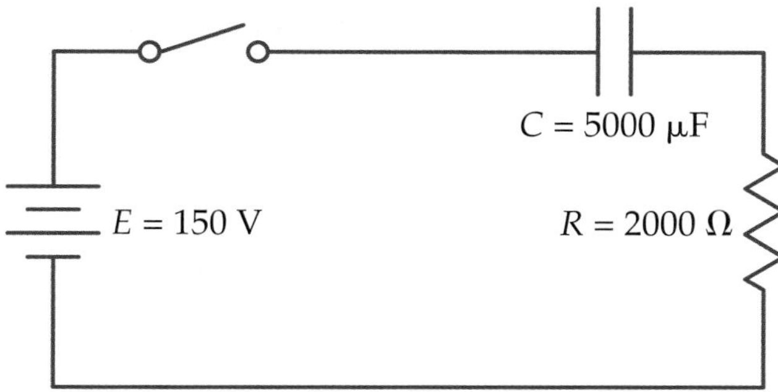

Figure 14-18. How long will it take for the capacitor to charge to battery voltage?

Then, use the RC time constant formula:

$$\begin{aligned}\tau &= RC\\ &= 2000\ \Omega \times 0.005\ \mathrm{F}\\ &= 10\ \text{seconds}\end{aligned}$$

Now, find the total charging time for this capacitor in this circuit:

$$\begin{aligned}\mathrm{T} &= 5\tau\\ &= 5 \times 10\\ &= 50\ \text{seconds}\end{aligned}$$

It will take the capacitor 50 seconds to charge to 150 V. The amount of time varies according to resistance and capacitance values.

In this example, it takes a while for the voltage to build across the capacitor. Although the current started to flow in the circuit immediately after the switch was closed, it took almost a full minute for the capacitor to become fully charged and for the current to go back to zero. Now, let us look at how a capacitor responds in an ac resistive circuit.

Capacitors in AC Circuits

Figure 14-19 shows a graph of both the voltage across the capacitor and the current in the test circuit in Figure 14-18. From these graphs of current (I) and voltage (E), you can see that the current leads the voltage. If you replace the battery in Figure 14-18 with an ac power supply, you will see that the current will always lead the voltage by 90°.

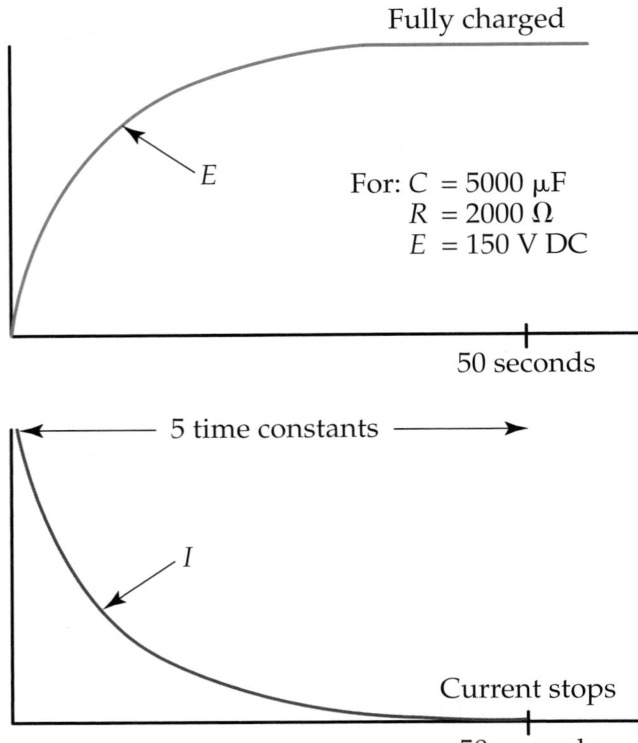

Figure 14-19. These graphs show what happens to the voltage and the current during the time the capacitor is charging.

Therefore, in a circuit with a capacitor, current leads voltage. Recall that the opposite condition exists for circuits with inductors: the voltage leads the current. The phrase "ELI the ICE man" is a simple device for remembering this. See **Figure 14-20.**

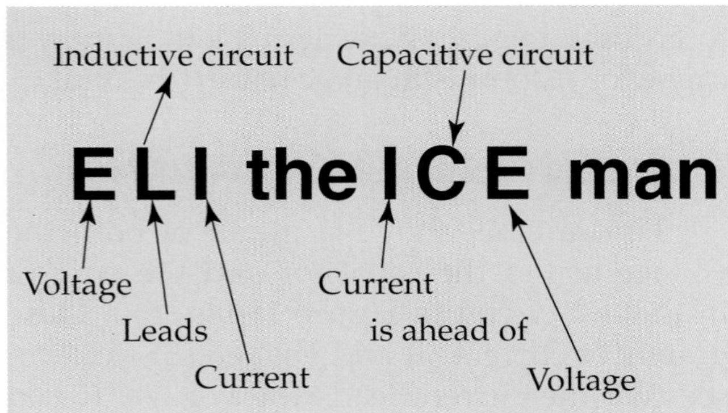

Figure 14-20. "ELI the ICE man" will serve as a memory aid in your study of inductive (*L*) and capacitive (*C*) circuits.

Capacitive Reactance

Capacitors also show reactance in an ac circuit. The opposition offered by capacitors to alternating current is called ***capacitive reactance.*** Capacitors resist the flow of current in a circuit. The number of electrons that flow back and forth across its plates in an ac circuit depends on two things:

- Value of the capacitor.
- Frequency of the circuit.

The larger the capacitor value, the greater the number of electrons that can be exchanged with every alternation of the sine wave. Also, the higher the frequency of the circuit, the greater the number of electrons that will change from plate to plate in a given amount of time.

The circuit shown in **Figure 14-21** can demonstrate capacitive reactance. Notice that the bulb in circuit *A* is brighter than the bulb in circuit *B*. This is caused by capacitive

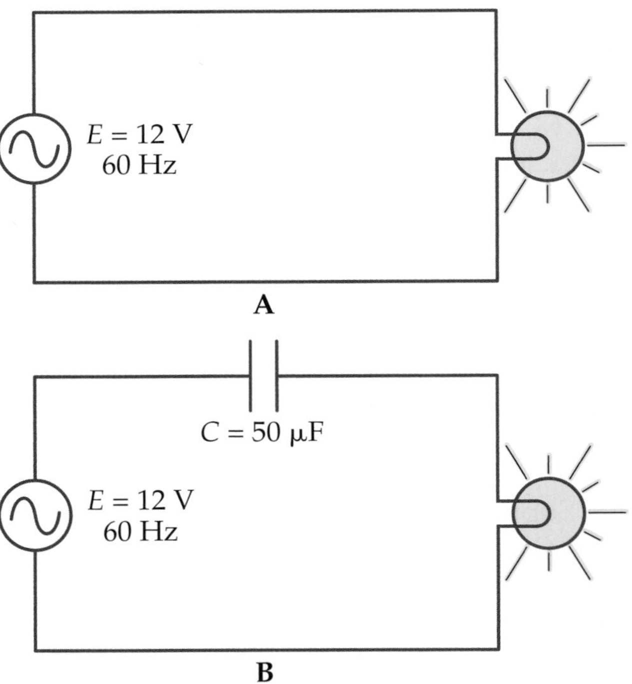

Figure 14-21. The effects of capacitive reactance. A—The lamp burns brightly because there is no capacitor in the circuit. B—A capacitor in an ac circuit reduces the brilliance of the lamp.

reactance (X_C). The formula for calculating capacitive reactance is

$$X_C = \frac{1}{2\pi fC} \quad (14\text{-}7)$$

where π is the constant 3.14, *f* is the frequency of the applied ac voltage in hertz, and *C* is the capacitance in farads. If the capacitance is given in another unit, it must be converted to farads before using this formula.

Look again at circuit *B* in Figure 14-21. The capacitance is measured in microfarads, so we must convert to farads to find the capacitive reactance (X_C) of this circuit:

$50\ \mu F \times 10^{-6} = 0.00005\ F$

Use the value 0.00005 F in the formula:

$$X_C = \frac{1}{2\pi fC} = \frac{1}{2 \times 3.14 \times 60\ Hz \times 0.00005\ F} = \frac{1}{0.01884} = 53\ \Omega$$

The capacitive reactance of this circuit is 53 Ω. Since the capacitor offers resistance to this ac circuit, the bulb does not burn as bright as it does in Figure 14-21A, which offers no resistance at all.

To make the bulb burn brighter in Figure 14-21B, you would need to raise the frequency or the capacitance. Doing so will lessen the capacitance reactance, or opposition to current flow, of the circuit. Look what happens when the frequency is raised to 120 Hz.

$$X_C = \frac{1}{2\pi fC} = \frac{1}{2 \times 3.14 \times 120\ Hz \times 0.00005\ F} = \frac{1}{0.03768} = 26.5\ \Omega$$

The capacitive reactance has dropped to 26.5 Ω, thus offering less resistance than the original circuit.

Example 14-6:

What is the capacitive reactance (X_C) of circuit in Figure 14-21B if the capacitance of the capacitor is 3 F and the frequency is 50 Hz?

$$X_C = \frac{1}{2\pi fC} = \frac{1}{2 \times 3.14 \times 50\ Hz \times 3\ F} = \frac{1}{942} = 0.001\ \Omega$$

The capacitive reactance in this circuit is 0.001 Ω.

Example 14-7:

What is the capacitive reactance (X_C) of circuit in Figure 14-21B if the capacitance of the capacitor is 150 μF and the frequency is 300 Hz?

First, convert 150 μF to farads:

$150\ \mu F \times 10^{-6} = 0.000150\ F$

Then, use the formula for capacitive reactance:

$$X_C = \frac{1}{2\pi fC} = \frac{1}{2 \times 3.14 \times 300\ Hz \times 0.000150\ F} = \frac{1}{0.2826} = 3.5\ \Omega$$

The capacitive reactance in this circuit is 3.5 Ω.

Vector Addition

A capacitor's reactance can be added to the resistance in an ac circuit using vector addition. For example, **Figure 14-22** shows a series circuit that has a 100-V, 60-Hz ac power supply, a 220-μF capacitor, and a 5-Ω resistor. First, find capacitive reactance (X_C):

$220\ \mu F \times 10^{-6} = 0.000220\ F$

$$X_C = \frac{1}{2\pi fC} = \frac{1}{2 \times 3.14 \times 60\ Hz \times 0.000220\ F} = \frac{1}{0.082896} = 12\ \Omega$$

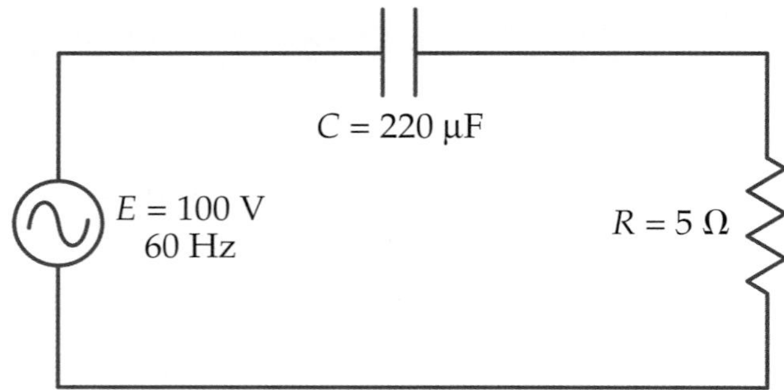

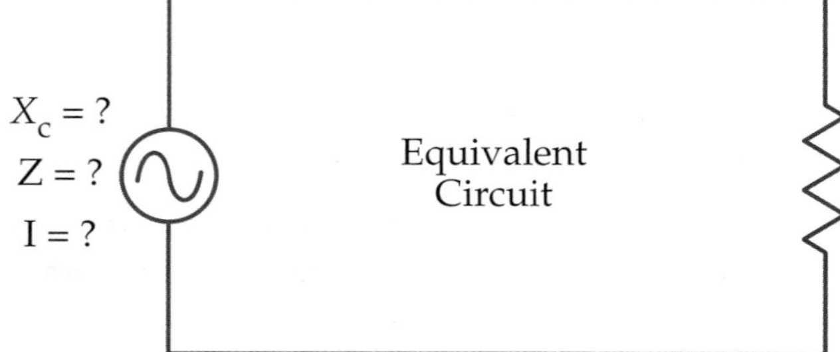

Figure 14-22. Find the various unknowns in this capacitive circuit. The unknowns include capacitive reactance (X_C), impedance (Z), and current flow (I).

Next, find impedance (Z) using the Pythagorean Theorem. Like inductive reactance, capacitive reactance can be used to find impedance with the following formula:

$$Z^2 = X_C^{\,2} + R^2 \tag{14-8}$$

$$= 12\ \Omega^2 + 5\ \Omega^2$$

$$= 144\ \Omega + 25\ \Omega$$

$$= 169\ \Omega$$

$$\sqrt{Z^2} = \sqrt{169\ \Omega}$$

$$Z = 13\ \Omega$$

Finally, find the current (I):

$$I = \frac{E}{Z}$$

$$= \frac{100\ \text{V}}{13\ \Omega}$$

$$= 7.7\ \text{A}$$

This circuit has a total impedance (Z) of 13 Ω and a current (I) of 7.7 A.

Example 14-8:

Find the capacitive reactance (X_C), the impedance (Z), and the current (I) for the circuit shown.

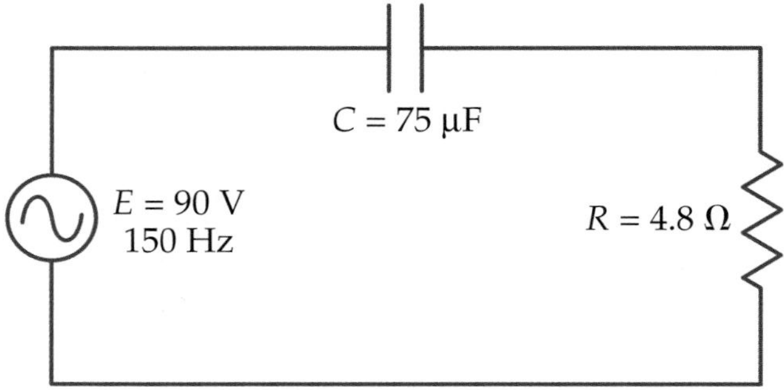

Start by finding the capacitive reactance:

$$75\ \mu\text{F} \times 10^{-6} = 0.000075\ \text{F}$$

$$X_C = \frac{1}{2\pi fC}$$

$$= \frac{1}{2 \times 3.14 \times 150\ \text{Hz} \times 0.000075\ \text{F}}$$

$$= \frac{1}{0.07065}$$

$$= 14.2\ \Omega$$

Next, find impedance (Z):

$$Z^2 = X_C^{\,2} + R^2$$

$$= 14.2\ \Omega^2 + 4.8\ \Omega^2$$

$$= 201.6\ \Omega + 23.0\ \Omega$$

$$= 224.6\ \Omega$$

$$\sqrt{Z^2} = \sqrt{224.6\ \Omega}$$

$$Z = 15\ \Omega$$

Finally, find the current (I):

$$I = \frac{E}{Z}$$

$$= \frac{90\ \text{V}}{15\ \Omega}$$

$$= 6\ \text{A}$$

This circuit has a capacitive reactance (X_C) of 14.2 Ω, a total impedance (Z) of 15 Ω, and a current (I) of 6 A.

Voltage Drops

By using vector addition, we found that the impedance of the circuit shown in Figure 14-22 is 13 Ω. Then, using Ohm's law and substituting impedance for resistance, we found that the current in the circuit is 7.7 A.

Since this is a series ac circuit, the same 7.7 A must flow through both the capacitor and the resistor. The voltage drops across the capacitor and resistor can be found using Ohm's law and this value for current:

$$V_C = IX_C$$
$$= 7.7 \text{ A} \times 12\ \Omega$$
$$= 92.4 \text{ V}$$

$$V_R = IR$$
$$= 7.7 \text{ A} \times 5\ \Omega$$
$$= 38.5 \text{ V}$$

Note that if we add these voltages as numbers, the answer is not 120 V. However, if we use vector addition, we get the correct answer.

$$E^2 = V_C^{\ 2} + V_R^{\ 2}$$
$$= 92.4 \text{ V}^2 + 38.5 \text{ V}^2$$
$$= 8537.8 \text{ V} + 1482.3 \text{ V}$$
$$= 10{,}020.1 \text{ V}$$
$$\sqrt{E^2} = \sqrt{10{,}020.1 \text{ V}}$$
$$E = 100 \text{ V}$$

This answer is the same as the applied voltage of 100 V.

Summary

- The opposition that is offered by inductors to any change in current in a circuit is called *inductive reactance.*
- The symbol for inductive reactance is X_L and is measured in ohms.
- The formula for finding inductive reactance is
  $X_L = 2\pi fL$
- In an inductive circuit, the voltage leads current by 90°.
- When current and voltage are out of phase, the degree to which they are out of phase can be measured by how far apart in degrees they are as they cross the reference line (x-axis).
- A vector is used to represent the magnitude and direction of a quantity.
- Two vectors are added by joining the tail of one to the head of the other.
- In both RL and RC circuits, voltages must be added by using the Pythagorean Theorem and treating the quantities as vectors.
- Impedance is the total opposition that components in a circuit provide to an ac circuit at a given frequency.
- The symbol for impedance is Z and is measured in ohms.
- Impedance is found using vector addition.
- You can wind the coils of a transformer so the primary and secondary are either in or out of phase with each other.
- The opposition that is offered by capacitors to any change in current in a circuit is called *capacitive reactance.*
- The symbol for capacitive reactance is X_C and is measured in ohms.
- The formula for finding capacitive reactance is
  $X_C = \frac{1}{2\pi fC}$
- In a capacitive circuit, the current leads the voltage by 90°.

Test Your Knowledge

Do not write in this book. Write your answers on a separate sheet of paper.

1. The opposition to ac by an inductor is called ______.
2. In what units is the value for inductive reactance given?
3. How many electrical degrees are there in one complete revolution?
4. The voltage on a sine wave is at its maximum at ______ degrees and ______ degrees.
5. At what number of degrees would you expect the counter emf to be the highest?
6. What is meant by impedance?
7. Impedance in a circuit is usually labeled ______ in a vector diagram.
8. Write the formula for the impedance of a circuit with inductive reactance.
9. Draw a vector diagram and tell how vectors can be used in ac circuits.
10. To add voltages in inductive and capacitive reactance circuits, we must use ______.
11. Dots are used on schematics with transformers to show whether the terminals are ______ or ______.
12. What does a capacitor do to current in a dc circuit?
13. It takes about ______ time constants to charge a capacitor to the voltage of a dc circuit power supply.
14. Find the RC time constant for a 30,000-Ω resistor and a 200-µF capacitor connected in series. Show your work.
15. Suppose an RC circuit has one capacitor and one resistor. Also suppose that it has a capacitive reactance of 4 Ω, a resistance of 3 Ω, and an applied voltage of 12 V. What is the current in this circuit, and what are the voltage drops across the capacitor and the resistor? Check your work using the Pythagorean Theorem. Show all work.

Activities

1. The electronic calculator has made the analysis of circuits with impedance a lot easier. How is a calculator able to show you the square of a number or square root of a number, multiply, divide, and do all of the other functions so easily?
2. See if you can locate a nomograph for inductive reactance and capacitive reactance circuits. Show how they work.
3. What is a saturable core reactor? How does it work?
4. How do you determine the value of a resistor that can be safely used to discharge a capacitor? What stops the resistor from overheating if you exceed its wattage rating when discharging capacitors?
5. How does an electric shaver make use of a transformer? What is done to its windings in terms of phasing? What stops it from overheating when it is not loaded?
6. Perform some experiments that prove what a capacitor does to the current in a dc circuit.
7. Take three different sizes of capacitors and charge them. Measure the amount of time necessary for the capacitors to lose their charge. Use three different sizes of resistors to discharge them and measure the time needed. What do the results tell you?

LCR Circuits

Learning Objectives

After studying this chapter, you will be able to do the following:

- Calculate the impedance of an LCR circuit.
- Determine the voltage and current of an LCR circuit.
- Describe what is meant by a resonant circuit.
- Explain how a radio is tuned.

Technical Terms

LCR circuit
multicasting
resonance

In this chapter, we will look at LCR circuits. An ***LCR circuit*** contains inductance (L), capacitance (C), and resistance (R). We will study their behavior and look at some of the useful things we can do with LCR circuits.

Simple LCR Circuits

Figure 15-1 shows an LCR circuit connected to a 50-V ac power supply with a frequency of 100 Hz. The value of the inductor is 0.159 H, the capacitor is 21.2 µF, and the resistor is 7 Ω. Since this is an ac circuit, we will have to find the capacitive reactance and the inductive reactance in order to calculate the impedance of the circuit. To solve the circuit for inductive reactance, use the formula 14-1 from Chapter 14:

$$X_L = 2\pi fL$$
$$= 2 \times 3.14 \times 100 \text{ Hz} \times 0.159 \text{ H}$$
$$= 100\ \Omega$$

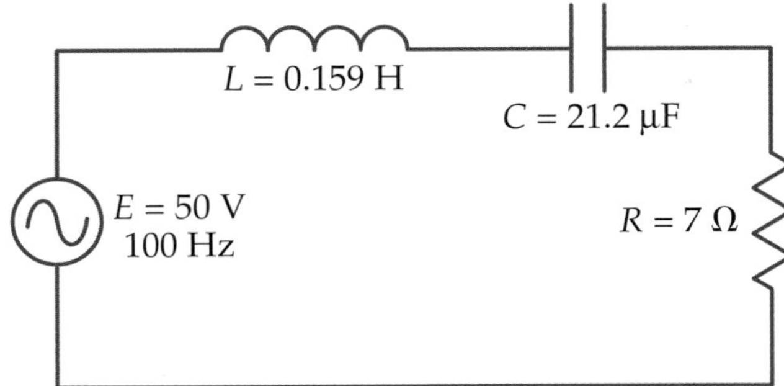

Figure 15-1. A typical LCR circuit contains an inductor (*L*), a capacitor (*C*), and a resistor (*R*).

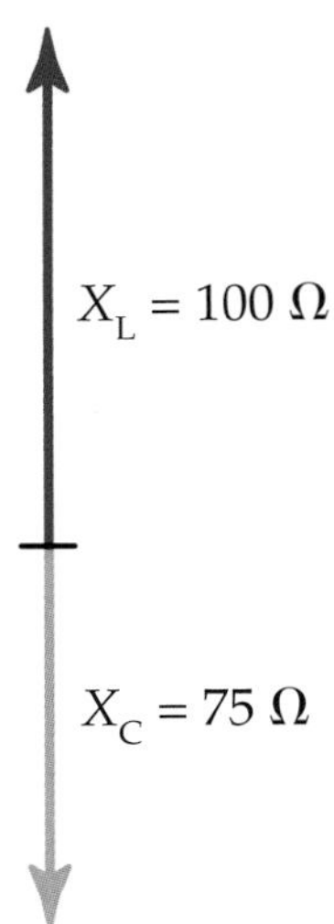

Figure 15-2. Vectors for inductive reactance (X_L) and capacitive reactance (X_C) are drawn in opposite directions because they represent opposing actions.

To solve the circuit for capacitive reactance, use formula 14-7 from Chapter 14. Remember to first convert to farads when the value of the capacitor is given in microfarads:

$$21.2\ \mu F \times 10^{-6} = 0.0000212\ F$$

$$X_C = \frac{1}{2\pi fC}$$

$$= \frac{1}{2 \times 3.14 \times 100\ Hz \times 0.0000212\ F}$$

$$= \frac{1}{0.013314}$$

$$= 75\ \Omega$$

Finding Impedance

To find the total impedance of the circuit shown in Figure 15-1, we need to add the reactance and the resistance as vectors. Note in **Figure 15-2** that the vectors for X_L and X_C are drawn in opposite directions. You can see the reason for this if you recall our earlier study of capacitors and inductors. In Chapter 14, we found that for an inductor, the voltage leads the current by 90° (a right angle), and for a capacitor, the current leads the voltage by 90°. These two events are opposing actions, as you can see in **Figure 15-3.** Notice that the current through an inductor and through a capacitor is 180° out of phase and that current flows in opposite directions. This is why the vectors for X_L and X_C are drawn in opposite directions.

Therefore, due to the 180° angle between the vectors, you can subtract the magnitude of the vector that points down from that of the one that points up to find the total reactance (X_T).

$$X_T = X_L - X_C \qquad (15\text{-}1)$$

This is the same as placing the head of one vector at the tail of the other and counting the difference between the vector lengths, **Figure 15-4.**

Earlier we found that the inductive reactance of the circuit in Figure 15-1 is equal to 100 Ω and capacitive reactance is equal to 75 Ω. Therefore, we can use the total reactance formula:

$$X_T = X_L - X_C$$

$$= 100\ \Omega - 75\ \Omega$$

$$= 25\ \Omega$$

The total reactance of the inductor and capacitor is 25 Ω, the same amount as when we counted the difference between the vector lengths.

To find impedance, we now must add total reactance (X_T) to the resistance (*R*) from the resistor. Using vectors, place the tail of one vector at the head of the other, **Figure 15-5.** This is a right triangle, so we can use the

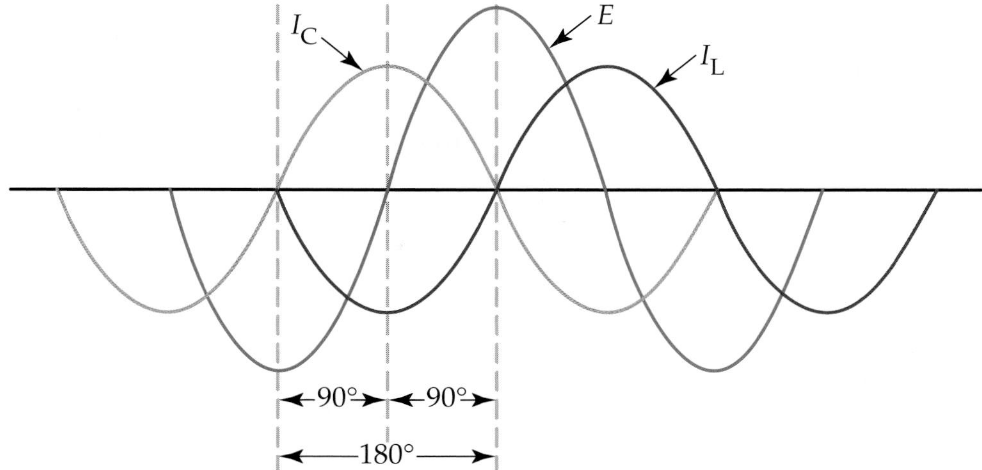

Figure 15-3. Current through a capacitor leads voltage by 90°. Current through an inductor lags voltage by 90°. The current through an inductor and through a capacitor is 180° out of phase.

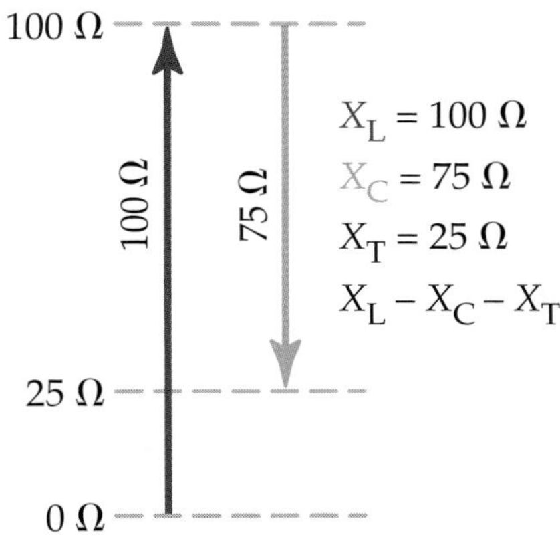

Figure 15-4. Total reactance (X_T) can be found by placing the head of one vector at the tail of the other and counting the difference between the vector lengths.

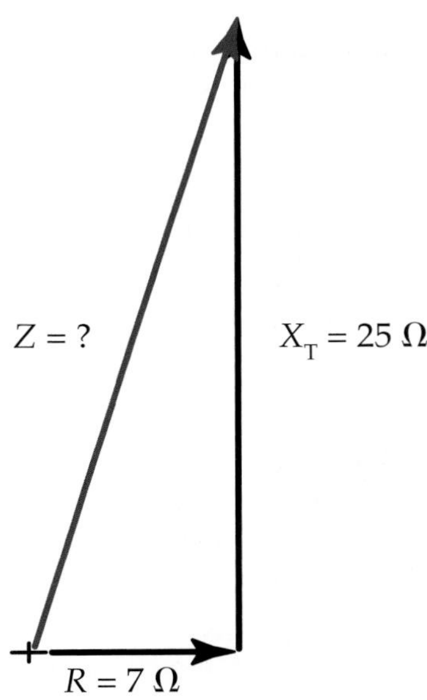

Figure 15-5. To find the impedance of an LCR circuit, draw the vector diagram shown and use the formula given in the text. The impedance is 26 Ω.

Pythagorean Theorem to solve for impedance using the following formula:

$$Z^2 = R^2 + X_T^2 \qquad (15\text{-}2)$$

Using the resistance and total capacitance value from the circuit in Figure 14-1, we have

$$Z^2 = 7\ \Omega^2 + 25\ \Omega^2$$

$$= 49\ \Omega + 625\ \Omega$$

$$= 674\ \Omega$$

$$\sqrt{Z^2} = \sqrt{674\ \Omega}$$

$$Z = 26\ \Omega$$

Finding Current and Voltage

You have found the total impedance of the circuit and know the applied voltage, **Figure 15-6.** You can use Ohm's law to find the current flowing through the circuit:

$$I = \frac{E}{Z}$$

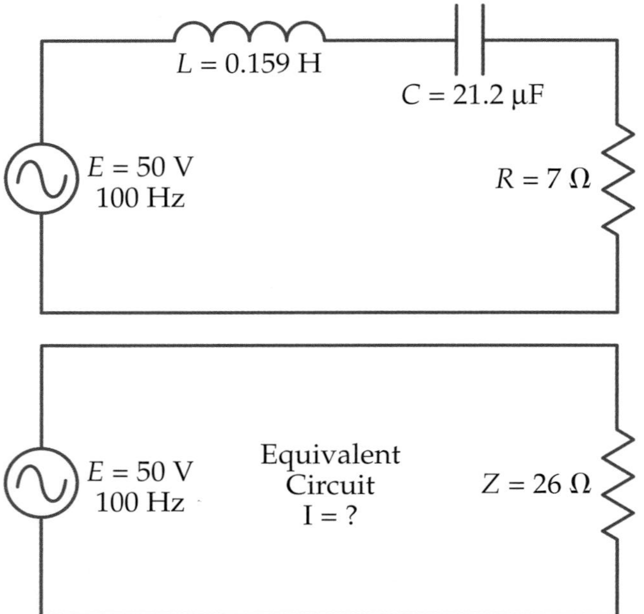

Figure 15-6. To find the current flowing in this LCR circuit, divide the applied voltage by the impedance. The current is 1.92 A.

Insert the known values:

$$I = \frac{50\text{ V}}{26\ \Omega}$$

$$= 1.92\text{ A}$$

Once you know the current in the circuit, you can use Ohm's law to find the voltage across each part of the circuit:

$$V_L = I \times X_L$$

$$= 1.92\text{ A} \times 100\ \Omega$$

$$= 192\text{ V}$$

$$V_C = I \times X_C$$

$$= 1.92\text{ A} \times 75\ \Omega$$

$$= 144\text{ V}$$

$$V_R = I \times R$$

$$= 1.92\text{ A} \times 7.0\ \Omega$$

$$= 13.4\text{ V}$$

Note that circuits containing reactance often produce voltages through induction or capacitance that are higher than the applied voltage. The apparent difference is sometimes called an *imaginary difference.* The voltage across the inductor and the voltage across the capacitor are opposite to each other (180° out of phase). If the voltages are added as ordinary numbers, the sum is much higher than 50 V. These voltages must be added as vectors, because vector addition takes into account the fact that these voltages are not in phase. The following formula is used:

$$E_T^2 = V_R^2 + (V_L - V_C)^2 \qquad (15\text{-}3)$$

Therefore,

$$E_T^2 = 13.4\text{ V}^2 + (192\text{ V} - 144\text{ V})^2$$

$$= 179.6 + 2304$$

$$= 2483.6$$

$$\sqrt{E_T^2} = \sqrt{2483.6}$$

$$E_T = 50\text{ V}$$

Total voltage (E_T) is equal to 50 V, just as it is in Figure 15-6.

Example 15-1:

Find the voltages across each of the components in the circuit shown.

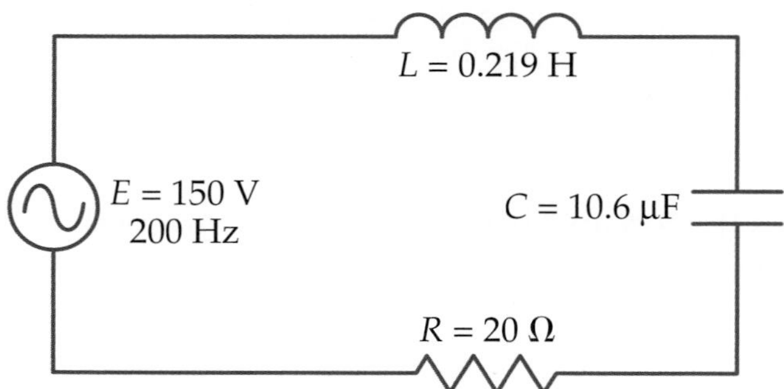

To begin, you must find the total reactance (X_T) by calculating inductive reactance and capacitive reactance. Find the reactance of the inductor:

$$X_L = 2\pi fL$$

$$= 2 \times 3.14 \times 200\text{ Hz} \times 0.219\text{ H}$$

$$= 275\ \Omega$$

Calculate the reactance of the capacitor. Remember that capacitance is given in microfarads.

$10.6\ \mu F \times 10^{-6} = 0.0000106\ F$

$$X_C = \frac{1}{2\pi fC}$$
$$= \frac{1}{2 \times 3.14 \times 200\ Hz \times 0.0000106\ F}$$
$$= \frac{1}{0.013314}$$
$$= 75\ \Omega$$

Find the total reactance (X_T) offered by the inductor and the capacitor:

$$X_T = X_L - X_C$$
$$= 275\ \Omega - 75\ \Omega$$
$$= 200\ \Omega$$

Use the Pythagorean Theorem to solve for total impedance:

$$Z^2 = R^2 + X_T^{\ 2}$$
$$= 20\ \Omega^2 + 200\ \Omega^2$$
$$= 400\ \Omega + 40{,}000\ \Omega$$
$$= 40{,}400\ \Omega$$
$$\sqrt{Z^2} = \sqrt{40{,}400\ \Omega}$$
$$Z = 201\ \Omega$$

Using the impedance value, calculate current in the circuit:

$$I = \frac{E}{Z}$$
$$= \frac{150\ V}{201\ \Omega}$$
$$= 0.75\ A$$

Find the voltage across each component:

$$V_L = I \times X_L$$
$$= 0.75\ A \times 275\ \Omega$$
$$= 206\ V$$

$$V_C = I \times X_C$$
$$= 0.75\ A \times 75\ \Omega$$
$$= 56\ V$$

$$V_R = I \times R$$
$$= 0.75\ A \times 20.0\ \Omega$$
$$= 15\ V$$

We can check our values by using the following equation to find total voltage:

$$E_T^{\ 2} = V_R^{\ 2} + (V_L - V_C)^2$$
$$= 15\ V^2 + (206\ V - 56\ V)^2$$
$$= 225\ V + 22{,}500\ V$$
$$= 22{,}725$$
$$\sqrt{E_T^{\ 2}} = \sqrt{22{,}725}$$
$$E_T = 150\ V$$

Using methods for adding vectors and Ohm's law, we have taken an LCR circuit and solved for the current through the circuit and the voltage across each component in the circuit.

Resonance

In circuits that contain both an inductor and a capacitor, the reactance produced by each of these two devices opposes each other. Sometimes, however, the amount of reactance offered by the capacitor equals that of the inductor. This is called ***resonance.*** When this happens, the total reactance of the circuit is zero, and the impedance of the circuit is equal to the resistance of any resistors in the circuit.

Look at the circuit in **Figure 15-7.** Analyze the circuit step by step. Start by finding the total reactance (X_T):

$$X_L = 2\pi fL$$
$$= 2 \times 3.14 \times 104.5\ Hz \times 0.387\ H$$
$$= 254\ \Omega$$

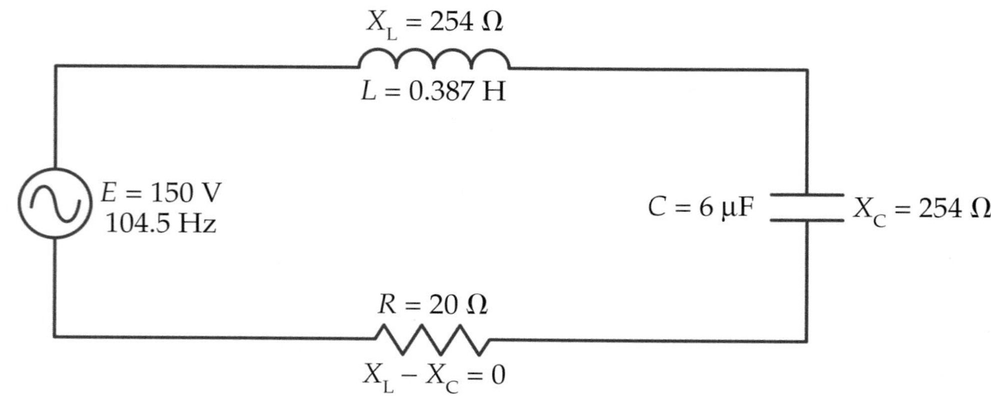

Figure 15-7. Resonance occurs when inductive reactance (X_L) equals capacitive reactance (X_C). When the difference in these vector lengths is totaled, the result is zero.

$$6.0\ \mu F \times 10^{-6} = 0.0000060\ F$$

$$X_C = \frac{1}{2\pi fC}$$

$$= \frac{1{,}000{,}000}{2 \times 3.14 \times 104.5\ Hz \times 0.0000060\ F}$$

$$= \frac{1}{0.003937}$$

$$= 254\ \Omega$$

$$X_T = X_L - X_C$$

$$= 254\ \Omega - 254\ \Omega$$

$$= 0\ \Omega$$

Since $X_L = X_C$, we can say that this circuit resonates at a frequency of 104.5 Hz. Continue by finding the impedance of the circuit at resonance.

$$Z^2 = X_T^2 + R^2$$

$$= 0^2 + 20\ \Omega^2$$

$$= 0 + 400\ \Omega$$

$$= 400\ \Omega$$

$$\sqrt{Z^2} = \sqrt{400\ \Omega}$$

$$Z = 20\ \Omega$$

Use this value to find the current and the voltages across the circuit's components.

$$I = \frac{E}{Z}$$

$$= \frac{150\ V}{20\ \Omega}$$

$$= 7.5\ A$$

$$V_L = I \times X_L$$

$$= 7.5\ A \times 254\ \Omega$$

$$= 1905\ V$$

$$V_C = I \times X_C$$

$$= 7.5\ A \times 254\ \Omega$$

$$= 1905\ V$$

$$V_R = I \times R$$

$$= 7.5\ A \times 20\ \Omega$$

$$= 150\ V$$

Note that since the circuit is resonant at this frequency, $V_L = V_C$.

Resonant Frequency

The voltages across the inductor and the capacitor are much larger at resonant frequency than at other frequencies. For example, compare voltages in the last example, which were at a frequency of 104.5 Hz, with voltages of the same circuit, but at 200 Hz.

Frequency (Hz)	Voltage (*V*)	
	V_L	V_C
104.5	1905	1905
200	206	56

At resonant frequency, the voltage across the inductor and the capacitor are at their highest value. Look at **Figure 15-8.** Note how quickly the voltage drops as the frequency moves farther from the resonant frequency on the graph.

A common and useful way resonance is used is in radio circuits. The circuit shown in Figure 15-7 is resonant at a frequency of 104.5 Hz. You might say that this circuit is "tuned" to 104.5 Hz. When you change the station on your radio, you are changing the resonant frequency of the circuit inside the radio.

If the value of the capacitor in an LCR circuit is changed, the resonant frequency also changes. To experiment, change the value of the capacitor in Figure 15-7 from 6.0 μF to 20 μF, as shown in **Figure 15-9.** When the capacitance changes, so does capacitive reactance (X_C). Since inductive reactance (X_L) remains the same, X_C no longer equals X_L, and the circuit will no longer resonate at this frequency.

The frequency at which this circuit will now resonate can be calculated. Remember that in a resonant circuit, $X_L = X_C$. Also, remember that $X_L = 2\pi fL$ and $X_C = 1/2\pi fC$.

$$X_L = X_C$$

$$2\pi fL = \frac{1}{2\pi fC}$$

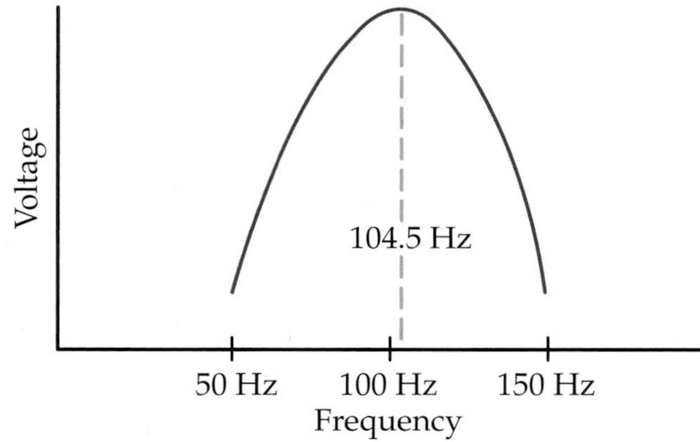

Figure 15-8. This graph shows the voltage across individual components in the circuit at different frequencies. Voltage is at a maximum at the resonant frequency, which in this case is 104.5 Hz.

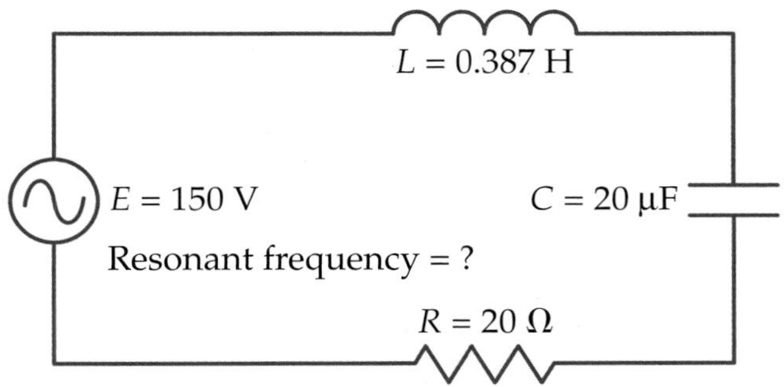

Figure 15-9. By changing the value of the capacitor in Figure 15-7, capacitive reactance changes. Therefore, the circuit will no longer resonate at 104.5 Hz, but at different frequency.

The value of *f* is the resonant frequency for this circuit. Plug in the known values and solve for *f*.

$$2 \times 3.14 \times f \times 0.387 \text{ H} = \frac{1}{2 \times 3.14 \times f \times 0.000020 \text{ F}}$$

$$2.43 \times f = \frac{1}{0.0001256 \times f}$$

$$2.43 \times f = \frac{7962}{f}$$

Next, multiply both sides of the equation by *f*:

$$2.43 \times f \times f = \frac{7962}{f} \times f$$

$$2.43 \times f^2 = 7962$$

Then, divide both sides by 2.43:

$$\frac{2.43 \times f^2}{2.43} = \frac{7962}{2.43}$$

$$f^2 = \frac{7962}{2.43}$$

$$= 3276.5$$

$$\sqrt{f^2} = \sqrt{3276.5}$$

$$f = 57.24 \text{ Hz}$$

If the capacitor is changed from 6.0 μF to 20 μF, we can make the circuit resonate at a frequency of 57.24 Hz. Therefore, if we want to tune a circuit to a certain frequency, all we have to do is change the value of the capacitor. This can also be done by changing the value of the inductor.

Tuning a Radio

To see how tuning works, look at the tuning circuit of a radio, **Figure 15-10.** The antenna picks up small voltages from radio waves transmitted through the air by many different stations. Each of these stations sends out waves at a different frequency. The antenna is shown as an inductor because most modern radio antennas are wound in a coil. Other types of antennas are found on land, cars, planes, and ships.

The purpose of the tuning circuit is to make the circuit resonant at the desired frequency. For example, if your favorite radio station broadcasts signals at a frequency of 680 kHz, you would adjust the capacitor in the tuning circuit until the circuit resonates at 680 kHz. The radio would then be tuned to receive your station's signal.

At the same time, small voltages with a frequency of 680 kHz develop very large voltages across both the antenna and the capacitor. These voltages are amplified by other components of the radio, and you hear your favorite station.

Tuning usually is performed by varying the capacitor in the circuit rather than the inductor. This is because it is relatively easy to make a variable capacitor.

Figure 15-11 shows a typical variable capacitor. It is made of a number of metal plates attached to a shaft called a rotor. Another set of plates, called the stator, is attached to the frame of the capacitor. Air is generally used as the dielectric material.

When the rotor shaft is turned, the plates on the rotor move between the plates on the stator. As the rotor is turned further, the distance between these plates and the rotor plates decreases. Recall from Chapter 4 that as

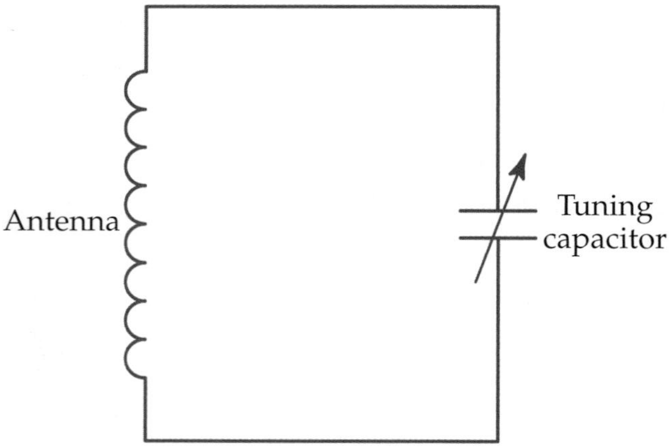

Antenna

Tuning capacitor

Figure 15-10. In a simplified tuning circuit, the antenna picks up radio waves and the capacitance is adjusted until the circuit is resonant and therefore tuned to a station's assigned broadcasting frequency.

Figure 15-11. Typical variable capacitor. Metal plates are turned by a dial to vary the amount of capacitance in the circuit.

the distance between the plates decreases, the capacitance increases. To increase the capacitance, you have to turn the rotor so that a larger area of the plates overlaps.

In radio models with a tuning knob or dial, the tuning knob is attached to the rotor shaft, which is extended through the front cover of the radio. Many are also attached to a pointer on a dial marked with the many different frequencies to help users find their favorite station.

Some digital radios are preset to accept the desired stations without having to turn a rotor on a capacitor. This type of radio is also capable of selecting stations by punching in their name. Some units have a "most played" grouping which you select. Others use a tuner from which you make your selection.

Project 15-1: Metal Detector

Used in conjunction with a tuned transistor radio, this simple metal detector will locate coins, keys, watches, and other metal objects lost in sand, dirt, or grass.

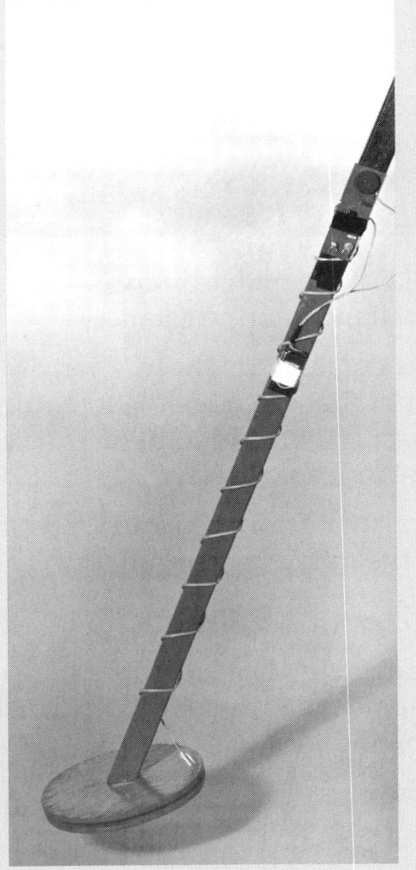

(Project by Ed Belliveau)

Gather the following items:

No.	Item
1	9-V battery
1	Capacitor, 680 pF
1	Capacitor, 0.001 µF
1	NPN silicon transistor (general purpose)
1	Potentiometer, 5 k
1	Resistor, 1 k
1	Resistor, 4.7 k
1	SPST switch
1	Coil made from 30′ of No. 26 wire
1	Wooden handle (e.g. broken hockey stick)
1	Plywood disk, 1/4″ thick, 6″ in diameter
2	Plywood disks, 1/4″ thick, 7″ in diameter
1	Circuit board
	Hookup wire, AWG 20

To build the metal detector, look at the schematic and follow the directions given.

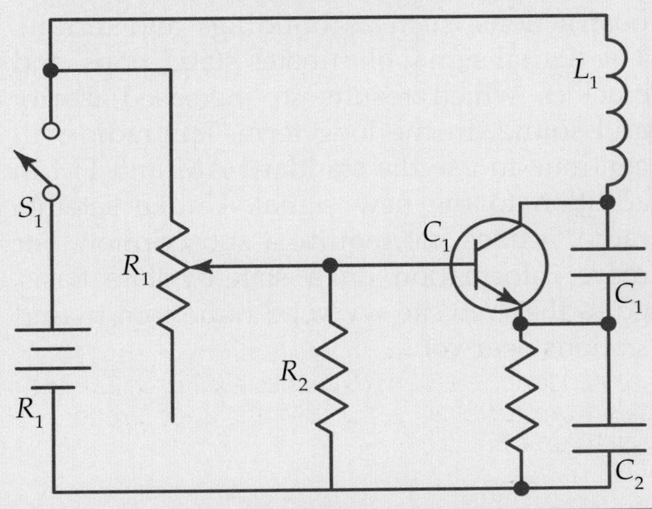

Project 15-1: Metal Detector *(Continued)*

Procedure

1. Using 1/4″ plywood, cut out one 6″ disk and two 7″ disks.
2. To make the coil form for the search coil, glue the three disks together with the small disk in the middle.
3. Glue the handle to the disks.
4. Wind 30′ of No. 26 wire on the coil form and wrap the lead wire up the handle.
5. Build the circuit as shown in the schematic. Mount the parts on a perforated board and solder all the connections. Q_1 can be any general-purpose transistor.
6. Connect the coil's lead wires to the circuit board.
7. Tape the circuit board and the 9-V battery to the handle.

To use the metal locator, follow these steps:

1. Place the metal detector on the surface to be tested.
2. Turn on the transistor radio and tune it to a high frequency station. Then move it slightly off the station, but make sure the station is still audible.
3. Turn on the metal detector and move the volume control until you can hear a squeal from the radio.
4. If you do not hear a squeal, repeat steps 2 and 3 at the next lower station. Keep doing this until you find the frequency with the loudest squeal.
5. Pass the coil slowly over the area to be searched. As the coil passes over buried metal, the inductance of the coil will change. This, in turn, will change the frequency of the squeal and tell you where to start digging for the metal you located.

High Definition Radio

The standard AM and FM radios are being supplemented by High Definition (HD) radio. These types of radio broadcast a digital signal that allows two or three signals to be sent at the same time. This is called ***multicasting.*** You choose which portion of the signal you wish to hear.

HD technology has fewer problems with obstructions, such as buildings and terrain. The digital signal eliminates static, pops, and crackles which results in increased clarity and sound. In the long term, HD radio will continue to use the standard AM and FM in addition to the new signal. Unlike satellite radio, it does not require a subscription. For more information on a state-by-state basis, go to the Web site www.hd-radio.com to find stations near you.

Summary

- An LCR circuit has inductance, capacitance, and resistance.
- To calculate impedance (Z) in an LCR circuit, first calculate the total reactance (X_T), and then add total reactance and resistance as vectors.
- To calculate current in an LCR circuit, use the following formula:

  $$I = \frac{E}{Z}$$
- To calculate voltages across the components in an LCR circuit, use the following formulas:

  $$V_L = I \times X_L$$

  $$V_C = I \times X_C$$

  $$V_R = I \times R$$

- In an LCR circuit, if the capacitive reactance (X_C) equals the inductive reactance (X_L), the circuit is resonant.
- At the resonant frequency, the voltages across the inductor and the capacitor are at their highest level.
- The purpose of tuning a circuit is to make it resonate at the frequency you desire.
- You can tune a radio by changing the value of a variable capacitor.

Test Your Knowledge

Do not write in this book. Write your answers on a separate sheet of paper.

1. What do the letters LCR stand for in an electrical formula?
2. What is the total impedance (Z) of a series circuit with an inductive reactance of 150 Ω, a capacitive reactance of 75 Ω, and a resistance of 10 Ω? Show your work.
3. Calculate the current for a circuit with an applied voltage of 10 V and an impedance value of 220 Ω. Show your work.
4. A circuit has the following voltage values: $V_L = 12$ V, $V_C = 9$ V, and $R = 5$ V. What is the applied voltage? Show your work.
5. In an LCR circuit, when the reactance of the capacitor equals the reactance of the inductor, the circuit is called a(n) ______ circuit.
6. In a resonant circuit, X_L will be equal to ______.
7. When a circuit is at a resonant frequency, will the voltage across the inductor and capacitor be at a maximum or minimum?
8. What happens when a circuit in a radio is tuned?
9. An antenna usually is shown in a circuit schematic as a ______ or an ______.
10. HD radio allows for ______, which means it can send one or more signals at the same time.

Activities

1. Find as many different kinds of radio coils as you can. Does the size or cost of a radio change the number of coils used? Why?
2. What are the sizes of the tuning capacitors used in radios?
3. Make a diagram of the tuning circuits used in a radio. What are the values for each?
4. Why do most radios use variable capacitors instead of variable inductors for tuning?
5. Does each city in the U.S. use the same set of frequencies for radio transmission? Are any of them the same or are they all different? Would the distance between cities make any difference which frequencies are used?
6. There are some nights when you can pick up radio stations hundreds of miles away from your home. How is this possible? Why does it happen at night? Why does it happen only on certain nights?
7. What agency of the government controls radio stations and the use of frequencies? Why must there be some type of control?
8. What international agency controls radio station frequencies when other countries are involved?
9. How would you go about learning how to operate a "ham" radio station? Who assigns call letters?

Can you identify the two round components in this picture?

16 Filters

Learning Objectives

After studying this chapter, you will be able to do the following:

- Explain the difference between a band-pass filter and a band-stop filter.
- Describe how to design a high-pass filter.
- Describe how to design a low-pass filter.

Technical Terms

attenuated
band-pass filter
band-stop filter
bandwidth
filter
filtering
high-pass filter
low-pass filter
pulsating direct current
pulses

In Chapter 15, we looked at LCR circuits and found that they are used as tuning circuits in radios. Tuning, however, is just one of many different ways LCR circuits are useful. This chapter covers LCR circuits and how they are used to filter the frequencies in a circuit. A ***filter*** is an electronic device that is used to eliminate a specified group of frequencies from an electronic signal. The action of doing that task is called ***filtering.***

There are a number of everyday uses for filters. When you turn on your TV, there must be a way to select each channel you want to watch. Just as important, you want to filter out other channels that might interfere with your viewing choice. This chapter shows the types of circuits that make this filtration possible, and it explains how these circuits filter out the unwanted frequencies. There are many different types of filters. Each has a specific function and is useful in certain applications. All of them, however, are LCR circuits. How the components are positioned with respect to the rest of the circuit determines the type of filter it is.

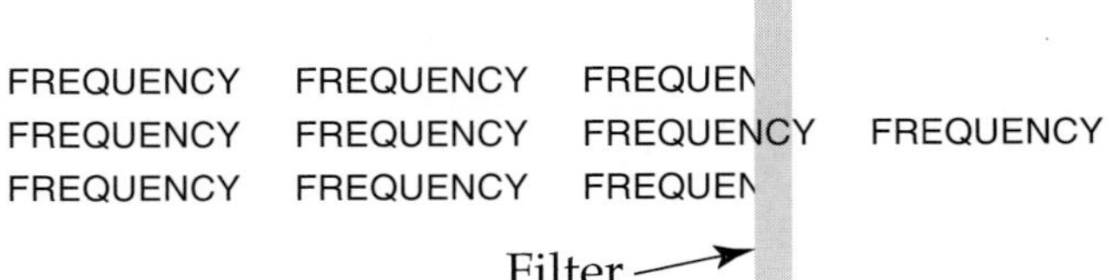

Figure 16-1. A band-pass filter allows only a narrow band of frequencies very close to the resonant frequency to pass through the circuit.

Band-Pass Filter

A ***band-pass filter*** allows only a narrow band above and below the resonant frequency (f_o) to pass through the circuit. The effect of this type of circuit is shown in **Figure 16-1.** As you can see, the band-pass filter blocks those frequencies outside of the desired range. A band-pass filter is constructed of inductors and capacitors. There are many configurations of band-pass filters.

Band-pass filters are used with transmitting equipment to limit the bandwidth of the transmitter's signal to the minimum amount that is necessary to send data at some desired speed. ***Bandwidth*** refers to the range of frequencies that are used to transmit a given signal. For example, the bandwidth that is capable of being heard by the human ear is about 20 kHz (20 Hz to 20,000 Hz). In a circuit designed for this bandwidth, the frequencies between the lower cutoff frequency (20 Hz) and the upper cutoff frequency (20,000 Hz) would be passed through a band-pass filter. The curve in **Figure 16-2** shows the range of frequencies passed. The lower cutoff frequency is represented by f_1 and the upper cutoff frequency is represented by f_2.

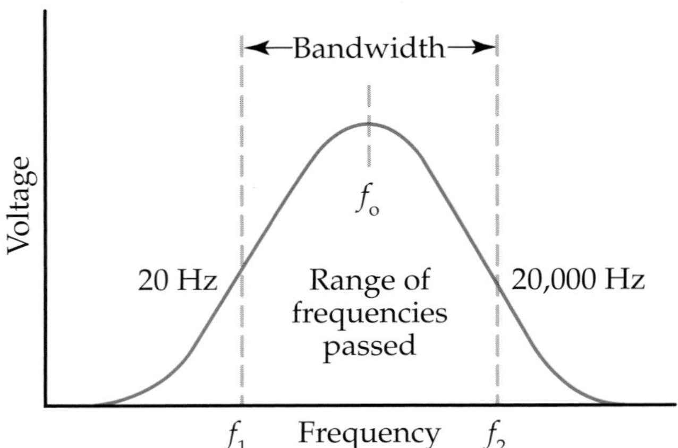

Figure 16-2. Curve showing the range of frequencies passed in a circuit designed to pass a 20 kHz bandwidth. Note that the center frequency is the resonant frequency of the circuit.

In actuality, no band-pass filter can truly block all the unwanted frequencies outside the desired frequency range. Some frequencies to the left of f_1 and others to the right of f_2 will tend to leak through the filter. These leaks represent noise when signals are transmitted and received. Noise is a problem when sending high-speed data transmissions over phone lines, cable lines, or satellite links. Imagine the loss of a single digit by a bank transferring billions of dollars using a computer. This has lead to a whole new industry and expensive electronics equipment constantly seeking a noise-free bandwidth for data transmission.

Once you know the frequencies you wish to pass and those you wish to attenuate (reduce or block), you can select the capacitors and inductors for the filter. It is not a simple matter of choosing the correct sizes and matching them against a chosen frequency. The process can be fairly complex and is outside the scope of this text.

Band-Stop Filter

A ***band-stop filter,*** on the other hand, stops frequencies very close to the resonant frequency from passing through the circuit, **Figure 16-3.** Remember that the band-pass filter allows a narrow band of frequencies to pass, and the band-stop filter prevents a narrow band of frequencies from passing.

A band-stop filter is useful when one certain frequency must be kept from entering a given circuit. For example, the frequency of the alternating current in your home is 60 Hz. This frequency lies in the audio range, which falls between 20 Hz and 20,000 Hz. If small voltages at this frequency entered a radio circuit, they would produce an annoying humming sound in the speaker.

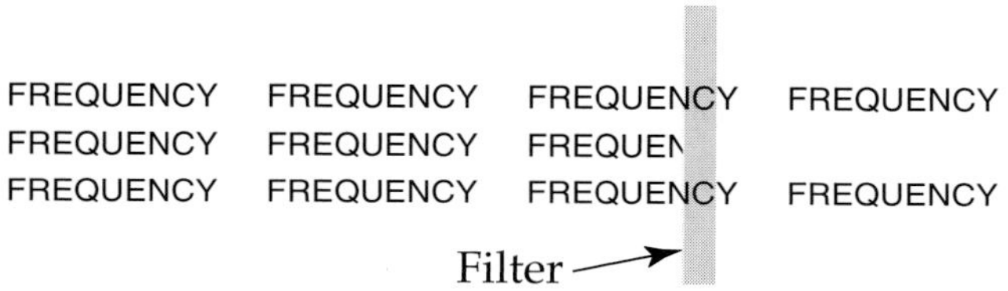

FREQUENCY FREQUENCY FREQUENCY FREQUENCY
FREQUENCY FREQUENCY FREQUEN
FREQUENCY FREQUENCY FREQUENCY FREQUENCY
Filter

Figure 16-3. A band-stop filter stops only a narrow band of frequencies very close to the resonant frequency.

The amplifier shown in **Figure 16-4** illustrates this use of a band-stop filter to stop any frequencies close to 60 Hz from entering the system. The amplifier has been protected against any 60-Hz voltage getting into it because the LCR circuit has been tuned to 60 Hz.

The input impedance of the amplifier is about 1000 Ω. That is, any signal coming into the amplifier is resisted by 1000 Ω. The filter is an LCR circuit with a resonant frequency of 60 Hz, and it is tuned to 60 Hz. Since the circuit is resonant, the only impedance for 60-Hz signals is from the resistor, which offers much less impedance than the amplifier.

The impedance from the resistor is only 10 Ω. If any 60-Hz signals get to terminals *A* and *B*, these signals take the path with the least resistance and travel through the filter, which means very little current at a frequency of 60 Hz flows into the amplifier. Signals at any other frequency, however, cause higher impedance in the filter since frequencies that are not resonant face greater resistance in LCR circuits. Those signals flow through the amplifier instead of through the LCR circuit.

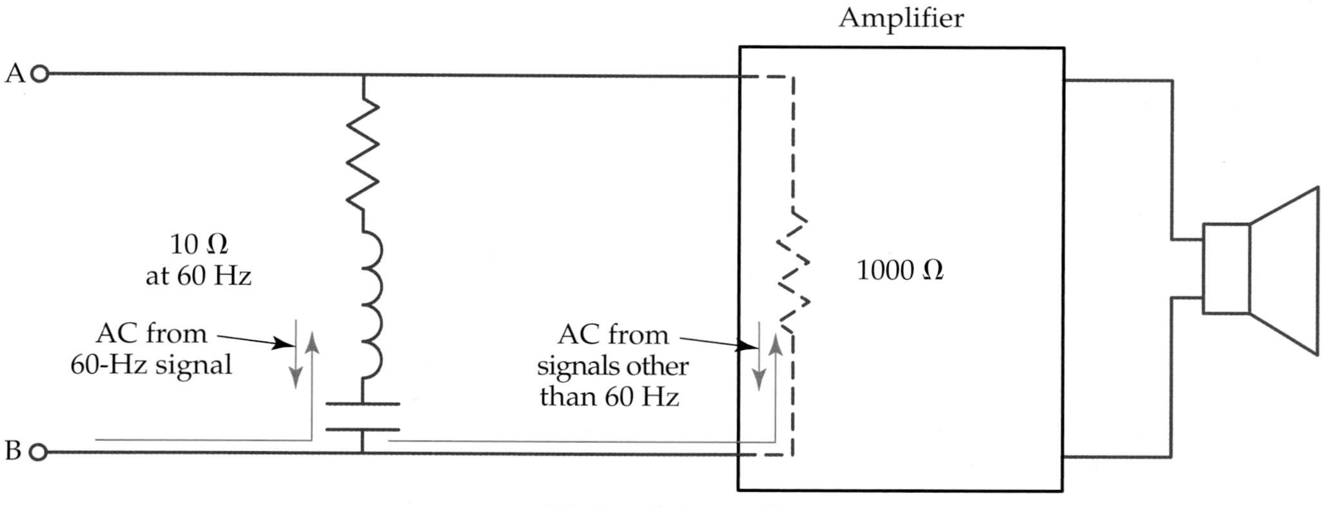

Figure 16-4. This band-stop filter intercepts a 60-Hz signal. Almost none of the 60-Hz signal passes through the amplifier.

High-Pass Filter

A ***high-pass filter*** is a circuit that allows only high frequencies to pass through it. See **Figure 16-5.** Any low frequencies that try to pass through this type of circuit will be ***attenuated.*** When a signal is attenuated, its amplitude is decreased. An attenuated signal becomes weaker, but it still has the same frequency. A simple high-pass filter can be made using only a capacitor, **Figure 16-6.**

To understand how a high-pass filter works, you have to remember how capacitors and inductors behave at different frequencies. The impedance of an inductor increases with high frequency. A capacitor, however, does just the opposite. Therefore, a high-pass filter is constructed of a capacitor and a resistor.

Note in **Figure 16-7** that as frequency gets higher, the reactance of the capacitor gets lower. As you will see in the following example, a capacitor with high reactance passes high frequencies but stops low frequencies from getting to the load resistor. High-pass filters can be useful in directing high frequency signals into the small tweeter speakers. When the frequency of the signal is high enough, it lowers the reactance of the capacitor enough to get past it.

High Frequency	High Frequency	High Frequency
Medium Frequency	Medium Frequency	Medium Frequency
Low Frequency	Low Frequency	Low Frequency

Figure 16-5. A high-pass filter is a circuit that passes a high-frequency signal and attenuates all other frequencies.

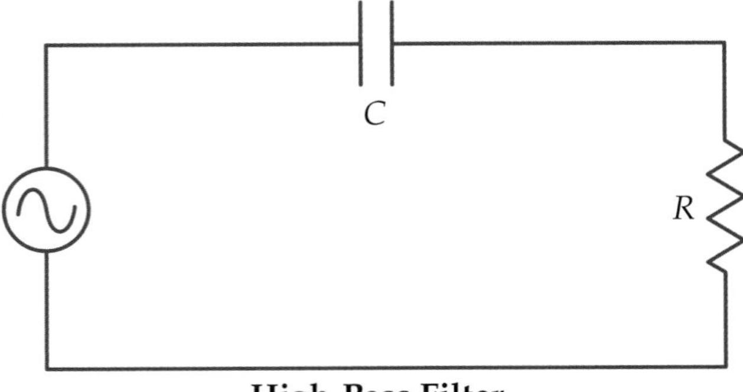

Figure 16-6. A simple high-pass filter.

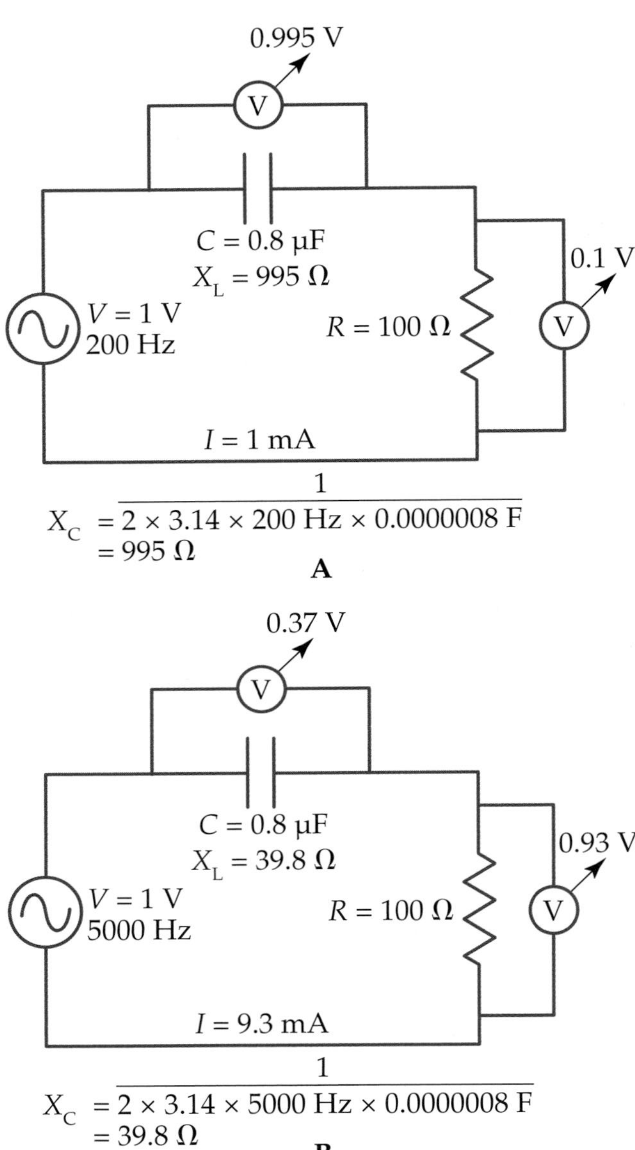

Figure 16-7. This high-pass filter is designed to stop frequencies lower than 500 Hz. A—A frequency lower than 500 Hz results in high capacitive reactance and a very low output across the load resistor. B—When 5000 Hz is applied to the circuit, the capacitive reactance is low and almost the entire amount of applied voltage can be measured across the load resistor.

Now, let us examine closely the circuits in Figure 16-7. To stop frequencies lower than 500 Hz, the circuit needs a capacitor whose reactance is higher than 100 Ω (the value of the load resistor) when the frequency is less than 500 Hz. This is achieved with a capacitor that has a capacitance of 0.8 μF.

Now, suppose a signal with a frequency of 200 Hz enters the filter, Figure 16-7A. The capacitive reactance (X_C) can be found using the equation from Chapter 14. Remember to first convert microfarads (μF) to farads (F):

$$0.8\ \mu F \times 10^{-6} = 0.0000008\ F$$

$$X_C = \frac{1}{2\pi fC}$$

$$= \frac{1}{2 \times 3.14 \times 200\ Hz \times 0.0000008\ F}$$

$$= \frac{1}{0.0010048}$$

$$= 995\ \Omega$$

Suppose that the 200-Hz signal applied to the circuit has an amplitude of 1 V. We can use Ohm's law to find how much of that voltage gets to the load resistor (R). First, we must find the current (I) in the circuit, which requires finding impedance (Z):

$$Z^2 = X_C^{\ 2} + R^2$$

$$= 995\ \Omega^2 + 100\ \Omega^2$$

$$= 990{,}025\ \Omega + 10{,}000\ \Omega$$

$$= 1{,}000{,}025\ \Omega$$

$$\sqrt{Z^2} = \sqrt{1{,}000{,}025\ \Omega}$$

$$Z = 1000\ \Omega$$

$$I = \frac{E}{Z}$$

$$= \frac{1\ V}{1000\ \Omega}$$

$$= 0.001\ A, \textit{ or } 1\ mA$$

Then, check and compare the voltages across the capacitor (V_C) and load resistor (V_R).

$$V_C = I \times X_C$$

$$= 0.001\ A \times 995\ \Omega$$

$$= 0.995\ V$$

$$V_R = I \times R$$

$$= 0.001\ A \times 100\ \Omega$$

$$= 0.1\ V$$

Most of the voltage drop at this frequency is across the capacitor. Only one-tenth of the voltage reaches the load resistor.

By circuit analysis, we have found that low frequencies are attenuated, or cut down, by a high-pass filter. In the following example, you will see what happens in this circuit at a high frequency.

Example 16-1:

Analyze the circuit in Figure 16-7B. This circuit has a frequency of 5000 Hz. Calculate how much of the voltage gets across the load resistor (R).

First, find the capacitive reactance (X_C):

$$X_C = \frac{1}{2 \times 3.14 \times 5000\ Hz \times 0.0000008\ F}$$

$$= \frac{1}{0.02512}$$

$$= 39.8\ \Omega$$

Now, find the impedance of the circuit:

$$Z^2 = X_C^{\ 2} + R^2$$

$$= 39.8^2 + 100^2$$

$$= 1584.0\ \Omega + 10{,}000\ \Omega$$

$$= 11{,}584.0\ \Omega$$

$$\sqrt{Z^2} = \sqrt{11{,}584.0\ \Omega}$$

$$Z = 107.6\ \Omega$$

Then, calculate the current:

$$I = \frac{E}{Z}$$

$$= \frac{1\text{ V}}{107.6\ \Omega}$$

$$= 0.0093\text{ A}$$

Finally, calculate the voltages across the capacitor (V_C) and across the load resistance (V_R):

$$V_C = I \times X_C$$

$$= 0.0093\text{ A} \times 39.8\ \Omega$$

$$= 0.37\text{ V}$$

$$V_R = I \times R$$

$$= 0.0093\text{ A} \times 100\ \Omega$$

$$= 0.93\text{ V}$$

Note that if the frequency of the signal is 5000 Hz, we get 0.93 V across the load resistor. This is almost the entire amount of applied voltage. In other words, this filter passed a frequency of 5000 Hz.

Adding an inductor in parallel with the load resistor improves this circuit, **Figure 16-8.** At high frequencies, Figure 16-8A, capacitive reactance (X_C) is very low. This means that most of the signal will get through the capacitor to point *A*. From point *A*, the signal has two possible paths, either through the inductor or through the load resistor. Since inductive reactance (X_L) increases as frequency increases, inductive reactance will be very high. Therefore, most of the signal takes the easier path and goes through the load resistor.

At low frequencies, Figure 16-8B, capacitive reactance (X_C) will be very high and most of the voltage drop is across the capacitor. The small amount of signal that does get to point *A* now can go through the inductor or the load. Inductive reactance (X_L) is very low at low frequencies. Therefore, most of the signal goes through the inductor, and very little goes through the load.

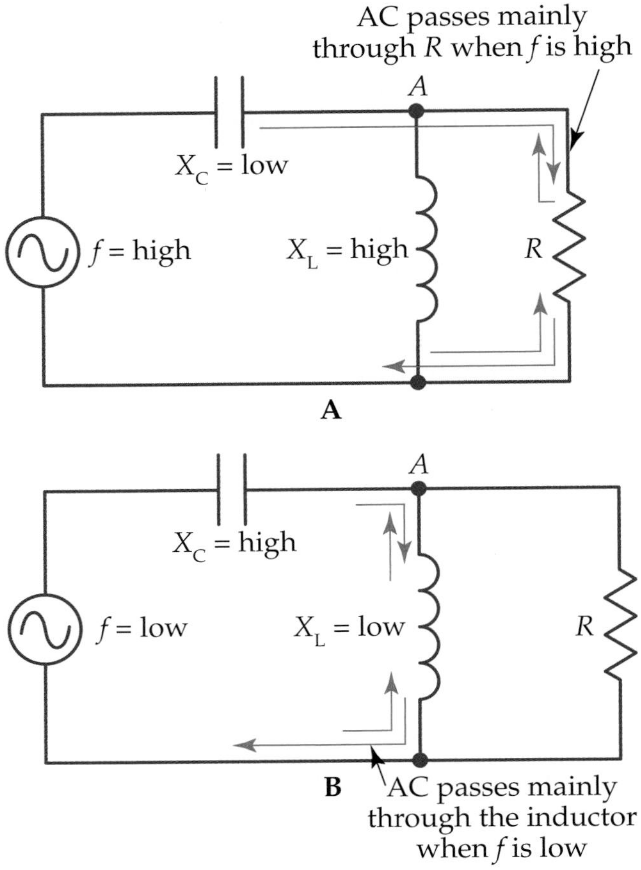

Figure 16-8. An inductor added to a high-pass filter improves performance. A—At a high frequency, most of the voltage passes through the load resistor. B—At low frequency, most of the voltage drop is across the capacitor and inductor.

This filter is the normal high-pass filter. The values of the capacitor and inductor depend on the frequency that needs to pass and on the load resistance.

Low-Pass Filter

A ***low-pass filter*** allows only low frequencies to pass through it and attenuates high frequencies. A low-pass filter works just the opposite to the way a high-pass filter works. Therefore, to make a low-pass filter, simply interchange the location of the inductor and capacitor in a high-pass filter, **Figure 16-9.**

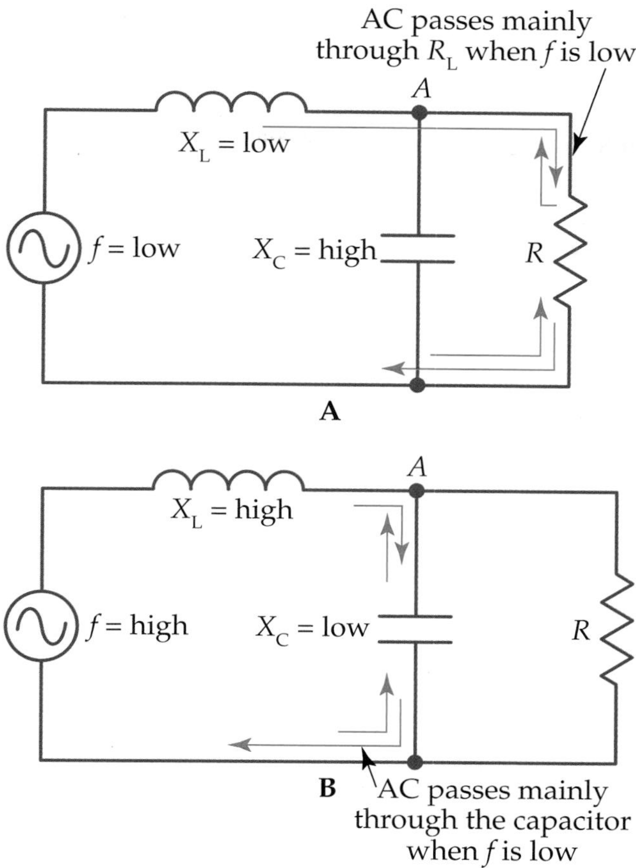

Figure 16-9. A low-pass filter. A—At a low frequency, most of the voltage drop is across the load resistor. B—At a high frequency, most of the voltage passes through the capacitor and inductor.

If a low frequency is applied to the low-pass filter, Figure 16-9A, the inductive reactance (X_L) is low, and the capacitive reactance (X_C), which is in parallel with the load resistance, is high. This means that low frequency signals will pass through the filter to the load resistor.

For high-frequency signals in low-pass filters, the inductive reactance (X_L) is high, and the capacitive reactance (X_C) is low, Figure 16-9B. Therefore, the small amount of voltage that does get by the inductor passes through the capacitor instead of through the load resistor. The high-frequency signals are blocked.

When designing a low-pass filter, the value of the capacitor and inductor depends on the range of frequencies you wish to block and pass. It also depends on the voltage being applied to the circuit and the value of the current. Since high-frequency and low-frequency are relative terms, you must know what values are being passed and which you want to block. Generally, low-pass filters can be useful in directing low-frequency signals into the large woofer speakers. Band-pass filters can also be created by combining a low-pass and a high-pass filter.

Power Supply Filter

Some devices, such as DVD players, amplifiers, and television sets, need a source of direct current. A power supply transformer is usually used to step down the voltage from the 120 V found in your home. Rectifiers, **Figure 16-10,** are circuits used to change the ac voltage to dc voltage.

You will learn more about rectifiers in Chapter 17. All you need to know now is that after the ac has passed through the rectifier, it is called ***pulsating direct current.*** See **Figure 16-11.** While the rectified current is dc,

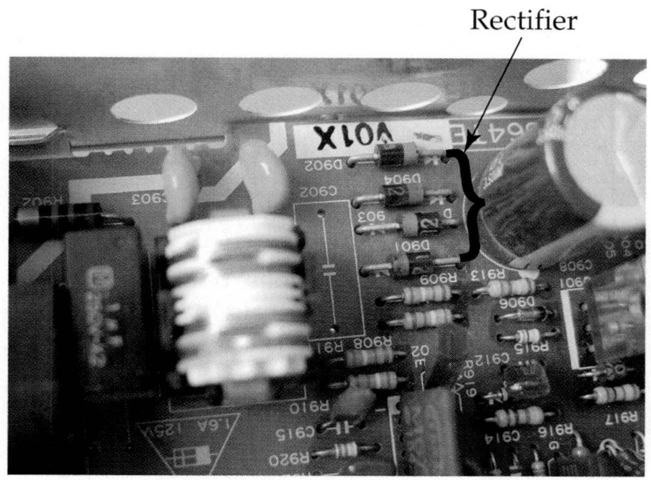

Figure 16-10. A rectifier converts ac voltages into dc voltages.

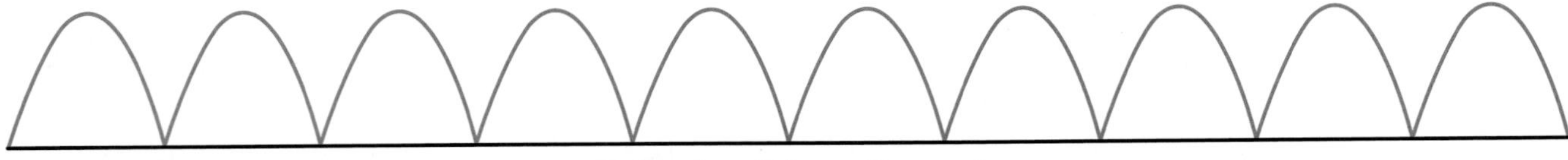

Figure 16-11. The rectifier changes alternating current to pulsating direct current.

the ***pulses*** at the top of the waves acts like ac. If this ac pulse gets into the amplifier section, it could cause an annoying hum.

To get rid of the ac pulsing, a filter is needed to level off the peaks and valleys of the output, **Figure 16-12.** One way to level out this pulsing is by adding a capacitor to the power supply's output. When the pulse rises to its peak value (point *A*), the capacitor charges to the peak value of the output voltage. When the pulse starts to drop in value (point *B*), the capacitor discharges through the load. This keeps the voltage across R_L at or near the peak value of the direct current. This cycle is repeated every time the pulsating current rises and falls.

By adding an inductor along with another capacitor, the output is dc with only a small amount of ripple, **Figure 16-13.** The circuit has filtered out the pulsating ac from a slightly rippled dc. This is called a Pi filter and is the most common type of filter used in power supplies.

Filter Arrangements

There are a number of different ways to arrange filters. Depending on the circuit, you may see those shown in **Figure 16-14.** These

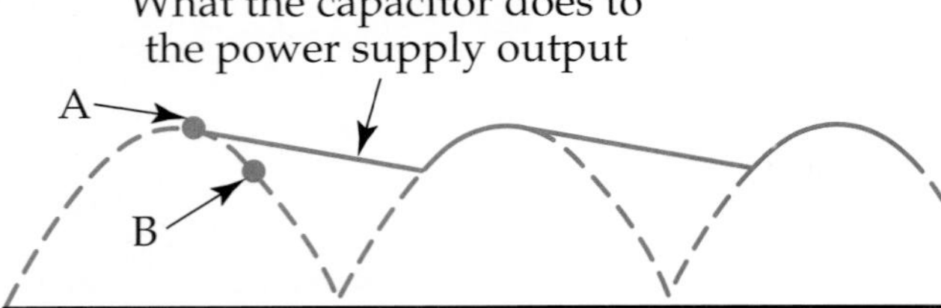

Figure 16-12. The charging and discharging of the capacitor filters pulsating dc into rippled dc.

Figure 16-13. Adding an inductor and another capacitor (Pi-Type Filter) reduces the rippled dc even further.

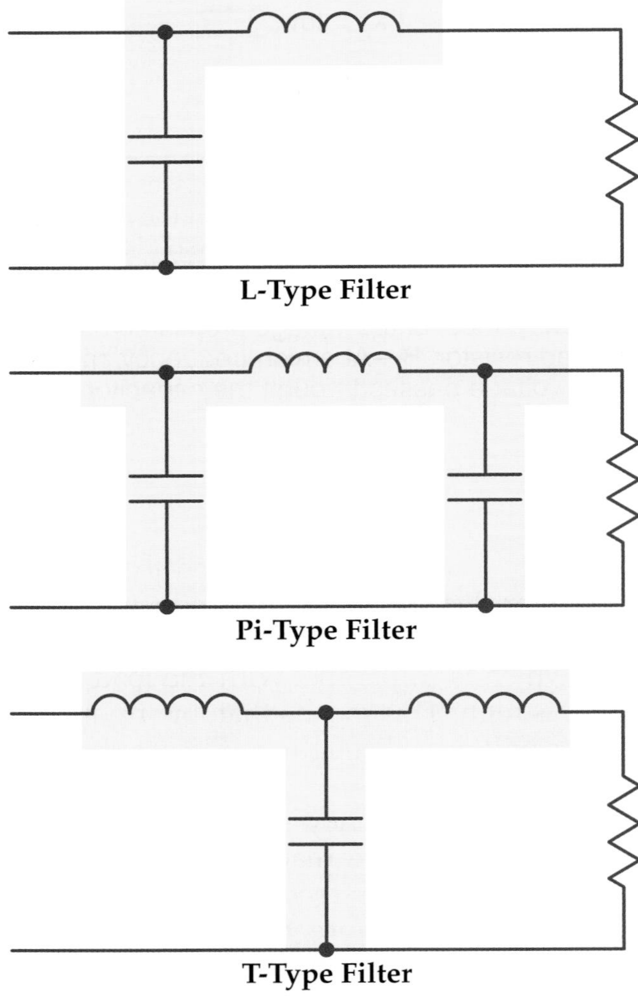

Figure 16-14. Special types of power supply filters are shown. They are named for the arrangement of components in the circuit.

arrangements are named after the shape (letter or symbol) of the circuit, such as *Pi, T,* and *L.* You may also see other combinations using many of these same components as filters, **Figure 16-15.** Some filters have additional components in their designs such as transformers and rectifiers. The important points to remember are that filters are designed to modify an incoming frequency, control the amount of direction of current flow, or modify or reshape the voltage.

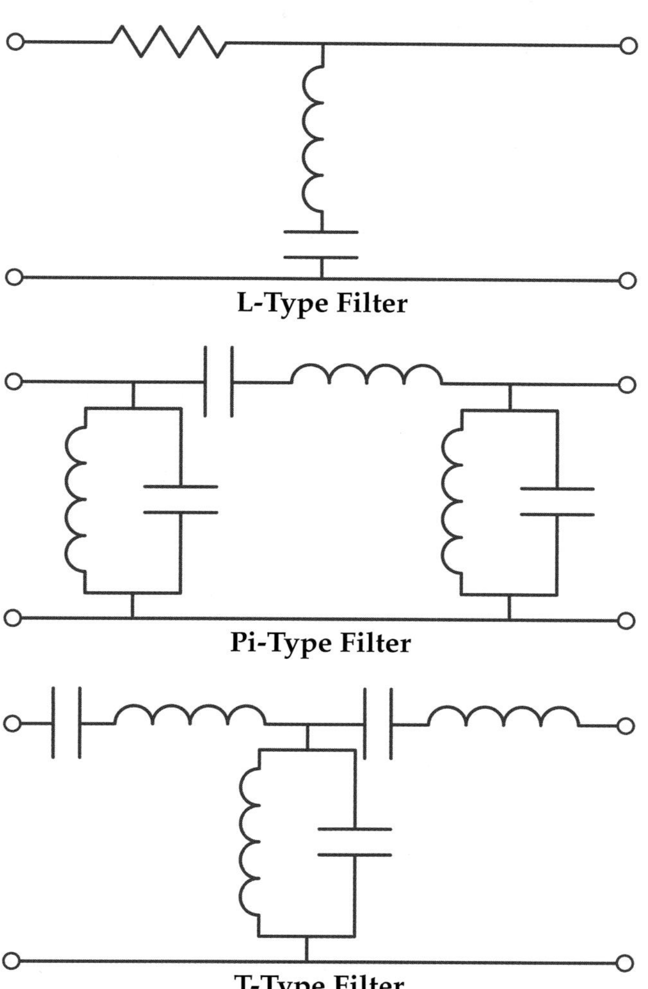

Figure 16-15. More complex arrangements of *L*, *Pi*, and *T* filters.

Summary

- The use of a circuit to separate different frequencies is called *filtering.*
- Filters are used to control which frequencies are allowed to pass and which frequencies are blocked in a circuit.
- A band-pass filter allows a narrow band of frequencies to pass through it.
- A band-stop filter stops a narrow band of frequencies from passing through it.
- A high-pass filter has a capacitor in series and an inductor in parallel with the load.
- A low-pass filter has an inductor in series and a capacitor in parallel with the load.

Test Your Knowledge

Do not write in this book. Write your answers on a separate sheet of paper.

1. A band-pass filter (passes, blocks) a narrow band of frequencies.
2. A band-stop filter (passes, blocks) a narrow band of frequencies.
3. When designing a low-pass filter, on what three things do the values of the capacitor and inductor depend?
4. Define *ripple.*
5. Using a(n) ______ will help cut down on ripple.

Activities

1. How does a radio station transmit frequencies? Why can't a message be heard as it goes through the air to a radio receiver?
2. What is the range of hearing for other animals?
3. Why is it possible to hear radio signals coming from the moon more easily than from some spot on earth that is only a few thousand miles away?
4. What are the bands that surround the earth and block radio signals called? Where are they located?
5. What are microwave signals? How are they transmitted?
6. How is it possible for the telephone company to use the same wire for more than one conversation? How is it possible that you can sometimes hear other voices when you are talking to someone on the telephone?
7. In some cities, you can pick up the audio from a local TV station on the radio. How is this possible?
8. All countries do not generate 60 Hz power for use in the home. Which ones do and which ones do not? What would happen if you took an electrical appliance designed for 60 Hz to a country operating at 50 Hz? What would happen if you used a 50 Hz electrical appliance in a country that uses 60 Hz?
9. How do ultrasonic machines work? How are they able to do things like welding without creating all the heat found in normal welding operations?
10. How do you determine the amount of inductance for a coil of wire? If you wanted a certain inductance, how would you determine how much wire is needed? What determines the size of cores used?
11. What kind of fire extinguisher should be used on electrical fires? Why?
12. Why are so many electrolytic capacitors used in filtering circuits?

Diodes

Learning Objectives

After studying this chapter, you will be able to do the following:

- Identify diodes and rectifiers.
- Explain the concept of a P-N junction.
- Determine the polarity of a diode and identify its anode and cathode.
- Discuss the methods of protecting diodes when soldering them into a circuit.
- Draw a full wave rectifier circuit and show the direction of flow through its leads.

Technical Terms

anode (A)
bridge rectifier circuit
cathode (K)
clipper potential (V_{CP})
depletion zone
diode
doping
forward bias
forward biased current (I_F)
hole flow
light emitting diode (LED)
organic light-emitting diode (OLED)
peak inverse voltage (PIV)
P-N junction
power supply
recovery time
reverse bias
solid-state devices
zener diode
zener knee
zener voltage (V_Z)

This chapter begins our study of ***solid-state devices.*** These components have replaced vacuum tubes in modern electronics. They include diodes, transistors, and silicon controlled rectifiers. Solid-state devices essentially do three things: stop, allow, and control the flow of electrons. This chapter covers diodes. A ***diode*** is a solid-state device that permits electron flow in one direction and blocks electron flow in the other direction.

> **Caution**
> An overload can easily ruin a solid-state device. Make sure the circuit does not exceed the voltage and current limits of the solid-state devices used in the circuit. Like fuses, if a solid-state device carries too much load, it opens. Unlike fuses, solid-state devices can be very expensive to replace.

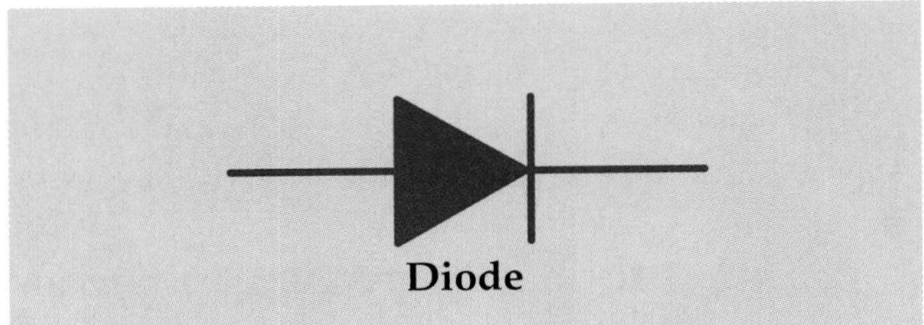

Figure 17-1. Schematic symbol for a diode.

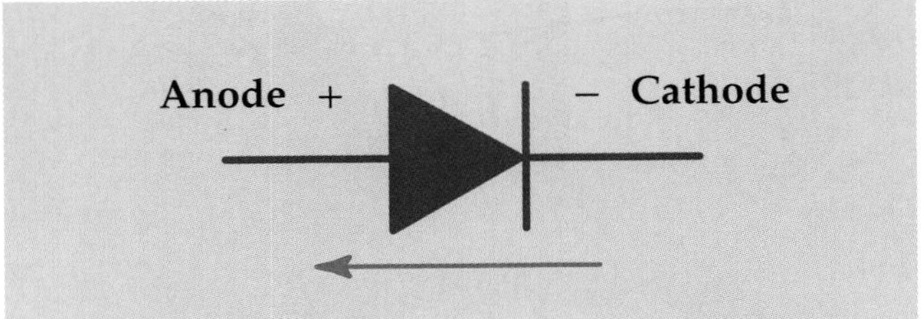

Figure 17-2. Current flows through a diode from the cathode (–) to the anode (+).

Diodes and Hole Flow

The schematic symbol for a diode is shown in **Figure 17-1.** A diode has a positive side, called an ***anode (A),*** and a negative side, called a ***cathode (K).*** In the symbol, the anode is the wide side of the triangle, and the cathode is the pointed side of the triangle with the line. A diode allows current to flow from the cathode to the anode, **Figure 17-2.** It does not allow current flow from the anode to the cathode.

A diode only works properly up to a certain limit. Once the current or the voltage exceeds a diode's limit, the diode breaks down. What happens in an overload is explained more thoroughly later in this chapter.

Diodes are typically made from either germanium or silicon. Germanium diodes carry smaller amounts of current than silicon diodes. Germanium diodes usually carry current somewhere in the milliamp range. Silicon diodes, sometimes called *rectifiers,* are larger and carry current that could easily be measured in amps.

To understand how diodes work, recall the explanation of electron flow from Chapter 1. Electrons are pushed through a circuit from one atom to the next. When an electron moves to the next atom, it leaves a hole where it was in the first atom. This hole is a space into which another electron could move. Solid-state devices, including diodes, work on the principle of ***hole flow,*** or the flow of these spaces that electrons leave behind. Hole flow is in the opposite direction of electron flow. To help you understand the principle, hole flow is demonstrated in **Figure 17-3.**

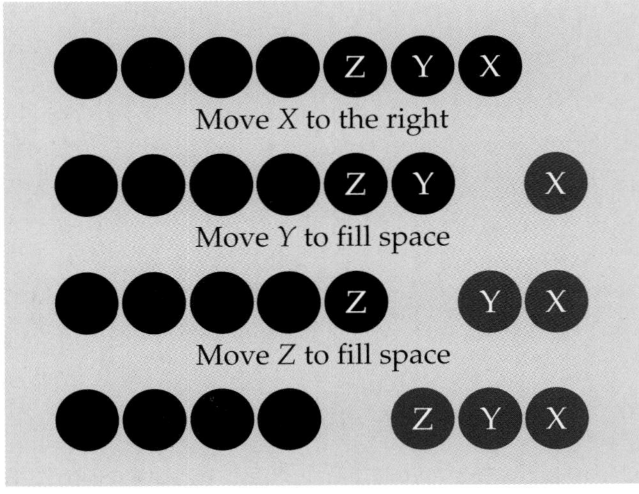

Figure 17-3. The electrons shown are moving to the right, and the holes are moving to the left.

Note that as *X* moves to the right, it leaves a hole to its left. *Y* moves to fill the hole and leaves a new hole in its place. As this continues and electrons move to the right, the hole moves to the left.

You learned in Chapter 1 that electrons are negatively charged. The holes the electrons fill are positively charged. The positive and negative charges and the principles of magnetism (like charges repel, unlike charges attract) are what make a diode work. Before you can learn how diodes work at an atomic level, you must learn about the P-N junction.

P-N Junction

A diode is made of P-type material, which contains holes, and N-type material, which contains electrons. You can think of the P-type material as positive and the N-type material as negative. The silicon or germanium used to make a diode can be made positive or negative through the process of ***doping.*** This process forces extra positive or negative charges into a material.

A ***P-N junction*** is an area of separation between P-type and N-type material in a diode. See **Figure 17-4.** A P-N junction is not a normal joint, such as a solder joint between two different materials. The diode is one piece of material, but part of it is negatively charged and part is positively charged. This is why the division between the P-type and N-type materials is called a *junction* and not a *joint* or *connection.*

When the diode is not in a circuit, a natural insulator or barrier is formed at the P-N junction, **Figure 17-5.** This is due to some of the electrons joining with some of the holes and some of the holes joining with some of the electrons near the P-N junction. The joined electrons and holes form a voltage potential across the P-N junction and block more electrons from joining other holes. They repel the other charges on each side of the junction, creating an area around the junction called the ***depletion zone*** where there are not many free electrons or holes.

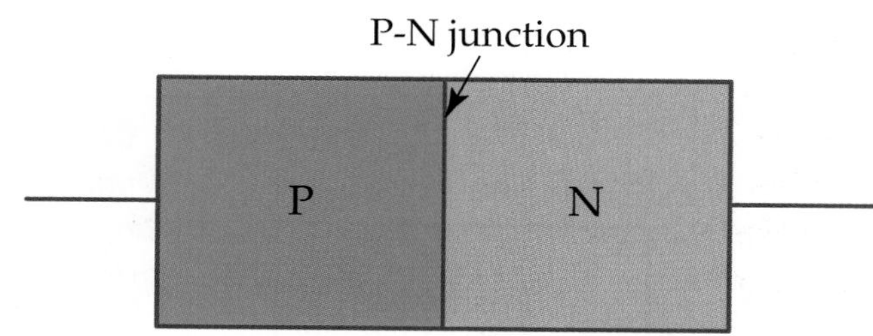

Figure 17-4. A P-N junction is the line of separation between the positive and negative materials that make up a diode.

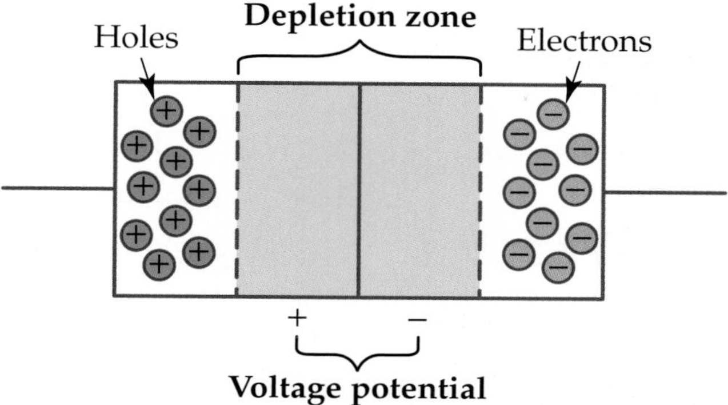

Figure 17-5. A P-N junction that does not have voltage applied to it forms a depletion zone, where there are not many free electrons or holes. A potential voltage exists across the P-N junction.

Forward Bias

When a diode is placed in a circuit and voltage is applied, the depletion zone is changed. However, it is not always changed in the same way. The polarity of the power source determines what happens within the diode. Suppose that the N-type material is connected to the negative side of the battery and the P-type material is connected to the positive side of the battery, **Figure 17-6.**

When the diode is positioned in this way, it is called ***forward bias.*** Remember that like charges repel. The negative polarity of the battery connected to the N-type material repels the negative free electrons, pushing them toward the depletion zone, Figure 17-6A. The positive polarity of the other side of the

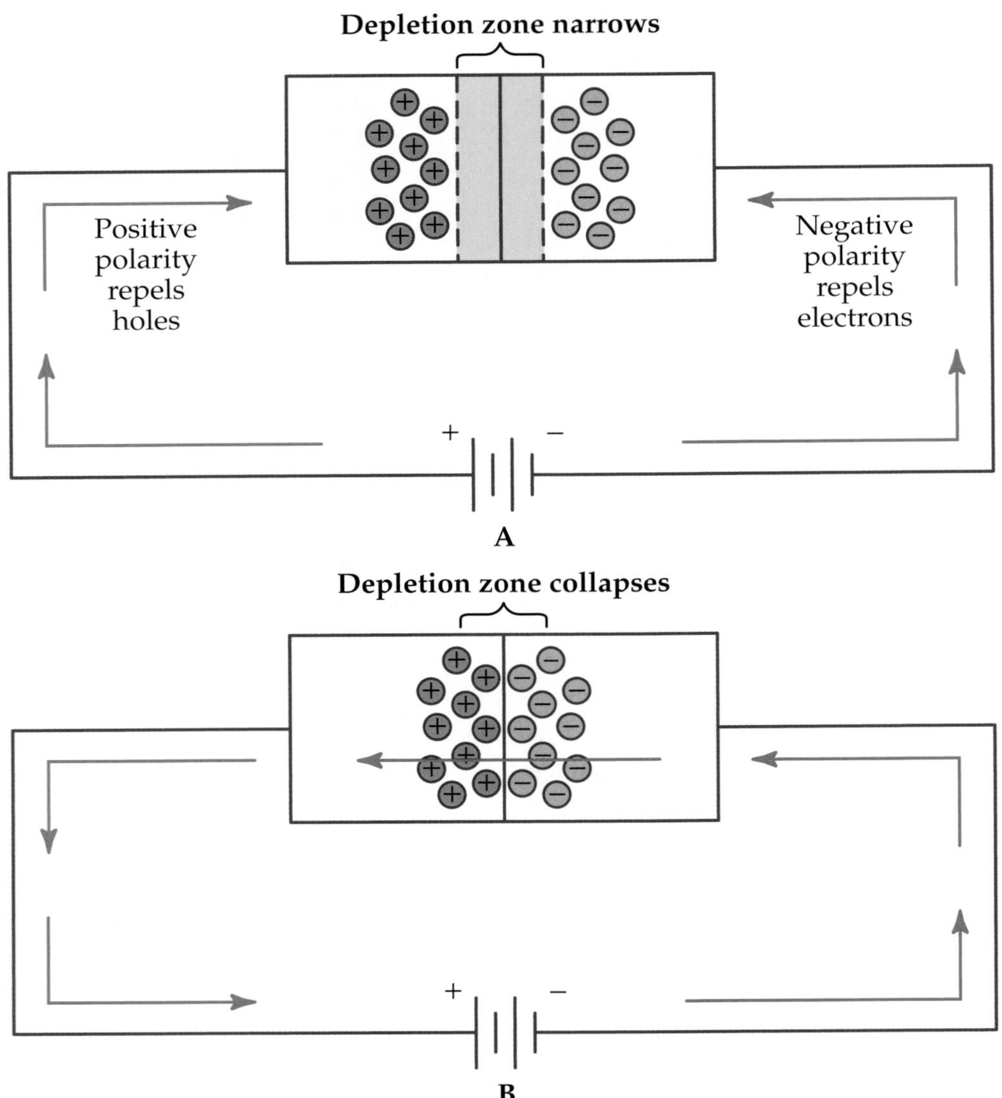

Figure 17-6. Forward biased diode. A—Electrons and holes are repelled toward the P-N junction. B—Current flows freely after the depletion zone has collapsed.

battery repels the positive free holes toward the depletion zone. When the free electrons and holes are pushed close to the neutral insulator, they narrow the depletion zone and begin to force the electrons that are joined with the holes toward the free holes in the P-type material. Eventually these electrons move out of the insulator, leaving free holes. The depletion zone and the insulator collapse, and current is allowed to flow through the diode, Figure 17-6B.

When a diode is forward biased, current is written I_F, and voltage is written V_F. A diode only allows current to flow in one direction.

Reverse Bias

When a power source is connected in the circuit with the opposite polarity, it is called ***reverse bias.*** For reverse bias, the positive terminal of the battery is connected to the N-type material, and the negative terminal of the battery is connected to the P-type material. Since opposite charges attract, the negative battery terminal attracts the holes in the P-type material, and the positive battery terminal attracts the electrons in the N-type material. This draws electrons and holes away from the junction, expanding the depletion zone, **Figure 17-7.** When this happens, current cannot flow through the diode.

It is possible to make current flow through a reverse-biased diode when too much voltage is applied. Sometimes, diodes are made for the purpose of letting current flow when the voltage reaches a certain level. However, most diodes are not made for this purpose, and overloading a diode can easily destroy it.

Peak Inverse Voltage

A diode has two important ratings: forward-biased current and peak inverse voltage. ***Forward biased current (I_F)*** is the current when the diode allows electron flow to complete the circuit. ***Peak inverse voltage (PIV)*** is the amount of voltage that can be applied in the reverse direction without damaging the diode. These ratings differ for all the different kinds of diodes and are frequently plotted on a graph. The curve on the graph is called a diode's characteristic curve, **Figure 17-8.** When plotting characteristic curves for diodes, voltage is shown on the horizontal axis (X), and current is shown on the vertical axis (Y).

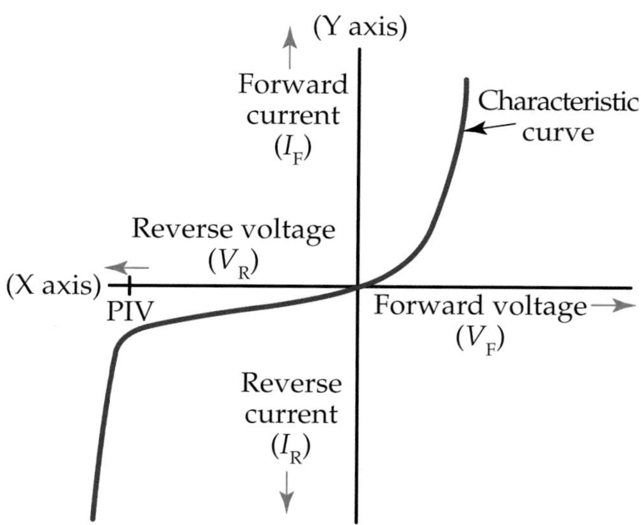

Figure 17-8. Forward current (I_F), forward voltage (V_F), reverse current (I_R), and reverse voltage (V_R) values for a diode can be plotted on a graph. The curve on the graph is called a diode's characteristic curve.

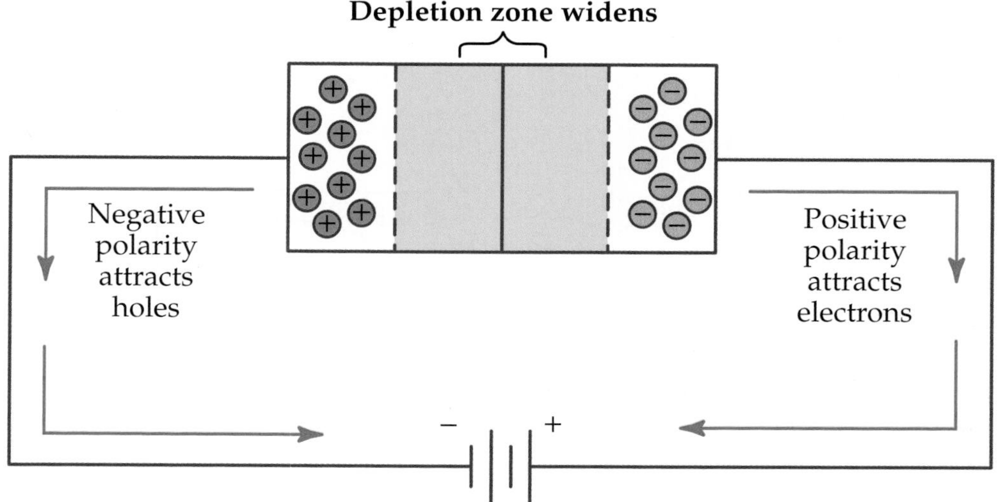

Figure 17-7. When a diode is reverse biased, electrons and holes are attracted away from the P-N junction. This expands the depletion zone and strengthens the diode's resistance to current flow.

There are four areas, called quadrants, used to plot curves, **Figure 17-9.** The four quadrants go counterclockwise around the center point where the two axes cross. The four quadrants are also assigned values based on their location. The locations are labeled and read as X axis, Y axis. For example, the first quadrant location is labeled +, + because the X axis and Y axis in this quadrant is in a positive region.

Characteristic curves for diodes fall into the first quadrant (+, +) and third quadrant (–, –), **Figure 17-10.** Forward-biased values are plotted in the first quadrant. Reverse-biased values are plotted in the third quadrant. By studying the characteristic curve of a diode, you can tell the following:

- The forward biased voltage (V_F).
- PIV and breakdown to destruction.
- Overload to destruction.

In **Figure 17-11,** notice that it takes about 0.3 V for germanium and 0.7 V for silicon diodes to be forward biased. Current flow does not take place immediately because of the internal resistance of the diode. Once the voltage reaches the bias level, there is a rapid increase in current. This is shown as I_F on the graph. Silicon diodes have more internal resistance than germanium diodes and take a higher voltage to reach its forward bias point. Silicon diodes also have a higher PIV rating.

Germanium diodes, however, can be turned on and off quicker than silicon diodes. That is why they are used for switching or signal tasks. Circuits for these applications have low voltages and low currents usually in the μA range. Silicon diodes are used in high-voltage and high-current applications.

The ***recovery time*** of a diode is the time it takes to turn on, turn off, and turn back on again. It is represented by the letters t_r. Germanium diodes have a fast recovery time that allows them to perform that task millions of times in one second. Although silicon diodes have a much slower recovery time, their high-current carrying ability makes them great for 60 Hz tasks, such as being used in ac-to-dc power supplies.

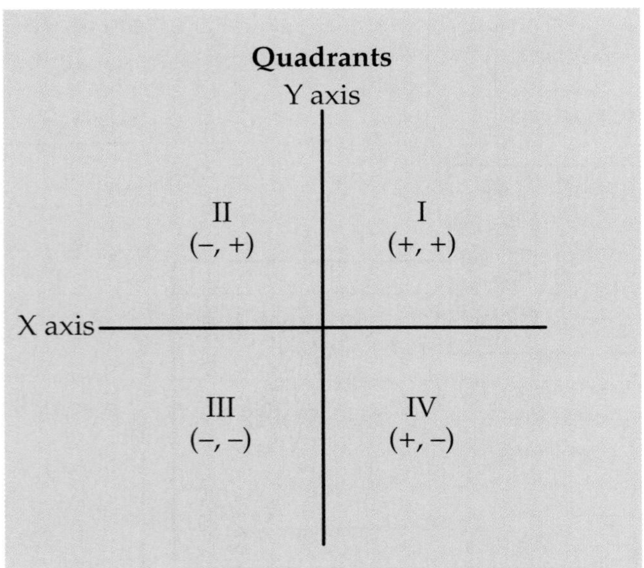

Figure 17-9. There are four areas, or quadrants, used on a graph to plot characteristic curves. They are assigned X axis, Y axis values based on their location.

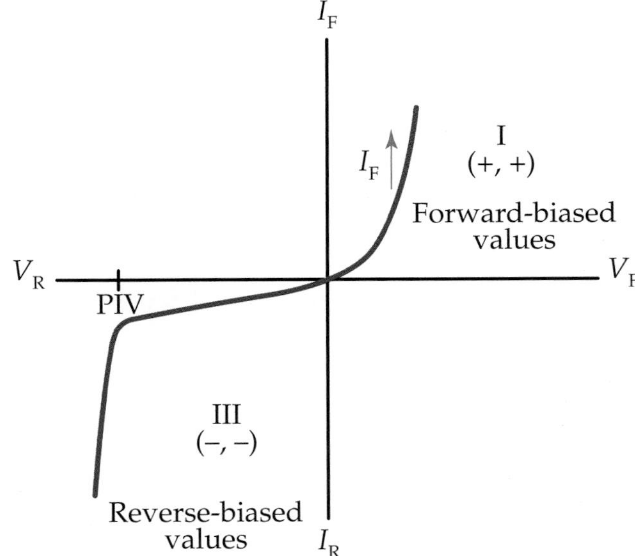

Figure 17-10. Characteristic curves for diodes occupy quadrant I and quadrant III.

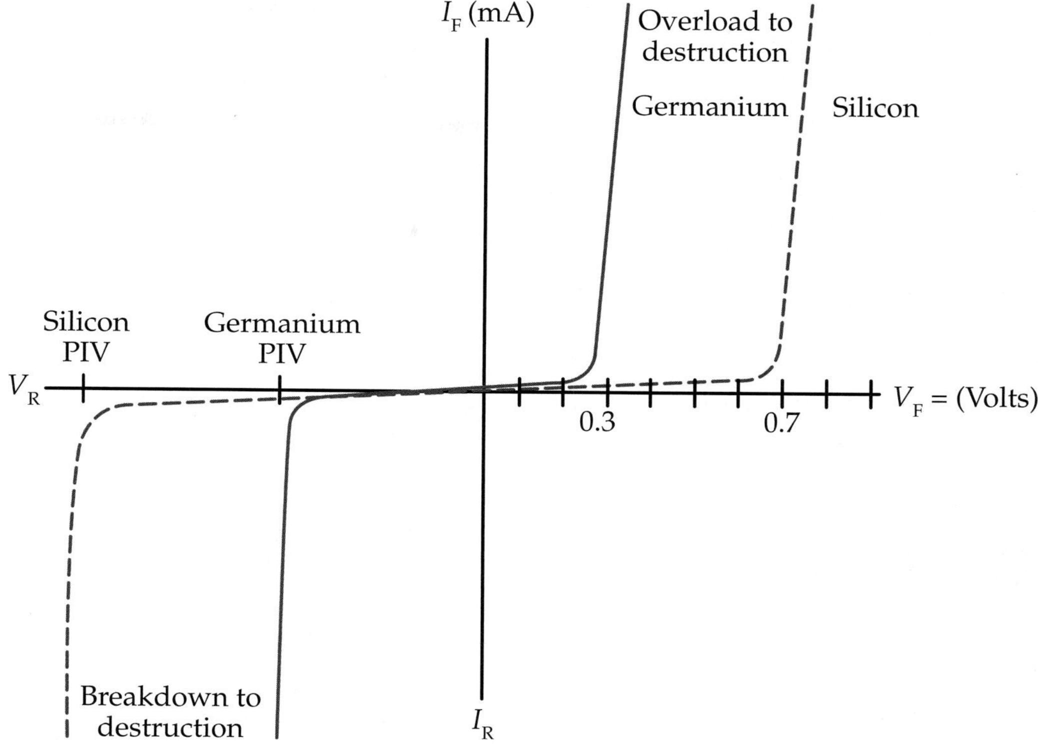

Figure 17-11. The characteristic curve of this graph shows that it takes about 0.3 V for germanium and 0.7 V for silicon diodes to be forward biased. A germanium diode has a smaller PIV rating than a silicon diode.

Identification Numbers and Color Code

The I_F and PIV ratings of a diode depend on the size of the diode. There is a wide range of diode sizes and ratings. To keep track of them, diodes are given numbers. Most diodes start with 1N, followed by other numbers. Examples include: 1N40, 1N363, 1N946 and 1N3750. Other letter and number combinations are also used by different manufacturers as their diode codes, such as RGP30G, NTE5332, and SK3092. Other solid-state devices, such as transistors, bridge rectifiers, thryristors, and SCRs, have similar code combinations. These numbers allow you to search for a specific device using a substitution manual.

Germanium diodes often have color-coded bands, **Figure 17-12.** The colors in this code represent the same numbers as in the resistor color code:

Color	Number It Represents
Black	0
Brown	1
Red	2
Orange	3
Yellow	4
Green	5
Blue	6
Violet	7
Gray	8
White	9

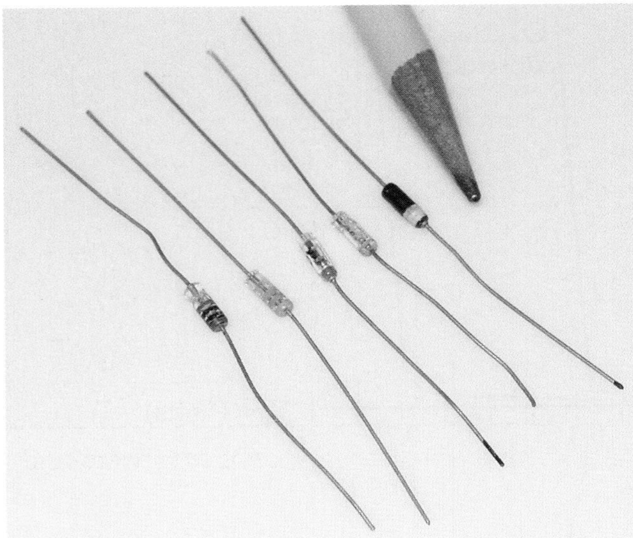

Figure 17-12. Germanium diodes usually have bands of color that help to identify them.

However, the code works slightly differently. To read the code, always turn the diode so the colored bands are to the left. Each code starts with 1N, followed by the numbers represented by the colored bands. Unlike the resistor color code, there is no multiplier. The third band just represents a digit instead of the number of zeroes to add to the first two digits. The colors on the diode in **Figure 17-13** are brown, orange, and red. This diode is a 1N132 germanium diode.

Example 17-1:

What is the identification number of the diode shown here?

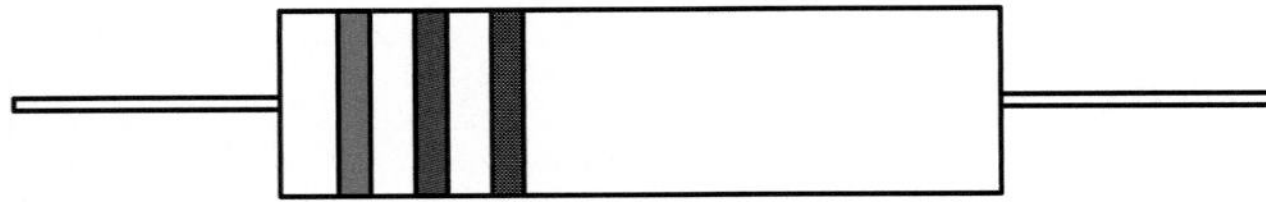

The diode is turned so the bands are on the left, and they can be read in order. The colors are green, violet, and black:

Green—5 Violet—7 Black—0

The identification number of the diode is 1N570.

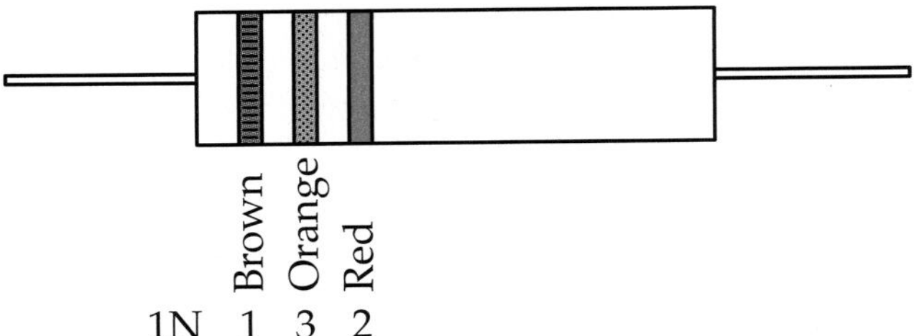

Figure 17-13. An example of a color-coded diode.

Resistance

Diodes do not have a fixed resistance. The resistance offered by a diode changes depending on the amount of voltage applied to the diode. **Figure 17-14** shows how to connect the ohmmeter test leads across the diodes or rectifiers to make resistance tests. As you move the range selector switch from R × 1 to R × 10, you are decreasing the voltage applied by the ohmmeter to the diode. Each diode or rectifier has different resistance values, depending on where the range selector switch is set.

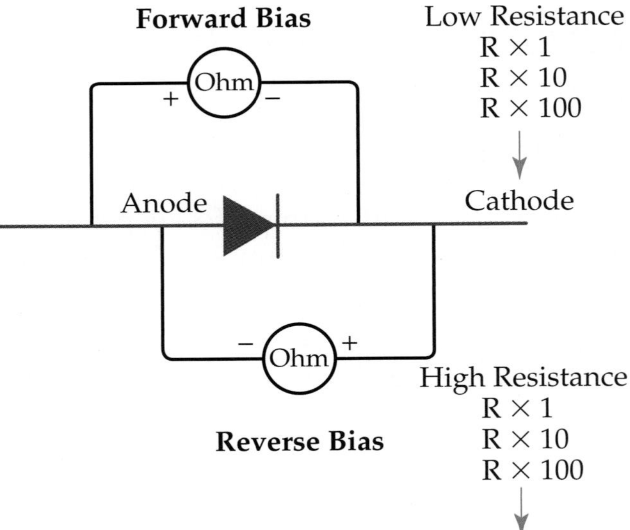

Figure 17-14. To make resistance checks on diodes, connect ohmmeter test leads as shown and set the range-selector switch to various positions on the meter.

	1N604		1N159		408A	
	Forward	Reverse	Forward	Reverse	Forward	Reverse
R × 1	13 Ω	∞	7 kΩ	∞	10 Ω	∞
R × 10	45 Ω	∞	37 kΩ	∞	72 Ω	∞
R × 100	200 Ω	∞	230 kΩ	∞	560 Ω	∞
R × 1000	1.1 kΩ	700 kΩ	1.5 kΩ	∞	4.1 kΩ	∞
R × 10 k	5 kΩ	420 kΩ	9 kΩ	Almost ∞	30 kΩ	∞
R × 100 k		160 kΩ	50 kΩ	7 kΩ	200 kΩ	Almost ∞
R × 1 M				6 kΩ	1.1 kΩ	70 MΩ

Figure 17-15. This table shows the results of resistance tests for three different diodes in forward and reverse bias on each range selector resistance setting.

Figure 17-15 shows a table of resistance values for three different diodes at the different settings of the ohmmeter. Note that the resistance is not constant for any of the diodes. Current increases or decreases, depending on the amount of voltage applied. How much the current changes is not fixed by Ohm's Law because, in some cases, the law does not apply to solid-state devices.

Diodes have different resistance values when connected in a circuit. As current flows through a diode, there is a change in temperature. As the surrounding temperature increases, the voltage value for the given amount of current changes. Some of these measurement changes will be very slight. Other changes will be very noticeable. These changes are taken into consideration for specific uses. For example, some diodes can be used for temperature sensing, while others can be used to protect equipment from current overloads.

Diodes in Circuits

The direction of current flow is important in diodes because they can be easily destroyed. It is essential to know the polarity, the I_F and PIV ratings, and how a diode will react in a circuit with other components.

Diodes in Series

Some circuits use two diodes in series, **Figure 17-16.** Notice that one of the diodes in Figure 17-16A has an I_F rating of 3 A, and the other has an I_F rating of 1 A. Remember

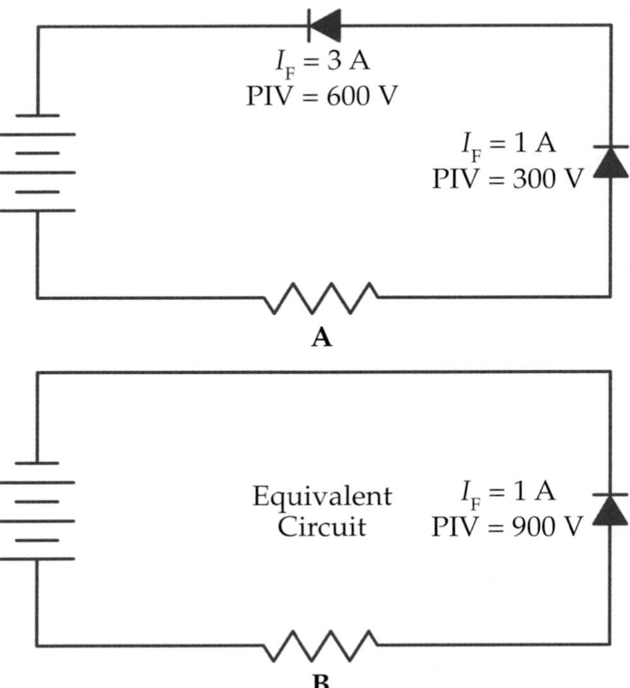

Figure 17-16. Diodes in series pass current that is equivalent to the diode of the lowest I_F rating. PIV is equal to the sum of the PIVs of each diode in series.

that components in series always have the same amount of current passing through them. Therefore, if more than 1 A were to pass through the circuit, this diode would be overloaded and destroyed. The amount of current that can safely pass through these diodes is 1 A. Diodes in series can only handle current equivalent to the I_F rating of the diode with the lowest rating.

When diodes are in series, the sum of the PIVs of the individual diodes is the PIV of the circuit. Look again at Figure 17-16A. The PIV of one diode is 300 V, and the PIV of the other is 600 V. The PIV of the circuit is found with the following formula:

$$PIV = PIV_{D1} + PIV_{D2} + \cdots \quad (17\text{-}1)$$

Therefore, the PIV for the circuit in Figure 17-16A is

$$PIV = PIV_{D1} + PIV_{D2}$$
$$= 300\text{ V} + 600\text{ V}$$
$$= 900\text{ V}$$

When diodes are in series, I_F remains as low as the smallest I_F rating of the individual diodes, and PIV increases to the sum of the PIVs of the individual diodes. Figure 17-14B shows the equivalent circuit.

Diodes in Parallel

Now, suppose these same diodes are connected in parallel, **Figure 17-17.** The I_F rating and PIV rating of diodes connected in parallel behave differently than diodes connected in series. The I_F value for the parallel hookup is the sum of the individual ratings.

$$I_F = I_{FD1} + I_{FD2} + \cdots \quad (17\text{-}2)$$

The individual I_F ratings of the diodes in Figure 17-17A are 2 A and 2 A. Therefore, the total I_F rating this parallel circuit is

$$I_F = I_{FD1} + I_{FD2}$$
$$= 2\text{ A} + 2\text{ A}$$
$$= 4\text{ A}$$

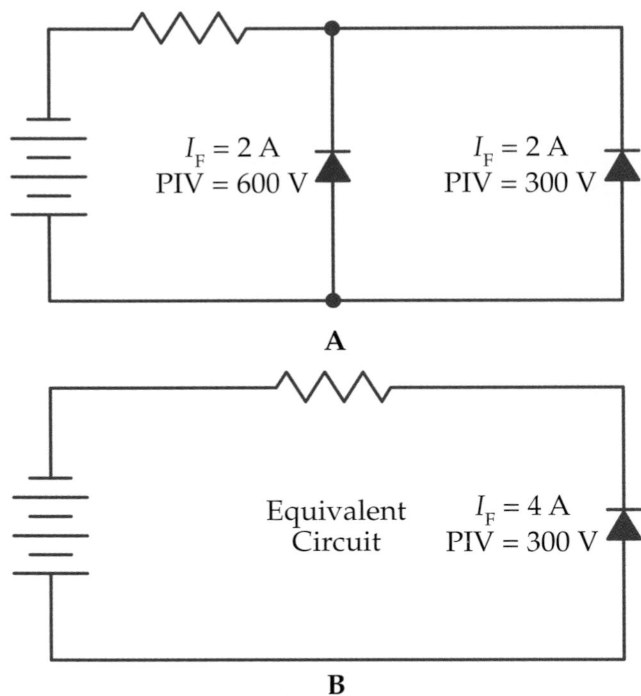

Figure 17-17. Diodes connected in parallel have a PIV equivalent to the diode with the lowest PIV rating. I_F is equal to the sum of the I_F of each diode in parallel.

This means that if the load in a circuit needs more current than a given diode can take, another diode can be added in parallel to handle the current load.

The PIV for a parallel hookup is the lowest PIV of the individual diodes. This is because the voltage drop across the parallel section of a circuit is the same for each branch. A voltage equivalent to the higher PIV would cause reverse current to pass through the branch of the circuit with the lower PIV. So, for the circuit in Figure 17-17A, the PIV is 300 V.

Therefore, when diodes are in parallel, I_F increases to the sum of the I_F ratings of the individual diodes, and PIV remains as low as the smallest PIV of the individual diodes. Figure 17-17B shows the equivalent circuit.

Practical Application 17-1: Determining Diode Polarity

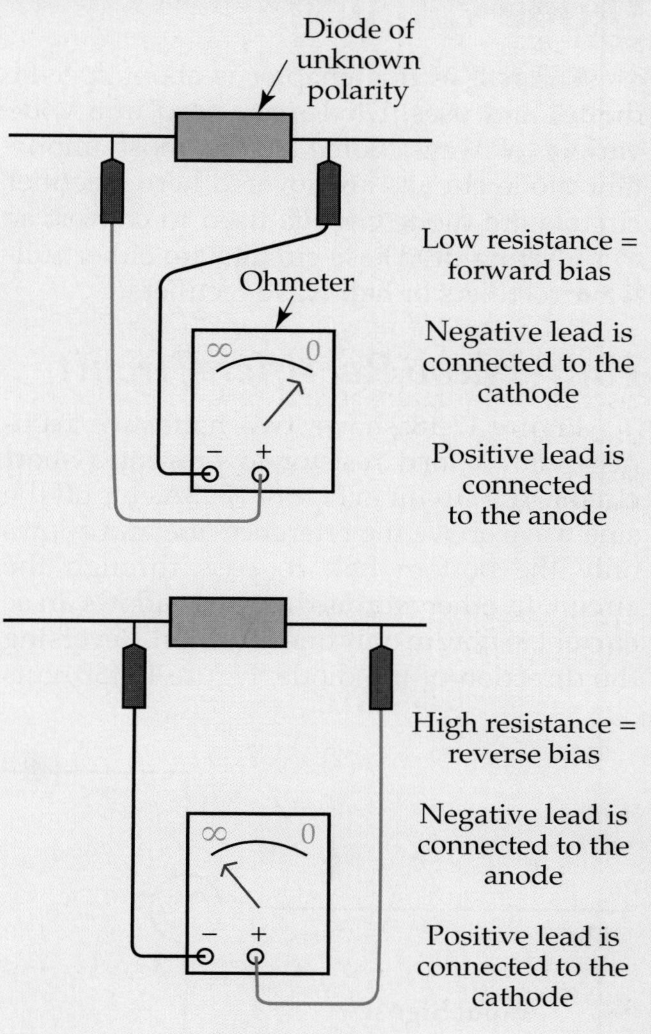

Sometimes the polarity of a diode is not marked. Before a diode is connected in a circuit, it is necessary to know its polarity. An ohmmeter can be used to determine which end of the diode is the cathode and which is the anode. Then the diode can be placed in the circuit with the correct polarity. Follow these steps to determine and mark the anode and cathode of a diode with unknown polarity:

1. Hook the positive lead from the ohmmeter to one end of the diode and the negative lead to the other end.
2. Record the measurement.
3. Reverse the ohmmeter leads. If the resistance reading is lower than the first reading, the diode is forward biased. If the resistance reading is higher, the diode is reverse biased. (Depending on the scale used, a high reading will be in the kilo to infinity range while a low reading will usually be somewhere between 5 Ω and 1000 Ω. A shorted diode will have a reading about the same in both directions.)
4. Connect the leads in the forward biased direction. The end of the diode connected to the negative lead of the ohmmeter is the cathode. The end of the diode connected to the positive lead of the ohmmeter is the anode.
5. Mark a minus sign (–) on the cathode.
6. Mark a plus sign (+) on the anode.

It is important when making ohmmeter tests that you know which of the terminals is plus and which is minus. Although the outside of the meter is marked with a plus and minus, this polarity might not be correct when compared with the battery. This is especially true in some, but not all, of the less expensive analog meters. The safest way to find out is to open the back of the meter and see where the leads from the battery are connected (plus to plus, minus to minus).

Diode Circuits

The rest of this chapter is about specific diodes and uses. Diodes are used in a wide variety of ways. Some of the most important diode circuits are covered here. Rectifier circuits are diode circuits used to convert ac to pulsating dc. These circuits are either full-wave rectifiers or half-wave rectifiers.

Half-Wave Rectifier Circuit

Figure 17-18 shows two half-wave rectifiers with a load resistor to prevent a short circuit. The circuit in Figure 17-18A cuts off the sine wave above the reference line and allows only the bottom half to pass through the circuit. In other words, this setup allows an ac current to flow in only one direction. Reversing the direction of the diode, Figure 17-18B, cuts off the sine wave below the reference line and allows only the top half to pass through the circuit. The diode still allows current in only one direction. When the diode is reversed, the direction of the current it allows is also reversed.

Full-Wave Rectifier Circuit

A full-wave rectifier forces all current to flow through the load in the same direction. Unlike the half-wave rectifier, there are not sections where the current is at zero. The diodes are wired to make both directions of ac pass through the load in the same direction.

There is more than one type of full-wave rectifier circuit. Two common full-wave rectifier circuits are covered here. The circuit in **Figure 17-19** is known as a ***bridge rectifier circuit.*** As you can see, four diodes are con-

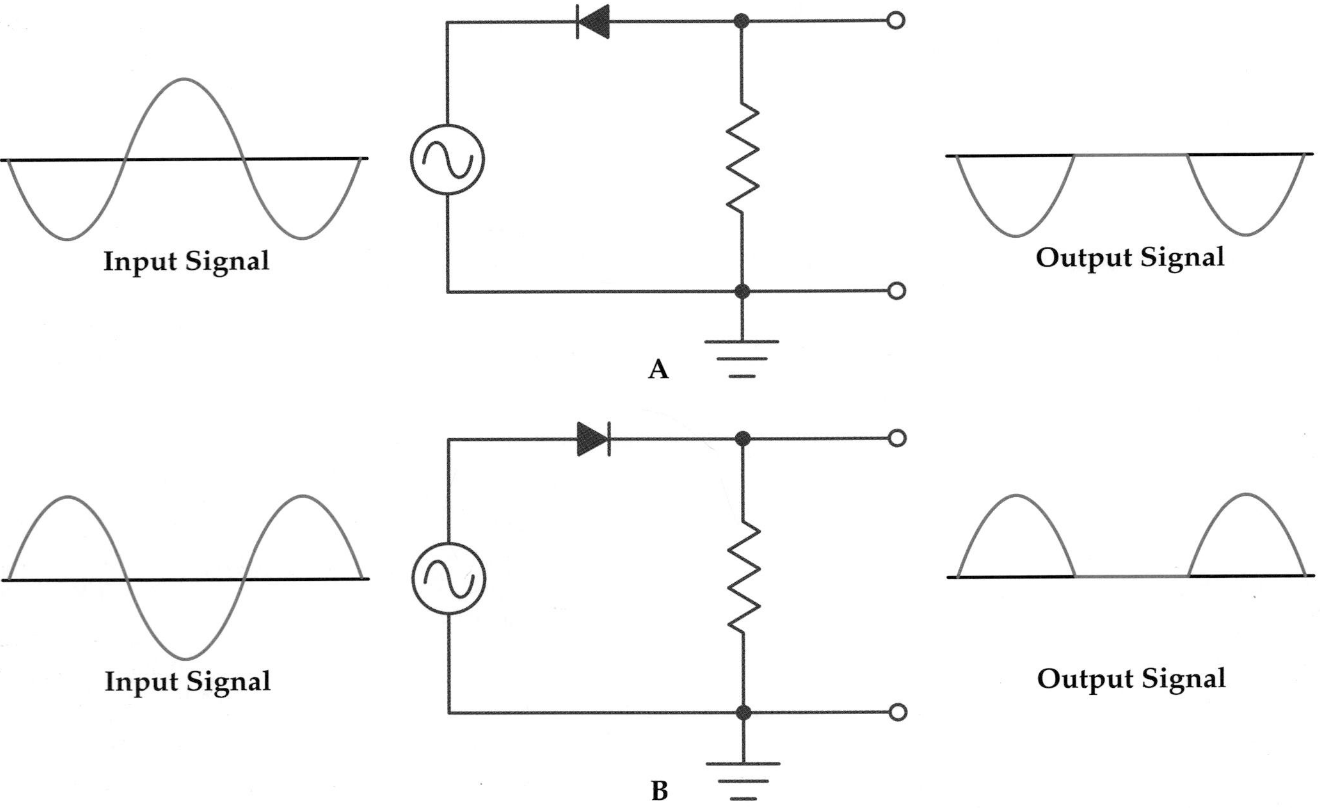

Figure 17-18. A half-wave rectifier circuit only allows current to pass in one direction. A—This configuration cuts off the sine wave above the reference line and allows only the bottom half to pass through the circuit. B—This configuration cuts off the sine wave below the reference line and allows only the top half to pass through the circuit.

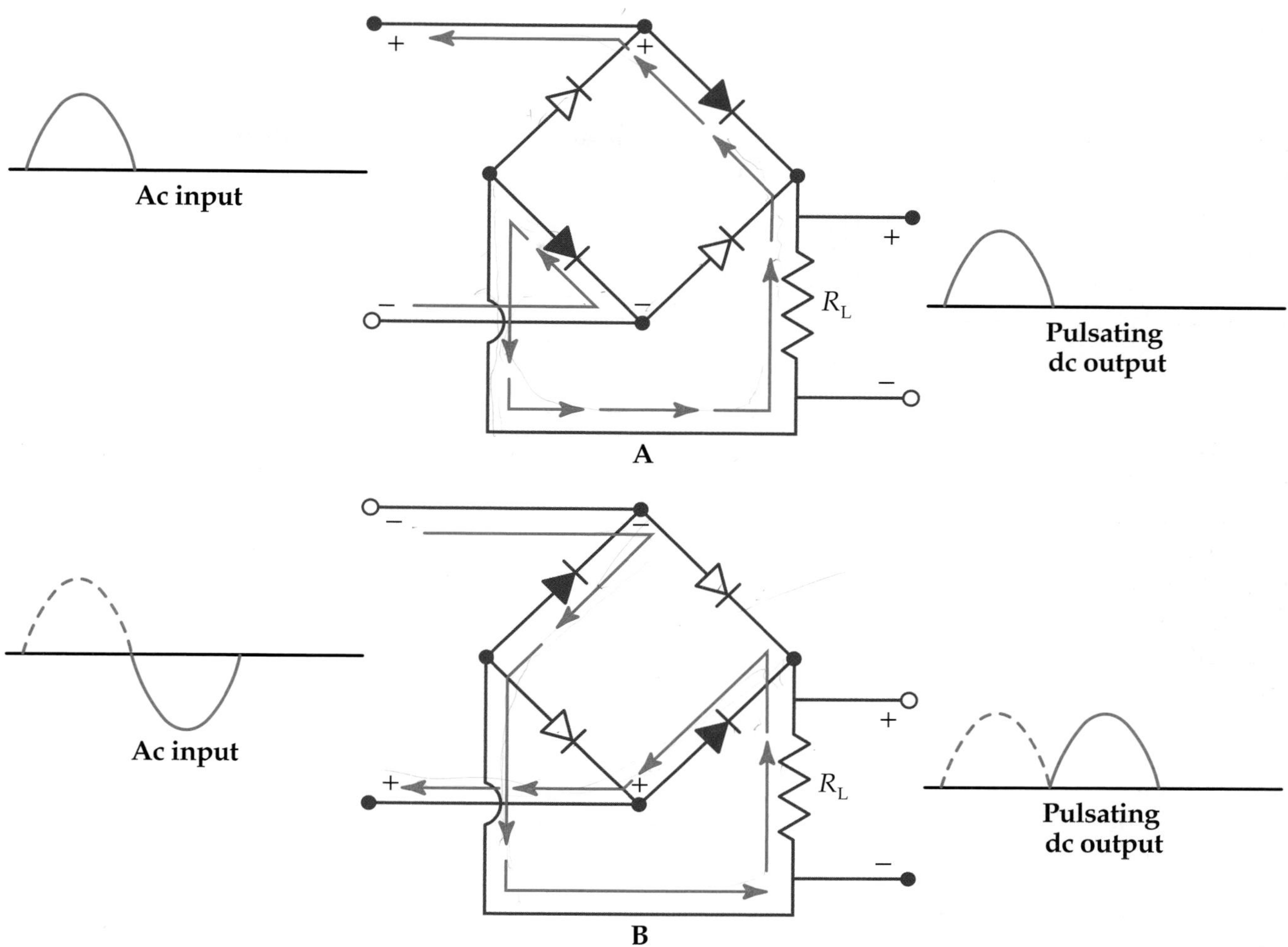

Figure 17-19. A schematic of a bridge rectifier circuit. Note the direction of forward bias through the four diodes. A—The first half cycle of the conversion from ac to dc is illustrated. Arrows indicate current flow. Note that two diodes are allowing current and two diodes are blocking it. B—The second half cycle of the ac-to-dc conversion. The ac has changed directions. The diodes blocking and allowing current have switched, and there is a new path for current flow. The current passes through the load in the same direction each time.

nected to the secondary of a transformer. The transformer is the ac power source for the circuit. The rectifiers change ac to dc by directing current through the load in only one direction. Study Figure 17-19 to better understand this process. The current that passes through the load is pulsating dc, which was discussed in Chapter 16.

Another common full-wave rectifier can be made using only two diodes and a center tap on the transformer secondary, **Figure 17-20.** The diodes are installed with opposite polarities on each side of the center tap circuit, and the load is attached directly to the center tap. If the load and center tap were removed, no current could flow in the circuit because the diodes would block it in both directions. However, the center tap circuit provides a path so that the current must only pass through one diode. In each direction of ac flow, only one diode allows the current to flow. The polarity of the diodes forces the current to flow through the center tap and load in the same direction, even though the current changes direction in the transformer.

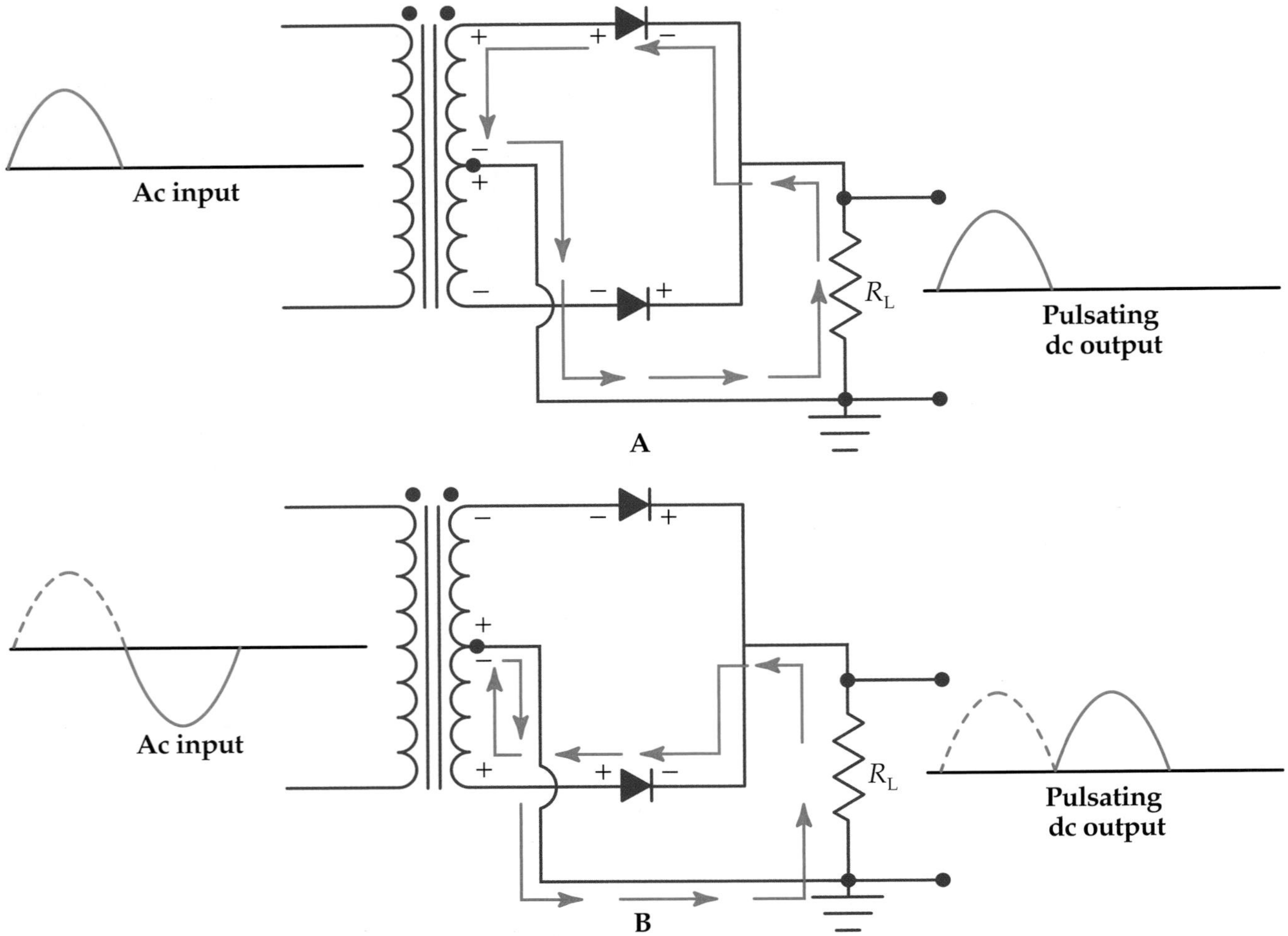

Figure 17-20. The center-tap full-wave rectifier circuit. A—The first half of a cycle. B—The second half of a cycle.

Because the transformer tap is in the center of the secondary windings, only half of the windings are used by each side of the rectifier circuit. This means that only half of the voltage is applied to the load each time the current changes direction. If the total voltage in the secondary of the transformer is 12 V, only 6 V is applied to the load, **Figure 17-21.** The advantage of this circuit is that the transformer runs cooler since only one-half of the secondary windings are in use at any one time. This can be useful if only half the voltage is needed. However, if all of the voltage in the secondary is needed in a circuit, the bridge rectifier circuit should be used.

Clipper Circuit

You may want to use a transformer with a diode to limit the circuit voltage. As the name indicates, a clipper circuit will clip (cut off) a portion of the voltage, **Figure 17-22.** For a silicon diode, the voltage will be clipped at 0.6 V. This voltage is called the ***clipper potential*** **(V_{CP}).**

Capacitors could also be inserted in series with the diode to further restrict the voltage. You can vary the size of the components to increase or reduce the transformer output depending on its use. Clipper circuits are used in electric guitars, for generating square waves, and limiting voltage spikes going to speakers.

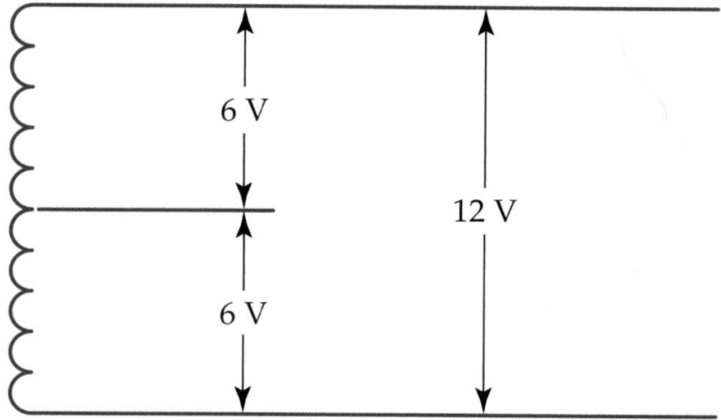

Figure 17-21. The applied voltage in a center tap full-wave rectifier circuit is half of the total voltage in the secondary of the transformer.

If you change the clipper circuit shown in Figure 17-22 by adding another diode to ground in the opposite direction, the circuit produces a square wave. The double shunt clipper circuit in **Figure 17-23** creates a clipped wave of about 1.2 V peak-to-peak (V_{PP}) when using a silicon diode.

Zener Diodes

A ***zener diode*** is a special type of diode used as a voltage regulator. Its symbol is shown in **Figure 17-24,** and its characteristic

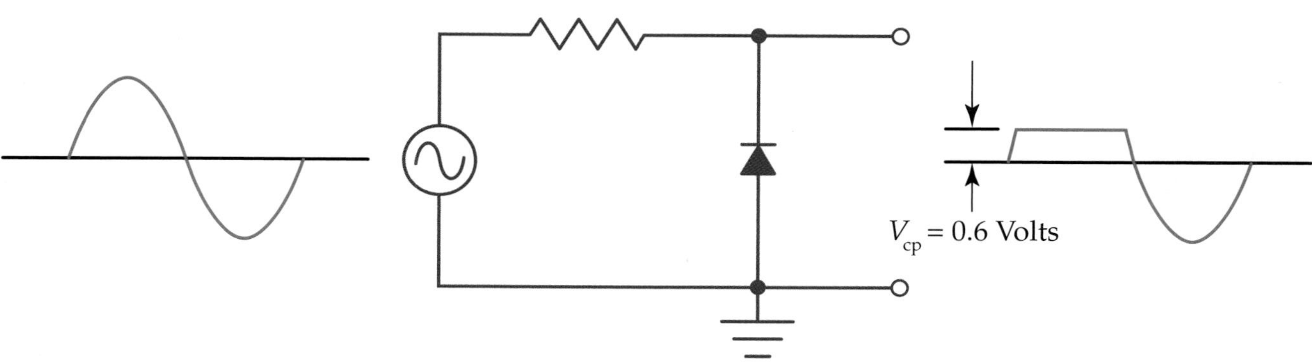

Figure 17-22. Clipper circuit and wave forms. The letters *CP* in the term V_{CP} stands for *clipper potential.*

Practical Application 17-2: Substituting Diodes

There is a wide variety of diodes. Different companies that make diodes may label them in different ways. If a diode needs to be replaced, finding the identical part to replace it can be difficult. With the variety of diodes, it is not always clear which diode would be an acceptable substitution. Even if the required type is known, it may be difficult to find since some diodes are not readily available in all electronics stores.

To help with substitutions, most companies publish a cross-reference catalog. This catalog makes it easier to find a replacement diode from another diode manufacturer. Be aware, however, that there may be a slight difference in their I_F and PIV ratings.

If you need to replace a diode and cannot find one with the exact ratings of the original diode in the circuit, use one with higher I_F and PIV ratings. Diodes with smaller ratings may be destroyed by too much current or voltage. Keep within the same family of diodes, such as germanium signal diodes and silicon power rectifiers. Make your selection from the information found in the cross-reference catalog.

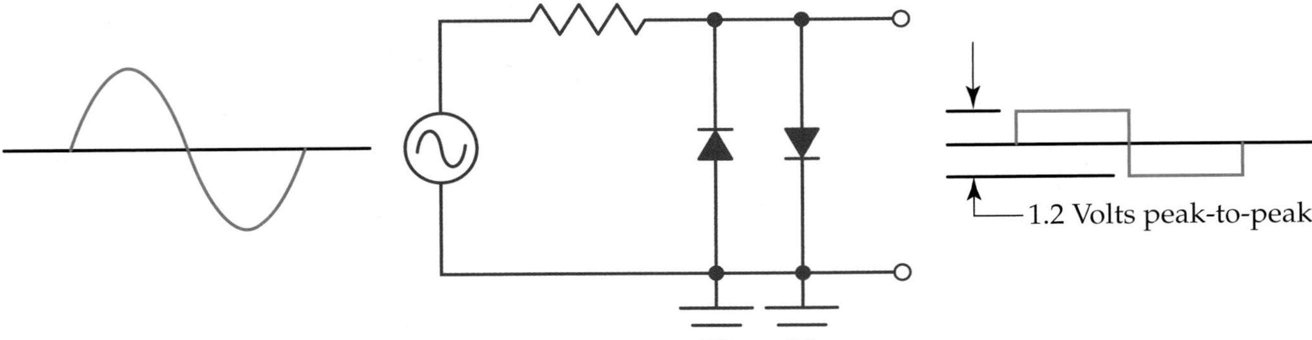

Figure 17-23. If we make a change in Figure 17-22 by adding another diode to ground in the opposite direction, we will get a square wave.

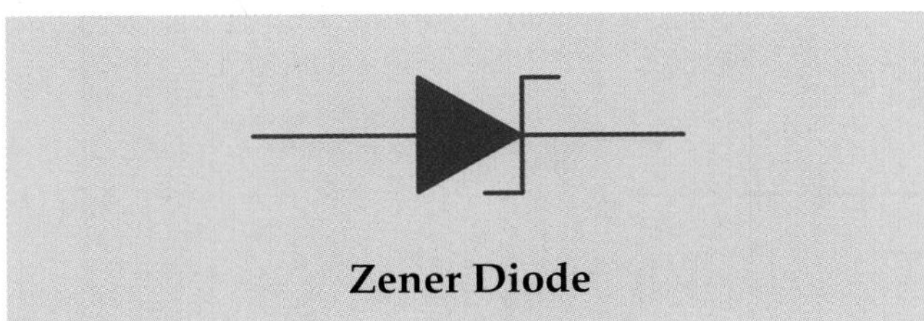

Figure 17-24. This symbol is used in drawings to represent a zener diode.

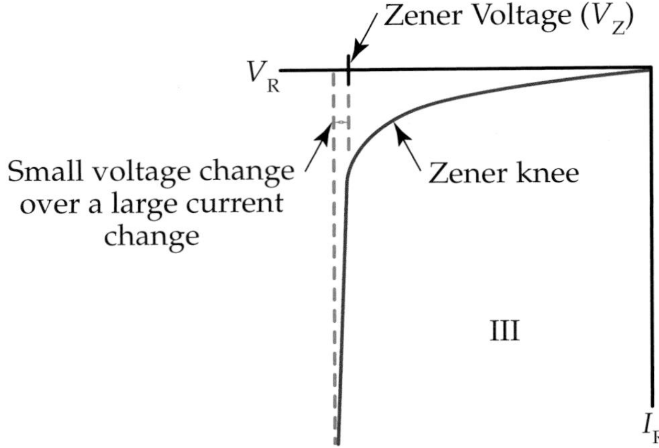

Figure 17-25. The curve on this graph shows that reverse current (I_R) increases sharply at zener voltage (V_Z). Notice that there is a range of current values for which reverse voltage (V_R) changes very little.

curve is shown in **Figure 17-25.** A zener diode is designed to be operated when it is reverse biased, which is why only the reverse portion of the curve is shown on the graph.

A zener diode is placed in parallel with the load, **Figure 17-26.** Recall that the voltage drop across different branches of a parallel circuit is always the same. The zener diode keeps the voltage drop fairly constant. Note in the characteristic curve in Figure 17-25 that there is a range of current values for which the voltage drop changes very little, even though the current changes very quickly. The zener diode is designed to operate within that range of current values.

The ***zener voltage (V_Z)*** is the voltage at which a zener diode begins to easily conduct without much change in voltage. The ***zener knee*** is the sharp curve in the graph that occurs at the zener voltage. As the current changes, so does the resistance of the zener diode, which keeps the voltage drop across the zener diode almost constant. This means that the voltage drop across the load is also almost constant.

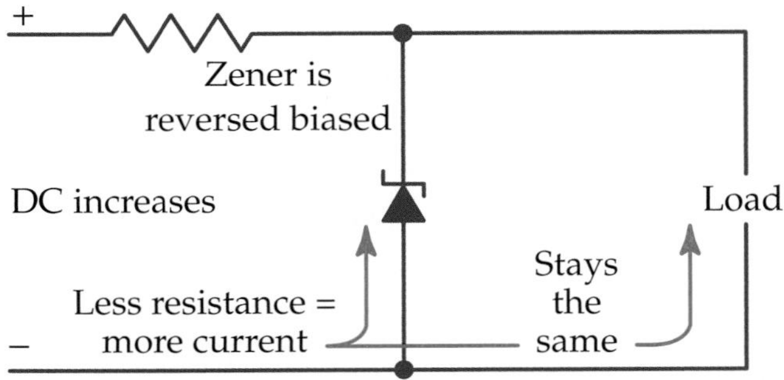

Figure 17-26. A zener diode is used in this circuit to keep the voltage drop across the load fairly constant.

The series resistor in the circuit drops any remaining voltage from the power supply. When a voltage regulator is designed, the value of this resistor is carefully calculated to make sure it can handle the required amount of voltage. The required zener voltage and power rating of the zener diode are also carefully calculated.

Light Emitting Diodes

A ***light emitting diode (LED)*** is a semiconductor device that gives off a tiny bit of light when current flows through it. Since an LED is a diode, current flows only when it is forward biased. The symbol for an LED is shown in **Figure 17-27.**

LEDs are used in many places where visibility is important. Traffic signal lights use a cluster of LEDs (red, yellow, and green). These LEDs have been installed to assist traffic control in most of our major cities, **Figure 17-28.** The benefit of using LEDs for traffic signal lights is they last longer and require much less maintenance than incandescent lights.

You will also find LEDs on the light bars used with police, fire, and emergency vehicles, **Figure 17-29.** LEDs allow these vehicles to be visible at great distances. Many of the major trucking firms use LEDs as brake lights and turn indicator lights. You will also find LEDs used on hundreds of different kinds of equipment for displays and as indicator lights, **Figure 17-30.**

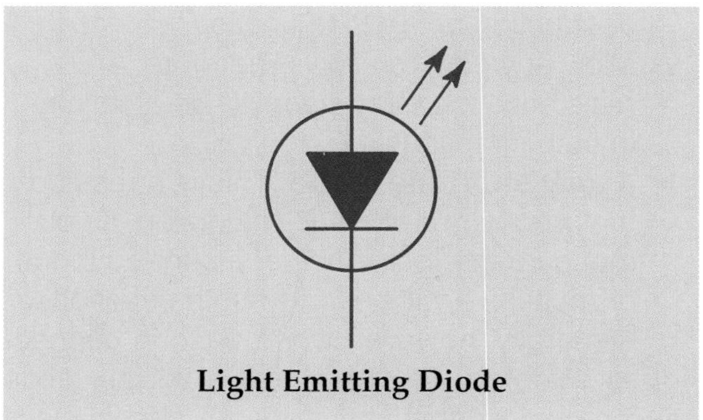

Figure 17-27. The schematic symbol for a light emitting diode (LED).

Figure 17-28. This traffic signal light uses LEDs for high visibility and low maintenance.

Figure 17-29. This Minnesota fire truck has LED light bars.

Seven or more segment LEDs are used to create numeric and alphabetical readouts when the proper voltage is applied, **Figure 17-31.** One big advantage of LEDs is that they can be turned on and off thousands of times without failing.

A relatively new area of research is the ***organic light-emitting diode (OLED).*** This new technology uses ultraviolet light to solidify polymers in the creation of a plastic-like film only about 8 mm of a desired color. The intent is to create OLEDs that are flexible enough to be rolled up, yet still give off light.

Figure 17-30. A red LED is used to indicate that this cell phone is charging.

Figure 17-31. Multiple segment LEDs can display numbers and letters.

The research includes various types of lighted displays, including plasma TV screens, digital camera displays, and cell phone screens. The problems with OLEDs include their ease of being damaged by moisture and their lack of durability. You will be hearing more about them in the years to come.

Power Supplies

A ***power supply*** is a circuit that provides the correct voltages to an electronic device. The frequency and voltage of the electricity provided by a power company is always the same. However, not all devices operate at that frequency and voltage. A power supply is used to change the frequency, the voltage, and sometimes the type of electricity to what is needed for a specific device.

In Chapter 16, you were introduced to filters used in power supplies. Here we will explain that subject in greater detail. **Figure 17-32** shows a block diagram of a basic power supply. Each block represents a different function or circuit in the power supply. As you can see, a basic power supply consists of a transformer, a rectifier circuit, a filter circuit, and a voltage regulator circuit. Some power supplies do not have all four components. The parts that are present depend on the purpose of the power supply.

Practical Application 17-3: Soldering Diodes

You must be very careful when soldering diodes, especially germanium diodes. The heat from a soldering iron could destroy the diode. To avoid this, use a heat sink between the diode and the circuit board on the lead being soldered. Overheating diodes or rectifiers can cause one of two things to happen.

- **It may open the diode.** This permanent destruction of the P-N junction will not allow any electron flow. Therefore, a resistance measurement will be at infinity in both directions.
- **It may short the diode.** This is a result of the breakdown of the P-N junction. If you measure the resistance of a shorted diode or rectifier, it will be low in both directions.

Generally, however, when a diode connection is soldered without a heat sink, the diode will open. Overloading a diode also causes these problems.

Project 17-1: Power Supply

A 12-V power supply can be used for anything requiring 12 V as its source of power. This includes burglar alarms, toys, some electrical games, and video security systems. It can be built at a reasonable cost. The amount of filtering could be improved, but it is not really necessary. This power supply does not use a voltage regulator.

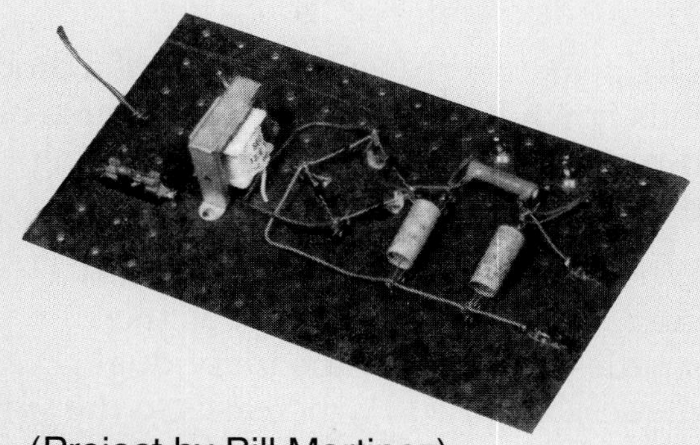

(Project by Bill Martinez)

When constructing this project, keep the leads short and compact so that the assembly can be placed in a small case. Buy a pre-packaged bridge rectifier or use four separate diodes. Use whatever capacitors you can find to make the capacitance around 6000 μF. Note that they are electrolytic capacitors. To build this power supply, obtain the parts in the table.

No.	Item
1	110-V/12-V transformer
1	Bridge rectifier (or 4 diodes)
2–4	Electrolytic capacitors to total 6000 μF at 25 WVDC
1	SPST switch
1	Fuse
1	Line cord and plug
1	Case
2	Output terminals

Assemble this project by looking at the schematic and following the directions given.

Procedure

1. Lay out the parts and determine the size of the case.
2. Mark the positions of the switch, output terminals, and other attachments on the case.
3. Drill all holes in the case.
4. Bend the case to shape. Paint it or cover it with contact paper.
5. Mount the parts in the case.
6. Wire the circuit according to the schematic.
7. Install the fuse in series with the transformer primary.

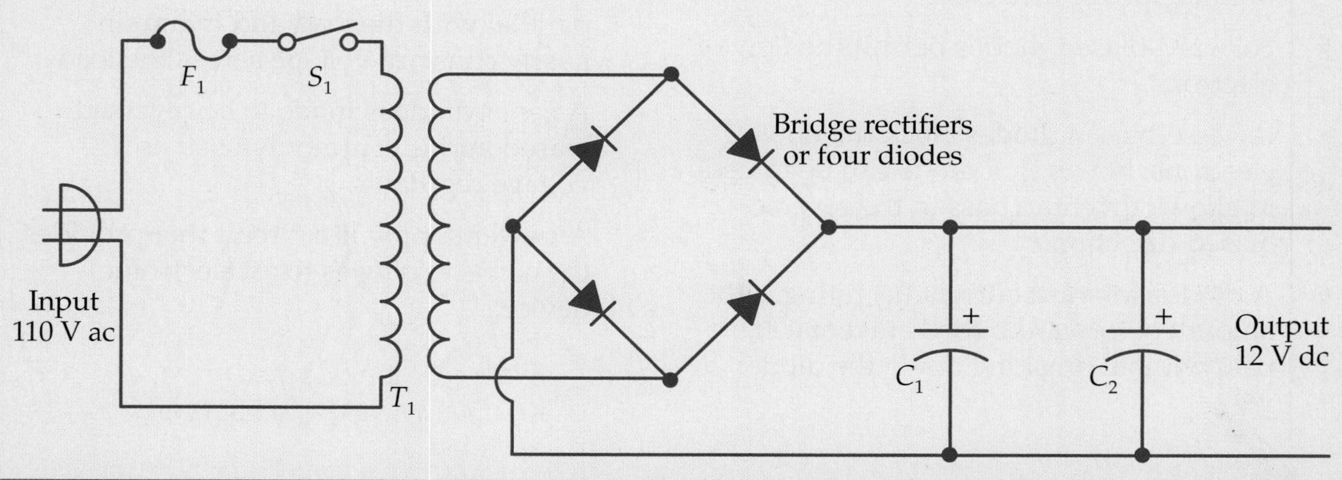

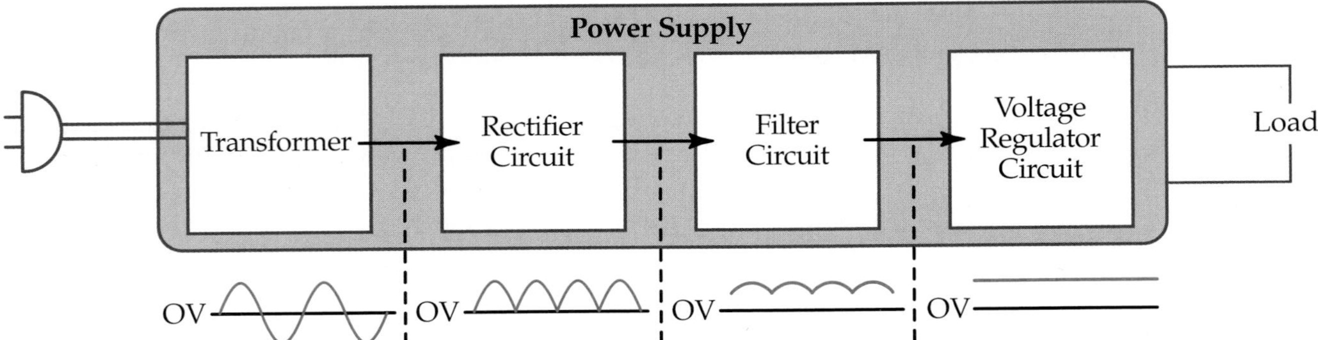

Figure 17-32. Block diagram of a basic power supply. The changes made to the voltage waveform by the components of each block are shown.

The *transformer* is used to step the voltage up or down to the appropriate level and to isolate the load circuit from the power source. This isolation can help prevent a person from being shocked by the secondary circuit. The *rectifier circuit* is used when ac is being changed to dc. The *filter circuit* is usually a capacitor used to filter the ripple out of the rectified current, which is pulsating dc. Finally, the *voltage regulator* makes sure that the voltage provided to the load stays constant, even if conditions in the power supply or load change.

Summary

- A diode is a solid-state device used to block the flow of electrons in one direction and permit the flow of electrons in the opposite direction.
- Forward-biased diodes permit the flow of electrons.
- Reverse-biased diodes block the flow of electrons; however, a diode can be forced to allow current to pass in the reverse-biased direction.
- A diode's forward current (I_F) rating is the amount of forward biased current it can pass without breaking down the diode.
- The peak inverse voltage (PIV) of a diode tells how much voltage in the reverse bias direction it can take without breaking down the diode.
- When diodes are in series, forward current (I_F) remains as low as the smallest I_F rating of the individual diodes, and the PIV increases to the sum of the PIVs of the individual diodes.
- When diodes are in parallel, forward current (I_F) increases to the sum of the I_F ratings of the individual diodes, and the PIV remains as low as the smallest PIV of the individual diodes.
- Rectifiers turn ac into pulsating dc.
- Clipper circuits clip off part of the sine wave of the voltage in a circuit.
- Voltage regulators are connected in parallel with the load and maintain nearly constant voltage across the load.
- A zener diode is made to be reversed biased and is commonly used as a voltage regulator.
- A power supply is a circuit that provides the correct voltages to an electronic device.

Test Your Knowledge

Do not write in this book. Write your answers on a separate sheet of paper.

1. Draw the symbols for a diode and a zener diode.
2. The positive end of a diode is the ______, and the negative end is the ______.
3. Diodes are usually thought of as carrying current in the ______ range, while rectifiers carry current in the ______ range.
4. The area of separation between the P-type material and the N-type material in a diode is called the ______.
5. For a diode to conduct, it must be properly ______.
6. PIV stands for ______.
7. What would the color code of a 1N420 diode look like?
8. What results would you expect to get if you were to make a resistance check on a diode?
9. How much current will two diodes in series be able to conduct?
10. If a diode must be replaced, what is the best thing to do if the exact values could not be found for a replacement part?
11. Before soldering, a diode should be protected with a ______.
12. A ______ turns ac into pulsating dc.
13. Give one use for a zener diode.
14. List at least three important uses for LEDs.

Activities

1. Obtain as many different types of diodes as possible. Label each lead. Are all of them identified the same way?
2. Why is a diode called a solid-state device?
3. Take a diode apart. Can you see how it has been made? If you were to use a microscope, what other things could you see?
4. Why are there so many different types of diodes? How is a manufacturer able to control the process for making so many different types?
5. How do you know which type of diode to use in a circuit? Make a list ranging from those carrying low current to those carrying high current.
6. How does a manufacturer make diodes with glass cases? How do they get the glass around it? How are the colored bands put on the glass without breaking it?
7. Try an experiment with a couple of diodes. Measure the amount of heat it takes to ruin each type. Plot the results to show the differences.
8. How much overload current will a diode take before it is ruined? Does it make any difference whether it is silicon or germanium?
9. Using rectifiers, build a full wave bridge. How many ways can you filter the output?
10. How many different circuits can you find using zener diodes? Do they fall into any one type?

11. Take a zener diode apart. Does it look any different from a rectifier or standard type diode?
12. How much voltage is needed to break down a zener diode? How much overload will it take before it is ruined?
13. Why is research being conducted on OLEDs? What are the newest uses that have come out of this research?
14. What caused the conversion of so many products to using LEDs? What are some of the latest uses and why are they important? What is the dollar estimate of the money now being spent on LED products? Is this going to increase or decrease in the future? Why?

Transistors

Learning Objectives

After studying this chapter, you will be able to do the following:

- Draw the direction of current flow and label the bias voltage on PNP and NPN BJTs in the common-emitter configuration.
- Draw common-emitter, common-base, and common-collector circuits.
- Explain the gain of a transistor and its importance.
- State the difference between a BJT and a FET.
- State the difference between a JFET and a MOSFET.
- Identify PNP and NPN transistors.

Technical Terms

amplifying transistors
base
bipolar junction transistor (BJT)
channel
collector
common-base amplifier
common-collector amplifier
common-emitter amplifier
cutoff voltage
depletion mode
drain
emitter
enhancement mode
field effect transistor (FET)
gain
gate
impedance
insulated gate field effect transistor (IGFET)
junction field effect transistor (JFET)
metal oxide semiconductor field effect transistor (MOSFET)
phototransistor
pinch-off voltage
source
substrate
switching transistors
transistor

Continuing the study of solid-state devices, this chapter covers transistors. A ***transistor*** is a solid-state device used for switching and amplifying the flow of electrons in a circuit. **Figure 18-1** shows the first transistor, which was invented at Bell Laboratories by John Bardeen, Walter H. Brattain, and William Shockley at the end of 1947. Although it looks rough by today's standards, its principle of operation is still the basic idea behind many solid-state devices in use today, **Figure 18-2.** In this chapter you will learn about the structure and operation of two major types of transistors: the bipolar junction transistor (BJT) and the field effect transistor (FET).

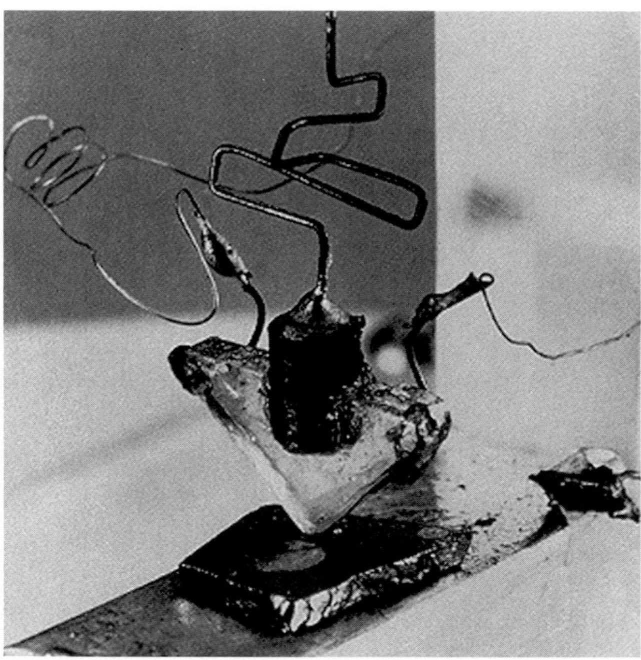

Figure 18-1. The first transistor was designed to amplify by passing electrical signals through a solid semiconductor material.

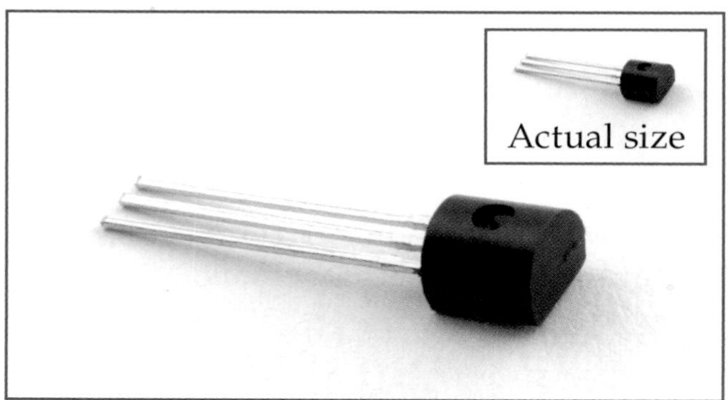

Figure 18-2. Today's transistors look more compact and attractive, but the operating principle is still based on the idea behind the first transistor.

Bipolar Junction Transistors

The most common type of transistor is the ***bipolar junction transistor (BJT).*** The makeup of a BJT transistor is similar to a diode because it is doped to contain P-type and N-type material. Recall that P-type material has a majority of holes and that N-type material has a majority of electrons.

A BJT has either P-type material between two N-type regions or N-type material between two P-type regions, **Figure 18-3.** It therefore has two P-N junctions instead of one. Notice that the names of these transistors, NPN and PNP, reflect their design. Also notice that the regions in a BJT are called ***collector, base,*** and ***emitter.***

A transistor can be thought of as two diodes with opposite polarities, **Figure 18-4.** Notice the arrangement of the diodes in the NPN transistor and the PNP transistor, Figure 18-4A. Recall from Chapter 17 that a forward biased P-N junction conducts easily. Therefore, for an NPN transistor, the emitter-base junction, which is represented by the lower diode, must always be forward biased, Figure 18-4B.

The other diode faces in the opposite direction; this second P-N junction has the opposite polarity. It represents the base-collector junction and must be reverse biased if the emitter-base junction is forward biased. The same biasing is true for the PNP transistor, also shown in Figure 18-4B. However, notice that the battery polarity differs from that in the NPN circuit. This is because the arrangement of diodes is different for the PNP transistor. This is why a PNP transistor cannot

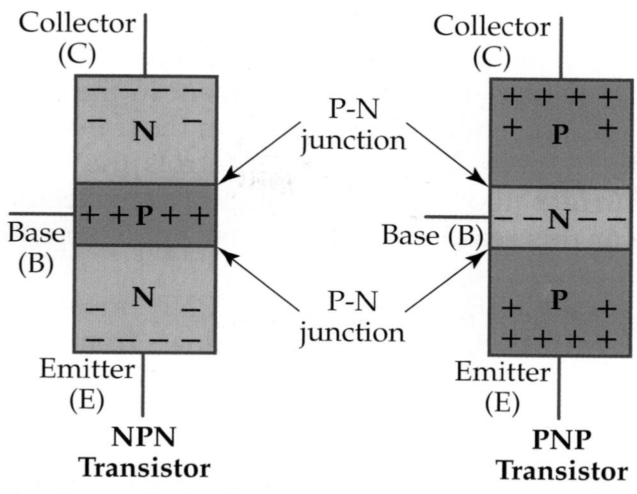

Figure 18-3. PNP and NPN are the two types of bipolar junction transistors. Bipolar junction transistors have two P-N junctions.

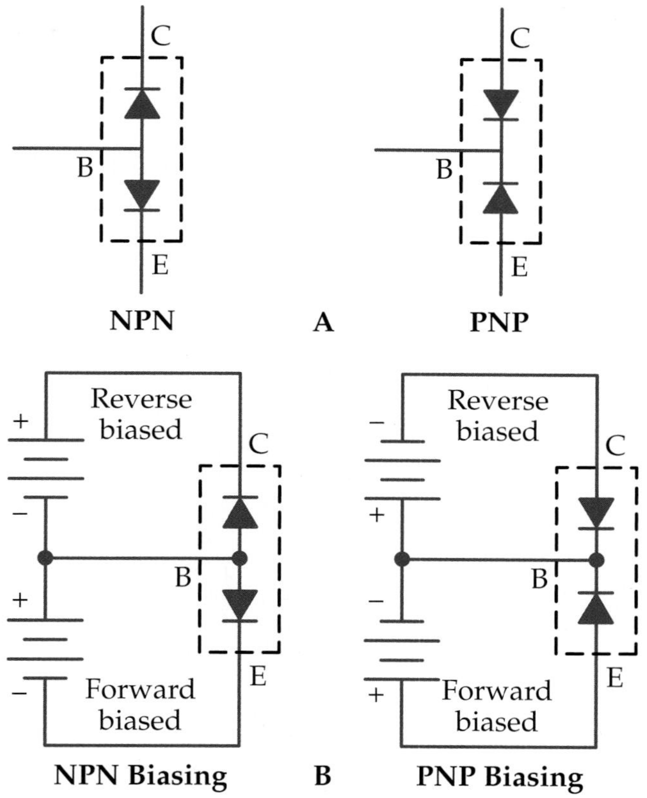

Figure 18-4. A BJT can be thought of as two diodes with opposing polarity. A—The direction the diodes face depends on whether the transistor is NPN or PNP. B—Biasing for an NPN and PNP transistor.

replace an NPN transistor in a circuit. Its P-N junctions would be incorrectly biased.

The schematic symbols for NPN and PNP transistors are shown in **Figure 18-5.** The base lead is always in the middle. The emitter lead is always connected to the arrow. **Figure 18-6** identifies the leads. Notice that the difference between an NPN and a PNP transistor's symbol is the direction of the arrow in the emitter. The arrow always points to the N-type material. The letters *NPN* can help you remember which symbol goes with which type of transistor. In an *NPN* transistor, the arrow is *N*ot *P*ointing i*N*.

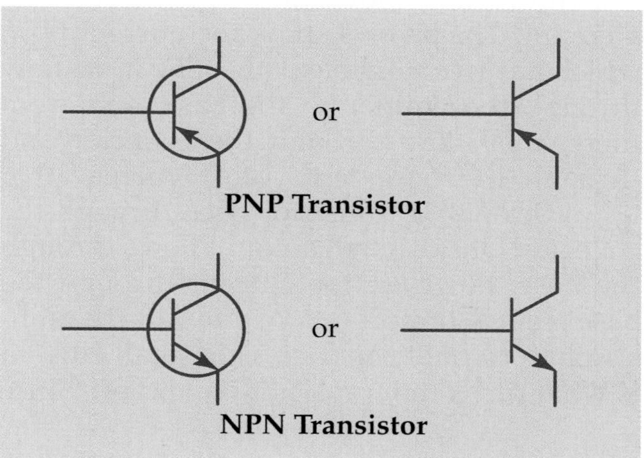

Figure 18-5. The schematic symbols for PNP and NPN transistors. Either way of drawing them is acceptable.

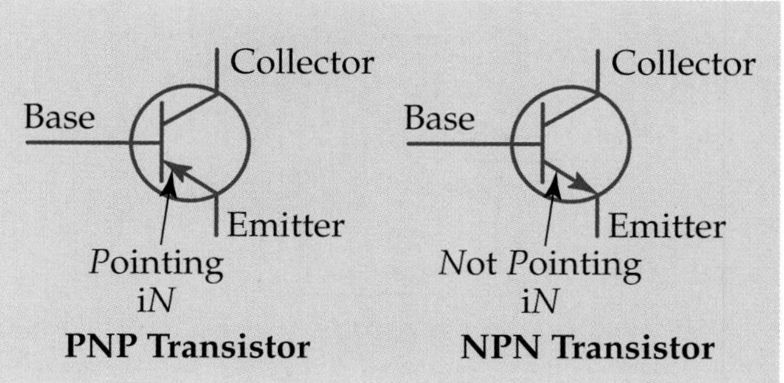

Figure 18-6. The leads of the BJT transistors are identified. Notice that the emitter is always the lead with the arrow.

BJT Operation

When an NPN transistor is correctly biased, current flows into the transistor through the emitter, which emits the current carriers to the base. The base is always the center region. It is important because it controls the amount of current that flows through the transistor, allowing no electrons, some electrons, or many electrons to pass. The collector collects them from the base, and the main flow of current goes out through the collector. The emitter is manufactured to have many free current carriers. It is designed to contain more free carriers than the collector and to conduct easily across the collector-base junction.

Figure 18-7 shows an NPN transistor in a circuit. The arrows show the current flow, which has two possible paths. Electrons flow through the emitter to the base, and from there, they flow through the collector and through the base lead. Most current that flows through a transistor goes through the collector. Only a small amount flows through the base; however, the current through the base lead is important. Without it, the transistor would not conduct. This small current is what turns the transistor on and off. Since base current is required to turn the transistor on, all BJTs are considered *normally off* devices. If there is no current in the base lead, there will be no current in the transistor at all. This small current controls the large current that passes through the collector. For this reason, BJTs are considered current-controlled devices.

In a circuit containing a PNP transistor, **Figure 18-8,** the electrons flow from the collector to the emitter, which is shown by the arrows that represent current. Comparing the current flow of both the NPN and PNP transistors, the flow of electrons always enters the point of the arrow, **Figure 18-9.**

Characteristic Curve

Two characteristic curves of a BJT are shown in **Figure 18-10.** The characteristic curves show the relationship between the collector current (I_C) and the collector-emitter voltage (V_{CE}) at different levels of base current (I_B). The red curve shows this relationship when I_B is 100 μA. Notice that the base current remains the same throughout the entire curve. At first, the collector current increases sharply as the collector-emitter

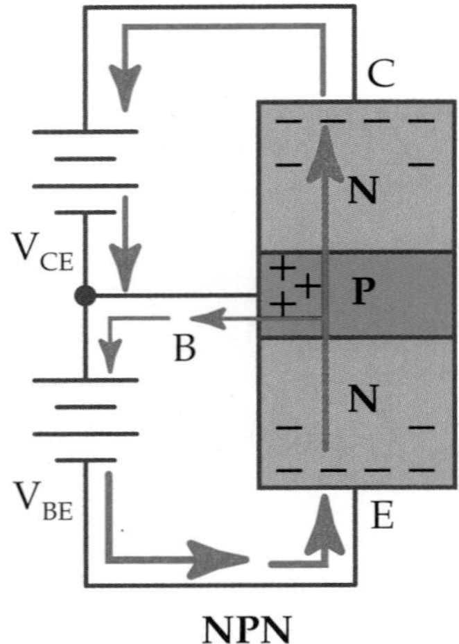

Figure 18-7. An NPN transistor in a circuit. The NPN transistor circuit has two batteries and two current paths passing through the emitter. The path through the collector carries larger current flow than the path through the base.

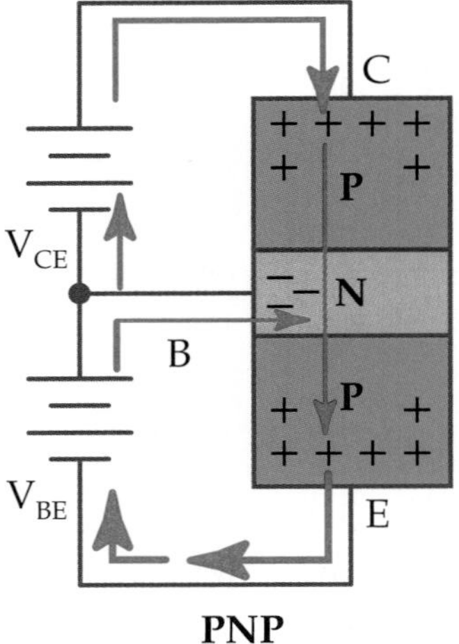

Figure 18-8. A PNP transistor in a circuit. The PNP transistor circuit has two batteries and two current paths passing through the emitter. Current in a PNP transistor flows in the opposite direction than in an NPN transistor.

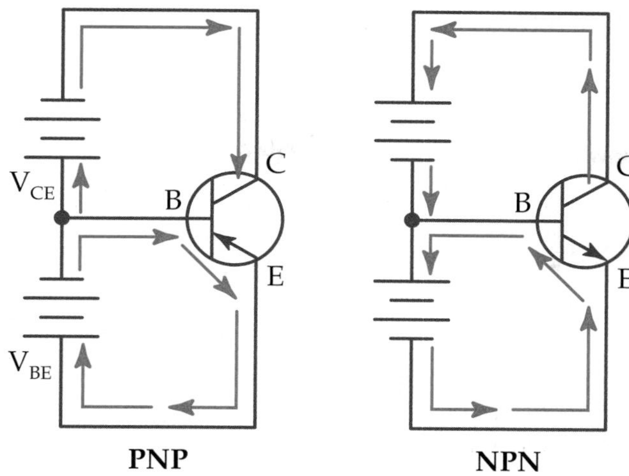

Figure 18-9. Notice how the flow of electrons always enters the point of the arrow.

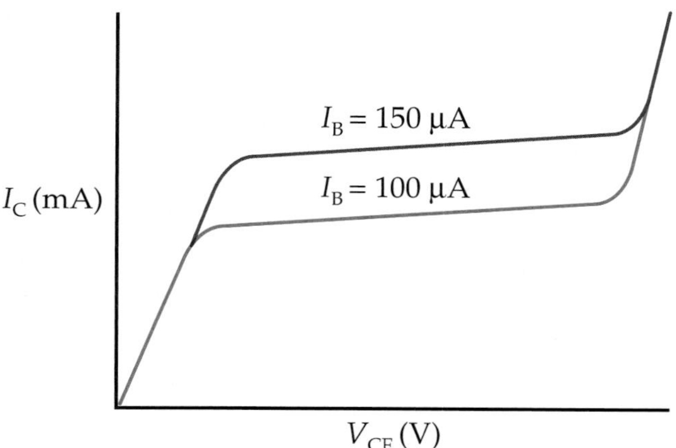

Figure 18-10. The characteristic curve of a BJT shows how the collector current (I_C) is controlled by the base current (I_B).

voltage increases. However, the current level quickly becomes constant. It is at this level of current that the BJT usually operates.

If the collector-emitter voltage becomes high enough so that the transistor breaks down, the collector current will increase rapidly again. This is shown by the sudden increase at the right side of the curve.

The blue curve shows what happens when I_B is 150 μA. Notice that if I_B increases by only a few microamps, the collector current changes by several milliamps. Next, we will look at how to calculate total current flow.

Total Current

Note that the emitter handles the current flow to both the base and collector, **Figure 18-11.** The amount of current that flows through the emitter (I_E) is equal to the sum of the current through the base (I_B) and the current through the collector (I_C):

$$I_E = I_B + I_C \quad (18\text{-}1)$$

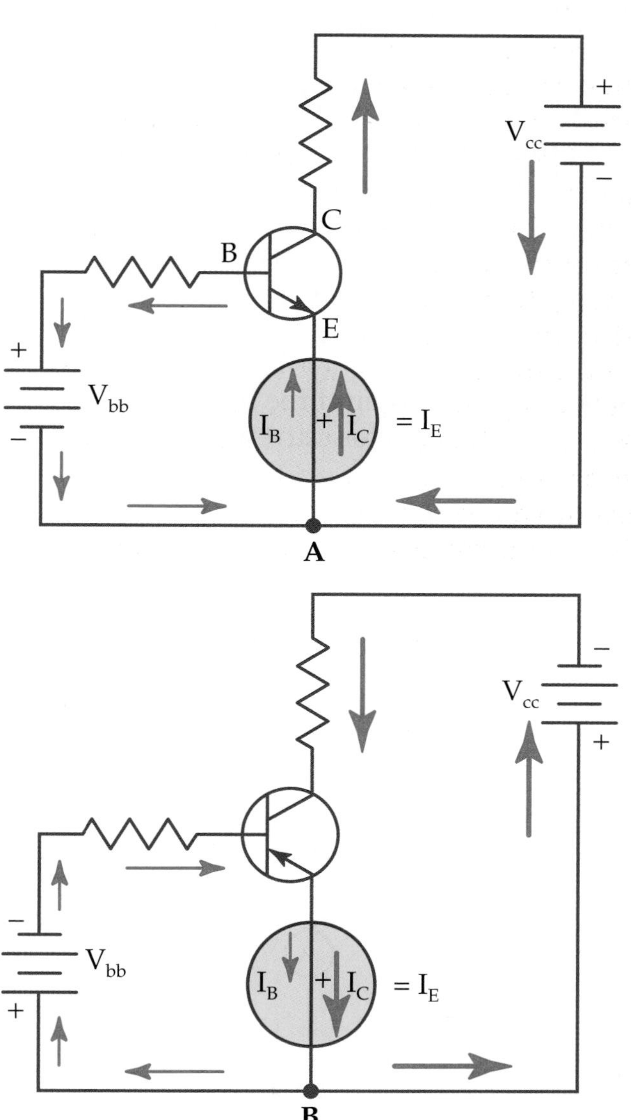

Figure 18-11. The current in the emitter can be found by adding the current in the collector and the current in the base.

Therefore, if the base current is 40 μA and the collector current is 3 mA, **Figure 18-12,** the emitter current can be found using the formula:

$$I_E = I_B + I_C$$
$$= 40\ \mu A + 3\ mA$$

To solve this equation, the values must be measured in the same units. Change 3 mA into microamps.

$$I_E = 40\ \mu A + 3000\ \mu A$$
$$= 3040\ \mu A, \textit{ or } 3.04\ mA$$

The current through the emitter is 3.04 mA.

Example 18-1:

Suppose the emitter and collector currents are known. The emitter current (I_E) is 6.4 mA, and the collector current (I_C) is 6.1 mA. Find the base current using this formula:

$$I_E = I_B + I_C$$
$$6.4\ mA = I_B + 6.1\ mA$$

Since you want to find the base current (I_B), subtract the collector current (I_C) from both sides.

$$I_E - I_C = I_B + I_C - I_C$$
$$6.4\ mA - 6.1\ mA = I_B + 6.1\ mA - 6.1\ mA$$
$$6.4\ mA - 6.1\ mA = I_B$$

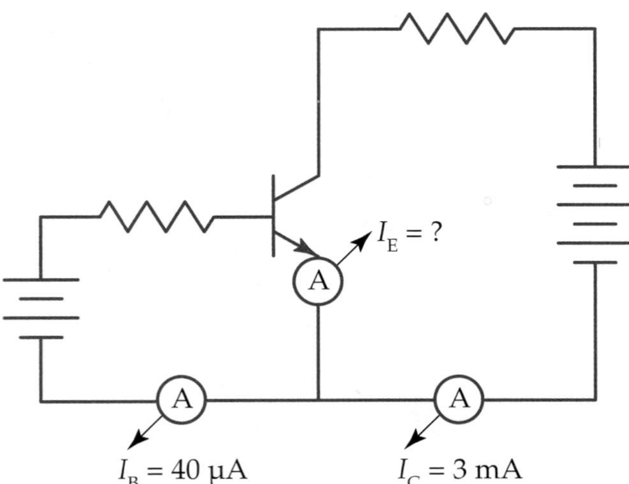

Figure 18-12. The base current (I_B) in this circuit is 40 μA and the collector current (I_C) is 3 mA. Can you find the value of the emitter current (I_E)?

By reversing the sides, you have

$$I_B = 6.4\ mA - 6.1\ mA$$
$$= 0.3\ mA$$

The base current is 0.3 mA.

Transistor Functions

Transistors have two basic functions: *switching* and *amplifying*. Some transistors are designed either to switch or to amplify. Some are designed to do both.

Switching

Switching transistors are used in circuits that are either on or off, similar to lighting circuits in your home or school. Most lights are either on or off, and this depends on the position of the light switch. A switching transistor is like the light switch: the circuit is either on or off, depending on the state of the switching transistor.

Switching circuits are used extensively in computers. When you write out a math problem involving division, you follow a specific set of steps. The computer does the same thing, only it does it faster and with greater accuracy than you can do it. Every math problem requires a series of switching. The computer interprets your keystrokes or voice commands and performs that request by switching transistors on and off thousands of times in a single second. This process also uses integrated circuits (ICs) and other time saving devices. You will learn about ICs in Chapter 20.

Switching transistors are also used with automobile computers. Every new car has somewhere between one to fifty computers. These computers, packaged within modules, make checks on the various safety systems, such as air bags and brakes. They receive this information from sensors and send this data to the main computer called the *engine control unit (ECU)*. The ECU uses this information to make sure the vehicle is running as efficiently

Project 18-1: Answer Box

Switching is used in the design of an answer box. This answer box is similar to those used on television quiz shows. The answer box can be used to prove which contestant pushed the button first. When a question is asked of the contestants, the first one to push the button is the one who gets the chance to answer the question.

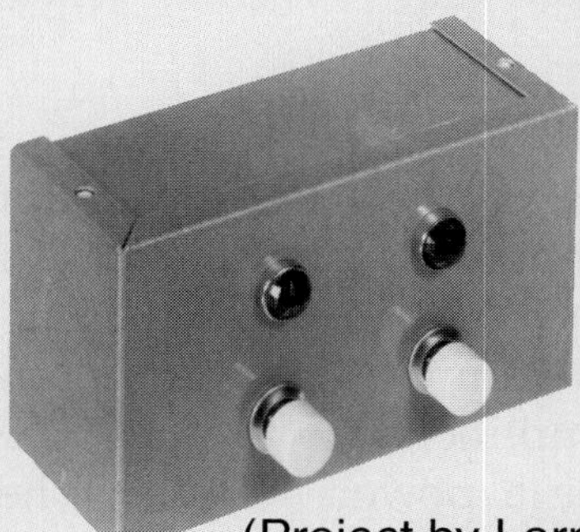

(Project by Larry Wood)

The circuit for the answer box has a transistor connected to each of two buttons, but only one transistor can conduct at any given time. Transistors can be activated in a small portion of one second. Because of their speed, when both buttons are pushed, a race will take place causing the base of the winner's transistor to be activated. Therefore, the lightbulb of the first person to push the button always lights first. Even if the second person hits the button a fraction of a second later, that person's lightbulb will not light.

Obtain the following materials. Assemble them according to the schematic diagram and the construction procedure given.

No.	Item
1	Battery, 6 V–9 V
2	Lightbulbs, 6 V–9 V
2	NPN transistors
2	Resistors, 2.2 kΩ
2	Push-button switches
1	Steel box, 4″ × 6″, 18-gauge
1	Circuit board, 4″ × 6″
	Connecting wires

Procedure

1. Connect the positive leads of the lightbulbs using wire or alligator clips.
2. Connect a resistor to the negative lead of each lightbulb.
3. Connect each resistor to one terminal of a switch.
4. Connect the other leads of the switches to the bases of the two NPN transistors.
5. Connect the emitters together. Also, attach a wire to the emitters, forming a ground.
6. Connect the collector of the transistors to the point where the resistor meets the lightbulb of the other wire.
7. Attach the power supply's positive terminal to both lightbulbs. Connect the negative terminal to ground.
8. Make a metal box large enough to hold the circuit board and the complete circuit (approximately 4″ × 6″ × 1 1/2″).
9. Drill four 5/8″ holes in the box and install the lightbulbs and the push buttons.
10. Solder all terminals on the circuit board and secure the board in the box. Use the answer box in conjunction with flash cards, questions from the teacher other quiz program.

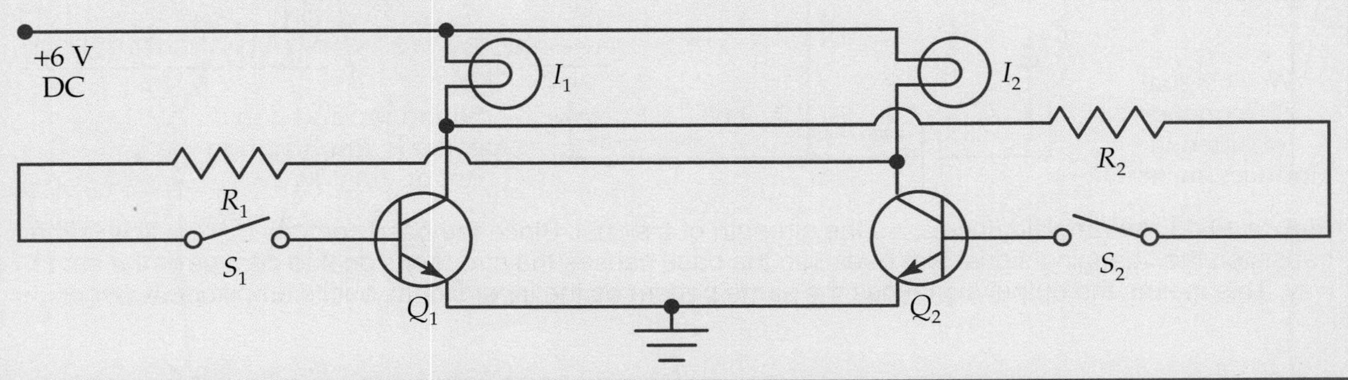

as possible. For example, the ECU uses the switching transistors within its processor to calculate the amount of fuel and the timing for injecting the fuel into each cylinder.

In addition, the ECU uses switching circuits to store data on the faults that occur during everyday driving. When you bring your car in for regular maintenance, the mechanic attaches a monitoring device to a connector on the ECU. This device reads and displays the data that the switching circuits in the ECU have stored.

Amplifying

Amplifying transistors are used in places where the amount of signal is important. This type of transistor is like a dimmer light switch. It allows the amount of current and power in the circuit to be adjusted. A small base current is used to control a much larger emitter-to-collector current. This type of transistor is used in a home theater system that needs to amplify an incoming signal, **Figure 18-13.** The amplitude of a radio signal is small when it is broadcast, and it must be increased to be heard over a speaker.

The amplifying transistor can go through a range of increasing power, similar to increasing a lightbulb's brightness with a dimmer switch. The radio signals must go through a few transistors to increase the size of the signals, which increases their volume. These transistors are the final power type that feed the signal into the speaker, which produces the sounds that are heard. There are hundreds of different power amplifier transistors. However, the vast majority are of the silicon variety capable of carrying current in the ampere range to the speakers.

Gain

Amplifiers make the input signal larger without changing the frequency of the signal. Both the voltage (E) and the current (I) can be increased, making this a gain in power (since $P = EI$). Amplifiers are the only devices that can provide a power increase. The increase an individual transistor can provide is called ***gain.*** Current gain, voltage gain, and power gain must all be calculated separately. This text only covers current gain in dc circuits.

In a dc circuit, the current gain of a transistor, which is represented by the Greek letter beta (β), is the ratio of the collector current (I_C) to the base current (I_B):

$$\beta = \frac{I_C}{I_B}$$

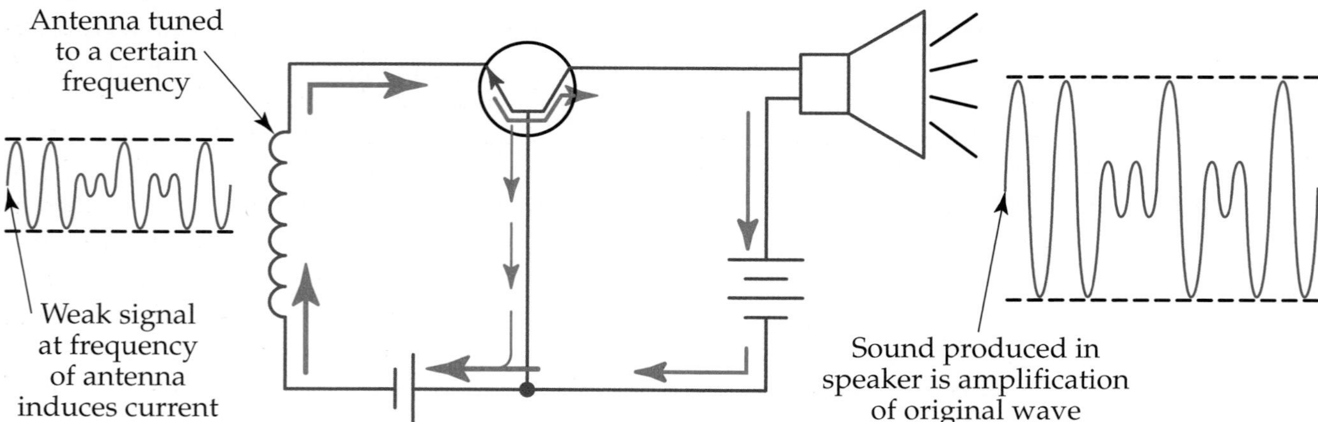

Figure 18-13. An amplifier increases the strength of a signal. Since the base controls current flow in the transistor, the changing signal that goes into the base causes the stronger signal to change in the same way. This means the output signal has the same pattern as the input signal, but its amplitude is larger.

Gain (β) does not have a unit, such as amps. Instead, it is a measure of the degree that the transistor increases the current. Current gain for an amplifier circuit may be anywhere from 30 to 50,000. However, the most common range for current gain is between 30 and 200.

Note

When calculating current gain, remember to make sure the units are the same. If they are not, convert the values so that their units have the same prefix, then perform the calculation.

A PNP transistor in a circuit has a collector current (I_C) of 3 mA and a base current (I_B) of 40 μA. Current gain can be found by using the formula for current gain.

$$\beta = \frac{I_C}{I_B} = \frac{3\ \text{mA}}{40\ \mu\text{A}}$$

Convert 3 mA to microamps.

$$\beta = \frac{3000\ \mu\text{A}}{40\ \mu\text{A}} = 75$$

This means that the ratio of the collector current to the base current is 75:1. In other words, if the collector current is 75 μA, the base current would be 1 μA, or the collector current would be 75 times greater than the base current. It does not mean that the collector current is more than the base current by 75.

If current gain and the base current are known, the collector current can be found by multiplying the base current by the gain. See Example 18-2. The actual amount of collector current is determined by the load. The load may draw 4 mA for some time, and then draw 6 mA later. The gain of the transistor is still 75, but the current will vary depending on the music, speed change, or some other variable for which it is used.

Example 18-2:

Suppose current gain (β) is 75 and base current (I_B) is 23 μA. Find the collector current.

The collector current is 75 times greater than the base current. Start with the original formula to see why.

$$\beta = \frac{I_C}{I_B}$$

$$75 = \frac{I_C}{23\ \mu\text{A}}$$

Multiply both sides by the base current (I_B), 23 μA, to solve for the collector current (I_C).

$$75 \times 23\ \mu\text{A} = \frac{I_C}{23\ \mu\text{A}} \times 23\ \mu\text{A}$$

$$75 \times 23\ \mu\text{A} = I_C$$

This is the same thing as

$$I_C = 75 \times 23\ \mu\text{A}$$

Now, solve the equation collector current.

$$I_C = 75 \times 23\ \mu\text{A} = 1725\ \mu\text{A}, \textit{ or } 1.725\ \text{mA}$$

The collector current is 1.725 mA.

BJT Configurations

There are multiple ways a transistor can be connected in a circuit. One type of configuration refers to how a particular lead is connected to ground or to a common power rail. For example, when the base lead is connected to the ground, it is called a ***common-base amplifier.*** Other configurations are connecting the emitter to ground, called a ***common-emitter amplifier*** and connecting the collector to ground, call a ***common-collector amplifier.*** **Figure 18-14** shows all three configurations.

The common-emitter amplifier are used when voltage gain and current gain is desired. The common-base amplifier is used when

Project 18-2: Flashing Light

BJTs can be used to make a small flashing light. Build the circuit shown and place it in a small box or soap dish.

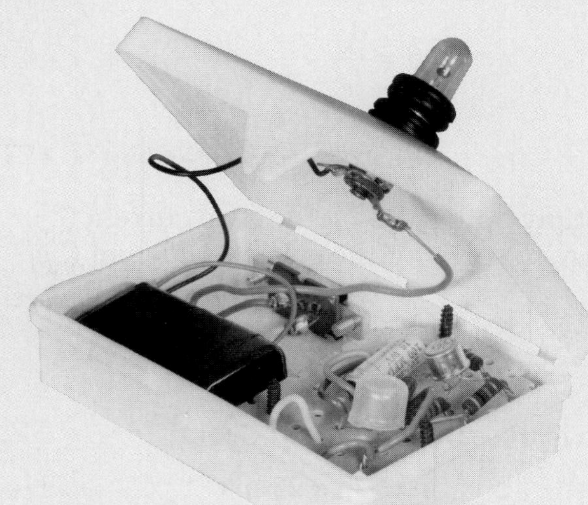

(Project by Dick Butcke)

To build it, obtain the following materials and assemble them according to the schematic diagram and the construction procedure given.

No.	Item
1	DC battery, 9 V
1	Electrolytic capacitor, 100-µF, 15-WVDC
1	DC lamp, 4 1/2 V
1	GE 2 transistor (Q_1)
1	GE 8 transistor (Q_2)
1	Resistor (R_1), 100 kΩ
1	Resistor (R_2), 100 Ω
1	Resistor (R_3), 2.7 kΩ
1	Resistor (R_4), 470-Ω
1	SPST switch
1	Perforated circuit board

Procedure

1. Lay out the wiring of the schematic on a breadboard.
2. Make sure that the flashing light works.
3. Transfer the wiring to the perforated board.
4. Solder and tape all the joints.
5. Place the flasher in the box or soap dish.

The flashing light can be used inside a papier-mâché animal and placed on a table. If you celebrate Halloween, you have another chance to use this flashing light. Turn it on and place it inside a jack-o'-lantern.

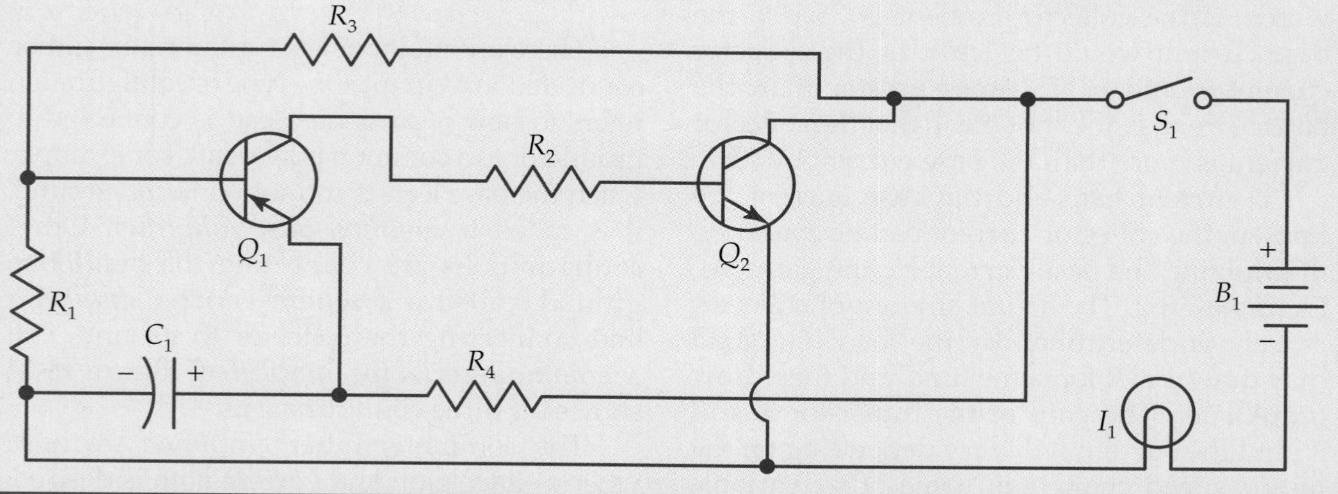

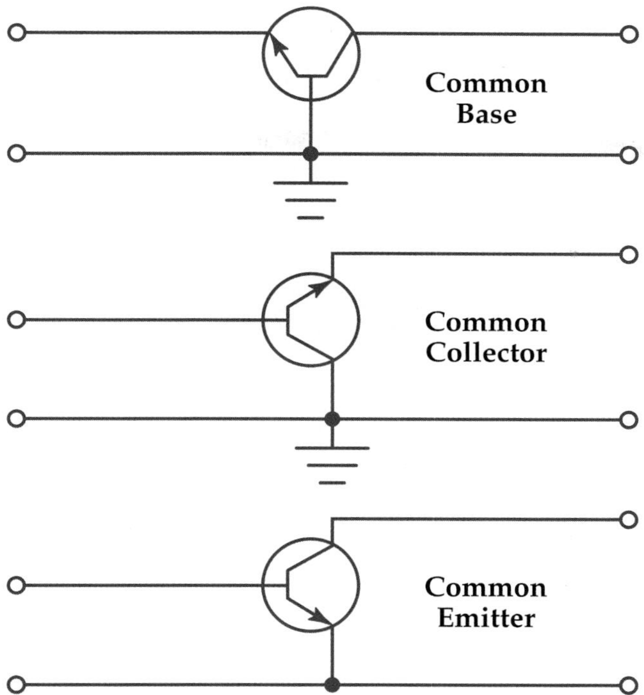

Figure 18-14. The three configurations for a BJT in a circuit are shown here.

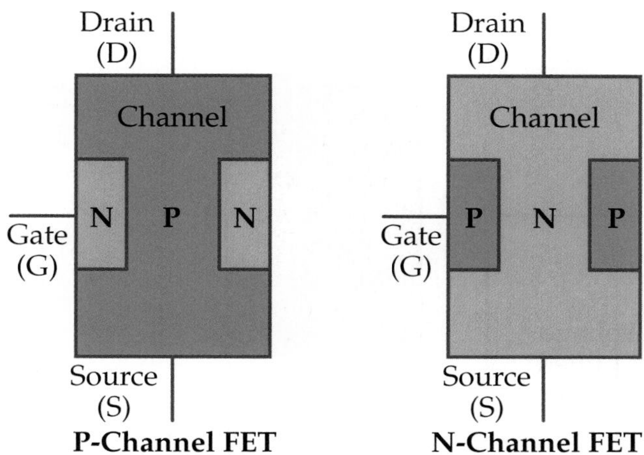

Figure 18-15. Field effect transistor (FET) construction. The channel of a FET carries circuit current. Notice that the channel in a P-channel FET is made of P-type material, and the channel in an N-channel FET is N-type material.

high voltage gain without current gain is desired. When only current gain is desired, the common-collector amplifier is used.

Field Effect Transistor

A ***field effect transistor (FET)*** is another type of transistor. FETs perform about the same function as BJTs, but their construction and operation differ in some important ways. FETs are unipolar, or have one pole. This means that the current travels through only one type of material, either P-type or N-type. The part of the FET that carries the current is called the ***channel.*** The channel in a P-channel FET is made of P-type material, and the channel in an N-channel FET is N-type material, **Figure 18-15.**

A FET has three leads: the ***source,*** the ***drain,*** and the ***gate.*** The source and the drain are at either end of the channel. The source is similar to the emitter of a BJT, the drain is similar to the collector, and the gate is similar to the base. The gate is material is opposite that of the channel material. Like the base in a BJT, the gate circuit controls the current that passes through the channel. However, unlike the base, the voltage applied to the gate (not the current) controls the current flow between the source and drain. For this reason, an FET is considered a voltage-controlled device.

There are two main types of FETs: junction field effect transistors (JFETs) and metal oxide semiconductor field effect transistors (MOSFETs). Both of these types can be either N-channel or P-channel FETs.

JFET

A ***junction field effect transistor (JFET)*** is a FET that allows current to flow without voltage on the gate, **Figure 18-16.** It is therefore considered a "normally on" device. The gate is turned on to block the flow of electrons that are usually allowed to pass through the JFET. Most of the time, a voltage is applied to keep the JFET from conducting. It is kept in the blocking position until current flow is allowed to pass by reducing the voltage. Without an applied voltage, there would be a

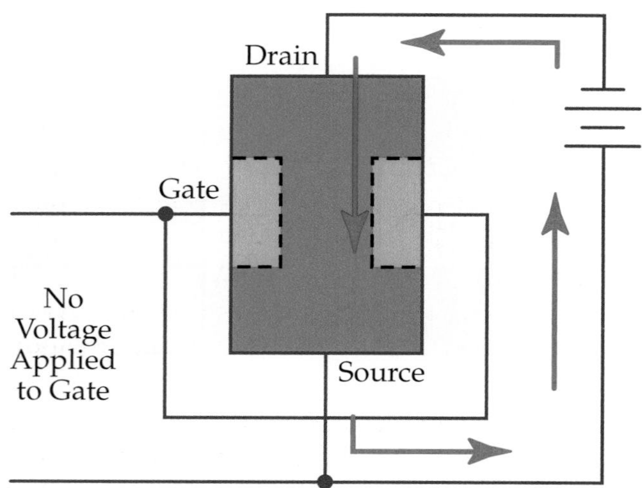

Figure 18-16. With zero input on the FET, current output is high because nothing is blocking the flow through the channel.

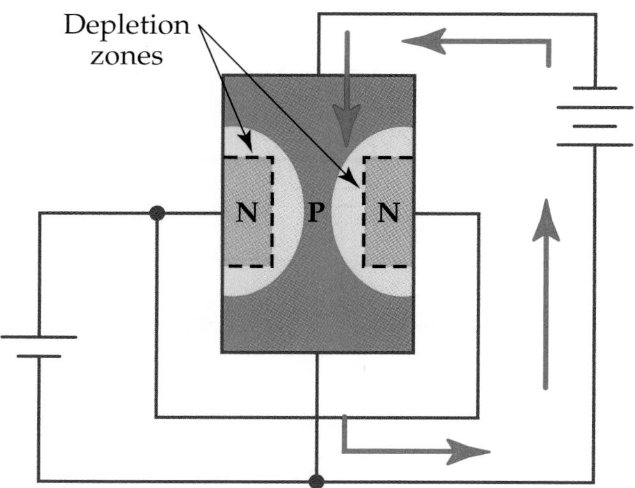

Figure 18-17. Biasing of a P-channel JFET. Increasing voltage on the input sets up a depletion zone that blocks the flow of current through the channel.

good chance that the too much uncontrolled current would flow and the JFET would be destroyed.

A P-channel JFET has N-type material on both sides of the channel. By applying voltage to the gate, the N-type material controls the amount of current that can flow through the channel. When reverse-biased voltage is applied to the gate, a depletion zone is created around the N-type material, **Figure 18-17.** The amount of voltage applied determines the size of this depletion zone. The current cannot flow through the depletion zone, so this zone makes the channel narrower. Recall that the width of a conductor affects how much current can flow through it. As the voltage applied to the gate increases, the channel width decreases and fewer electrons can flow from the source to the drain. In this way, the voltage applied to the gate controls the amount of current that flows through the channel. **Figure 18-18** shows this effect on an N-channel JFET.

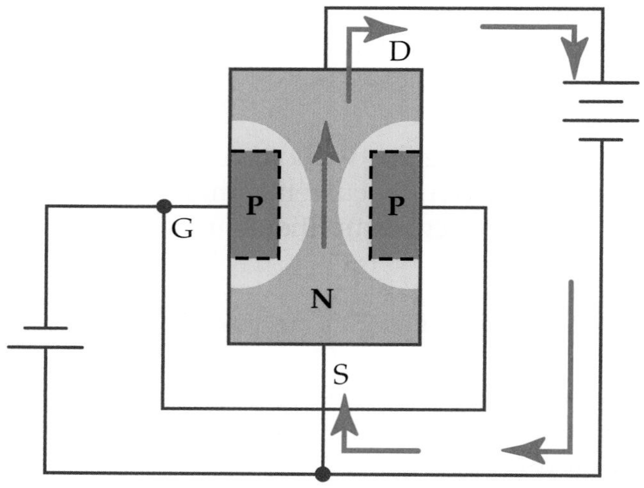

Figure 18-18. Biasing of an N-channel JFET.

JFET symbols

The symbols for N-channel and P-channel JFETs are shown in **Figure 18-19.** The only difference between these two symbols is the direction in which the arrow points. As in BJTs, the arrow points toward the N-type material.

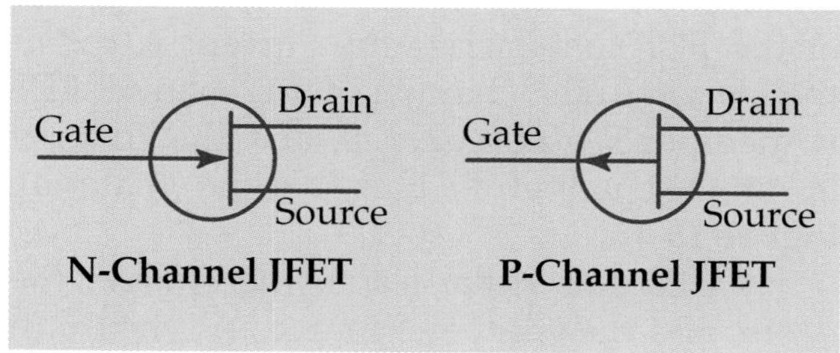

Figure 18-19. Symbols are shown for N-channel and P-channel JFETs. Only the arrow changes direction.

FET characteristic curves

Figure 18-20 shows the characteristic curves for a JFET. Notice that the characteristic curves show the relationship between drain current (I_D), drain-source voltage (V_{DS}), and gate-source voltage (V_{GS}). Each curve represents a different gate-source voltage (V_{GS}).

You have already learned that when no gate-source voltage (V_{GS}) is applied (V_{GS} = 0), the drain current (I_D) is the highest. If you look at the characteristic curve for this V_{GS} value, you will see that the drain current at this point equals 10 mA. This value is known as the "drain-to-source current with gate shorted," or I_{DSS}.

Now, recall that the gate-source circuit is reverse biased. As this voltage increases (becomes more negative), drain current (I_D) decreases. This is shown in the characteristic curves for V_{GS} = –1, V_{GS} = –2, V_{GS} = –3, and V_{GS} = –4. Notice that as V_{GS} increases (becomes more negative), I_D decreases.

If V_{GS} reaches a negative value that is high enough, the depletion zones of the JFET come together, thus blocking current flow through the channel, **Figure 18-21.** The value of V_{GS} at which this happens is called the ***cutoff voltage*** of the transistor. Looking at the characteristic curves in Figure 18-20, you can see that, for the JFET represented, this value is –4 V_{GS}. Keep in mind that this value varies among different JFETs.

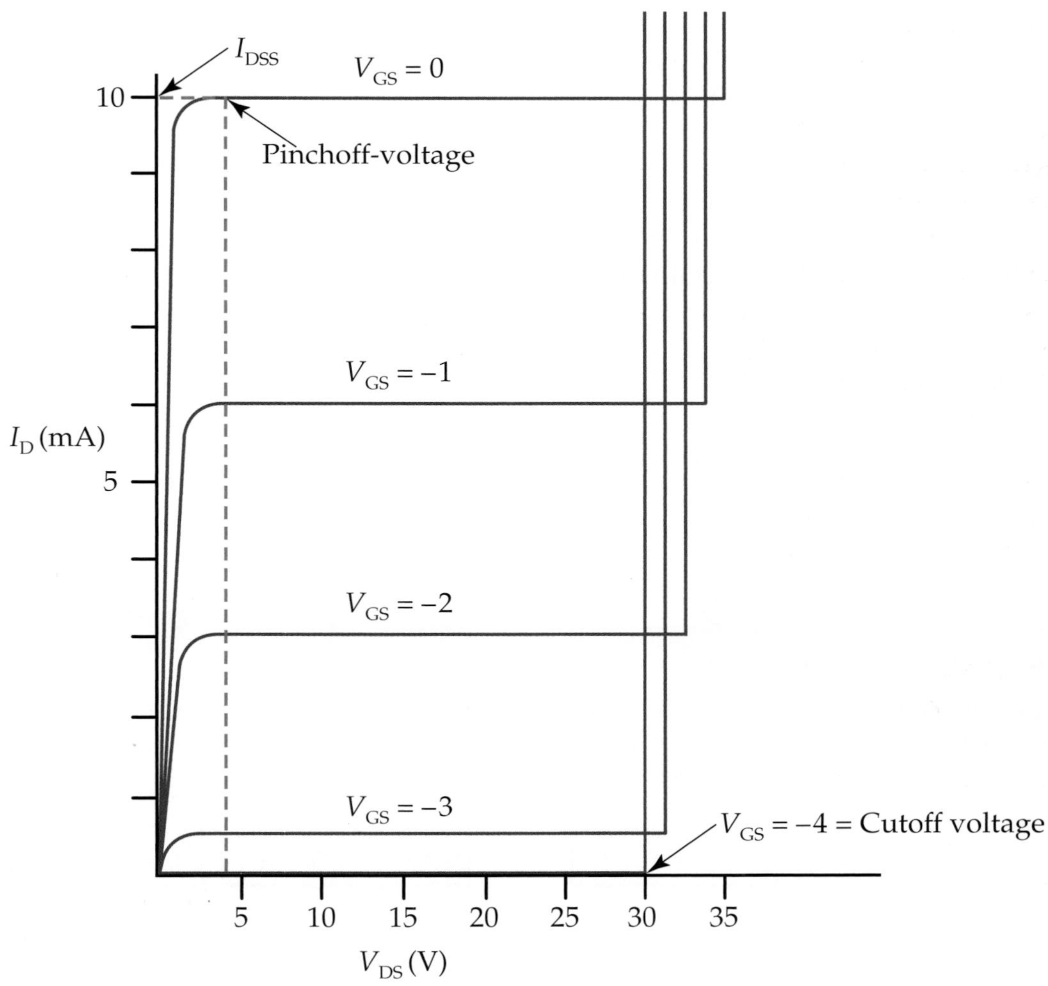

Figure 18-20. The characteristic curves of a JFET whose cutoff voltage is –4 V. The V_{GS} values are negative because the gate-source junction is reverse biased.

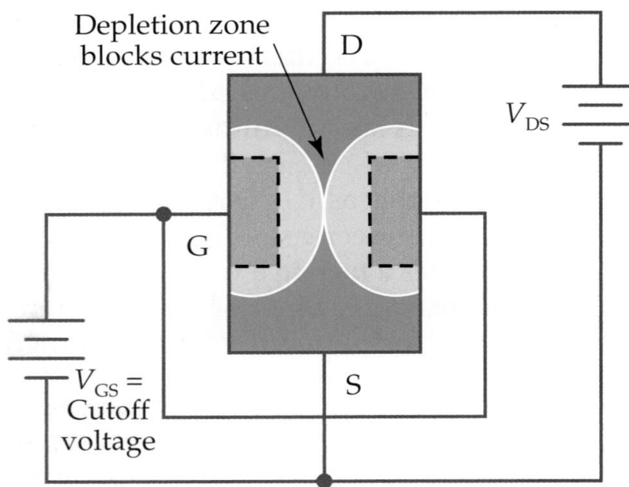

Figure 18-21. When the cutoff voltage is applied to the gate-source junction, the depletion zone expands and blocks current flow through the channel.

Notice that for each level of gate-source voltage (V_{GS}), the drain current (I_D) levels off (stops increasing) even as the source-drain voltage (V_{DS}) continues to increase. For example, when V_{GS} equals zero ($V_{GS} = 0$), the drain current increases until the drain source voltage (I_D) is equal to 4 V. At this point, called the ***pinch-off voltage,*** the drain current reaches 10 mA and holds this value until breakdown occurs. Breakdown is shown on the right side of the characteristic curve as a sharp rise in current.

MOSFET

Figure 18-22 shows a cross section of the two types of ***metal oxide semiconductor field effect transistor (MOSFET).*** The bottom section of material is called the ***substrate.*** The substrate is always made of the opposite type of material of the channel. A layer of insulation is located between the gate and the channel. A MOSFET may sometimes be called an ***insulated gate field effect transistor (IGFET)*** because of this insulation.

Like JFETs, some MOSFETs are "normally on" devices. This type of MOSFET operates in what is called ***depletion mode.*** The gate voltage is used to deplete the channel and control the current in the drain. This type of MOSFET requires the gate to be reverse biased.

The other type of MOSFET operates in ***enhancement mode.*** In this type, the source and the drain are not connected by a channel, so the MOSFET is normally off. The gate voltage is used to create and enhance the

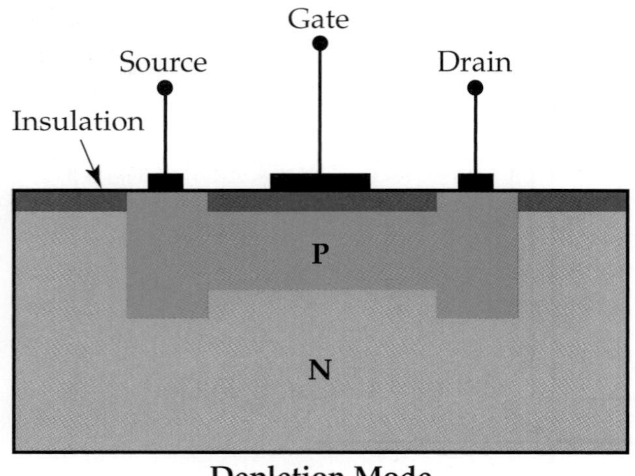

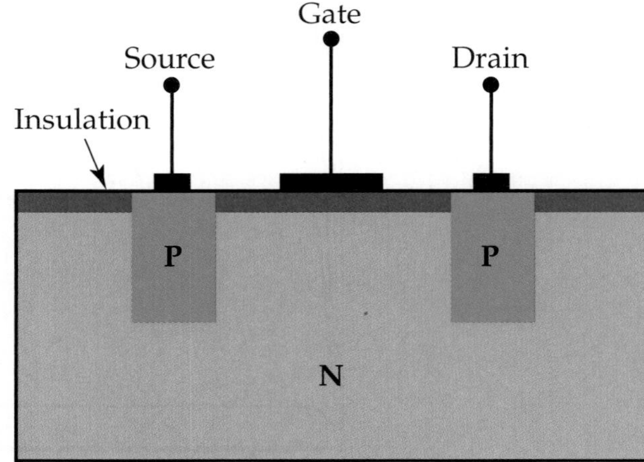

Figure 18-22. Depletion mode and enhancement mode P-channel MOSFETs. The insulation creates a capacitor between the gate and the channel.

channel and control the current in the drain. In the enhancement MOSFET, the gate is forward biased.

As you recall from Chapter 4, when a dielectric separates two conductors, they form a capacitor. The gate terminal and the channel in a MOSFET act as the conductors of a capacitor, and the insulation acts as the dielectric. The small capacitor in a MOSFET can be damaged very easily by surges or static charges. Even the capacitance of a human body or stray voltages from soldering can ruin a MOSFET.

Caution

Be extremely careful when working with a MOSFET. Solder all other components first, saving the MOSFET until last. Use heat sinks and avoid touching the leads (or wear protective gloves) because your body capacitance can ruin it. Ground all sources of static electricity before contacting a MOSFET.

The schematic symbols for MOSFETs are shown in **Figure 18-23.** Again, the only difference between P-channel and N-channel symbols is the direction in which the arrow points. In the symbols for the enhancement mode type, there is no solid line between the source and the drain, just as the source and the drain are not connected in the actual enhancement mode MOSFETs.

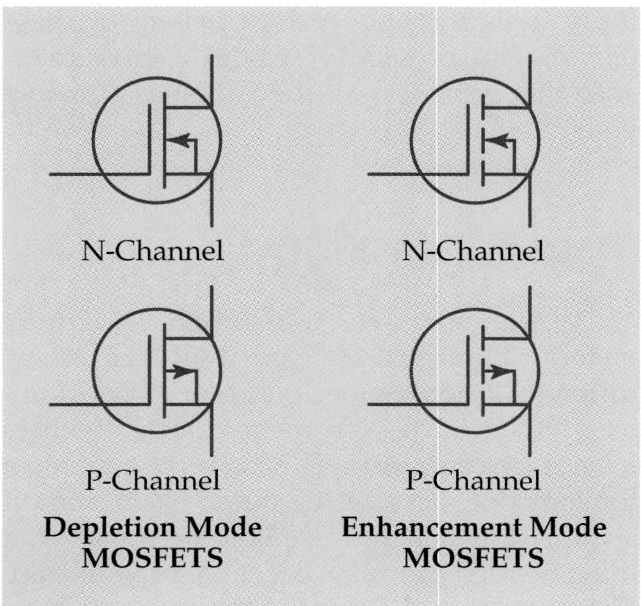

Figure 18-23. Schematic symbols for MOSFETs.

Advantages of the FET

The main advantage that FETs have over BJTs is the fact that any signal trying to enter the FET encounters very high impedance. ***Impedance*** is the combination of resistance and reactance. The FET is usually reverse biased to stop or reduce the current from flowing in the circuit. This explains why FETs are said to have high impedance.

MOSFETs are also high impedance when in the off condition. **Figure 18-24** shows a FET used as an amplifier connected to a radio tuning circuit. When you tune in a station, the tuning circuit develops the signal voltages of that station. These voltages, in turn, are amplified by the FET, which uses a small voltage at the gate to control a large current through the channel.

The MOSFET uses very little current in a circuit until it is needed. It also reduces the problem of thermal runaway, which is experienced with BJTs. When used in switching applications, MOSFETs may be turned on and off over 25 μs intervals carrying 40 A at less than 12 V. Therefore, they can be used to carry high current for short periods of time by selecting the correct size heat sink to keep the temperature low.

Current in a FET does not flow across a junction as it does in a BJT. This means that much less noise is produced by FETs. This characteristic makes them very useful in amplifier circuits. Since there is very little noise on the input, less noise is amplified to the output.

Phototransistors

Another type of transistor is called a ***phototransistor.*** Its schematic symbol is shown in **Figure 18-25.** Its three leads are the

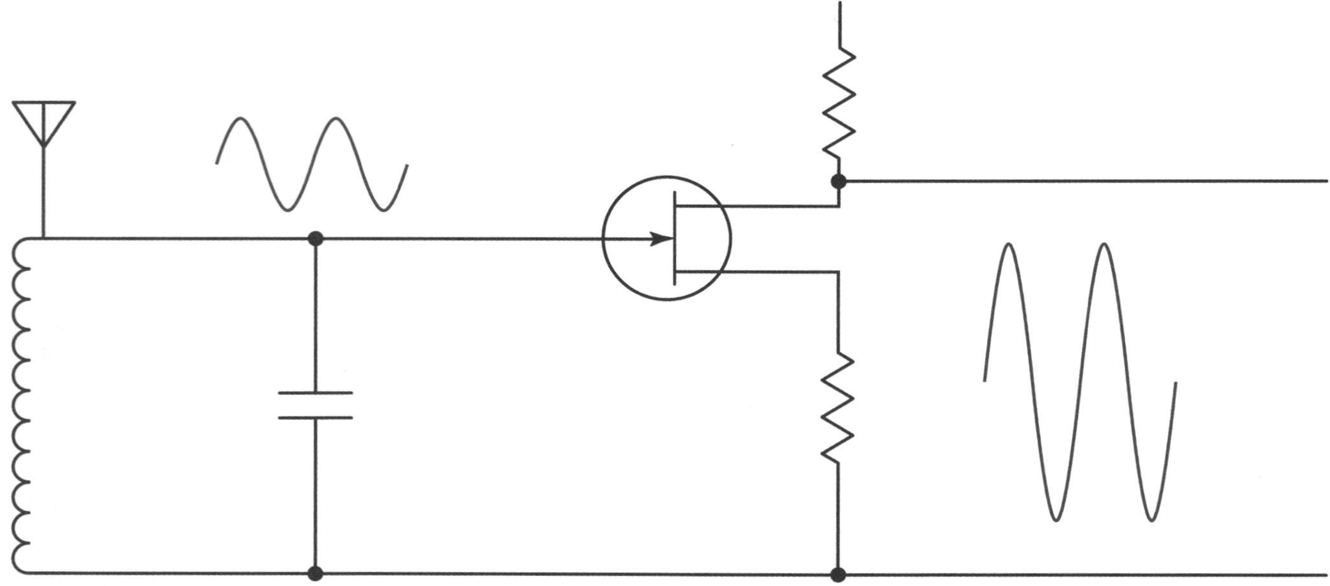

Figure 18-24. Using the FET in a circuit to amplify the radio signal.

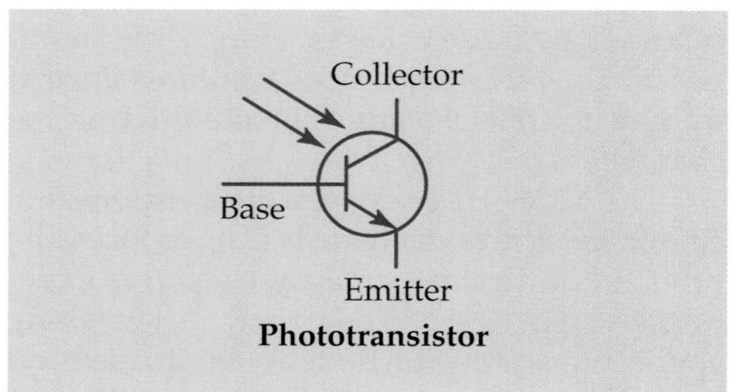

Figure 18-25. Schematic symbol for a phototransistor. The phototransistor is activated by light.

emitter, collector, and base, just like a BJT. The only difference between a phototransistor and a BJT is that a phototransistor uses light applied to its base as a power source, whereas a BJT uses the power source of a circuit. The base of a phototransistor is positioned underneath a tiny lens. As the amount of light that passes through the lens changes, the current through the transistor changes. The amount of light required to trigger a phototransistor falls in a range between 350 nm (nanometers) and 100 nm (nanometers) depending on the transistor chosen.

There is a whole series of light-operated electronic devices available, depending on the application. For example, phototransistors are used for dimming the headlights and the rearview mirror of some automobiles and to control outdoor lighting. To determine the phototransistor to use for a specific application, you must consider things such as the available wavelength of the light source, sensitivity, stability, and cost. Suppliers of these devices can provide you with a cross-reference that considers all the variables affecting your choice.

Transistor Data

When you see transistors on circuit boards, **Figure 18-26,** you will see many different kinds. Some look like small thimbles. Others have shiny metal cans. Still others have black containers. This presents a problem for someone just starting out in electronics. If a transistor fails and needs to be replaced, it must be replaced with the same type of transistor and with the leads of the new transistor in the same positions. To do this, you must

Figure 18-26. Transistors are scattered about this circuit board. Can you identify them?

be able to identify the leads on an unmarked transistor, and you must be able to use a parts catalog to ensure that the replacement transistor is the correct type.

Lead Identification

The base, emitter, and collector are not always labeled on a transistor. Look at **Figure 18-27.** To identify the leads, start by turning the transistor so you can see the leads. There are some standards for the leads that some manufacturers follow when making transistors. For example, if the transistor is made with a tab, the tab is closest to the emitter. If it is a semicircle, the base is in the middle with the emitter on the left when the straight side of the transistor faces up, Figure 18-27A. If the leads are different lengths, the emitter is usually the shortest, Figure 18-27B. If a red mark is used, it typically identifies the collector, Figure 18-27C. In some power transistors, the metal plate is also the collector, Figure 18-27D.

These are not universal ways to identify the leads. Some manufacturers do not abide by these standards. For this reason, the safest and best way to check transistor leads is to look at a parts catalog with the transistor's specifications.

Parts Catalogs

Parts catalogs and data sheets are valuable when a transistor needs to be replaced. These contain necessary information about transistors, including gain, operating temperature, cutoff frequency, and leakage current.

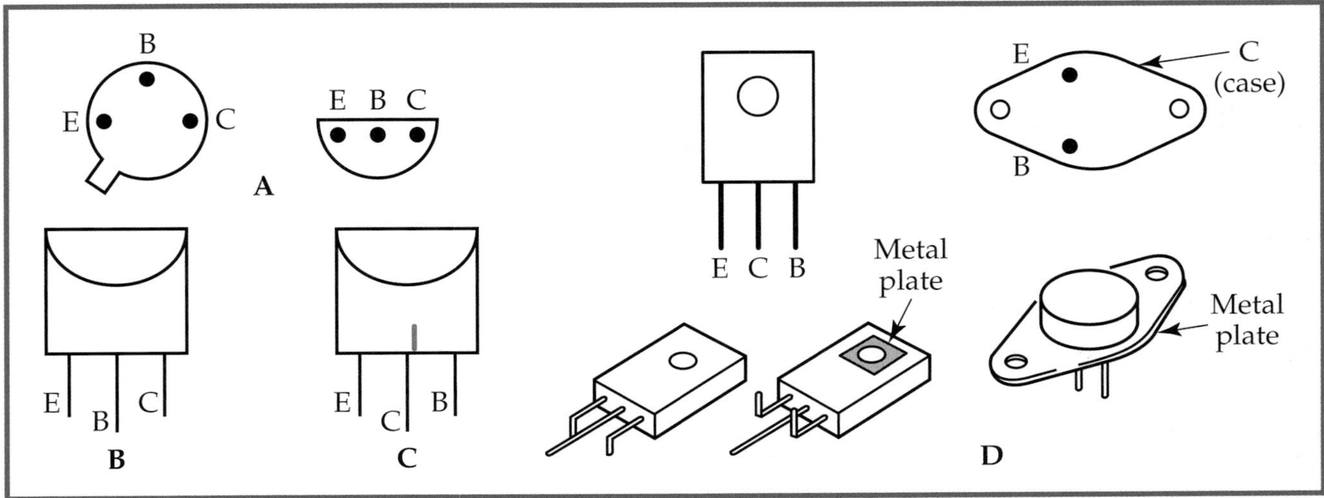

Figure 18-27. Shown are some ways to identify the leads on a BJT. A—For small signal types of transistors like the ones shown, the emitter is typically near the tab. The base is in the middle with the emitter on the left when the straight side of the transistor faces up. B—When the leads are different lengths, the emitter is usually the shortest. C—A red mark typically identifies the collector. D—Two types of power transistors with their leads identified. Power transistors usually are flat and plastic with leads on the bottom or back. The metal plate dissipates excess heat.

Some catalogs also list the type of case the transistor comes in. The shapes used to illustrate how to find the labels for the leads, like those shown in Figure 18-27, are just a few of the many case types available. Other information provided by catalogs includes:

- Replacement information.
- Temperature ratings.
- Power dissipation.
- Maximum currents.
- Maximum voltages.
- Matching characteristics.
- Static electrical characteristics.
- Dynamic electrical characteristics.
- Switching electrical characteristics.
- Switching test circuit.

Heat Sinks and Transistors

Like many electronic devices, when a transistor is being soldered into a circuit, it can easily be destroyed by the heat of the solder gun. As you know, a heat sink is temporarily added to the circuit to protect the device. The heat that transistors generate in operation can also destroy them. They need a way to dissipate heat, so many transistors have permanent heat sinks.

Practical Application 18-1: Identifying Transistor Type

Suppose the type of transistor in a circuit is unknown. Using an ohmmeter, it can be determined whether this transistor is a PNP or an NPN transistor. To begin, set the meter to the R × 100 range.make sure that the positive probe of the ohmmeter is connected to the positive end of the battery in the ohmmeter and that negative is connected to negative. To test the transistor's type, perform these steps:

1. Hook the positive probe from the ohmmeter to the base of the transistor.

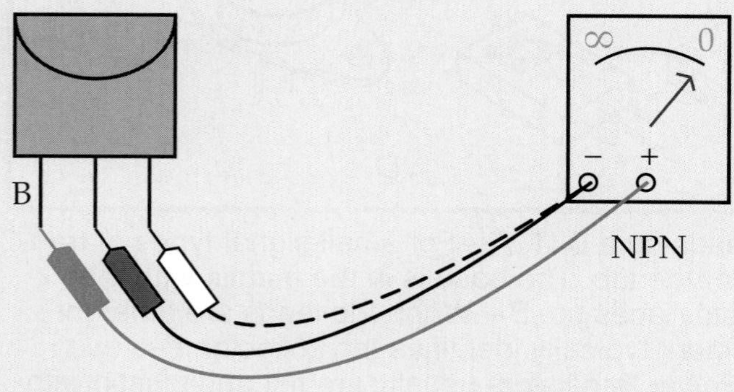

2. Connect the negative probe from the ohmmeter first to one of the other two transistor leads, then to the other.
3. Read the meter. If both readings show low resistance, this is an NPN transistor. If both readings show high resistance, hook the negative ohmmeter probe to the base of the transistor.
4. Connect the positive probe from the ohmmeter first to one transistor lead, then to the other.

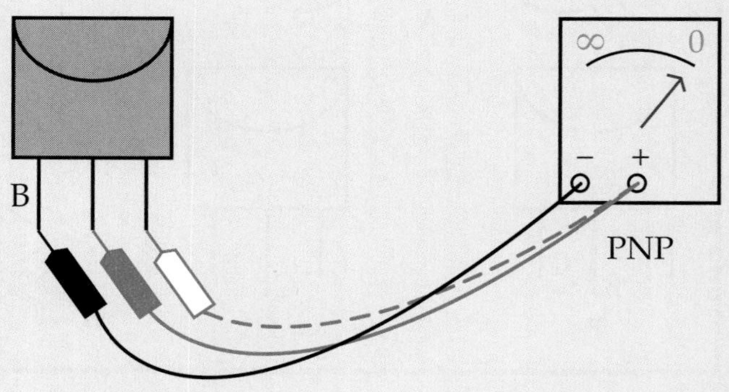

5. Read the meter. If both readings show low resistance, this is a PNP transistor.

Practical Application 18-2: Troubleshooting BJTs

As you learned earlier in this chapter, a BJT is like two diodes in series with opposite polarities. The base lead is in the middle of these two diodes. If there is a problem with a BJT, it can be found using an ohmmeter and information about how diodes behave.

Testing both polarities of the ohmmeter's probes across each pair of the BJT's leads can reveal any problems with the transistor. If one of the leads is the base and the first reading is high, when the ohmmeter probes are reversed, the reading should be low. The tests shown in the following figure should produce the results shown there.

When testing the collector and emitter leads, resistance should be high in both directions. The resistance is high both ways because there is a reverse-biased diode in both directions.

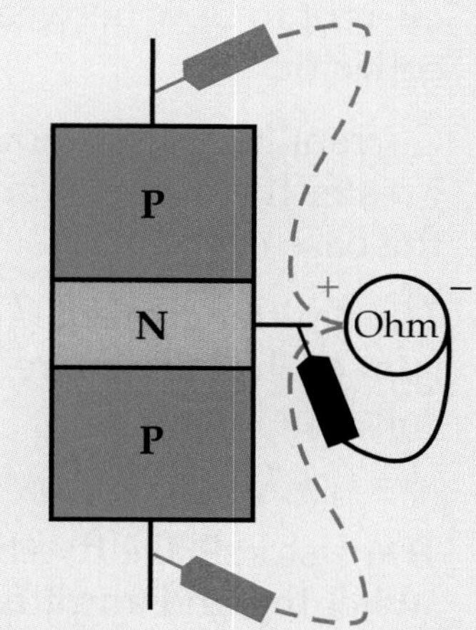

Both Readings Are Low

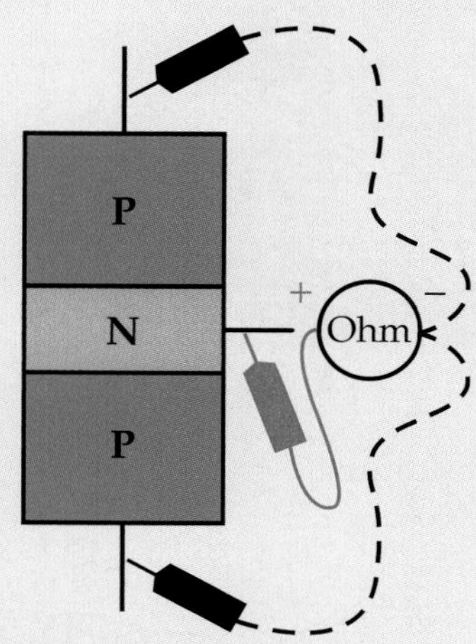

Both Readings Are High

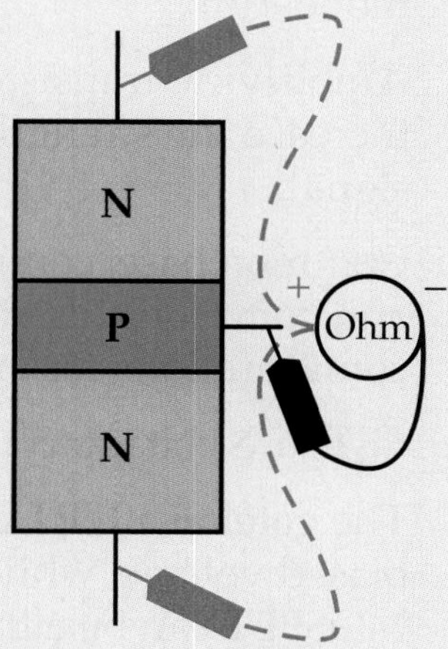

Both Readings Are High

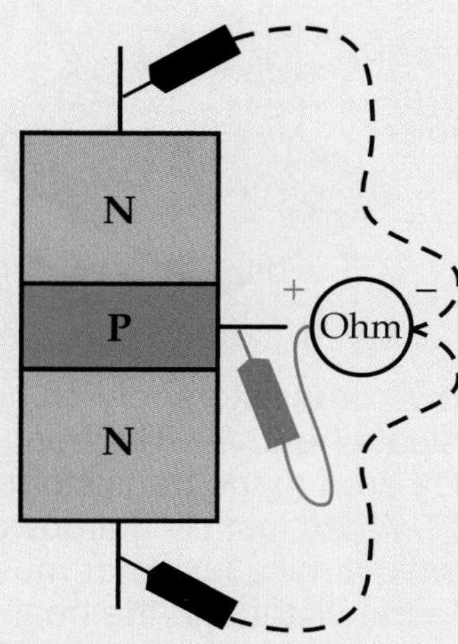

Both Readings Are Low

Power transistors, for example, carry enough current to create extra heat that the transistor cannot dissipate. However, if the transistor is attached to a heat sink, **Figure 18-28,** it will be able to get rid of the heat. This is not the same kind of heat sink used while soldering, but it has the same purpose. It takes heat away from the transistor so the transistor is not destroyed.

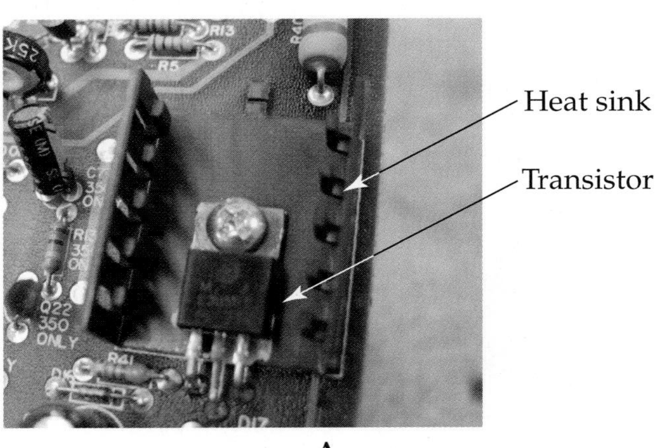

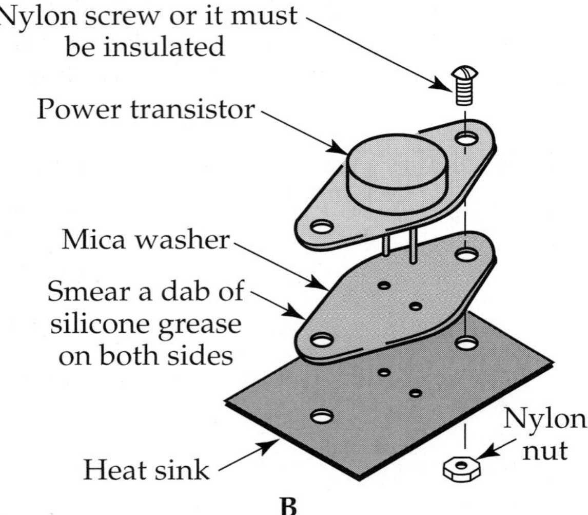

Figure 18-28. Power transistors rely on heat sinks to help dissipate heat. A—The heat sink adds more surface area to the transistor to help dissipate heat. B—When installing a power transistor on a heat sink, a mica insulator must be used between the transistor and the heat sink. The nuts and bolts also must be insulated if they were not made of nylon.

Summary

- BJTs are current-controlled devices.
- Two types of bipolar junction transistors are PNP and NPN.
- An NPN transistor has P-type material between two N-type regions.
- A PNP transistor has N-type material between two P-type regions.
- The base of a transistor must have the proper polarity, positive voltage for NPN or negative voltage for PNP, to allow current to flow between the emitter and collector.
- Current gain (β) is how many times greater the collector current (I_C) is than the base current (I_B).
- The emitter current (I_E) is equal to the sum of the base current (I_B) and the collector current (I_C):

  $I_E = I_B + I_C$
- Transistors have two basic functions: switching and amplifying.
- Switching transistors are used extensively in both personal computers and with computers found in all newer model cars.
- Amplifying transistors are used to increase the strength of a very small signal.
- Common base, common emitter, and common collector are the three configurations for an amplifying circuit.
- FETs are voltage-controlled devices.
- The gate on a JFET requires a reverse-biased voltage, while the gate on a MOSFET can be either forward or reverse biased.
- An ohmmeter can be used to determine an unknown type (PNP or NPN) of BJT by checking for bias voltage.

Test Your Knowledge

Do not write in this book. Write your answers on a separate sheet of paper.

1. The two basic functions of transistors are ______ and ______.
2. Draw the symbols for PNP and NPN transistors. Label the leads of each transistor and record the transistor type beneath each symbol.
3. A BJT transistor is a ______-controlled device.
4. In the symbols for BJTs, the lead that has the arrow is the ______.
5. The symbol for *current gain* is ______.
6. Draw a circuit with an NPN BJT in the common-emitter configuration.
7. A FET is a(n) ______-controlled device.
8. The part of a FET where current flows is called the ______.
9. The three leads of a FET are the ______, ______, and ______.
10. MOSFETs operate in either ______ mode or ______ mode.

Activities

1. Take apart NPN and PNP transistors. Is there any difference in construction between the two types? Can you take them apart without destroying them?
2. Find as many different shapes of transistors as you can. Identify the lead location and name for each type.
3. Solder a couple of transistors without using heat sinks. How much heat can the transistors take before they are ruined? Do the leads discolor in any way? Can you *see* when they are ruined?
4. Visit a plant or try to get a film that shows how a transistor is encapsulated. How is the process done? How do they identify which type of transistor is inside the plastic? Are they processed in batches or one at a time?
5. Why do manufacturers give guarantees on solid-state components? Do they last longer than those with tubes?
6. You often hear that if a solid-state device is going to fail, it will do so in the first 30 days. If not, it will last for years. Why?
7. Find some circuits that have common-emitter, common-base, and common-collector wiring. Are there any advantages for each type?
8. Make a list of the properties listed in a replacement catalog. What does each item mean or stand for? Why are there so many different categories?
9. Communications satellites are affected by their altitude. Some have solid-state devices which were ruined by things in the atmosphere. How is that possible?
10. What are majority and minority carriers?
11. Where are unijunction transistors used?
12. If you were to increase the temperature surrounding a transistor, how would it affect operation? Does the heat found in the desert or hot industrial plants affect the operation of a transistor? What effects does space travel have on a transistor?

New techniques for building circuit boards have continued to shrink the size of transistors and other components and improve their speed of operation at the same time.

Thyristors

Learning Objectives

After studying this chapter, you will be able to do the following:

- Draw the symbols for an SCR, a DIAC, and a TRIAC.
- Describe how the SCR, DIAC, and TRIAC are turned on and off.
- State the uses for an SCR, a DIAC, and a TRIAC.

Technical Terms

DIAC
silicon-controlled rectifier (SCR)
thyristor
TRIAC

At this point you should understand P-N junctions as they are used with diodes and transistors. Now, we will look at some of the other solid-state devices and how they work through the use of their P-N junctions. A ***thyristor*** is the name given to a broad group of devices that are made up of four or more alternating layers of P-type and N-type materials. Four layers create three P-N junctions, one at each region where the P-type and N-type materials are joined together.

Thyristors are most often found in circuits that have high-voltage or high-current requirements. In addition, they are made with the ability to lock into the conducting mode once they reach a specific voltage or current level. There are many types of thryristors. This chapter covers only the SCR, DIAC, and TRIAC.

Silicon-Controlled Rectifiers

The ***silicon-controlled rectifier (SCR)*** is a type of thyristor that usually is used in circuits where the current is very high. SCRs may be used in circuits carrying as much as several thousand amps. An SCR allows a circuit to be switched on and off thousands of times in one second.

The symbol for an SCR, **Figure 19-1,** looks like a diode symbol with one additional line. This line represents what is called its *gate*. The gate is very important in the operation of an SCR. **Figure 19-2** shows the physical structure of an SCR. Note that it is like a four-layer BJT. An SCR can be thought of as the combination of two BJTs, **Figure 19-3.** When the SCR is turned on, each BJT provides the necessary base current for the other.

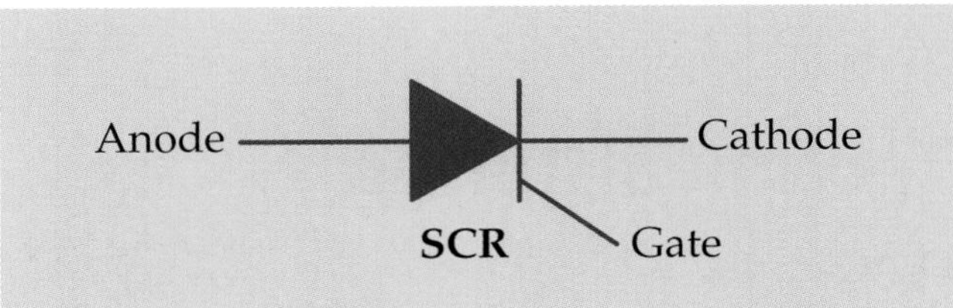

Figure 19-1. The SCR schematic symbol looks like the diode symbol, but it has an additional lead called the *gate*.

Turning on an SCR

To turn on the SCR, the gate circuit is used to send a pulse of electricity into the SCR. This current is usually relatively small and is positive with respect to the cathode. Once it is turned on, the SCR conducts from cathode to anode.

When the gate circuit is opened, the SCR continues to conduct. Removing the gate current does nothing. This means once an SCR is turned on, it will not turn off on its own. For this reason, an SCR is often called a *latch* or a *latching switch*.

Turning off an SCR

SCRs have a minimum forward-biased current that keeps them turned on. One way to turn off an SCR is to lower the current in the circuit to less than this minimum current. This

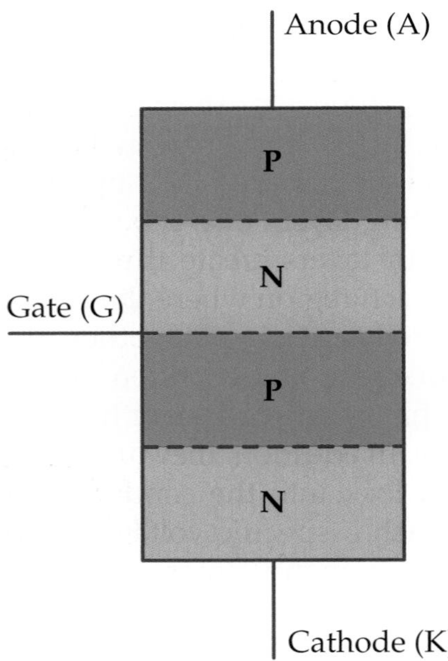

Figure 19-2. An SCR has four layers of P-type and N-type material.

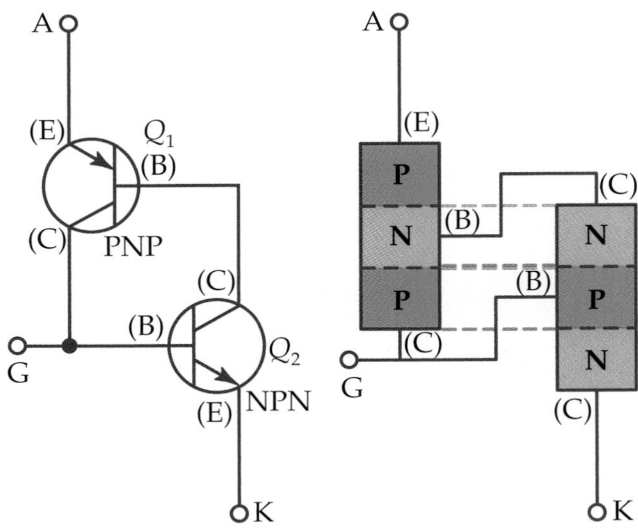

Figure 19-3. When the gate provides base current for the NPN transistor, the NPN transistor provides base current for the PNP transistor. Once they are both turned on, each provides the necessary base current for the other.

can be done with a switch in the load circuit, **Figure 19-4.** However, if the load current is too high, opening the circuit may cause an arc when the switch is opened.

A second way to turn off the SCR is to connect a push-button switch in parallel with the SCR, **Figure 19-5.** When the switch is closed, the SCR stops conducting because it is shorted out of the circuit. When the push button is released, the SCR remains off and can only be turned on again if the gate is triggered.

Turning off an SCR in AC Circuits

When an SCR is used in an ac circuit, it is easy to turn off. The SCR is automatically shut off during each reverse-biased half of the ac cycle. Usually, the SCR in an ac circuit is triggered on at some point during every forward-biased half cycle and automatically turned off at the beginning of each reverse-biased half cycle.

Since the SCR can be triggered on at any point during the forward-biased half of the cycle, it has many uses in circuits that control heavy currents. It can be triggered on early so that it is on for most of the half cycle, **Figure 19-6.** It can also be triggered late in the half cycle so that it is off for most of the cycle, **Figure 19-7.** This allows complete control at ANY point on each half cycle that current is allowed to flow. It also means that the current will be shut off at the end of that half cycle when it becomes reverse-biased (goes negative).

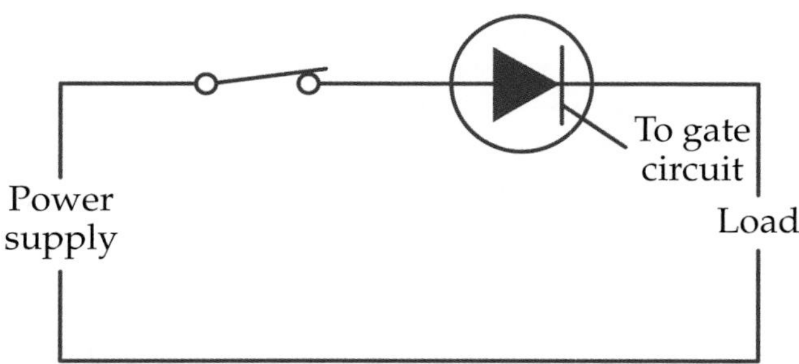

Figure 19-4. A normally closed switch can be wired in series with the SCR. To turn off the SCR, the switch is opened.

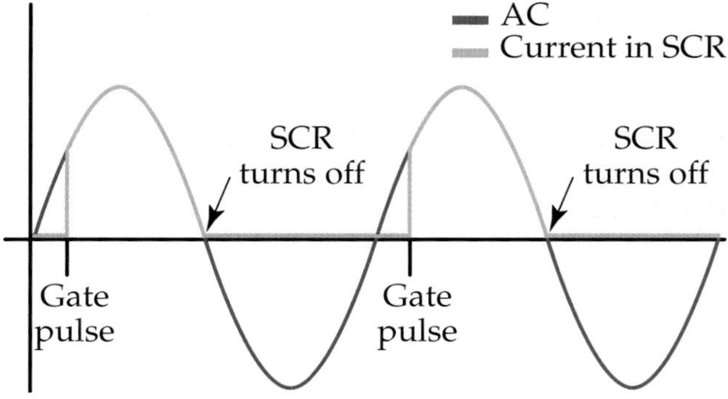

Figure 19-6. The SCR can be turned on early in the cycle. In this case, the SCR is on for most of the positive half of the cycle.

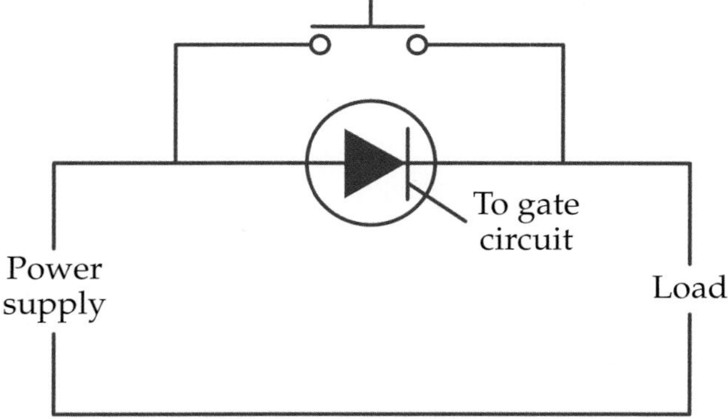

Figure 19-5. Another way to turn off an SCR uses a normally open push-button switch in parallel with the SCR. Closing the switch shorts the SCR out of the circuit.

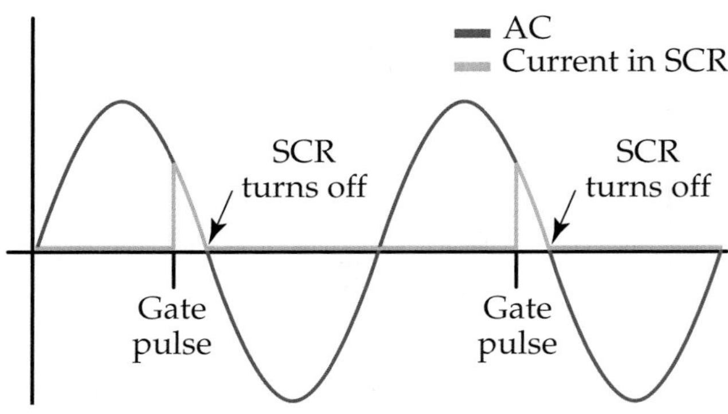

Figure 19-7. The SCR may also be turned on late in the positive half of the cycle.

Project 19-1: Power Tool Speed Control

You can build the speed-control unit shown as a valuable exercise in working with ac circuits that have resistors, rectifiers, and SCRs.

(Project by Jim Shelton)

By turning the knob, you can vary the speed of any series universal (brush-type) motor that draws less than 400 W. This is the type of motor found in most drills, table saws, hand grinders, and sewing machines.

Caution

Do not use the speed-control unit for other types of motors.

This speed-control unit allows you to slow power tools when working with different kinds of material. Have you ever tried to drill holes in a piece of steel and the drill bit squealed and burned? This indicates that the drill is running too fast. Installing this speed-control unit and using it to slow the drill can solve this problem. Obtain the following items:

No.	Item
2	Silicon rectifiers (D_1, D_2), Motorola HEP 162
1	Resistor (R_1), 2.5 kΩ, 5 W
1	Potentiometer (R_2), 500 Ω, 2 W
1	Resistor (R_3), 1.5 kΩ, 0.5 W
1	SPST switch, 5A
1	SCR, Motorola HEP 302
1	Aluminum heat sink for SCR, 3 1/2″ × 1 1/2″
1	Three-prong panel-mount ac receptacle
1	Three-wire grounded line cord
1	Aluminum minibox, 3″ × 4″ × 5″
	Connecting wiring and cord strain relief
	Heat sink compound

To build the speed-control unit, look at the schematic and follow the procedure.

Procedure

1. Discard the insulating washers furnished with the SCR.
2. Before you install the SCR on the heat sink, apply a thin coat of heat sink compound to the surface of the minibox that will contact the top of the heat sink.

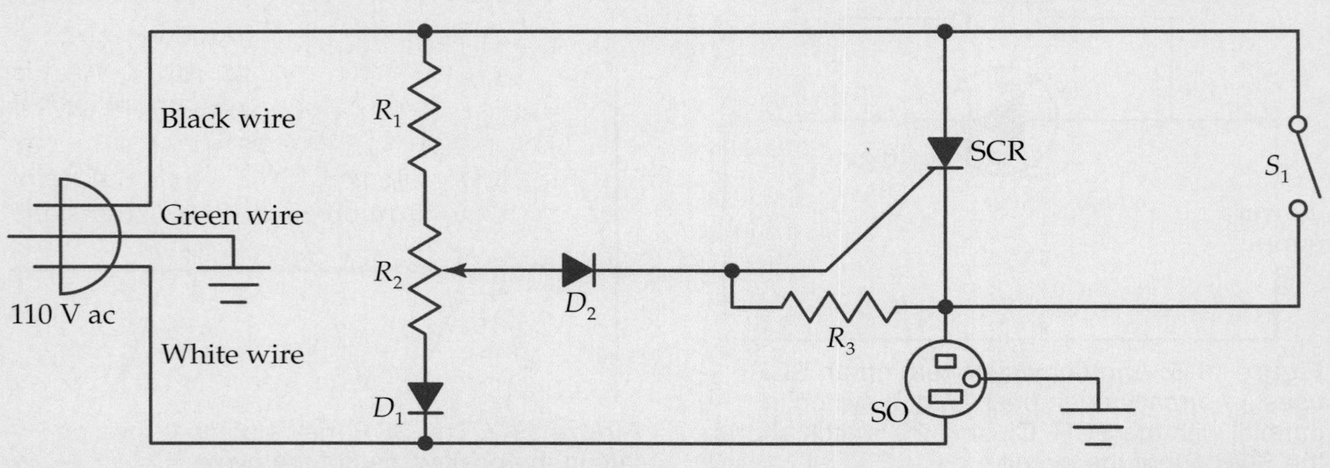

Project 19-1: Power Tool Speed Control *(Continued)*

3. Mount the heat sink/SCR assembly in the minibox. Make a direct electrical connection between the minibox and the heat sink, but do not overtighten the nut.
4. Mount and connect the other components. Make sure that the rectifiers and other parts do not touch the bottom half of the minibox, which could cause a short. The receptacle is wired to ensure against incorrect connection, which could otherwise create a safety hazard.
5. Connect the line cord.

Be sure to use a three-wire grounded line cord and three-prong panel mount socket to prevent electrical shock in case the insulation within the power tool breaks down. Also, ensure that your bypass switch and wiring are able to carry at least 5 A ac.

After you have completed the speed-control unit, plug the line cord of the unit into a 110-V ac receptacle. Plug your power tool line cord into the panel mount socket of the speed-control unit. You should be able to control the power tool up to 70% of its maximum speed. If you need full speed, simply flip the bypass switch.

DIACs

A DIAC uses the information that you learned about the diode. ***DIAC*** is the short name given for **DI**ode for **A**lternating **C**urrent. It is similar to two parallel diodes joined together in opposite directions, as shown in **Figure 19-8.** Notice that a DIAC consists of two anodes (A1 and A2) but no cathode. Since the DIAC conducts in both directions, neither lead has a cathode label.

A DIAC operates similar to two opposed zener diodes hooked up in series. Recall that a zener diode reaches a breakdown point before it conducts. The breakdown voltage for many DIACs is around 30 V. DIACs are used most often to trigger a TRIAC or an SCR in many types of circuits.

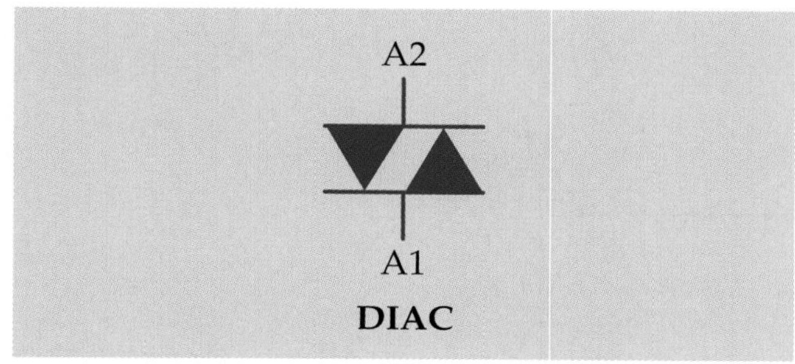

Figure 19-8. The DIAC schematic symbol looks like two diodes arranged in parallel and in opposite directions.

Turning on a DIAC

Once a DIAC reaches its breakdown voltage, a sharp drop in the diode resistance occurs, which causes a rapid increase in current flow. The rapid current flow switches the DIAC into conduction mode. The DIAC latches in the on position as long as there is current flow.

Turning off a DIAC

The DIAC stays in conduction mode in either direction until the current flow falls below the threshold value required to keep it in that state. Once the holding current falls below the threshold value, the DIAC stops conduction and returns to its high-resistant state.

TRIACs

A TRIAC uses the information that you learned about the SCR. ***TRIAC*** is the short name given for **TRI**ode for **A**lternating **C**urrent.

Project 19-2: Color Organ

The color organ shown is an easy-to-build project that uses one SCR. It varies the brightness of a low-wattage bulb in keeping with changes in the audio level from a radio or stereo.

(Project by Ed Belliveau)

No.	Item
1	Plastic box, approx. 3″ × 6″ × 1 1/2″
1	Circuit board, approx. 2″ × 5″
1	Audio output transformer
1	Potentiometer (R_1), 5 kΩ
1	Control knob
1	Resistor (R_2), 1 kΩ
1	Capacitor (C_1), 0.001 μF
1	SCR
1	Receptacle, 110 V
1	Lamp cord with plug
	Connecting wiring

You can mount this simple circuit in any handy plastic box and use a knob on the outside to control the potentiometer. Obtain the items indicated in the table.

To build the color organ, look at the schematic and follow the directions.

Procedure

1. Mount transformer, resistor, capacitor, and SCR on the circuit board.
2. Place the board in the plastic box.
3. Install the receptacle in the plastic box.
4. Install the potentiometer and control knob.
5. Attach the lamp cord and connecting wiring.
6. Solder the connections.

The transformer output is the input for this circuit. Connect the audio input side of the transformer into the speaker of your radio. Connect the lamp cord for a low-wattage bulb into the receptacle in the side of the plastic box. Plug your color organ lamp cord into a 110-V receptacle. Turn on the radio and watch the results.

If you want to experiment further, mount the lamp behind different colored plastic sheets. You can even make a four-sided, four-color lamp for this use.

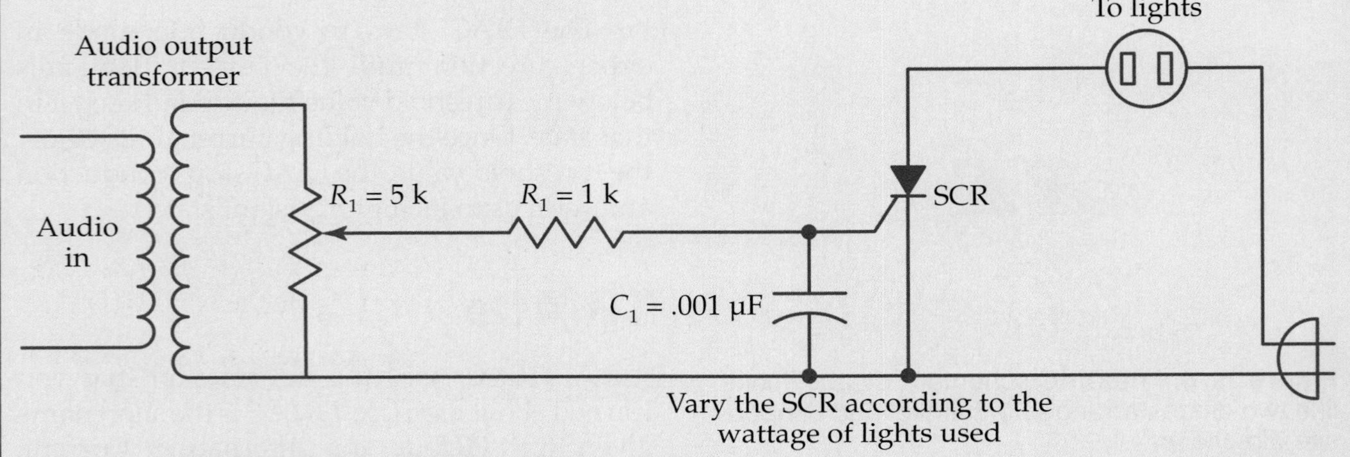

It is similar to two SCRs joined together in parallel and facing opposite directions. However, both gates are tied together as shown in **Figure 19-9.** This is often called being connected in inverse parallel. You will also notice that a TRIAC consists of two anodes (MT1 and MT2), one gate, but no cathode. For this reason, a TRIAC has sometimes been called a *bidirectional silicon switch.*

Uses for a TRIAC

A TRIAC, similar to the SCR, acts as a high-speed switch. It is used as a voltage regulator, a control for alternating current devices (such as speed control for various motors), a control for digitally-controlled electrical appliances, and as a dimmer control for lights.

Turning on a TRIAC

Since both gates are tied together, the single gate is bidirectional. That means the TRIAC can be turned on with either a negative or a positive voltage applied to the gate. Once it is triggered on, just as with the SCR, the TRIAC will continue to conduct without an applied gate voltage. The advantage to this setup is that a very low voltage can be used to trigger a device carrying a large current, just like the SCR. The big difference between the TRIAC and the SCR is that the TRIAC can be triggered with either a negative or a positive voltage.

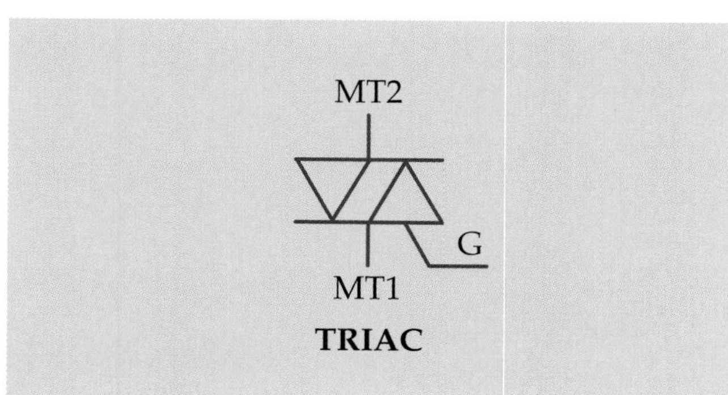

Figure 19-9. The TRIAC schematic symbol looks like a DIAC with a gate.

Turning off a TRIAC

Once a TRIAC is turned on, it is a simple matter to turn it off. Turning off the TRIAC occurs every time the alternating current gets near the end of each half-cycle. At that point, the current falls below the threshold value required to keep the TRIAC conducting. Thus, the TRIAC shuts off until the gate is triggered to turn it back on.

> **Note**
> When a TRIAC fails, it may be shorted in one direction, but will operate normally in the opposite direction.

Summary

- A thyristor is a semiconductor device that has four or more alternating layers of P-type and N-type material.
- Thyristors are typically used in circuits that have high-voltage or high-current requirements.
- A thyristor locks into conduction mode once it reaches a specific voltage or current level.
- A silicon-controlled rectifier (SCR) is a four-layer device that has an anode, a cathode, and a gate.
- An SCR is used in circuits carrying as much as several thousand amps and can be switched on and off thousands of times in one second.
- The SCR is turned on by applying a pulse of electricity to its gate.
- A DIAC is similar to two parallel diodes joined together in opposite directions.
- DIACS have only two terminals (anode 1 and anode2) and can conduct in either direction.

- A DIAC turns on when its breakdown voltage is reached and it stays on as long as the current flow is above the threshold value required to keep it in that state.
- A TRIAC is similar to two SCRs connected in reverse parallel.
- The TRIAC consists of two anodes (MT1 and MT2) and one gate.
- TRIACs can be turned on with either a negative or a positive voltage applied momentarily to the gate.
- A TRIAC turns off when the current falls below the threshold value required to keep the TRIAC conducting.

Test Your Knowledge

Do not write in this book. Write your answers on a separate sheet of paper.

1. In what types of circuits are thryristors most often found?
2. Draw the symbols for an SCR, a DIAC, and a TRIAC and label their terminals.
3. An SCR is turned on by applying a pulse of electricity to its ______.
4. For a DIAC to conduct, it must reach its ______ voltage.
5. A DIAC is often used to trigger a(n) ______ and a(n) ______.
6. A TRIAC can be turned (off, on) by applying either a negative or a positive voltage applied momentarily to the gate.
7. A TRIAC will stop conducting when the current falls below its ______ value.

Activities

1. What other types of thryristors are there? Create a table that lists their names, symbols, and uses.
2. How many different size of SCRs are there? What is the minimum amount of current they will carry? What is the maximum current?

Integrated Circuits

Learning Objectives

After studying this chapter, you will be able to do the following:

- Explain why the IC is important for technology.
- State the differences between linear and digital circuits.
- Draw two symbols for integrated circuits.
- Draw the symbols for basic logic gates.
- Create truth tables for logic gates.
- Describe the procedure for making a printed circuit board.

Technical Terms

AND gate
binary
digital ICs
hybrid IC
integrated circuit (IC)
inverter
large-scale integration (LSI)
linear ICs
logic family
logic gate
medium-scale integration (MSI)
monolithic IC
NAND gate
NOR gate
NOT gate
OR gate
photomask
small-scale integration (SSI)
substrate
thick-film IC
thin-film IC
truth table
ultra large-scale integration (ULSI)
undercut
Very large scale integration (VLSI)
XNOR gate
XOR gate

An ***integrated circuit (IC)*** is a single electronic device built on a semiconductor chip capable of holding hundreds of thousands of microscopic components, **Figure 20-1.** It contains all the necessary components of the circuit including resistors, transistors, and capacitors and all the necessary internal circuitry. The IC also provides external connections within working products to allow them to function. Without the IC, it would have been impossible to develop computers and thousands of other tools that represent our present day standard of living. ICs are one of the greatest technological advancements of the twentieth century.

Advantage of Integrated Circuits

The main advantage of ICs is that so much electronic circuitry can be put in a very small package, **Figure 20-2.** ICs are not intended to be repaired because of their size and the way they are made. Fixing them would be nearly impossible. The electronic parts are completely sealed inside of the IC, and any attempt to open the box can ruin the parts inside. Since they are so inexpensive, ICs are made to be replaced. It is somewhat fast and easy to disconnect or unplug an IC and replace it with a new one.

Figure 20-1. A hard plastic case protects the complex circuitry of the integrated circuit. (Courtesy of National Semiconductor Corporation)

Figure 20-2. This unsealed IC shows its many components and interconnections.

Integrated circuits are inexpensive even though they are difficult to manufacture. This is because a manufacturer can make so many ICs at once. Special machines produce ICs so fast and accurately that the price can be kept quite low.

ICs have special functions because they are so small. For example, their small size makes them ideal for use in hearing aids, miniature security cameras, and cell phones.

Types of Integrated Circuits

There are two general types of ICs: linear and digital. ***Linear ICs*** are found in many analog devices. They are designed to accept a continuous range of signals. Examples of circuits that use linear ICs are filters and amplifiers.

Digital ICs work with signals that are discrete. Discrete signals are not part of a continuous range of values. They are distinct values, such as 0 V and 1 V and may represent states, such as off and on. Digital ICs are found in calculators and computers. Some equipment have circuits that use both linear and digital ICs.

IC Classification

ICs are classified according to how many components they contain. In 1964, when the first ICs were commercially produced, an IC contained up to 11 components. This is called ***small-scale integration (SSI).*** By the late 1960s, ICs contained up to 99 components, which is now considered ***medium-scale integration (MSI). Large-scale integration (LSI)*** was reached by 1970, and these ICs contained from 100 to 9999 components. ***Very large-scale integration (VLSI)*** contains from 1000 to 99,999 components and was reached in the early 1980s. Sometimes the term ***ultra large-scale integration (ULSI)*** is used to describe ICs that contain 100,000 to over 1,000,000 components or to emphasize the complexity of an IC.

Another way to classify ICs is by logic family. A ***logic family*** describes the physical makeup of the IC. For example, the CMOS logic family of ICs use CMOS technology for the physical makeup of its ICs. The following is a list of some logic families:

- CMOS (Complementary Metal Oxide Semiconductor).
- ECL (Emitter Coupled Logic).
- NMOS (N-Channel Metal Oxide Semiconductor).
- PMOS (P-Channel Metal Oxide Semiconductor).
- TTL (Transistor-Transistor Logic).

IC Symbols

Two symbols for ICs are shown in **Figure 20-3.** Because the circuit inside the IC container is so complex, the internal circuit of the IC is never drawn. Instead, you must refer to schematic diagram and data sheets for information about the numbered leads and internal components. You will also have to obtain data sheets or use cross-reference books to understand the numbers printed on top of the IC, **Figure 20-4.** Since there are literally millions of different ICs, you will never know what is inside them without deciphering their code numbers. The schematic will tell you what the IC can do.

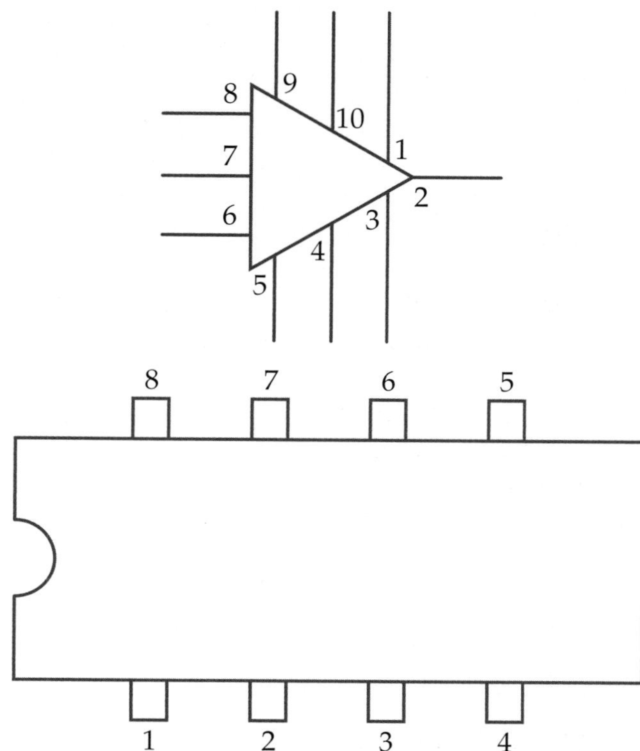

Figure 20-3. IC symbols used in circuit drawings can have many different shapes. The circuits inside the IC do not need to be drawn on schematic diagrams.

IC Manufacturing

Circuits can also be divided according to how they are manufactured. ICs are either monolithic or hybrid. A ***monolithic IC*** is made on a single piece of silicon. A ***hybrid***

Figure 20-4. The label on top of the IC allows you to find data about the circuitry inside. The pins are the leads used to make the electrical connections in the circuit.

IC is one which is made by combining two or more monolithic ICs. A hybrid could also be made by combining monolithic with discrete components, such as inductors and additional transistors.

Monolithic IC

In the process of making a monolithic IC, the first step is drawing the circuit. The circuit drawing is huge when compared to the size of the finished IC, **Figure 20-5.** This is done so that the drawing can be reduced and still maintain accurate spacing between parts. If parts get too close and touch, there would be a short within the IC, and it would not work properly.

A camera is used to reduce the drawing to 400 times smaller than its original size. The camera can shrink the lines on the drawing, which represent conductor paths (traces) and electronic parts, down to a width of 0.0001″. The technology allows oaver 1 million transistors and other components to be placed on a tiny rectangular area of a silicon wafer, **Figure 20-6.** Silicon wafers are typically 3″ to 8″ in diameter and can hold many ICs, or chips.

Figure 20-5. A camera will be used to reduce this integrated circuit drawing so that its image can fit on a single chip.

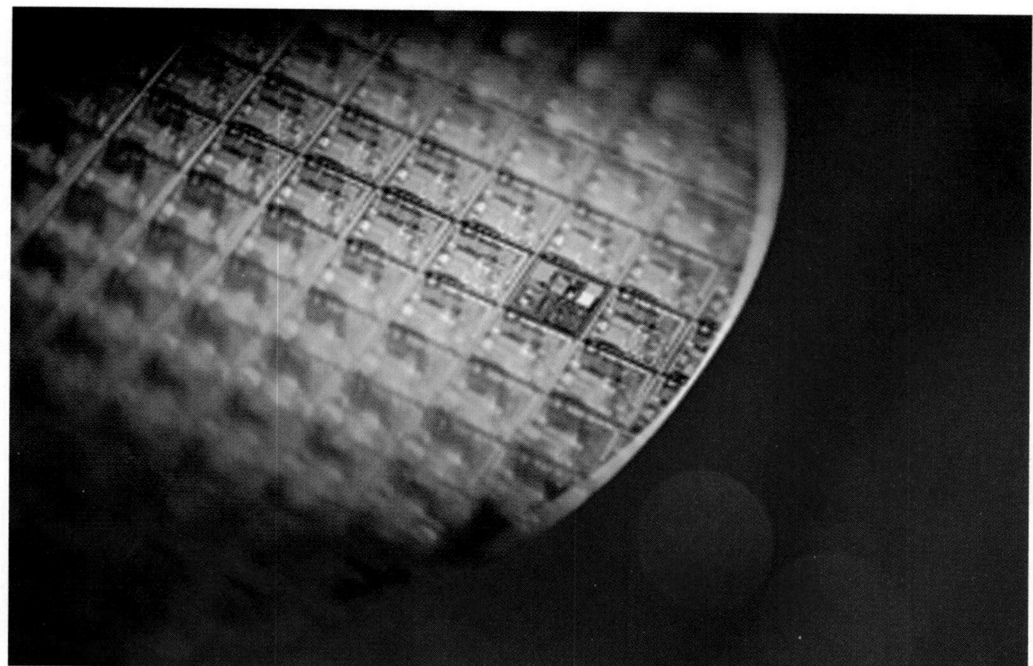

Figure 20-6. Each square on this silicon wafer is an IC.

To make the completed IC, a combination of special furnaces, ultraviolet light, and photomasks are used to add the impurities, which are the N-type and P-type materials that form the diodes, transistors, and other components. First, the wafer is covered with a protective layer of material, and the photomask is placed over it. A ***photomask*** has the pattern of the different parts of the IC cut out of it and is used to transfer the pattern to the IC. The whole thing is exposed to ultraviolet light, which hardens the exposed protective layer, while the parts that are covered remain soft. See **Figure 20-7.**

Chemicals are used to etch away the soft parts of the protective layer and of the material beneath it. If N-type material is etched away, P-type material is diffused over the IC to fill the areas surrounding the N-type material that is left. This process is repeated with N-type and P-type material until the components are formed. Several photomasks may be used.

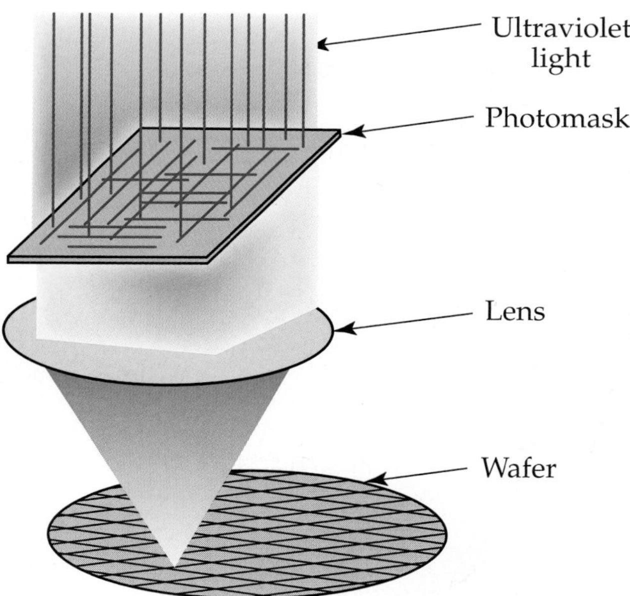

Figure 20-7. A photomask and ultraviolet light are used to transfer circuit patterns to each IC on the wafer.

Next, the interconnections are added using small amounts of aluminum, gold, or copper. This conducting material is applied in a manner similar to the N-type and P-type materials. A layer of aluminum is placed completely over the IC, and then the unwanted parts are etched away.

Finally, the wafer is washed and the ICs tested. A computer keeps track of the failed ICs. A diamond saw is used to cut out the individual ICs, **Figure 20-8.** The failed ICs are discarded. The good ICs are mounted onto a framework. Leads are bonded to the IC, and then the IC is enclosed in permanent packaging. All of this must be done in a sterile environment, because even one piece of dust can ruin an IC.

Hybrid IC

There are two types of hybrid ICs: thin-film and thick-film. A ***thin-film IC*** is made by depositing very thin layers of vapor on an insulating material known as a ***substrate.*** The process used to do this is called *sputtering.* It consists of depositing atoms two to three layers thick of silver and various oxides on a substrate, covering it with a resist material, applying a mask, and exposing it to a light source. The light hardens the material not covered by the mask. Unexposed resist material is washed off and the IC is baked to harden it.

A ***thick-film IC*** is made by depositing multiple layers of paste-like material on a substrate. A screen printing process is used to deposit the paste in layers for the various components. The material is fused together in a high-temperature oven at about 1562° F (850° C.) This process, called *firing,* is used for ICs that must be made with a much shorter manufacture time. Traditional silicon wafer ICs require three or more months to complete the process from design to production.

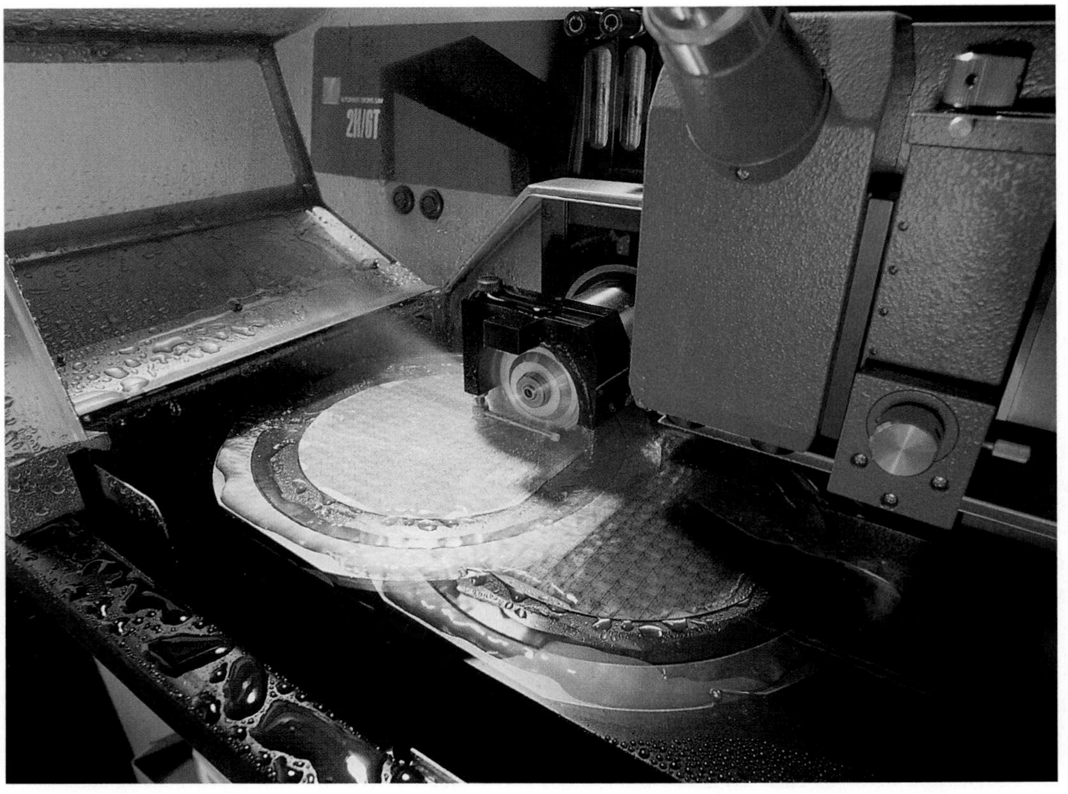

Figure 20-8. A diamond saw is used to cut out the individual ICs from the wafer.

With the development of new thin pastes, the comparison between thick-film and thin-film ICs shows very little difference in their dimensions. The chief difference lies in the higher cost of thin-film because of the materials and equipment used in their manufacture. However, thin-film ICs have fewer connections made by humans as part of the manufacturing process. This increases their reliability. Also, because of the photo process, they have better resolution when compared to thick-film ICs.

Scientists are researching ways to fabricate transparent ICs made from organic compounds. This would eliminate the need for expensive metals, wafers, substrates, and many of the chemical processes. In essence, because of the material with which they would be made, they would be truly "throw away" devices.

Logic Gates

A ***logic gate*** is a circuit whose output depends on the combination of digital signals applied to its inputs. Input and output values are one of two states: off or on. Logic gates are found extensively in computers, calculators, and all kinds of circuits that specialize in on and off conditions.

There are many types of logic gates. This section covers a limited group of these devices. However, if you continue your study of electronics, you will be exposed to a much wider grouping of logic gates.

A couple of examples will help you understand their importance. Suppose you had a programmable thermostat to control the heat and air-conditioning in your home. The thermostat has a three-position switch marked HEAT, OFF and, COOL. It also has two additional switches: TEMP and TIME. During the summer, you set one switch to COOL. You set the TEMP switch to 78° F and the TIME switch for 6 am. You also set the TEMP switch for a higher setting of 80° F at 8 am. You have another setting of 78° F at 4:30 pm. The thermostat is really part of a logic circuit that tells the air-conditioning unit to turn on and off at various times with those temperatures during the course of the day.

A more critical example of logic gate use is found on a gas-fired hot water heater. It has a circuit that checks whether the gas is on or off. It also checks whether the pilot light is on or off. This logic circuit shuts off the main gas valve if the pilot light goes out. This is a life-saving circuit that prevents you from creating an explosion. If the logic circuit had not shut off the gas, a spark from the light switch in a gas-filled house could be deadly.

Binary

Binary is a number system that uses only the numbers *0* and *1*. Using TTL (Transistor-Transistor Logic) ICs, these binary values are equal to the following voltages:

0 = 0 V to 0.8 V

1 = 2 V to 5 V

Since binary only uses two digits, *0* and *1*, it is useful for use with logic gates, or electrical circuits that can be either off or on. In logic gates, the binary number *1* represents the presence of a signal, and the binary number *0* represents the absence of a signal. This translates to the options in the following table:

Binary number	State	Gate logic
0	Off	False
1	On	True

AND Gate

The ***AND gate*****, Figure 20-9,** is a logic gate that requires a signal at every input to have a signal at the output. **Figure 20-10** shows

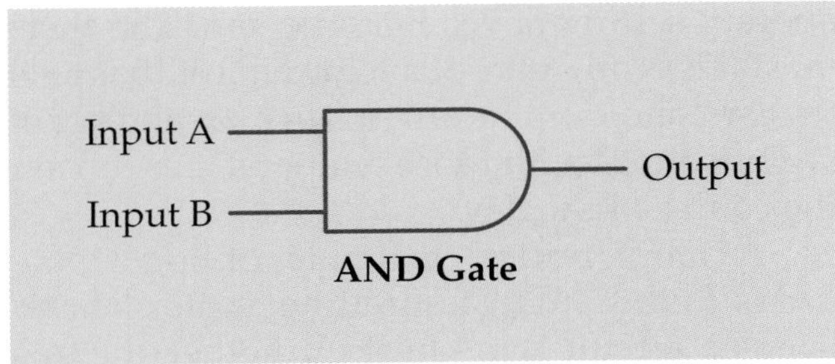

Figure 20-9. The schematic symbol for the two-input AND gate.

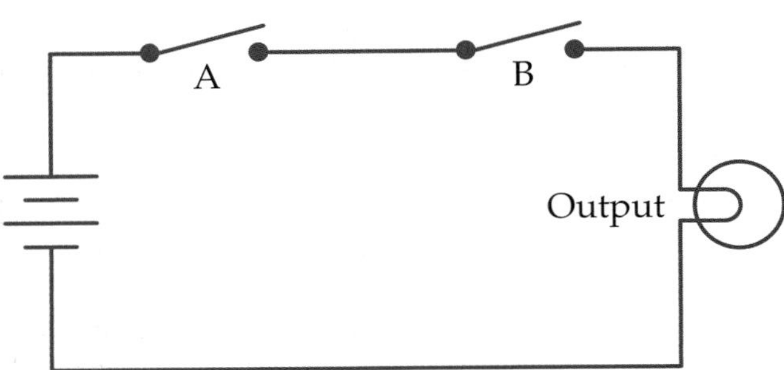

Figure 20-10. This circuit is equivalent to the AND gate. The switches represent the input leads, and the lamp represents the output.

a circuit that is equivalent to the AND gate. Both switches must be closed for there to be a signal in the lamp. Once both switches have been closed, the lamp lights. If either one of the two switches is left opened, the lamp will not light. This logic can be depicted by the following table:

AND Gate		
Input A	Input B	Output
0	0	0
1	0	0
0	1	0
1	1	1

This table is known as a ***truth table.*** It shows all possible combinations of inputs and the outputs they produce. Each row of the table represents the various options for signals on the input leads. For example, the first row shows that if there is no signal at input A and input B, there will be no signal at the output. Notice that the only time a signal is present at the output is when a signal is present at both inputs.

When several AND gates are used together, they are normally found packaged inside of an IC, **Figure 20-11.** Each of the input and output leads are wired to one of the pins on the outside of the unit. Notice that there can be more than two input leads for a logic gate. To help you identify the location of each lead, manufacturers provide a diagram of the IC on its accompanying data sheet. Normally, this diagram is not shown on the outside case of the IC. However, each integrated circuit has some type of identification mark to help you locate pin one. For example, when the notch, like that shown on the IC in Figure 20-11, is positioned to the left, pin one is at the bottom left of the IC. The rest of the pin numbers follow sequentially in a counterclockwise direction.

OR Gate

To get an output from the ***OR gate,*** **Figure 20-12,** a signal must be present at one

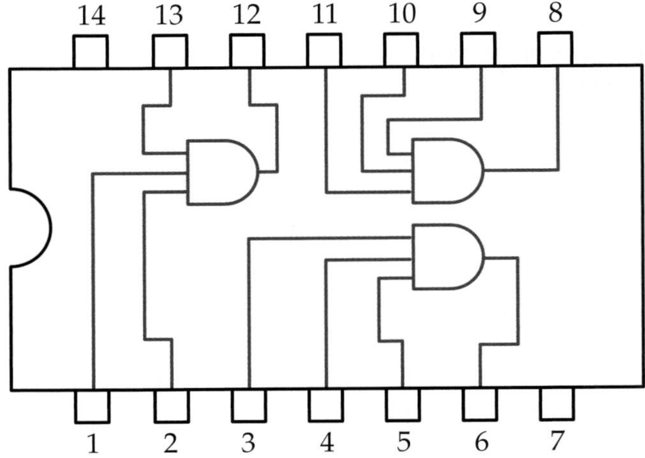

Figure 20-11. Three AND gates are contained within this IC.

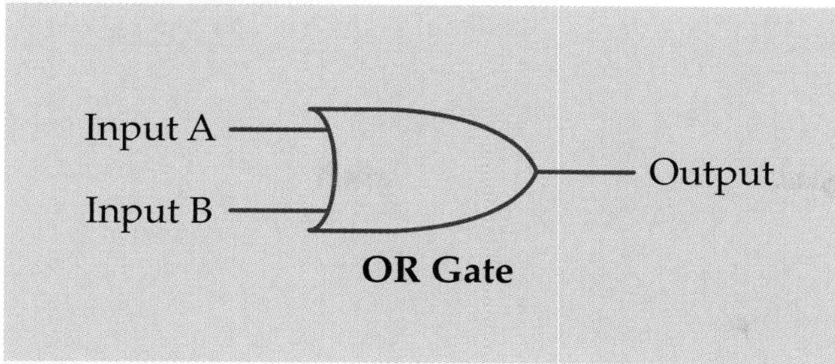

Figure 20-12. The schematic symbol for the two-input OR gate.

or more of its input leads. The following is the truth table for the OR gate:

OR Gate		
Input A	Input B	Output
0	0	0
1	0	1
0	1	1
1	1	1

Notice that any time there is an input signal on one of the OR input leads, an output signal is present. The only time that there is no signal at the output of the OR gate is when there is no input signal at either input leads.

Figure 20-13 shows a circuit that is equivalent to the OR gate. Only one of the switches needs be closed for there to be a signal in the lamp. If both switches are closed, the lamp still lights. If neither of the two switches is closed, the lamp will not light.

NOT Gate

The *NOT gate,* **Figure 20-14,** is also called the ***inverter*** because it inverts, or does the opposite of, the input signal. The truth table for the inverter is as follows:

NOT Gate	
Input	Output
0	1
1	0

Notice that if the input lead does not have a signal present, the output will have a signal. If a signal is present at the input lead, the output will not have a signal.

The AND and OR gate sections showed equivalent circuits made of switches and lamps. This was done to simplify how these solid-state logic devices operate. However, the actual circuitry involved in these devices is made of transistors, diodes, resistors and many other components. They are extremely small and are usually built and packaged inside an IC chip. However, their main function is

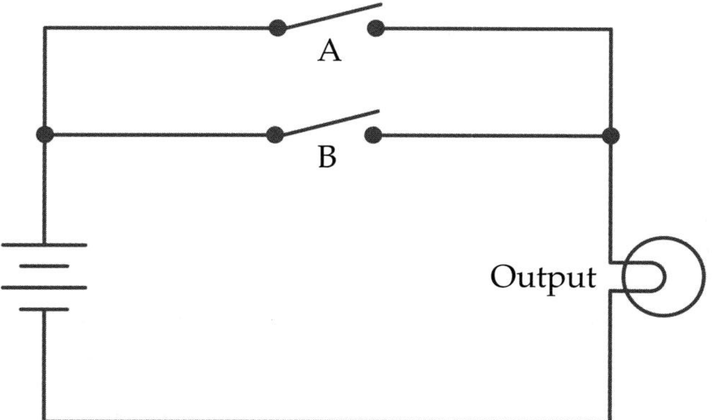

Figure 20-13. This circuit is equivalent to the OR gate.

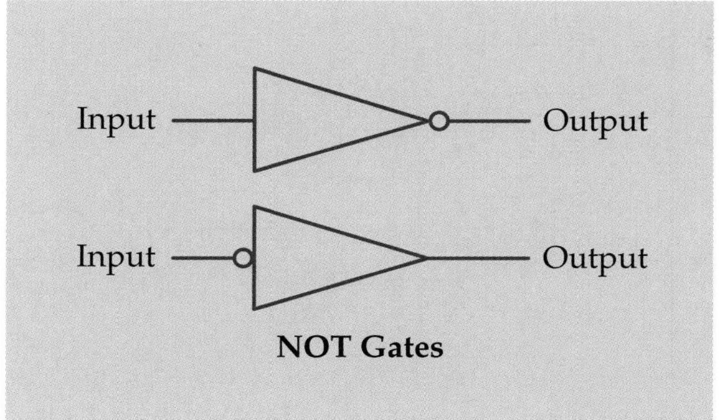

Figure 20-14. Two ways of showing the schematic symbol for the NOT gate.

switching from a zero state to a one. Again, that means from OFF to ON or the reverse ON to OFF. Each lead of the IC is connected to a terminal. The individual legs on the IC are also called *pins*.

NAND Gate

When the NOT gate is added to the AND gate, they become the ***NAND gate***, **Figure 20-15.** Note that this symbol combines the AND symbol with the circle portion of the NOT symbol. The following is the truth table for the NAND gate:

NAND Gate		
Input A	**Input B**	**Output**
0	0	1
1	0	1
0	1	1
1	1	0

The only way to prevent an output signal in the NAND gate is to have a signal present at each input lead.

NOR Gate

When the NOT gate is combined with the OR gate, they become the ***NOR gate***, **Figure 20-16.** Notice that the circle portion of the NOT symbol is combined with the OR symbol to form the NOR symbol. The following is a truth table for the NOR gate:

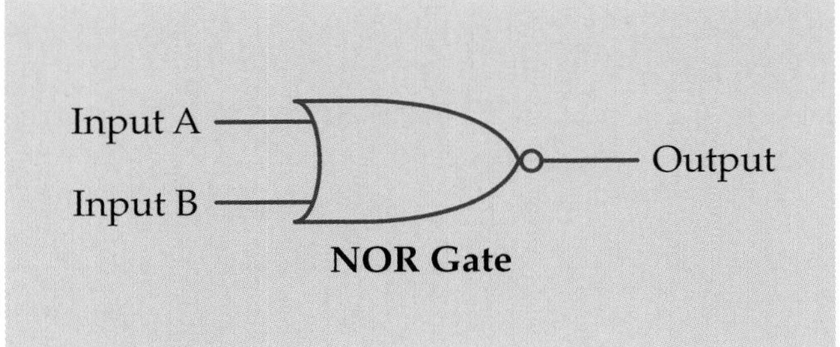

Figure 20-16. The schematic symbol for the two-input NOR gate.

NOR Gate		
Input A	**Input B**	**Output**
0	0	1
1	0	0
0	1	0
1	1	0

The only way to get an output signal from the NOR gate is to have no signal at any input lead. If there is a signal at one of the input leads, the output will not have a signal.

XOR Gate

The ***XOR gate***, **Figure 20-17,** or *exclusive or* gate, is similar to the OR gate. The difference is that if more than one lead has a signal,

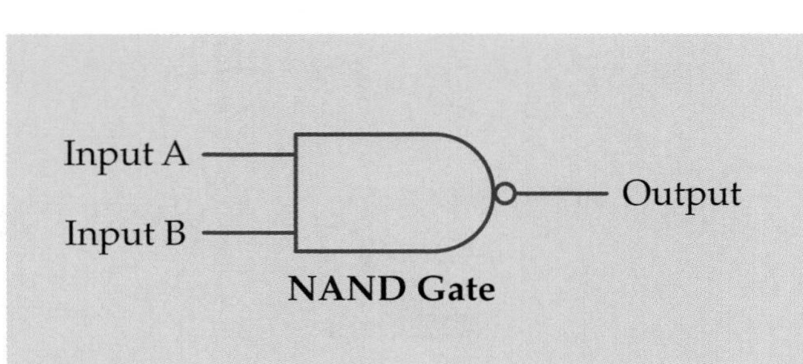

Figure 20-15. The schematic symbol for the two-input NAND gate. The NAND gate is called the NOT AND gate by some people.

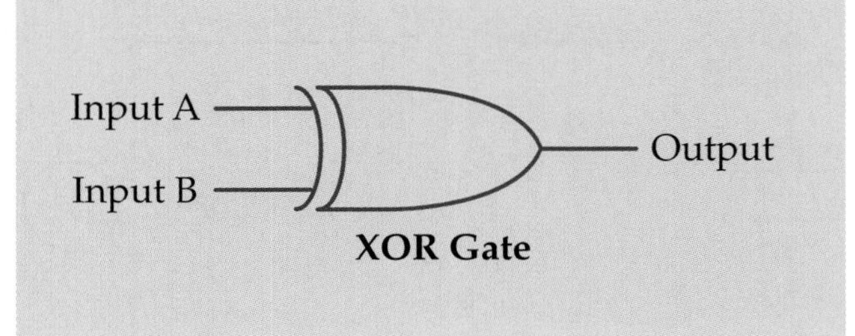

Figure 20-17. The schematic symbol for the two-input XOR gate.

there will be no output signal. In other words, there must be an input signal at exclusively *A* or exclusively *B* for an output signal to be present. The following is the truth table for the XOR gate:

XOR Gate		
Input A	**Input B**	**Output**
0	0	0
1	0	1
0	1	1
1	1	0

XNOR Gate

The ***XNOR gate,*** **Figure 20-18,** is the combination of the NOT gate and the XOR gate. This means that it has the opposite output from the XOR gate. The following is the truth table for the XNOR gate is as follows:

XNOR Gate		
Input A	**Input B**	**Output**
0	0	1
1	0	0
0	1	0
1	1	1

There are sets of international symbols used for all these gates. Although they are rarely used in this country, you may want to examine how they compare to the ones you studied in this chapter.

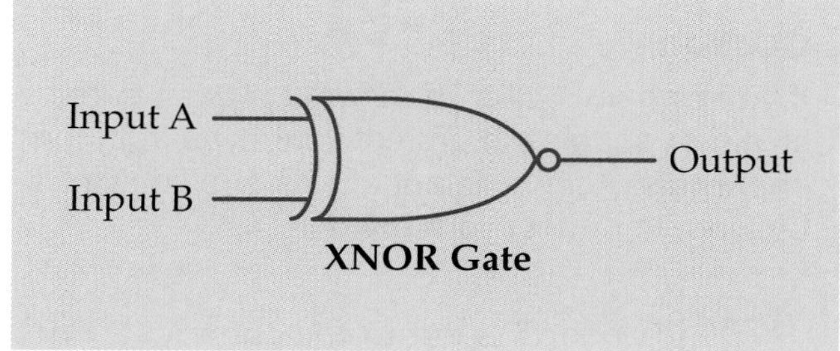

Figure 20-18. The schematic symbol for the two-input XNOR gate.

Printed Circuit Boards

When you do not need a circuit as small as an IC, you can make your own printed circuit (PC) board. Making a PC board starts with a board made of fiberglass mat and epoxy resin or one of the new polystyrene materials coated with copper. The circuit is laid out on the board using special etch-resistive ink or tape. The ink is put on the copper in the desired shape for the finished conductors, **Figure 19-19.** Most of these conductors are the wires of the circuit.

The next important step is to lay out each component that will be attached to the board exactly where you want it. Once the copper is removed, it is difficult to make repairs. If you missed a trace, you will have to use a small piece of wire to connect that portion of the circuit that was skipped.

When the PC board has been laid out, the next step is to remove the copper from the places where it is not wanted. This is done by placing the PC board in chemicals that etch, or eat away, the unwanted copper. The ink over the wires of the circuit board resists being etched away.

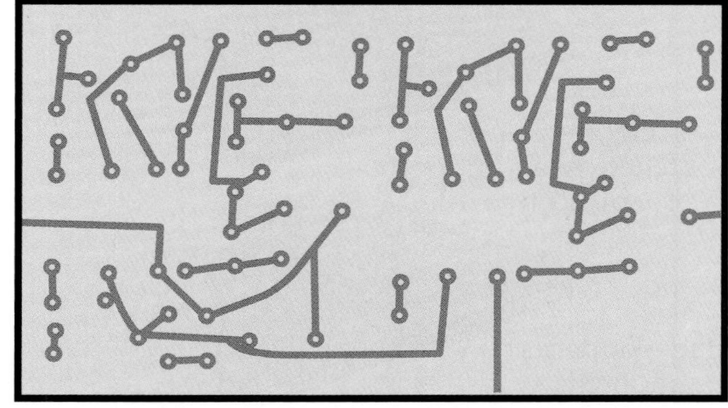

Figure 20-19. Special tape is used to cover the copper on this PC board. You can even mark your name on the board with special ink.

Caution

If you etch your own PC boards, wear a safety shield or goggles to protect your eyes from the chemicals. Acid resistant gloves should also be worn to protect your hands.

The PC board is left in the chemicals only until the uncovered areas of copper are etched away. If the board is left in the solution too long, the conductors will be ***undercut***, **Figure 20-20.** Undercutting occurs when the chemicals start to etch away the copper under the protective coating. This can cause the circuit to be open at the place it is undercut. In some cases, the undercut conductor cracks easily, which also opens the current path. Once the copper is etched properly, the board is washed, and the ink or tape is removed.

When you have a good PC board, you can solder the parts to your printed circuit. The main advantage of PC boards is that they use less wire than traditional circuits. This reduces the size of the circuit, allowing it to fit into a smaller package or chassis box. If you would like to use a PC board but do not want to make your own, there are inexpensive, generic PC boards available in electronics stores.

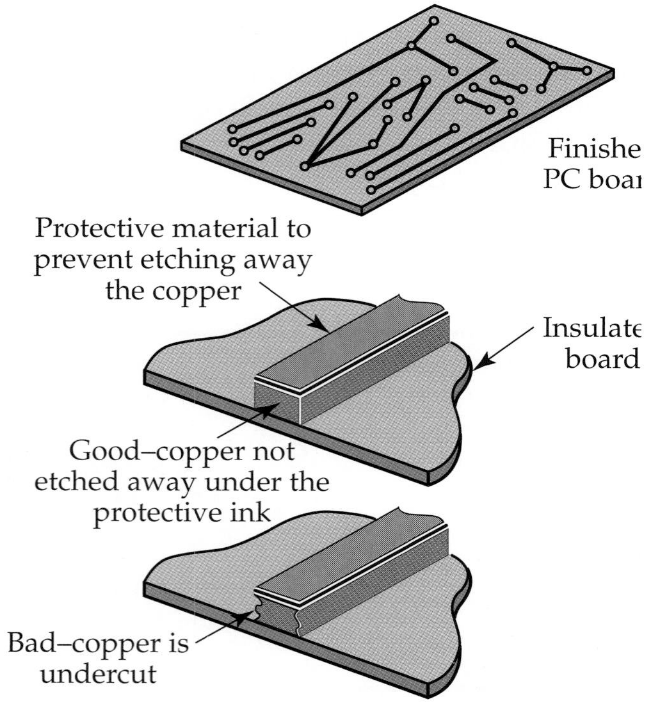

Figure 20-20. Close-up views show two conditions of a finished PC board. The undercut copper was left in the etching solution too long.

Summary

- An integrated circuit is a complete circuit contained within a compact package.
- An IC can contain thousands of components.
- Linear ICs are found in many analog devices and are designed to accept a continuous range of signals.
- A digital IC works with signals that are discrete, such as off or on.
- A monolithic IC is made on a single piece of silicon.
- A hybrid IC is made by combining two or more monolithic ICs or by combining monolithic with discrete components.
- Logic gates are circuits used in digital ICs.
- The output of a logic gate depends on the combination of digital signals applied to its inputs.
- Several types of the logic gates are AND, OR, NOT, NAND, NOR, XOR, and XNOR.
- Truth tables show all the possible combinations for signals at the input leads of logic gates and their resulting output.

Test Your Knowledge

Do not write in this book. Write your answers on a separate sheet of paper.

1. *True or False?* The size of an integrated circuit is part of the reason for the importance of ICs in electronics.

2. Digital ICs work with signals that are ______.
3. Linear ICs work with a(n) ______ of signals.
4. Draw two symbols for ICs.
5. How are diodes and transistors added to a thick film circuit?
6. Draw the symbol for the AND gate.
7. Show the truth table for the AND gate with only two input leads.
8. Show the truth table for the OR gate with only two input leads.
9. Draw the symbol for the XOR gate.
10. The copper on a printed circuit board may be ______ if it is left in the etching solution too long.

Activities

1. How is the copper clad attached to the printed circuit board material? How can it be put on one side only, or on both sides?
2. Make a list of the different types of solid-state devices. How many can you find? Show the symbols for each one.
3. If you are given the number of an IC, how can you tell what it will do? Does each manufacturer label them in a different manner? How do you know where to start counting for each lead?
4. Logic gate symbols are different for different companies. How do you know which one to follow and use? Are some of the symbols the same for different companies?
5. Make a list of at least ten different logic symbols. Does each have a truth table?
6. How can you tell which company made a printed circuit board? Do they use special symbols?
7. You sometimes hear the words *consumer products* being used in the electrical field. What does this mean?

The IC manufacturing process at this National Semiconductor plant is controlled by computers. (Photo courtesy of National Semiconductor corporation.)

Fiber Optics

Learning Objectives

After studying this chapter, you will be able to do the following:

- Diagram the process used by fiber optics to transmit electrical signals.
- Compare and contrast fiber-optic cable and copper-core cable.
- Give one advantage of radio waves over fiber optics.
- Explain why spliced fiber-optic cables must be perfectly aligned.

Technical Terms

binary code
demodulation
fiber optics
modulation
multiplexing
optical detector
photons
repeater station

Fiber optics is becoming a standard method for transmitting data and voice communications. It has many advantages over copper-core cable and radio waves. However, it requires a skilled technician and special tools for its installation.

Fiber Optics Technology

Fiber optics refers to communication using light pulses transmitted through a solid glass fiber. This fiber is about the same thickness as a human hair, about 0.003″ (three thousands of an inch) to 0.005″ (five thousands of an inch) thick, **Figure 21-1.** This is equal to 76.2 microns to 127 microns.

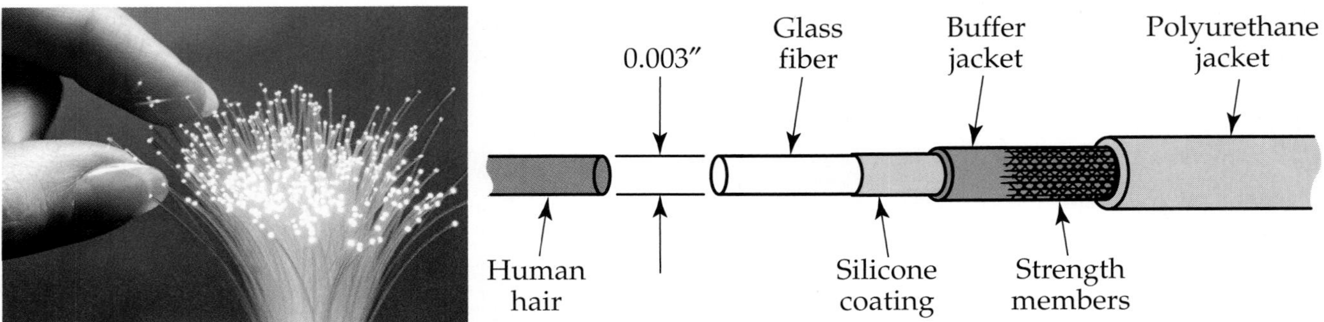

Figure 21-1. Fiber-optic cable consists of glass fiber surrounded by protective layers.

Unlike the glass in most windows, the glass used for fiber-optic cable is strong, yet it is flexible enough to bend. The glass fiber is covered by a silicone coating, and then wrapped in a buffer jacket, strength members, and finally a polyurethane jacket. These layers protect the glass fiber from damage. They also prevent the light pulses from escaping from the fiber wall. While the vast majority of fiber is glass, some use Plastic Optical Fiber (POF), a thicker (about 0.030″) fiber-optic material. However, POF is used mostly with low-speed networks.

Glass is an insulator, so an electrical signal cannot be transmitted along a piece of glass fiber. However, a glass fiber can carry pulses of light. Electrical signals are changed into binary code and sent to a transmitter. Recall that binary is a number system that uses only the numbers *0* and *1*. This number system works well with electrical circuits that can either be off or on. A ***binary code*** uses the on-off patterns to represent data. The transmitter of a fiber-optics system uses the *on* and *off* binary electrical signals to trigger either an LED or a laser diode. The output of both diodes is pulses of light known as ***photons.*** These photons are transmitted through the glass fiber to a receiver, **Figure 21-2.**

The LED must be placed at a perfect 90° angle with the center of the fiber. Light signals tend to bounce off the walls of the core as they travel from the transmitter to the receiver. The steeper the angle at which a signal enters the core, the more often it will reflect off the inner wall of the fiber. The greater the bounce, the greater the loss in the signal before it reaches the receiver, **Figure 21-3.**

There are other causes of signal loss. They include impurities in the glass, power of

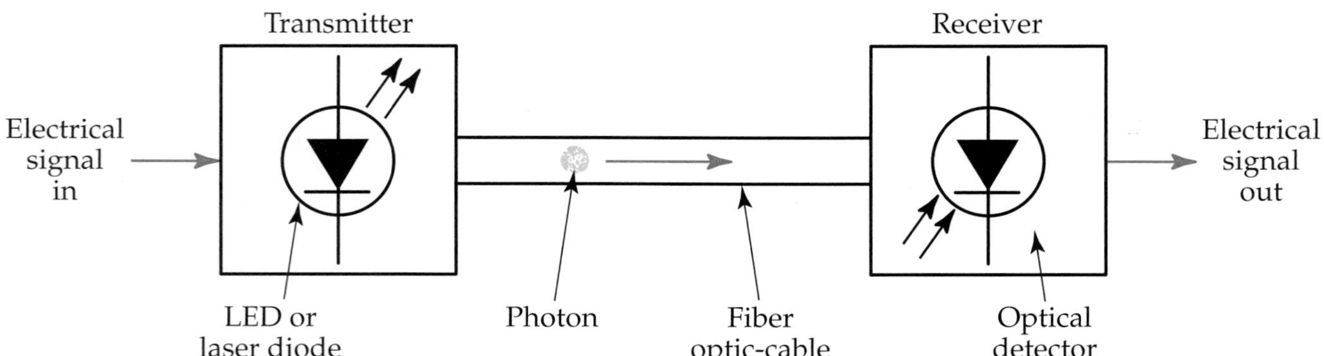

Figure 21-2. Electrical signals are converted to photons by a transmitter. The photons travel along the glass fiber until they are detected by the receiver and converted back into electrical signals.

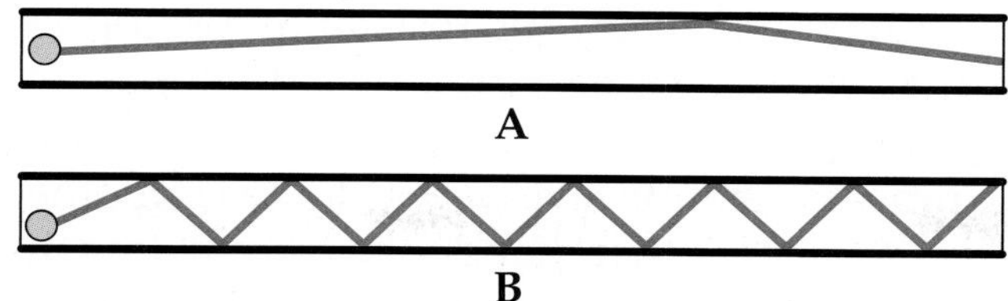

Figure 21-3. The LED or laser diode of the transmitter must be placed at a perfect 90° angle with the center of the fiber to minimize signal loss. A—A photon transmitted from a perfectly aligned LED or laser diode. B—A photon transmitted from an incorrectly aligned LED or laser diode.

the signal, and severe bends in the fiber. To keep losses to a minimum, the purest glass is generally found where the cable runs are the longest. Light pulses can also be lost through the edge of the fiber. New coatings and core materials are being developed to prevent some of this loss.

When the photons reach the receiver, they strike an ***optical detector.*** The detector converts the photons into an electrical signal. Two diodes commonly used as detectors are the positive intrinsic negative (PIN) diode and the avalanche photo diode (APD). Both diodes convert the light pulses to electron pulses.

If the photons travel farther than 60 km (kilometers), they must pass through a ***repeater station.*** The repeater station receives and resends the signal, boosting its strength to prevent the loss of any part of the message.

A glass fiber can carry more than one signal at a time. In fact, 50,000 voice messages can travel at the same time through a single glass fiber. This is done using a time shift technique, which produces signals at different frequencies and places them on the fiber-optic cable at different times. This time shift technique uses a combination of modulation, multiplexing, and demodulation.

An example of this time shift technique might look like that in **Figure 21-4.** The example in this figure shows three different

(1) **Modulation**

Simultaneous conversations are split into different frequencies and split into parts.

Hello Judy. This is John.

At what time are we meeting?

May I speak with Mr. Wilmington?

(2) **Multiplexing**

Conversions are converted into photons and sent out on a single cable with slight delays.

Wilmington Mr. with meeting we are John is this speak I may time what at Judy hello

(3) **Demodulation**

Signals are detected by receiver and converted back to their original frequencies and sentence order.

Hello Judy. This is John.

At what time are we meeting?

May I speak with Mr. Wilmington?

Figure 21-4. Time shift process, which uses a combination of modulation, multiplexing, and demodulation.

simultaneous conversations. The process of splitting the data, changing its frequency, and adding it to a carrier signal is known as ***modulation.*** Converting the data back to its original frequency and order is known as ***demodulation. Multiplexing*** is sending two or more separate signals down a single channel at a higher data rate. This is done to save the cost of using multiple channels.

Advantages of Fiber-Optic Cable

Fiber-optic cable has many advantages over copper-core cable and radio transmissions. These include the number of signal transmissions it can carry, resistance to electrical and atmospheric interference, and reduced signal loss.

A fiber-optic cable with only 12 glass fibers can carry about the same number of signals as a copper-core cable with 900 pairs of copper wire. This copper-core cable would have a 3″ diameter and would weigh 16,000 pounds per kilometer. The fiber-optics cable would have 3/8″ diameter and would weigh only 132 pounds per kilometer. With this size and weight advantage, the cost to lay fiber-optic cable is less than the cost of laying the larger copper-core cable.

Another advantage of fiber optics is that glass fibers are not affected by lightning. Since there is no electrical charge in a light pulse, it is immune to electrical surges caused by lightning. Also, light waves are not affected by electromagnetic fields found around power lines that can affect electrical signals in copper-core cable.

Some of the same advantages hold true for fiber-optic cable when compared to radio transmissions. Radio transmissions are very sensitive to weather and atmospheric conditions. Sunspots, clouds, and electrical storms can cause signal loss. Fiber-optic cable, on the other hand, is immune to these conditions, unless the cable itself is damaged.

Over distance, the resistance of copper wires causes signal loss. For this reason, signals traveling in copper-core cable require a repeater station every one to two kilometers. On the other hand, the signal loss in fiber-optic cable is about 0.5 decibels per kilometer. This is why a fiber-optic transmission needs a repeater station only every 50 to 60 kilometers.

Radio transmissions have an advantage over fiber-optic cable in distance because a radio wave can travel much farther than a fiber-optic signal. However, as mentioned, radio waves have other distinct disadvantages.

Disadvantages of Fiber-Optic Cable

The speed of light in a vacuum is about 300,000 kilometers per second. This is equal to about one foot per nanosecond. However, the speed of a photon is reduced inside a glass fiber. This loss in speed is caused by extremely small impurities in the glass.

The early fiber-optic systems had problems with glass purity and water absorption in the glass fibers. This made the signal losses too high for efficient communication. However, by 1984, silica glass fibers were made that suffered losses of only 0.5 decibels per kilometer. There are still some problems caused by impurities and deflection of light inside the glass fibers. Researchers are working to reduce even further the impurities in the fibers that hinder the photons. As research continues, photon transmission speed is increasing as a result.

Splicing Fiber-Optic Cable

A fiber-optic cable may need to extend to great lengths to connect two distant places. For example, it may need to extend from a phone installation center to a business complex which is a greater distance than the length of a single cable. In this case, the fibers of each fiber-optic cable must be attached (spliced) together to make the needed cable length.

Splicing is also required when someone accidentally cuts a fiber-optics phone line. The glass fiber must be spliced together. The fiber-optic cables are clamped in position to prevent movement during the splicing operation, **Figure 21-5.**

> **Note**
> Large fiber-optic cables are usually buried in the ground by a construction group. The end of one of the cables is used to connect the phone lines in a new subdivision. The process of attaching them one at a time to provide phone service to each subscriber is also known as splicing.

All the color-coded cables are separated for matching the glass fibers. A holding rack makes it easier to locate the correct fibers for splicing, **Figure 21-6.** The glass fibers to be spliced are separated for easy identification and placed on a tray, **Figure 21-7.** The tray also serves as a way to control the amount the glass fibers are bent, **Figure 21-8.** Too much of a bend causes what is known as signal attenuation (loss of signal strength). Technicians

Figure 21-6. Holding rack containing glass fiber. (FiberTech)

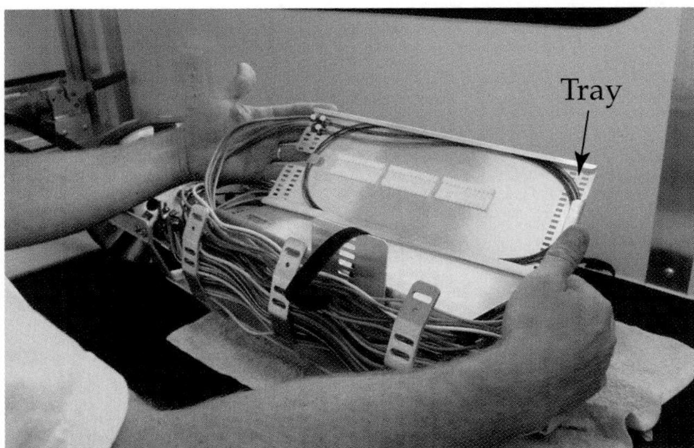

Figure 21-7. Individual trays which will hold the spliced fiber-optic cables. (FiberTech)

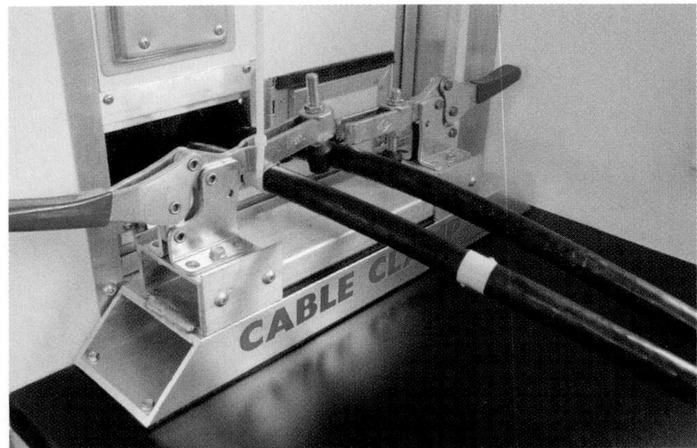

Figure 21-5. Repair station clamp locks cable pairs in position. (FiberTech)

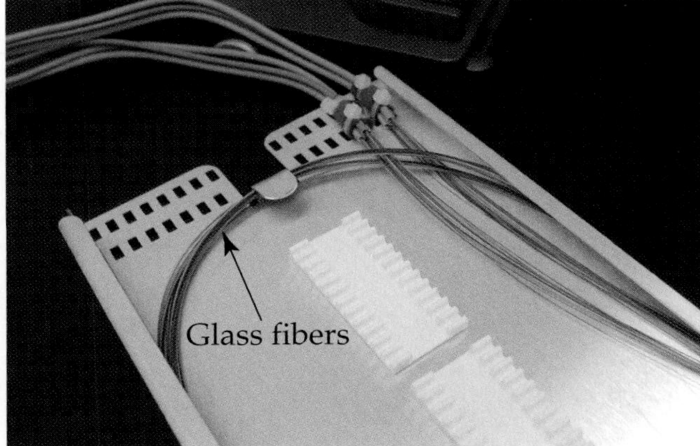

Figure 21-8. Individual pairs which are to be spliced. (FiberTech)

making the splices or laying fiber-optic cable must know the maximum amount that it can be bent. The glass fiber must not be damaged during manufacture or installation. If it is, light signals inside the cable will be diffused, and the cable will be useless. To prevent nicks, a machine is used to strip the insulation off the glass fiber, **Figure 21-9.**

It is critical that the ends of the glass fibers being spliced are in perfect alignment to prevent signal loss after they are joined. A machine is used to hold the glass fibers and ensure they are aligned, **Figure 21-10.** The glass fiber ends must be cut square with one another. Note the two pointed electrodes that are used to fuse the glass fibers together, **Figure 21-11.** A piece of protective heat shrink material is positioned away from the actual splice. The protective cover is lowered over the splice area. The screen shows the location of the fibers on the x- and y-axis for perfect alignment. Once the alignment is correct, the electrodes fuse the fibers together.

Once the splice is completed, the machine tests the signal loss. The technician raises the cover to perform a tension test. A good joint will not break.

The joined fiber is removed, and the heat shrink is moved over the joint, **Figure 21-12.** The covered joint is positioned in the machine, which heats it to shrink the heat shrink around the joint, **Figure 21-13.** The heat shrink protects the joint in the way the insulation protects the other portions of the fiber.

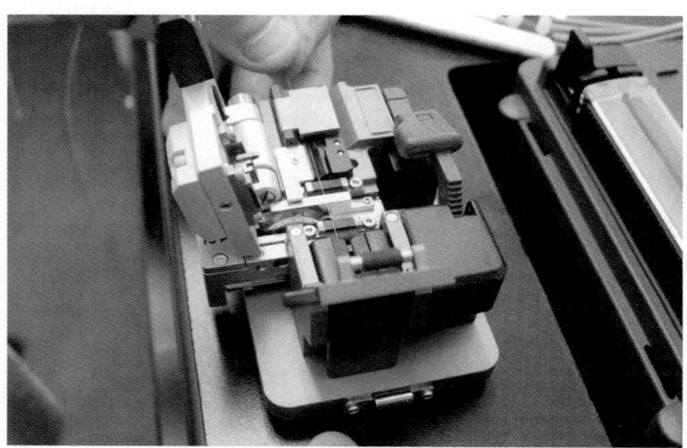

Figure 21-9. Stripper used to remove the insulation from the glass fibers. (FiberTech)

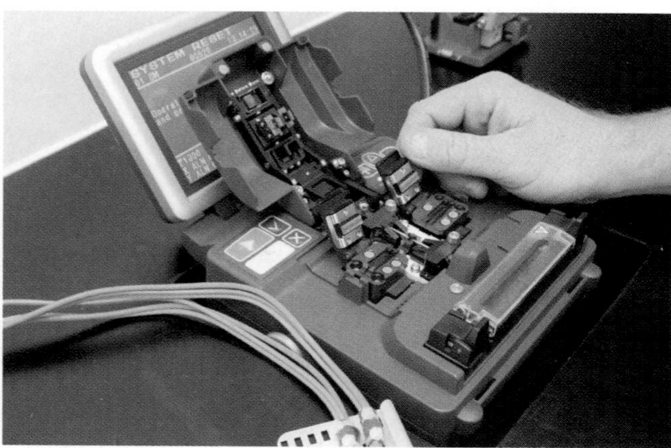

Figure 21-10. Machine used to splice the pieces of glass fiber together. (FiberTech)

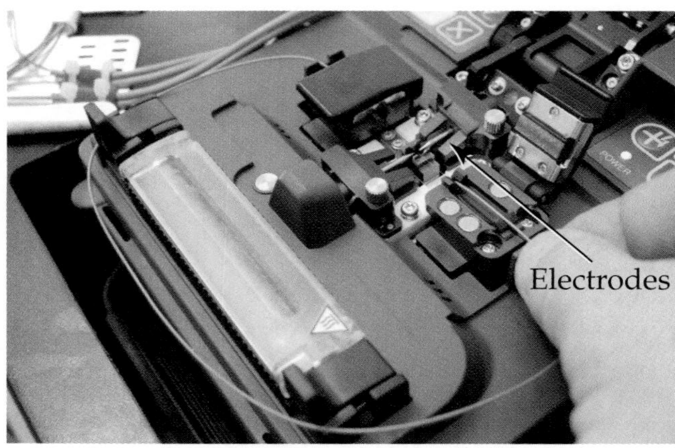

Figure 21-11. The ends of the glass fibers must be perfectly square. (FiberTech)

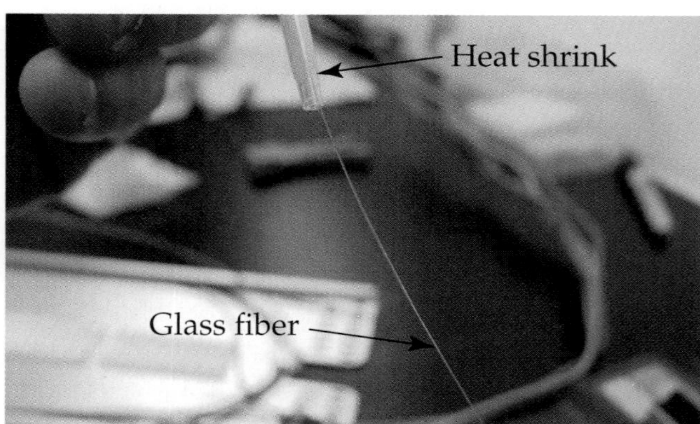

Figure 21-12. Heat shrink, placed on the cable earlier, is positioned over the fused glass fiber. (FiberTech)

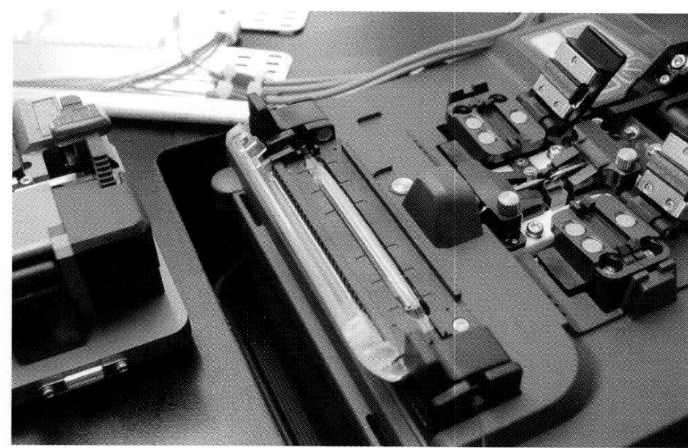

Figure 21-13. A heat shrink machine holds the pieces in place. (FiberTech)

The spliced fiber is then returned to its protective tray, **Figure 21-14.** The technician continues to match and fuse together the remaining fibers until the whole cable is ready to be installed. Splicing two fiber-optic cables together can be a long and tedious process because there may be hundreds of glass fibers to splice for one project.

Caution

Fiber-optics material has very sharp ends. Handle the ends with care so you do not stab yourself.

Project 21-1: Fiber-Optic Lamp

You can make your own fiber-optic lamp like that shown in the following picture. The size of the lightbulb you use is your choice, but it is best to keep the wattage low. This prevents the lightbulb from overheating the inside of the can.

(Project by Allan Witherspoon)

To dress up the project, cover the outside of the can with contact paper. If you decide to paint the outside of the can, use a smaller bulb to avoid burning your design. A base can be cut from wood to protect the table or surface on which you display your project. For further protection, glue felt or another soft material to the bottom of the wood base.

To make this project, gather the items listed in following table. Refer to the schematic as you follow the procedure.

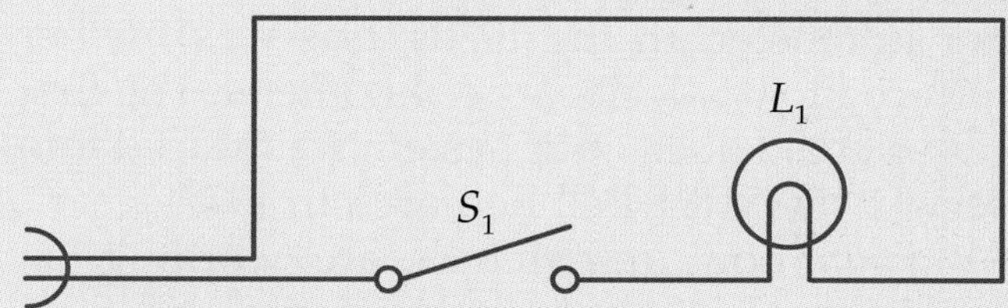

Procedure

1. Drill a hole in the side of the can for placement of your switch.
2. Wire the switch to the light cord and attach the end to the lamp holder.
3. Insert the lightbulb and position it inside the can.
4. Drill a small opening in the top of the can just large enough to hold the assembled fiber-optics material.
5. Position the lightbulb near the end of the fiber optics. You may want to wrap some electrical tape to help hold the fiber optics together inside the can. A colored bulb will provide you with options for a different color.
6. Place the contact paper over the outside of the can. Cut the piece of wood and attach it to the bottom of the can for added weight and balance.

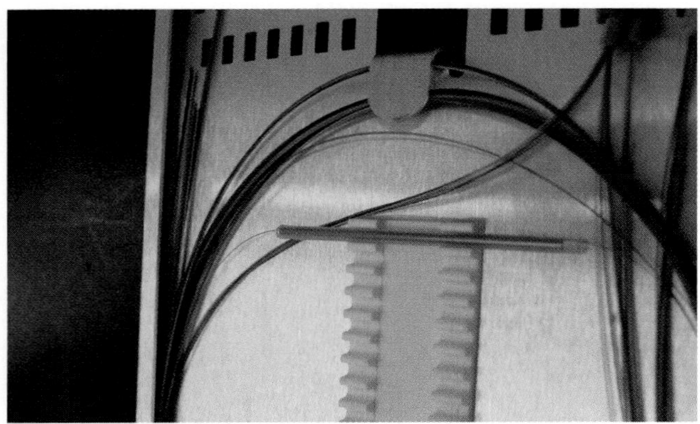

Figure 21-14. Spliced glass fibers are placed in a protective tray for easy identification. (FiberTech)

Future of Fiber Optics

Fiber optics has made great strides since first introduced in the 1950s. Refinements in the silica glass fiber during the 1970s decreased the signal loss to about 20 decibels per kilometer. In the 1980s, this was further reduced to 0.5 decibels per kilometer. The first commercial fiber-optic system was placed into service by AT&T Bell Research in 1977.

Even now, researchers are working on replacing glass fibers with plastics and fluoride glass compounds that last much longer than previous glass fibers. These new materials also have fewer problems with signal losses caused by distance and bending the cable. There are claims of decibel losses eight times less than those of silica fibers. However, there are major problems with temperature, moisture, and in the ability to heat these fibers and draw them to a constant size. A number of research centers have been established at several collegiate institutions. These institutions have Web sites that describe their findings.

In addition to the glass fiber, researchers are working on developing LED transmitters and optical receivers capable of over 1,000,000,000,000,000 (one quadrillion) pulses per second. This would increase the number of transmissions that can be carried over a single fiber and further reduce the size of the cable.

Summary

- Fiber optics is communication using light pulses transmitted through a solid glass fiber about the thickness of a human hair.
- Glass fiber is smaller and much lighter than the same amount of copper wire.
- The pulses of light that are transmitted through fiber-optic cable are called *photons.*
- The transmitter uses the binary form of an electrical signal to send the light pulses through the fiber-optic cable.
- The light pulses are received by an optical detector, which converts them back into an electrical signal.
- A fiber-optic signal can travel farther than an electrical signal in copper-core cable without needing a repeater station.
- Radio waves can travel farther than fiber-optic signals without needing a repeater station.
- When glass fibers are spliced together, they must be perfectly aligned to prevent signal loss.

Test Your Knowledge

Do not write in this book. Write your answers on a separate sheet of paper.

1. Describe the two purposes of the protective layers around the glass fiber.
2. *True or False?* A glass fiber can carry an electrical charge.
3. A(n) ______ or a(n) ______ is used to create the light pulses (photons) needed to carry a message through the fiber-optic cable.
4. The optical detector located at the receiver may be either a(n) ______ or ______.

5. List two advantages that fiber-optic cable has over copper-core cable.
6. A light pulse is boosted by a repeater station every ______ kilometers, whereas an electrical signal must be boosted every ______.
7. The technical term for signal loss is ______.
8. *True or False?* Radio waves can travel farther than a fiber-optic signal without needing a repeater station.
9. Explain how the purity of the glass used in fiber-optic cable can affect data transmission.
10. *True or False?* Fiber-optic cables that are being joined together are soldered just like copper wire.

Activities

1. When was the first fiber-optic cable invented? By whom was it invented?
2. What was the first fiber-optics material made from? What was it to be used for?
3. Why did it take so long for fiber optics to be used commercially?
4. What are some of the new uses for fiber-optic cable?
5. How do you get colored images out of fiber-optic cable?
6. How much fiber-optics material is used in the average year? What does it cost?
7. What are some of the challenges faced in the fiber optics industry?
8. Which country has the most fiber-optics material installed for public use?

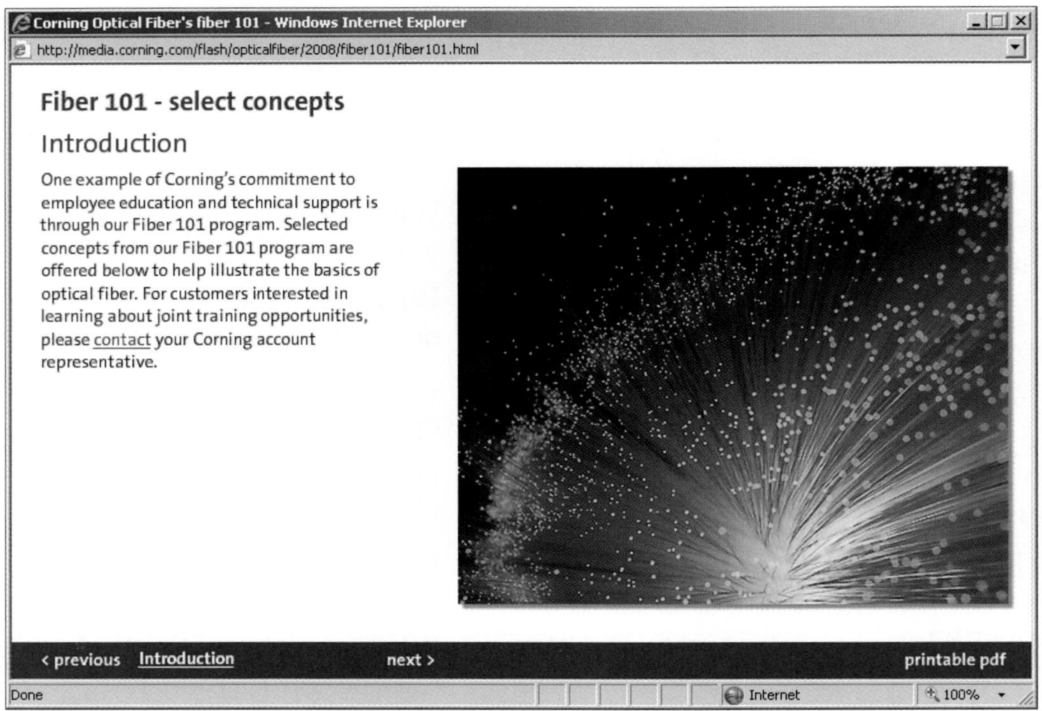

Check out the Fiber 101 tutorial from Corning Incorporated located at www.corning.com/opticalfiber/library/fiber_101/index. aspx. Other tutorials and videos about fiber optics can also be found in the Corning Web site.

Switches

Learning Objectives

After studying this chapter, you will be able to do the following:

- Identify a relay's coil and contacts in a ladder diagram.
- Explain the numbering system used in relay circuits.
- Explain the operation of each of the four types of normally open and normally closed timed relays.
- Explain the difference between a limit switch, pressure switch, and float switch.

Technical Terms

float switch
holding contacts
ladder diagram
limit switch
normally closed (N.C.)
normally open (N.O.)
pressure switch
relay
solenoid
timed relay

Switches are used in many ways in circuits. They can be used to control circuits or to move parts. Switches can be controlled by other circuits, magnetic fields, or physical contact. You discovered that solid-state devices, like transistors, are high-speed, low-voltage devices. In this chapter, we will study switching devices that are generally used with higher current and voltages.

Relays

A ***relay*** is an electrical device that consists of a coil, an armature, and electrical contacts, **Figure 22-1.** The coil provides the mechanical

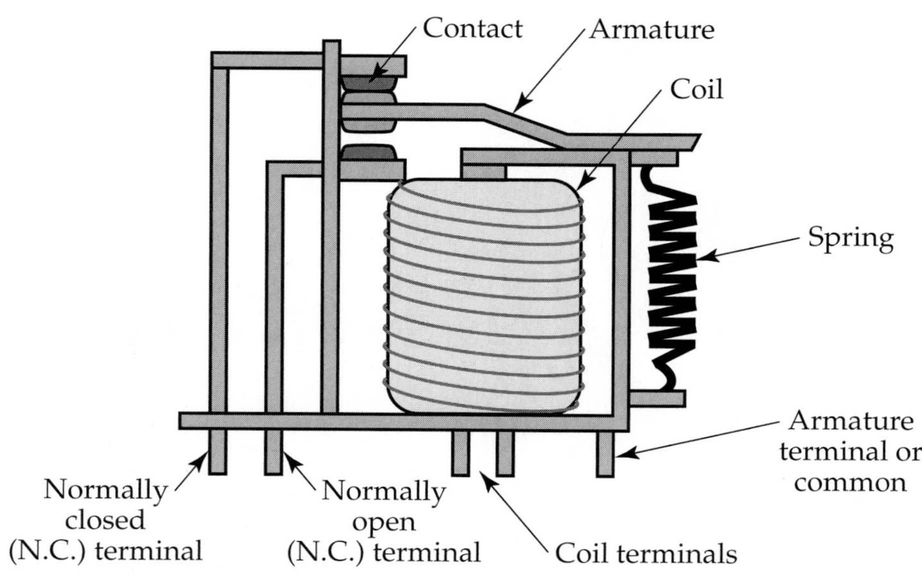

Figure 22-1. The parts of a relay.

action needed to open and close the electrical contacts. When the coil is energized, a magnetic field is created that is strong enough to move the relay's armature. This causes the contacts to open or close, depending on how the relay is wired. When the coil is de-energized, either gravity or a spring forces the armature back to its normal position.

Look at **Figure 22-2.** Notice that the coil is wired to a control circuit. The control circuit is typically powered by a low-voltage source and has a device to turn the control circuit off and on. Usually, this device is a push button. The push button may be controlled by a person or by another electrical device. The armature and set of contacts are part of a high-voltage circuit that provides power to a load. This circuit is called the *load circuit.* Notice that in this circuit, the relay is wired in a normally open (N.O.) position. This means that when the control circuit is not powered, the contacts in the load circuit are open.

A relay can have from one to twelve sets of contacts, each of which can complete a load circuit. The contacts can be either ***normally open (N.O.)*** or ***normally closed (N.C.).*** *Normally* refers to their condition when no current is flowing through the coil. A normally closed relay allows current to flow in the load circuit when there is no current in the control circuit. A normally open relay prevents current flow in the load circuit when there is no current in the control circuit.

Some relays are mounted inside a plastic box, **Figure 22-3.** In this way, the relay is protected from dirt and other particles getting between the opening and closing contacts and disrupting the circuit. With an open relay, you can see the armature move and the contacts open and close. You may be able to hear the contacts click.

The ANSI symbol in **Figure 22-4** is used in schematic drawings. **Figure 22-5** shows the circuit in Figure 22-2 also drawn using ANSI symbols. Ladder diagrams using three JIC (Joint Industry Conference) relay symbols are shown in **Figure 22-6.** Ladder diagrams are covered later in this chapter.

Control Circuit

The power to the coil usually comes from a transformer; although, it could come from any power source, **Figure 22-7.** The voltage

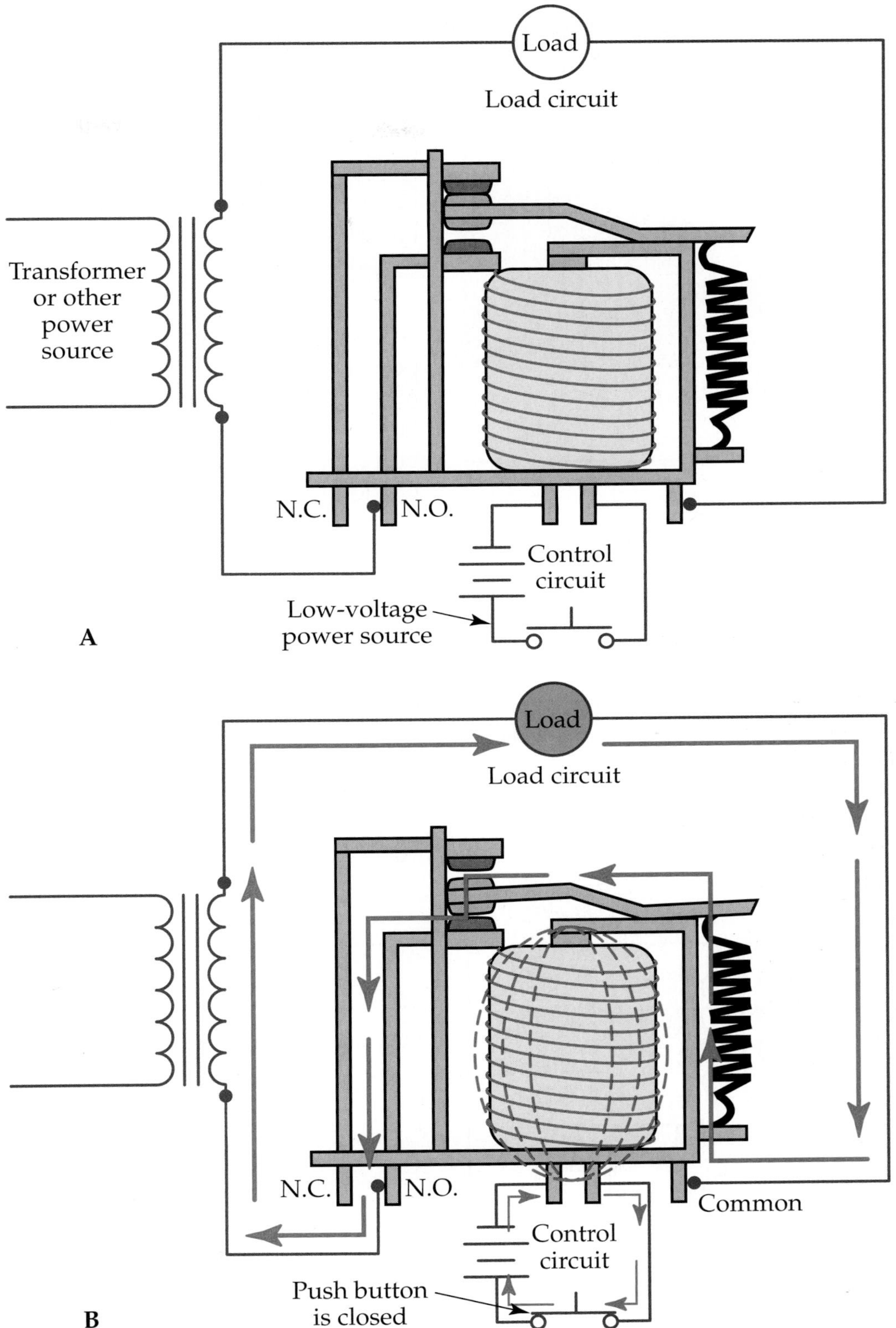

Figure 22-2. The coil of a relay is typically used as part of a control circuit and the contacts used as part of a load circuit. A—Coil is not energized. B—Energizing the coil closes the normally open contacts and opens the normally closed contacts.

Figure 22-3. A sealed relay.

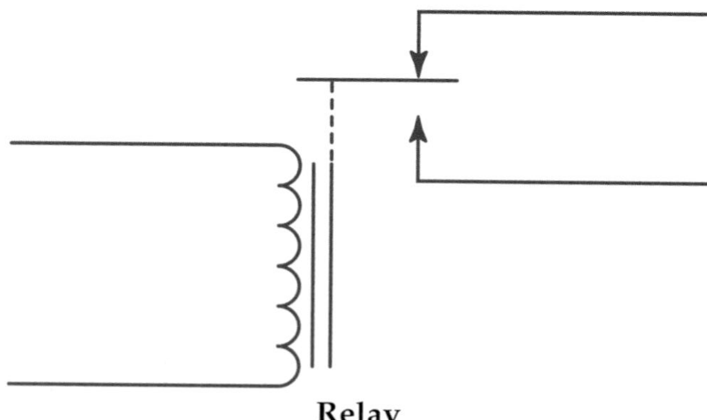

Figure 22-4. Symbol for the coil and contacts of a relay.

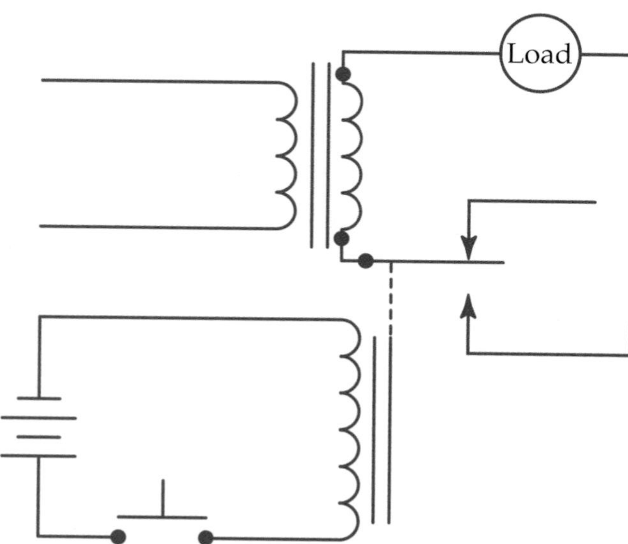

Figure 22-5. Relay symbols used in a schematic.

Symbol	Name
—○—	Coil of the control relay
—| |—	Normally open contacts
—|/|—	Normally closed contacts

Figure 22-6. Relay symbols used in ladder diagrams.

levels and type of current of the coil and power source must match. A 110-V ac coil must be attached to a 110-V ac power source. For a relay that requires dc, the current must be rectified. A simple ac circuit with a relay is shown in Figure 22-7A. A simple dc circuit with a relay is shown in Figure 22-7B.

The remaining discussion about relays focuses on ac relays. However, the same concepts apply to dc relays. The only difference in using a dc relay is that the components selected for the control circuit must be matched to the operating dc voltage level.

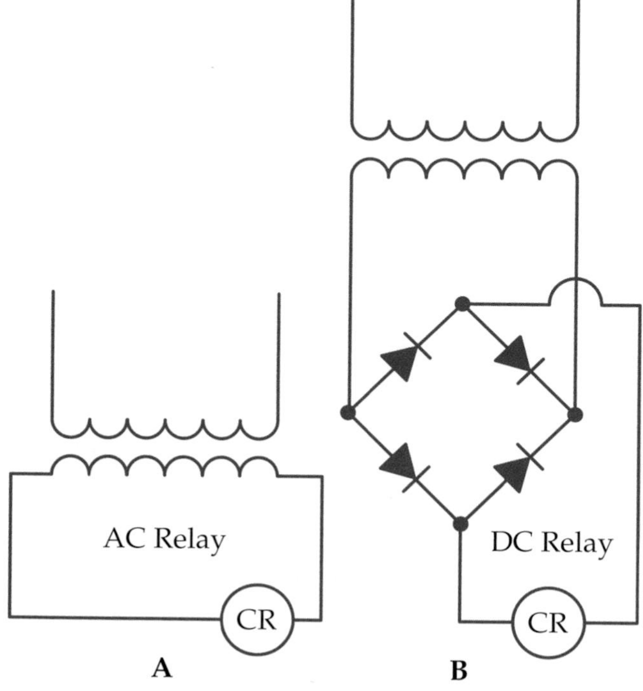

Figure 22-7. Power sources used for relays. A—AC circuit with relay. B—Rectified dc circuit with relay.

Control circuit components are typically located inside a control or relay panel. The control panel, **Figure 22-8,** holds a wide variety of electrical components such as contacts, relays, lights, timers, and other devices that control the operation of the load circuit. This is done to isolate the electrical devices and potential sources of electrical shock away from the operating portions of the machine they control.

Many times, a push-button switch is used to turn the control circuit on and off. The circuit must be protected from an electrical overload, so a fuse should be installed in the circuit. To be safe, most companies install fuses on each side of the power supply. A schematic diagram of a typical relay control circuit is shown in **Figure 22-9.**

Holding contacts

Except for timed contacts (discussed later in this chapter), contacts will retain their changed, non-normal position only while the coil is energized. For example, suppose a push button completes the control circuit. When the push button is released, the circuit is de-energized and the relay armature returns to its normal position. The push button must be constantly depressed as long as the contacts are needed in their changed positions.

Figure 22-8. A control panel holds a wide variety of electrical components such as contacts, relays, lights, timers, and other devices that control the operation of the load circuit.

Holding in the push button is inconvenient, but the relay can be used to avoid this problem. If a set of normally open contacts is connected in parallel with the push button and the push button is pressed, the relay coil energizes and closes these contacts, **Figure 22-10.** This creates a parallel path to the coil that goes around the push button. The coil stays energized even when the push button is released because the current goes around the open push button. The contacts in this parallel circuit are called ***holding contacts.*** The circuit now needs a way to de-energize the relay. De-energizing the relay can be accomplished with a stop button.

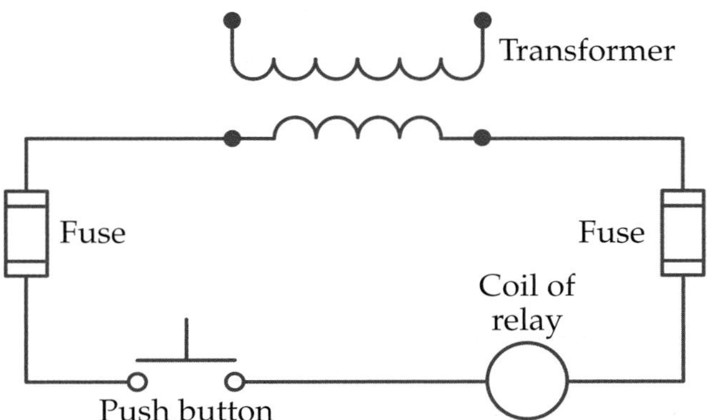

Figure 22-9. Fuses protect the control circuit.

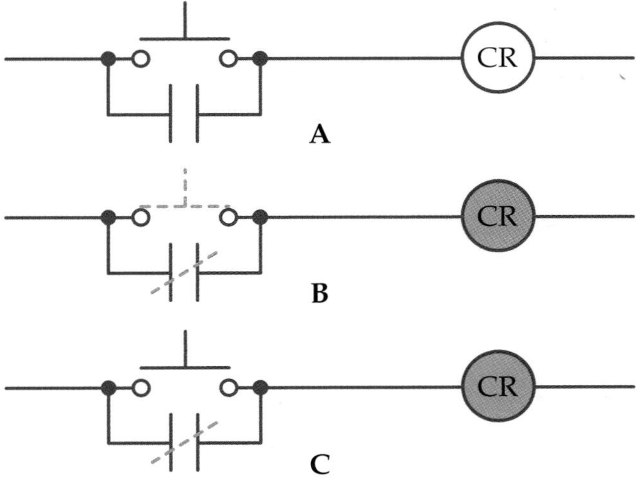

Figure 22-10. Designing a method to hold a circuit closed. A—Push button and normally open set of contacts. B—Pushing the button energizes the circuit and closes the contacts. C—Releasing the start button opens the switch but the circuit remains completed through the holding contacts.

Stop button

In a control circuit with holding contacts, there is a start button and a stop button, **Figure 22-11.** The holding contacts are placed in parallel with the start button, which is usually a normally open switch. The stop button, which is normally closed, is placed in series with the start button/holding contact combination and the coil, Figure 22-11A. When the start button is depressed, current flows through the normally closed stop button and energizes the coil. The energized coil closes the holding contacts, Figure 22-11B. When the start button is released, current flows through the holding contacts and through the normally closed stop button, which completes the circuit, Figure 22-11C.

To open the circuit, the stop button is pressed. This causes the coil to de-energize and the holding contacts return to their original state (open), Figure 22-11D. The only way to restart the current flow is by pushing the start button again.

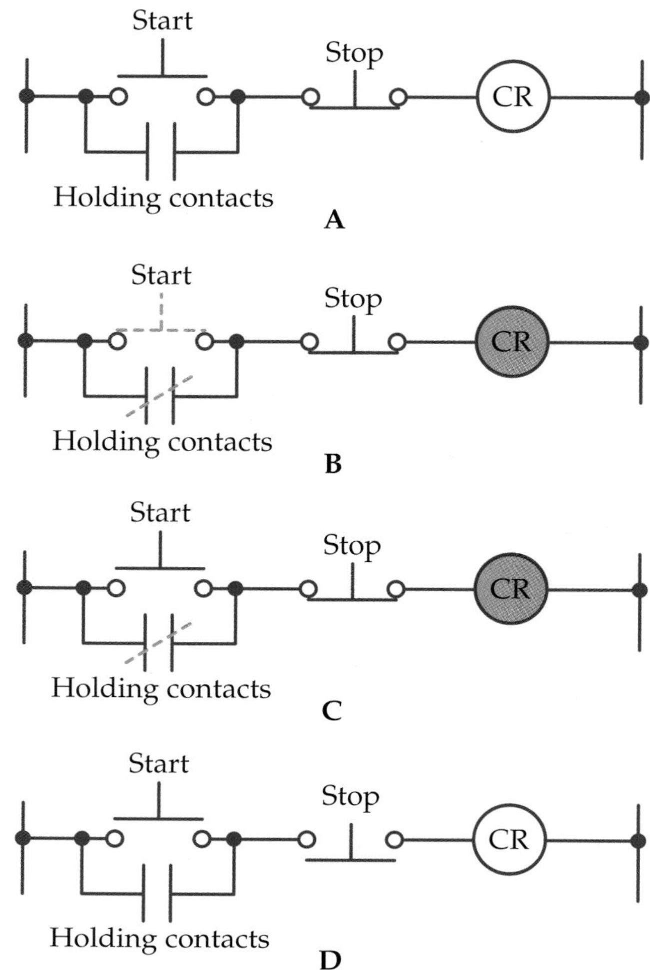

Figure 22-11. A stop button is added to break the circuit completed by the holding contacts. A—Initial state. Circuit is not energized. B—Start button is depressed to energize the circuit. C—Start button is released. Circuit remains energized. D—Stop button is depressed to de-energize the circuit.

Ladder Diagrams and Relay Circuits

Earlier in this chapter, you learned that a diagram called a ladder diagram uses three different symbols to depict the parts of a relay. These parts are the coil, normally closed contact, and normally open contact. You have also been introduced to sections of a ladder diagram in Figure 22-10 and Figure 22-11.

Practical Application 22-1: Testing Relays

With a continuity tester, you can check the position of the contacts to see whether they are normally open or normally closed. This check is necessary when the position of the contacts cannot be seen. A continuity tester can also show whether the contacts are making the proper connection. There are times when relay contacts are broken off or damaged by arcing. At times they may even weld themselves together.

When using a continuity tester to check a relay's contacts, make sure that no power is passing through the contacts. Place the tester leads to each terminal. Recall that the light of the continuity tester will turn on only if the contacts are closed. You will need to move the armature up and down by hand to make these checks. On a relay with a clear plastic case, you should be able to see the contacts easily. If not, you can test them on the terminal strip, found inside most large control panels.

A *ladder diagram* is a drawing which shows the logic of a circuit. A technician uses a ladder diagram when troubleshooting a system. The ladder diagram typically looks like a ladder. The rails of the ladder represent the main lines through which power flows. The rungs, or lines between the rails, represent circuits. Each line is usually numbered, **Figure 22-12.** Numbering starts at the top line of the left rail.

Many circuits have more than one relay. The relay and the contacts it controls do not have to be on the same line, or rung, in a ladder diagram. To identify the coil of a relay they are numbered and labeled CR. CR stands for a control relay. The order of numbering is typically, but not always, done from top to bottom. Notice in Figure 22-12 that the coil in line 2 is the first coil in the ladder diagram starting at the top. It is numbered 1CR. The next coil, proceeding downward, is on line 4 and is labeled 2CR. The number of the coil can be placed either in front of or behind its letter code. Always use only one numbering scheme throughout a schematic.

The contacts can be numbered in two ways: by contact number or location. If you pick up a relay and look at it, the contacts are numbered from left to right. There are also two terminals for the coil. The relay 1CR diagrammed in **Figure 22-13** is shown with eight contacts.

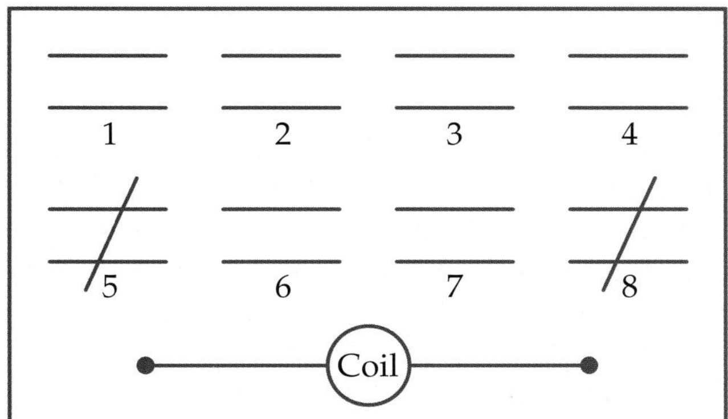

Figure 22-13. Labeling on a relay.

Six are normally open and two are normally closed. Those normally open numbered 1, 2, 3, 4, 6, and 7. Contacts 5 and 8 are normally closed. Labeling the contact on the diagram as 1CR-4 tells the electrician to use the fourth set of contacts. It also specifies relay 1CR for the particular load circuit being connected.

There is also a way of labeling the relay to show where the contacts are being used. This is done by placing a set of numbers on the drawing beside the coil. **Figure 22-14** shows relay 8CR. The numbers to the right of 8CR tell you on what line of the schematic the contacts are being used. They also show whether the contacts are open or closed. The line under the number 48 tells you that those contacts are normally closed.

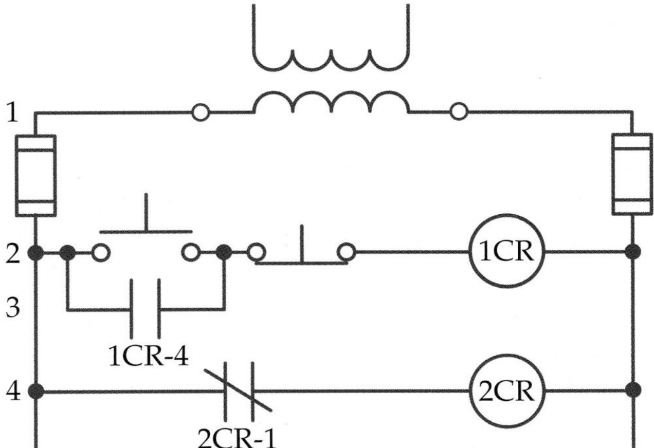

Figure 22-12. Each line of a ladder diagram is typically numbered.

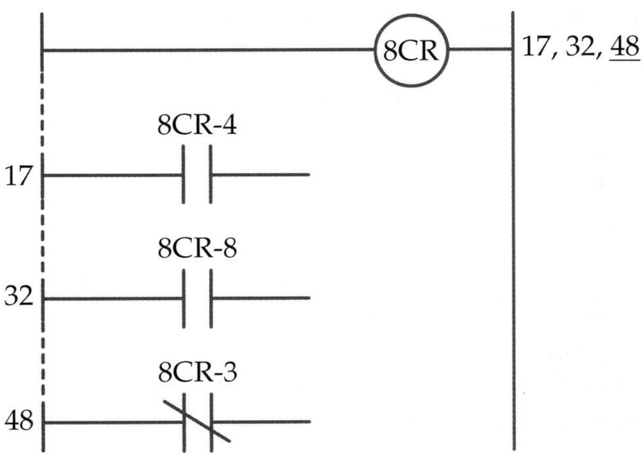

Figure 22-14. Labeling relay contacts within a ladder diagram.

Timed Relays

A ***timed relay*** provides a time delay, often as long as one minute, before the contacts change positions. The industrial (JIT) symbol for a timed relay is the same as the symbol for a normal control relay, except that it is identified by the letter *T*, **Figure 22-15.** The number of the relay can be placed either in front of or behind its letter code. Thus TR1 and 1TR could refer to the same relay. Again you should always use consistent labeling when drawing a ladder diagram.

Like other relays, the contacts on timed relays may be normally open or normally closed. However, the contacts in a timed relay can be timed to open or timed to close. Thus, the four possible positions for timed relay contacts are as follows:

- Normally open, timed closed.
- Normally open, timed open.
- Normally closed, timed open.
- Normally closed, timed closed.

The symbols for each of these are shown in **Figure 22-16.** The position of the switch shows whether the contact is normally open or closed, and the arrowhead points in the direction in which the contact is timed.

Timed relays are manufactured to operate with either ac or dc coils. These range from 12 V to 48 V. The higher voltages of 120 V and 240 V only operate with ac. Most of these relays can handle current load levels ranging from 4 A to 25 A. Depending on the relay chosen, the time range can vary from 0.05 seconds to 30 hours.

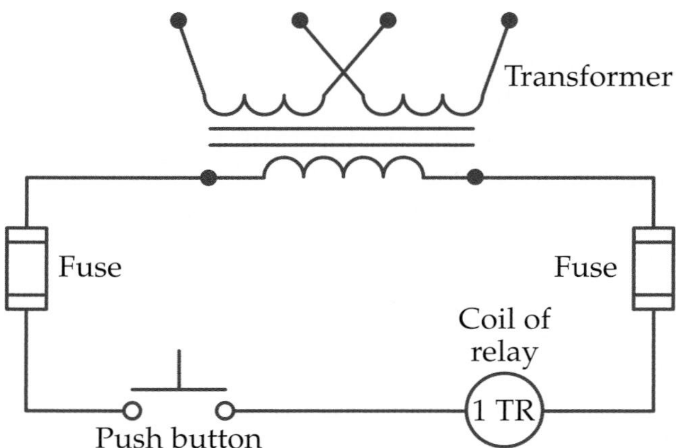

Figure 22-15. Timed relay coil in a ladder diagram.

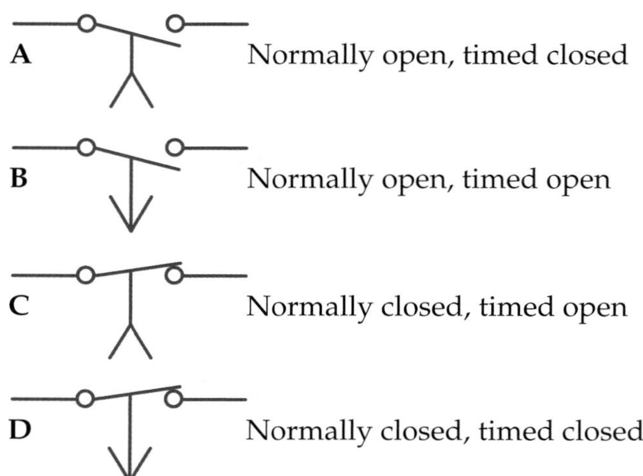

Figure 22-16. Symbols for timed contacts.

Normally Open, Timed Closed

A normally open, timed closed contact is open when no current is passing through its control circuit. When the timer relay coil is energized, the contact does not close immediately because it is timed to close. It begins timing once the coil is energized. When the timer reaches the preset time, the contact closes. **Figure 22-17** shows this process. An adjustment screw is turned to set the amount of time it waits to close. The contact remains closed until current to the coil of the relay is switched off. When the control circuit is opened, the contact opens immediately. A timed closed contact cannot also be timed open.

An example of a normally open, timed closed relay circuit is a burglar alarm system. When a door or window is opened, the control circuit is closed. This energizes the coil of the relay, but the contact is timed to give the

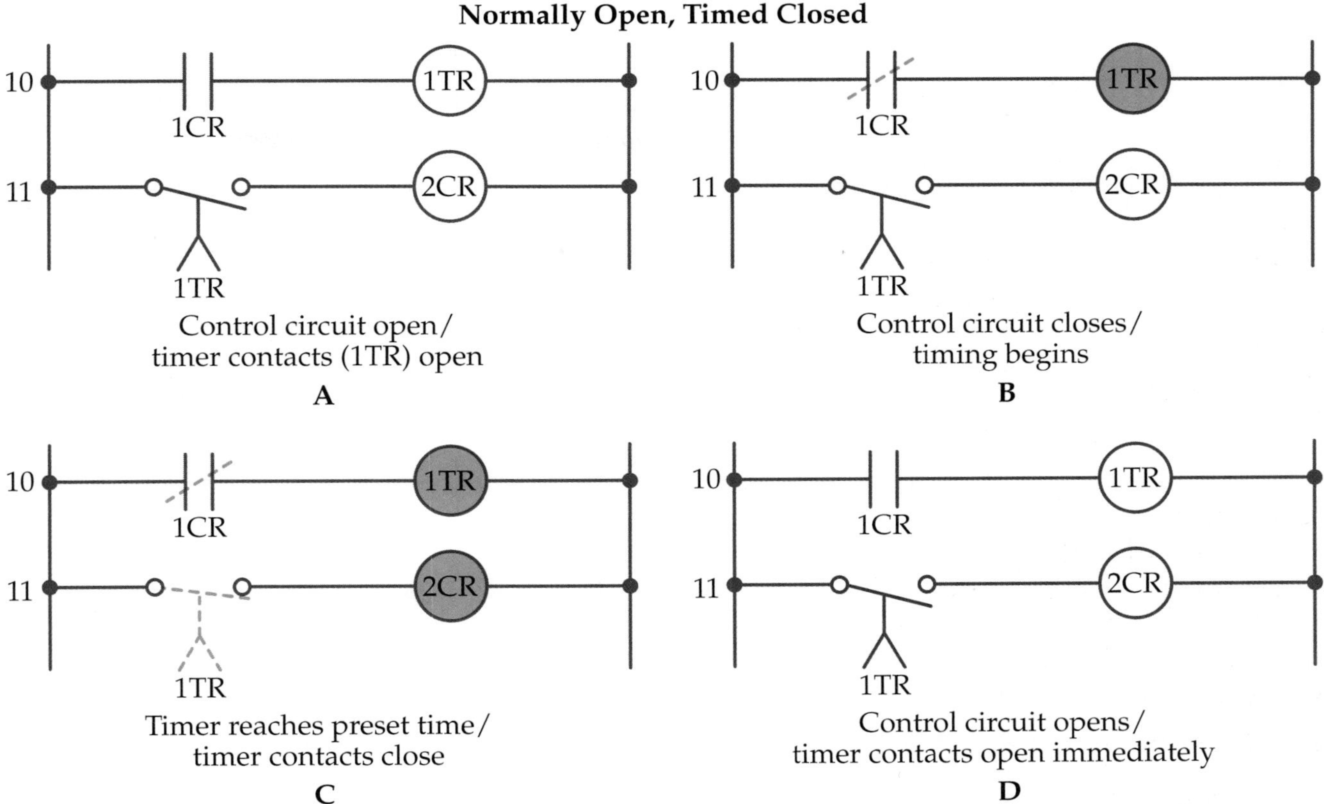

Figure 22-17. Normally open, timed closed circuit operation.

homeowner time to turn off the alarm. If the alarm is not turned off within a set amount of time, the normally open contacts close to complete the load circuit. This load circuit may automatically call the police or alarm company, sound a bell or siren, or turn on lights.

Normally Open, Timed Open

A normally open, timed open contact closes immediately when the timer relay coil is energized. The contact remains closed as long as there is power to the coil of the relay. As soon as the control circuit is opened, the contact begins timing. It remains closed as long as it is timing, but when the timer reaches the preset time, the contact opens. It stays open until the relay is energized again, and the process repeats itself. **Figure 22-18** illustrates this process.

A good example of a normally open, timed open contact on a relay is illustrated in the operation of the dome light circuit in a car. Have your ever shut your car door and gone into the house knowing that the dome light remains on? A timer relay is used to prevent your battery from running down. When the car door is closed, the control circuit is normally open, and the timer relay coil is not energized. This is similar to the circuit in Figure 22-18A. However, for our example, Coil 2CR would be the dome light.

When you open the door, the control circuit closes and the timer relay coil energizes. The timer relay contacts in the load circuit close and the dome light comes on, Figure 22-18B. When you close the door, the dome light stays on for a set amount of time, even though the control circuit is open and the coil is de-energized, Figure 22-18C. After

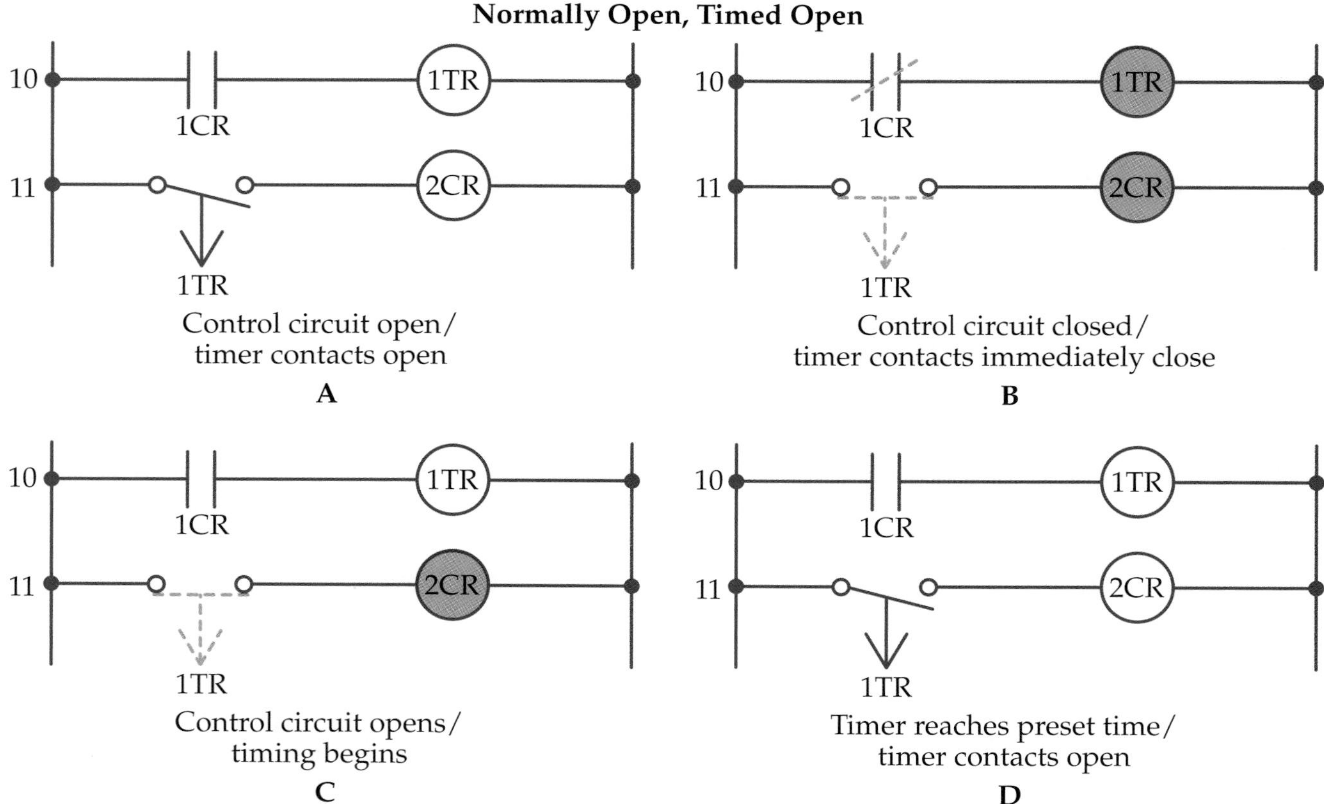

Figure 22-18. Normally open, timed open circuit operation.

a delay, the contacts in the load circuit open, and the dome light goes out, Figure 22-18D.

Normally Closed, Timed Open

Normally closed, timed open contacts are closed when no power passes through the control circuit. When power is applied to the control circuit, the contact remains closed but starts timing. When the timer reaches the preset time, the contact opens and remains open until the control circuit is de-energized. **Figure 22-19** illustrates this process.

A good example of a normally closed, timed closed circuit is in switching circuits that require a sequence of opening and closing over a regular, specified period of time. For example, a conveyor belt carrying a piece of hot plastic needs time for a vacuum former to stretch the plastic over a mold. At this time, the control circuit closes. The timer contacts remain closed, but the timing begins. The preset value allows enough time to stretch the plastic and for it to cool. After this time, the timed contacts open. The machine advances the formed part so it can be removed from the mold.

Normally Closed, Timed Closed

Normally closed, timed closed contacts are closed when no power passes through the control circuit. The contacts open immediately when the coil is energized. As soon as the control circuit is opened, the contact begins timing. It remains open as long as it is timing, but when the timer reaches the preset time, the contacts close. It stays closed until the relay is energized again, and the process repeats itself. **Figure 22-20** illustrates this process.

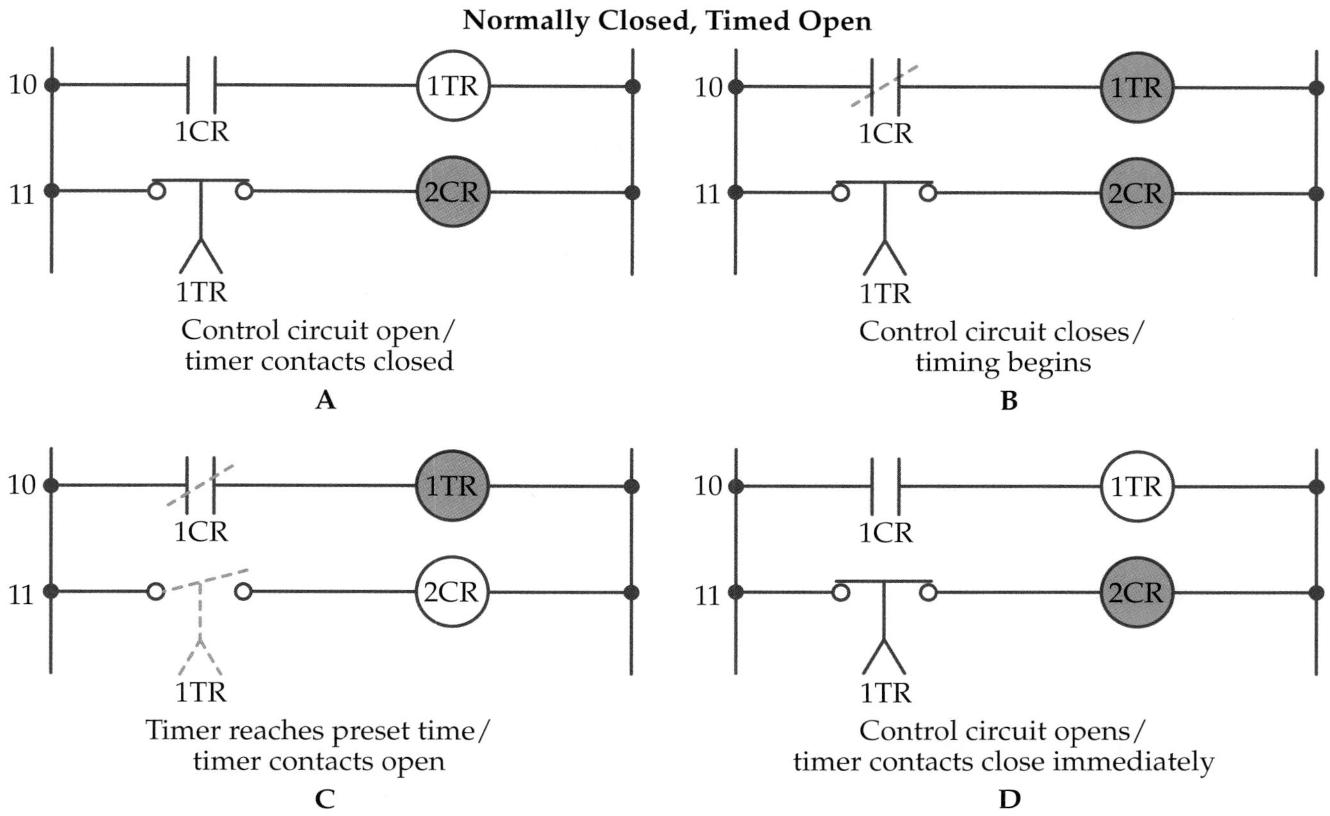

Normally Closed, Timed Open

10 1CR 1TR
11 1TR 2CR

Control circuit open/
timer contacts closed

A

10 1CR 1TR
11 1TR 2CR

Control circuit closes/
timing begins

B

10 1CR 1TR
11 1TR 2CR

Timer reaches preset time/
timer contacts open

C

10 1CR 1TR
11 1TR 2CR

Control circuit opens/
timer contacts close immediately

D

Figure 22-19. Normally closed, timed open circuit operation.

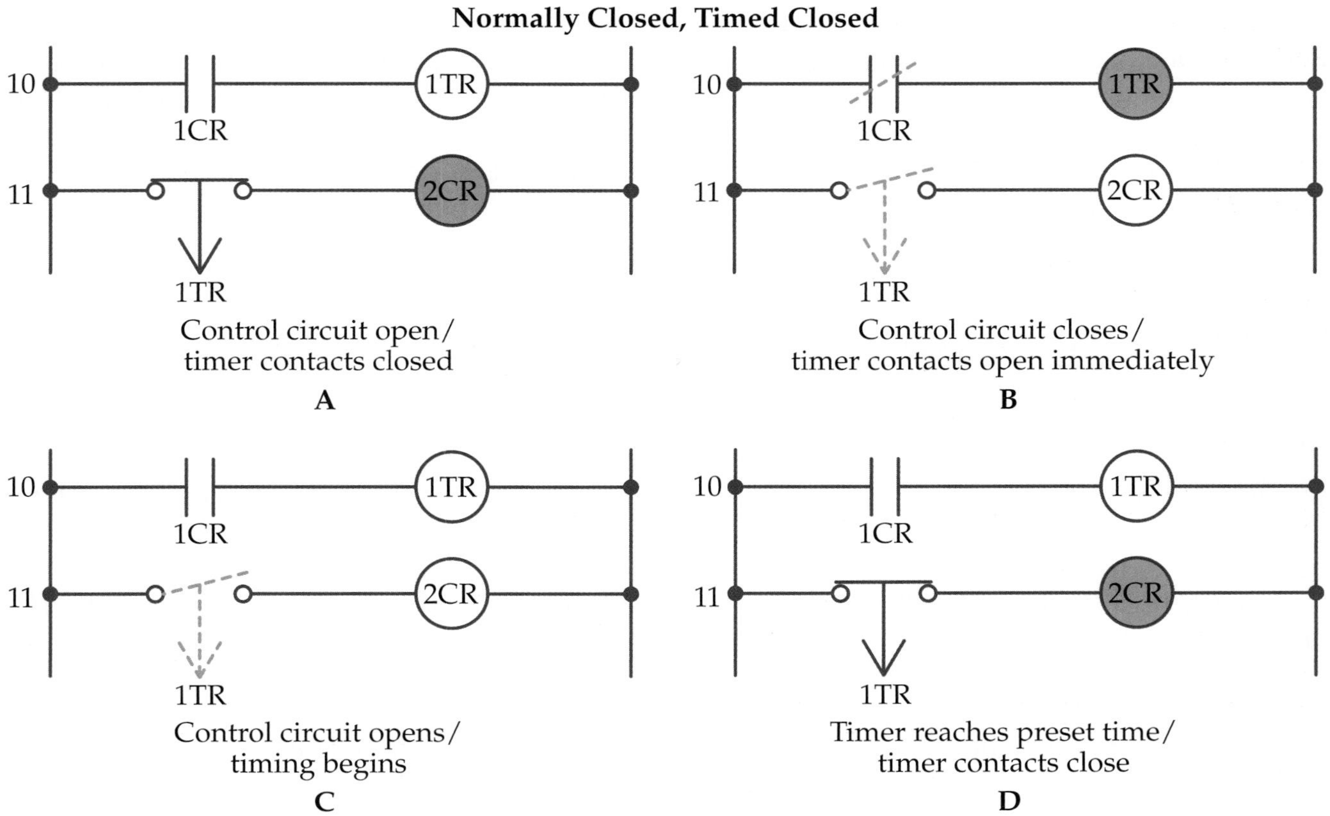

Normally Closed, Timed Closed

10 1CR 1TR
11 1TR 2CR

Control circuit open/
timer contacts closed

A

10 1CR 1TR
11 1TR 2CR

Control circuit closes/
timer contacts open immediately

B

10 1CR 1TR
11 1TR 2CR

Control circuit opens/
timing begins

C

10 1CR 1TR
11 1TR 2CR

Timer reaches preset time/
timer contacts close

D

Figure 22-20. Normally closed, timed closed circuit operation.

A good example of a normally closed, timed closed circuit is the timing circuit in a clothes dryer. Before you start your clothes dryer, you must close the door and turn on the timer cycle. This is shown by the circuit state in Figure 22-20B. The door represents the timer contacts (1TR). The timer control circuit is shown in line 10. Sensors can also be used to measure the moisture remaining in the clothes.

When you reach the selected moisture level, the timing contacts (1TR) close and a buzzer goes off to indicate the clothes are dry. If you open the door, the process shuts off.

This process can also be done by discharging a capacitor. If you fail to open the dryer door, the timer circuitry will reset itself and after a period of time the buzzer goes off again. This happens because the capacitor recharges, the circuit resets, and the buzzer sounds. This process repeats itself until you open the door.

Solenoids

A ***solenoid*** has a coil and a plunger and is similar to a relay in that it operates on the principles of electromagnetism. The electromagnetic force causes the plunger to move. When the plunger moves, it can control something else that needs to move, such as part of an electronic lock or part of the sound mechanism in door chimes, **Figure 22-21.**

You will also find a solenoid in your dishwasher. When you push the button to begin the washing process, you activate a circuit that energizes a solenoid. When the plunger inside the coil of the solenoid moves, it shifts a valve that opens the water supply line. This fills the dishwasher with the amount of water based on the dishwasher setting, such as pots and pans, normal, or light china. When the solenoid is de-energized, the plunger returns to its original position, thus shutting off the water supply. Depending on where it is used, the plunger can be spring loaded or returned by gravity.

Other Switches

In addition to relays and solenoids, there are other ways to control electrical circuits. Rather than timing a circuit, it can be controlled through the use of mechanical switches. These switches use other moving parts or liquid levels to turn the circuit on or off. Switches are made to open and close by the movement of a machine or a part moving on a conveyor (limit switches). Other switches are controlled by pressure of a liquid or a gas (pressure switches) or the height of a liquid (float switches). In spite of all their differences, turning a circuit on or off is still the purpose of every switch.

Limit Switches

Like relays, a ***limit switch*** has an electrical contact that is either normally open or normally closed. Some outside mechanical source usually triggers a limit switch. The trigger source can be many different things, such as a machine component or a human operator. The symbol for a limit switch is shown in **Figure 22-22.**

A limit switch might be used on a conveyor belt in a factory's production line, **Figure 22-23.** A limit switch can be used to make sure none of the products fall off the end of the belt. The switch is placed near the end of the conveyor belt and is part of the motor circuit that runs the belt.

When A product hits the limit switch, the limit switch contact opens and stops the motor that runs the conveyor belt, Figure 22-23A. This means the conveyor belt stops, too. As soon as the product is removed from the limit switch, its contact closes, and the conveyor belt starts moving again.

To see the logic of this operation, look at the ladder diagram in Figure 22-23B. The limit switch is normally closed. When power is applied to the circuit, the coil of relay 3CR energizes and closes contacts 3CR-1. Closing contacts 3CR-1 completes the circuit to the motor, and the conveyor belt begins to move. As

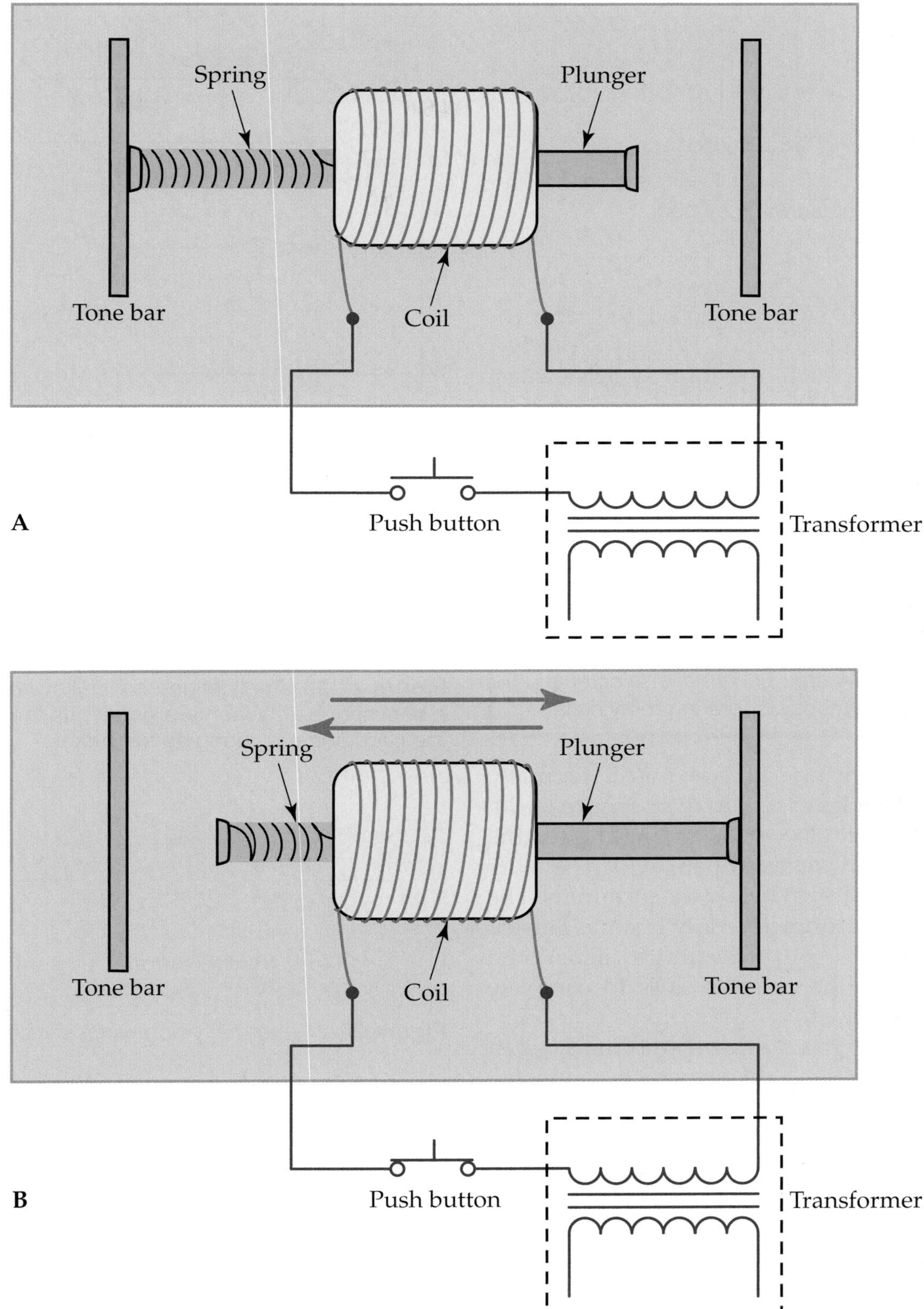

Figure 22-21. A solenoid operates on the principles of electromagnetism like the relay. A—The push button has not been pressed. The plunger is in a resting state. B—The push button is pressed. The plunger moves forward to strike the right tone bar and then returns to its original position, striking the other tone bar.

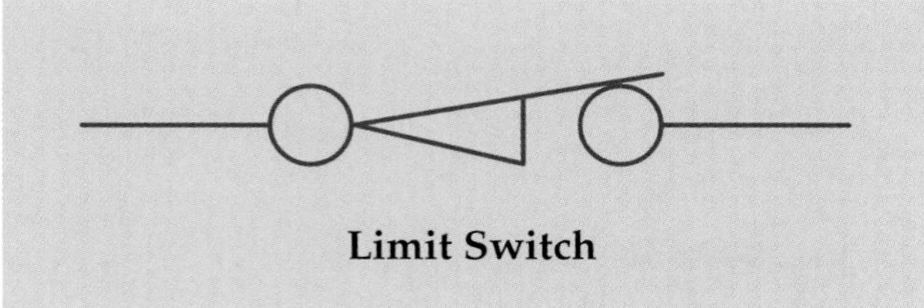

Figure 22-22. Limit switch symbol.

long as the limit switch is closed, the conveyor belt will continue to move. When the product hits the limit switch, the limit switch contact opens, Figure 22-23C, thus de-energizing 3CR. This opens contacts 3CR-1 in line 8, which stops the conveyor motor.

> **Note**
>
> Relay contacts are designed to withstand the constant opening and closing under an electrical load, such as what is required to start and stop a motor. The contacts in a limit switch are best used with much smaller electrical loads, such as those operating relay coils.

The schematic symbol for a normally open limit switch is shown in **Figure 22-24.** Remember, gravity or a spring holds the normally open switch open, which is why the symbol is drawn below the terminals. The circuit can be completed only if some outside force exerts pressure on the limit switch. When that happens, the contacts close to complete the load circuit.

A limit switch can have a normally open and normally closed set of contacts connected together mechanically. This ensures that if one set of contacts is open, then the other set will be closed. However, the mechanical connection does not allow both sets of contacts to be open or closed at the same time. This type of limit switch is shown in a circuit with a dashed line connecting the two contacts, **Figure 22-25.**

This limit switch may be used with a two-speed motor, which has a high-speed and low-speed winding. You do not want to energize both windings together. This switch ensures

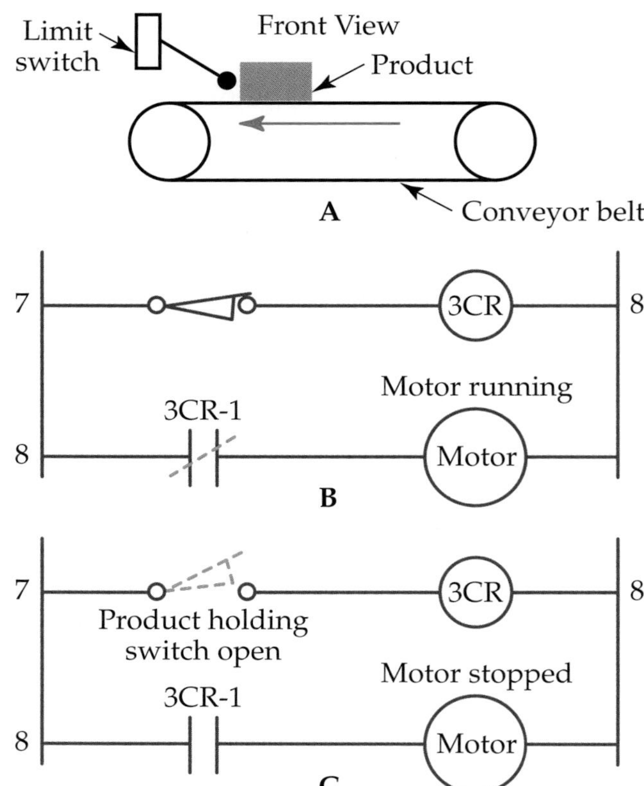

Figure 22-23. Application of a limit switch to stop a conveyor belt. When a product hits the switch, the circuit opens, stopping the motor.

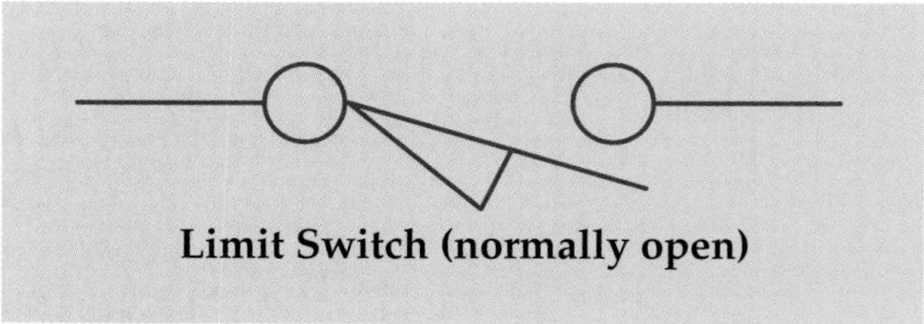

Figure 22-24. Normally open limit switch symbol.

that only one circuit—either the high-speed circuit or the low-speed circuit—is completed. When one circuit is turned on, the other circuit is simultaneously turned off.

Pressure Switches

Another common device found in a circuit is called a ***pressure switch*** or a ***vacuum switch.*** Its contacts can also be either normally open or normally closed, **Figure 22-26.** Changes in

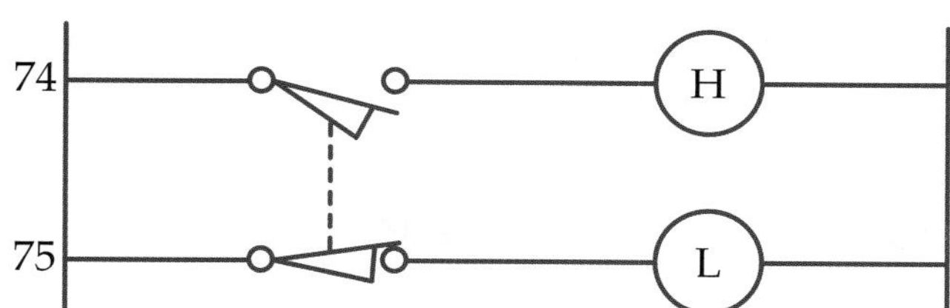

Figure 22-25. Using normally open and normally closed limit switches in combination to control high-speed and low-speed motor windings.

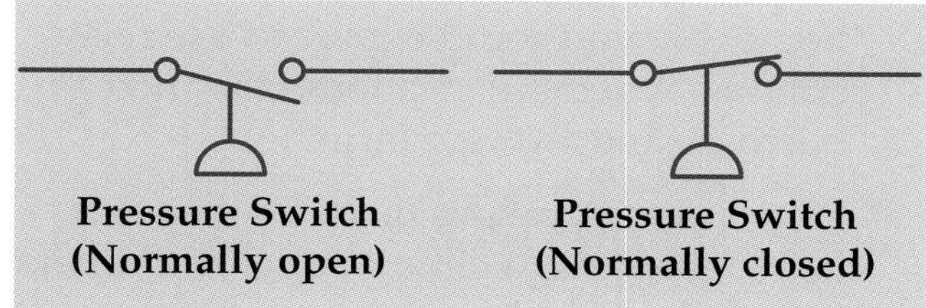

Figure 22-26. Symbols for pressure switches.

pressure are used to activate the switch that closes or opens the contacts.

Pressure switches are used frequently in air compressors. A normally closed pressure switch is installed on the air tank of the compressor. When the electrical switch is turned on, the motor runs the air compressor until the pressure is high enough to activate the pressure switch. At that time, the contacts on the pressure switch open, stopping the motor. Without the pressure switch, the motor would continue to run and the pressure would eventually rupture the tank or harm the pump. As the compressed air is used, the pressure falls until it reaches a level that closes the pressure switch. This activates the pump motor to recharge the tank.

Float Switches

A ***float switch*** is activated mechanically by a rising or falling water level. This type of switch has an arm connected to a ball float that rises with the surface of the water. When the water's surface falls far enough, the float pulls on the switch. This may either open or close the switch.

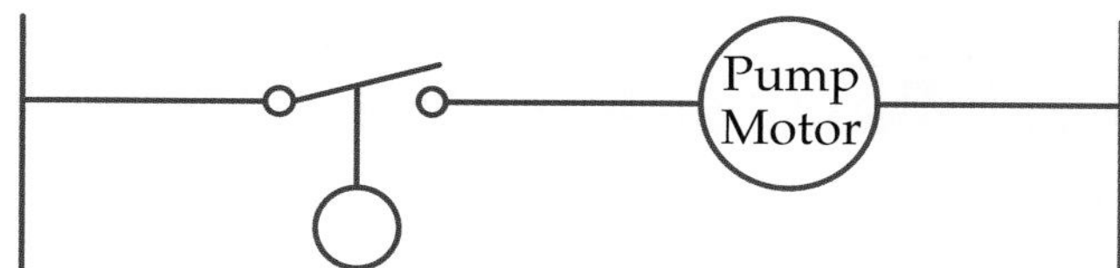

Figure 22-27. A float switch symbol in a ladder diagram.

Some wells with a pump that transfers water to a storage tank have a float switch, **Figure 22-27.** The float switch has settings that indicate when the pump should run and when the tank is full. A float that sits inside the storage tank is attached to the arm of the switch. When the water level drops, the float drops, allowing the switch to close. This completes the circuit to the pump, which then delivers more water until the tank is again full enough so that the float pulls the switch open.

Summary

- A relay is a magnetic switch.
- A relay's coil becomes magnetic when current flows through it.
- The magnetic field surrounding the coil pulls the relay's armature.
- Relays in a ladder diagram are numbered, usually from top to bottom.
- Relay contacts in a ladder diagram are given corresponding numbers.
- A switch's contacts may be normally open or normally closed.
- A timed relay can be set to open or close after a specific amount of time.
- Limit switches are triggered by an outside, mechanical force.
- Pressure switches are triggered by changes in pressure.
- Float switches are triggered by changing liquid levels.

Test Your Knowledge

Do not write in this book. Write your answers on a separate sheet of paper.

1. The diagram below shows a description of relay 13CR. What would the numbers on the right of the relay tell you?

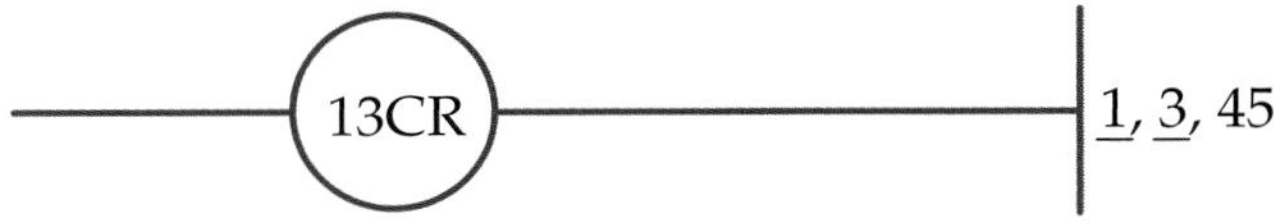

2. The letters CR stand for ______.
3. Draw the ladder diagram symbol for a normally open contact.
4. Explain why holding contacts are so important in a relay circuit.
5. What is needed to energize the relay coil 5CR in the circuit below?

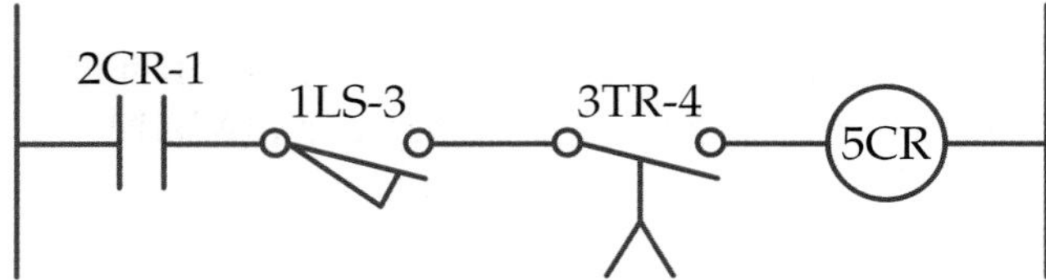

6. Normally open, timed closed contacts (close, do not close) immediately when the timer coil is energized.
7. Normally open, timed open contacts (close, do not close) immediately when the timer coil is energized.
8. Normally closed, timed open contacts (immediately open, remain closed) when the timer coil is energized.
9. Normally closed, timed closed contacts (immediately open, remain closed) when the timer relay coil is energized.
10. A solenoid operates on the principles of ______, just like a relay.
11. For each of the following switches, draw its normally open and normally closed symbols and write what triggers it.
    A. Limit switch
    B. Pressure switch
    C. Float switch

Activities

1. Make a list of the closest places near your home that you could expect to use relay circuits. Are these locations classified as business, industry, or farm?
2. Why are some relays made so they can be plugged into a circuit panel and others to be bolted in place? Where would you expect to find these different types used? Do they all carry the same amount of current?
3. Relays are generally classified as slow compared to so many high-speed devices, such as computers. Why would anyone still build circuits using relays? What are the advantages and disadvantages of each?
4. How can the contacts in a relay weld themselves together? What would happen to the circuit? Can the same thing happen in high-speed devices? Explain your answer.
5. Could a relay circuit be used in space flight where there is no gravity? Explain your answer.
6. Are other electricity and electronic devices affected by a lack of gravity? How do you prevent heat build-up on devices or components that require heat sinks while on earth?

23 Computers

Learning Objectives

After studying this chapter, you will be able to do the following:

- State the difference between hardware and software.
- Identify the main parts of a computer.
- State the functions of the main parts of a computer.
- State the difference between input and output devices.
- Explain the purpose of ASCII.
- Discuss how software is written and what it does.

Technical Terms

adapter cards
American Standard Code for Information Interchange (ASCII)
arithmetic-logic unit (ALU)
backup
bit
cache
central processing unit (CPU)
CMOS memory
computer
computer integrated manufacturing (CIM) cell
hard drive
hardware
inkjet printer
input devices
laser printer
local area network (LAN)
metropolitan area network (MAN)
modem
monitor
motherboard
network
output devices
parallel ports
peripherals
personal computer (PC)
personal digital assistant (PDA)
ports
programmable logic controller (PLC)
random-access memory (RAM)
serial port
server
software
wide area network (WAN)
workstations

A ***computer*** is a device that takes in information, processes it, and outputs the results. The information going into the computer is called *input*. The results are called *output*. Commonly used input devices include keyboards, touch-sensitive monitors, and scanners. The most used output device is a monitor. However, there are many other output devices that a computer uses. Some devices, such as hard drives, CD drives, and modems, work as both input and output devices.

The most commonly used computer is the ***personal computer (PC)***, **Figure 23-1.** A PC is a computer that is often used by a single person. There are also numerous computers that you normally do not even think about or see during your daily activities. For example, your car has numerous computers to control functions such as airbag deployment, engine performance monitoring, headlight control, and warning light activation. Computers are a part of electronic gaming systems, sewing machines, and many other devices in your home. Although there are many other examples of computers, this chapter focuses mainly on the PC.

Figure 23-1. A personal computer, which is typically referred to as a PC.

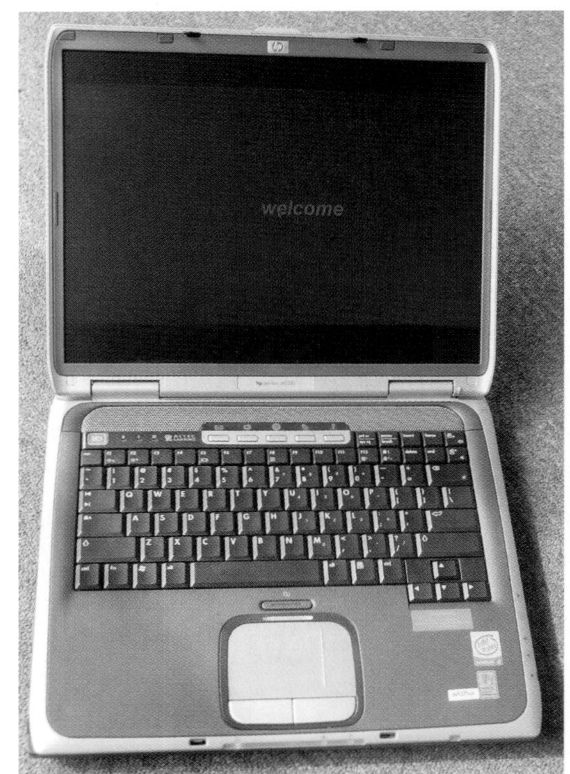

Figure 23-2. Laptop computer.

Types of Computers

The systems popular today are the mainframe, personal computer (PC), notebook, and personal digital assistant (PDA). Mainframes are the largest and most powerful of these systems. Universities, research and development groups, and very large industries may use a mainframe. Mainframes are used to handle extensive mathematical calculations and operations which involved very large database files containing millions or billions of pieces of data. For example, these databases may contain payroll records or census information.

PCs and notebook computers (also called *laptop computers*), **Figure 23-2,** are most commonly found in homes, schools, and businesses. PCs in general are thought of as a machine used by a single operator. They have many purposes, such as word processing, accounting, Internet exploration, and electronic mail (e-mail). They can also be used to store, edit, and view image files, such as those from a digital camera, or store and listen to music files.

The ***personal digital assistant (PDA)*** is a small, handheld, wireless computer, **Figure 23-3.** It can compute and store data, but because it is so small, it has significantly less memory than most PCs. PDAs may have MP3 and video playback ability, which means they can allow you to listen to music or to view a video. They can also be used to send and receive e-mail and act as a mobile phone. Many may have limited word processing, database, and spreadsheet capabilities.

> **Note**
> Computer technology changes very quickly. Manufacturers continue to reduce the size of integrated circuits and increase the speed of electronic devices used by computers. Compared to today's computers, those from just a few years ago are almost obsolete.

Parts of a Computer

You have already learned in earlier chapters about most of the components that make up a computer. These items include transformers, capacitors, diodes, transistors, and ICs. However, these small components make up larger parts of the computer, which are covered here. These parts are called the computer's ***hardware*** and include the power supply, central processing unit, hard drive, and all the parts you can see when using a computer. Parts that are attached to the computer are called ***peripherals.***

Power Supply

All computers have a power supply. Some computers run on ac power, some run on dc power, and some can be run from either source. Portable computers use batteries, which are a dc source of power. PCs, however, are normally plugged into an ac outlet.

A PC power supply, **Figure 23-4,** steps down the 110-V ac from the wall outlet to 5 V and 12 V and converts theses values to dc. The power supply makes sure the current remains at levels that the solid-state devices in the computer can handle without being damaged.

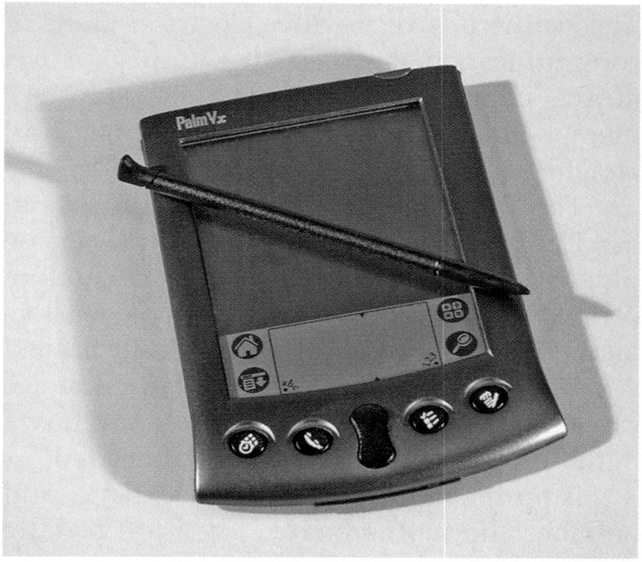

Figure 23-3. Personal Digital Assistants, commonly referred to as PDAs, are handheld devices that can allow you to access the Internet and send and receive e-mail.

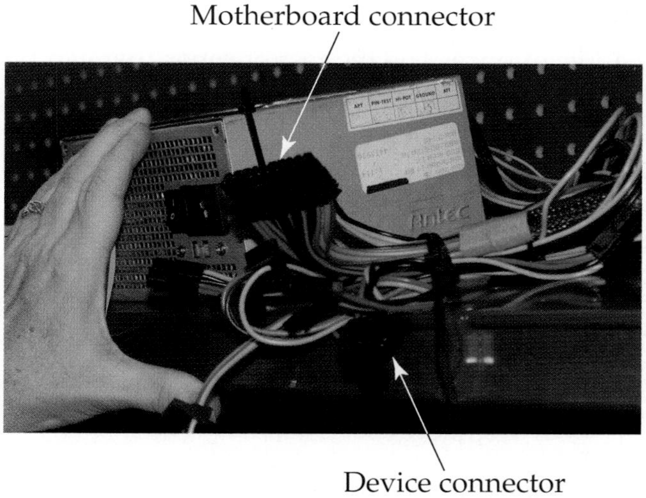

Motherboard connector

Device connector

Figure 23-4. A PC power supply steps down the 110-V ac from the wall outlet to 5 V and 12 V and converts these values to dc. It has one connector that attaches to the motherboard and several other connectors that attach to devices. (Courtesy of Theresa Miller)

The PC power supply has many connectors. One connector attaches to the motherboard. The other connectors plug into the devices that are part of the computer, such as hard drives and DVD drives.

Central Processing Unit

The ***central processing unit*** (***CPU***) receives all input and controls the other components of the computer. The CPU is an IC mounted on the computer's motherboard and is also called the *microprocessor*, or just the processor of a computer, **Figure 23-5.** It performs all of the computer's tasks, including performing calculations and manipulating data.

All input to the CPU is in the form of binary numbers. A ***bit*** is a binary digit, either a 0 or a 1. In fact, it is derived from the words ***bi**nary digi**t***. Pieces of information, such as letters, are encoded in strings of 0s and 1s. The 0s and 1s are transmitted through transistors as present or absent signals, as discussed in Chapter 20. The CPU is like any other IC in that it uses logic gates to do its calculations and determine what to do next.

Processors are measured based on how much information they can process at one time. For example, a 64-bit processor can handle 64 bits at once. The biggest number that can be held in 64 bits is 18,446,744,073,709,551,616 (2^{64}). A 64-bit processor can add and subtract numbers up to 2^{64} in a single operation.

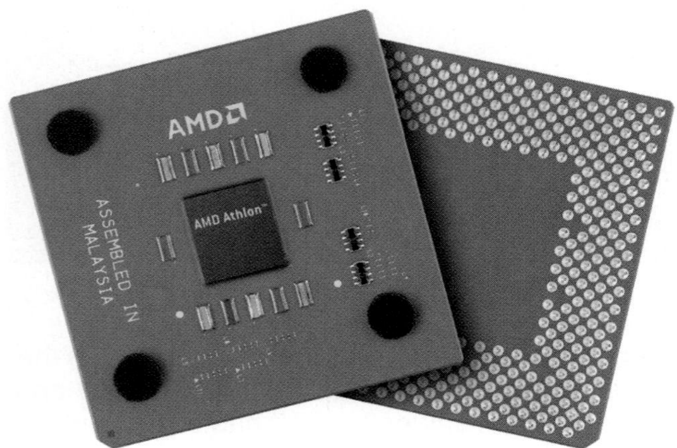

Figure 23-5. A central processing unit (Mobile AMD Athlon™ Processor). (Advanced Microprocessor Devices, Inc.)

The CPU contains an ***arithmetic-logic unit*** (***ALU***) and ***cache.*** The ALU performs the binary calculations, allowing the rest of the CPU to control the other tasks. The cache is a small amount of memory where the CPU stores information it must access frequently. This allows quick and easy access to that information, thus increasing the computer's processing speed.

Processing speed is measured based on the clock of the CPU. The clock sends a pulse of electricity through the CPU at regular intervals. Usually, the CPU can perform a task with each pulse, although some tasks may take several pulses. Early CPU processing speeds were around 1 MHz. This meant that there were one million pulses per second. This sounds like a lot, but today's processors are measured in gigahertz (GHz). Each gigahertz is one billion pulses per second.

Motherboard

The ***motherboard*** is the main printed circuit board of the computer, **Figure 23-6.** The motherboard provides electrical paths of communication between all the major parts of the computer. The power supply, CPU, memory, hard drive, and adapter cards are all attached to the motherboard.

The motherboard usually has several expansion slots where adapter cards can be plugged into the motherboard and therefore connected to the CPU. The expansion slots allow the adapter cards to be easily added without soldering. They also allow the adapter cards to be interchanged. The user can buy a better or more sophisticated one if needed.

Adapter cards are used to expand and enhance the motherboard. They are usually designed for one specific purpose. For example, a graphics adapter card offers better video technology to the user. A graphics adapter card may have its own RAM or even its own processor to support the high-speed graphics

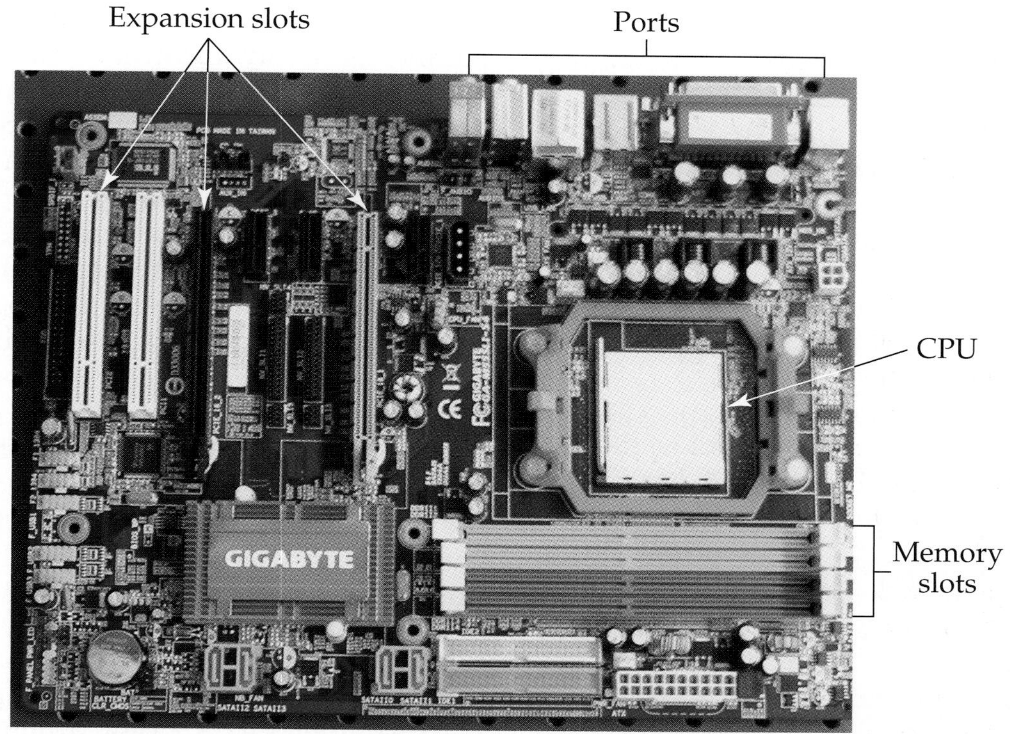

Figure 23-6. The motherboard contains ICs and slots for cards and memory. (Fry's Electronics)

found in many computer games, **Figure 23-7.** This allows the computer's CPU to run the rest of the program without being slowed down by supporting the graphics as well.

Many other computer parts are not on the motherboard, but need to be connected directly to it. These parts are called *peripherals,* and the places they are plugged into are called ***ports.*** There are several different kinds of ports. **Figure 23-8** shows the ports on the back

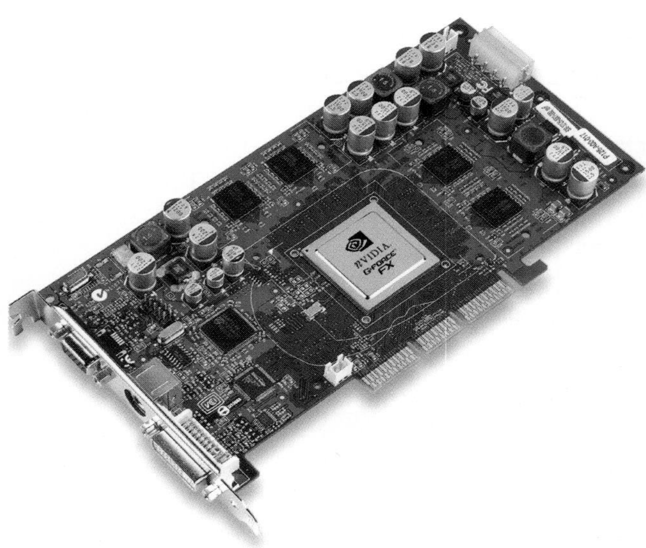

Figure 23-7. This graphics adapter card (NVIDIA® GeForce™ FX) has its own RAM and processor. (NVIDIA Corporation)

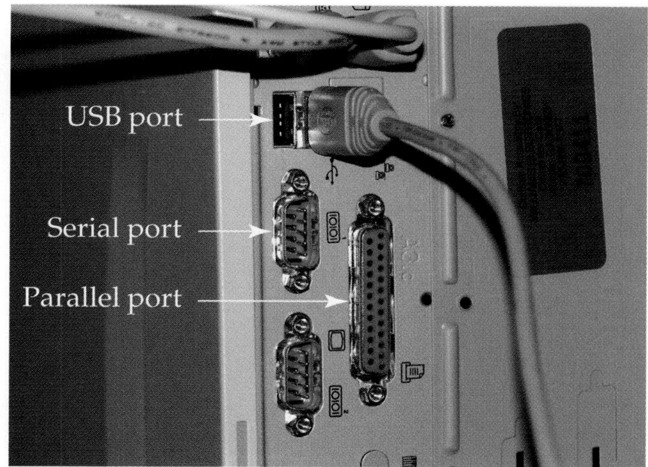

Figure 23-8. The ports on a computer are used to attach external devices called *peripherals.* (Courtesy of Nora Sullivan)

of a motherboard. The types and number of ports included on a motherboard vary among different motherboards.

A ***serial port*** typically has 9 or 25 pins. Bits travel through the wire and through the serial port in serial fashion, one bit at a time. A serial port may be used to connect a mouse to a computer. A serial port can also be called a *communications port* or an *RS-232 port*.

Computers are also equipped with ***parallel ports.*** Parallel ports have a connector with 25 holes. A parallel port usually connects a printer to a computer. The wires in the parallel port can transmit at least eight bits at one time. This increases the transfer speed and allows the computer to quickly transmit more information to a printer than is possible using a serial port.

Figure 23-9. Each end of a USB cable. The end on the left connects to the USB port on the computer and the other end connects to the USB device.

As the size of files increased, the speed of data transfer through serial and parallel ports decreased. High-speed data transfer is made possible with the universal serial bus (USB) port. USB cables connect computers to devices such as printers, scanners, and digital cameras, **Figure 23-9.** USB ports transfer around 480 Mbps (Megabits per second). Both the computer and the peripheral (the device connected to the computer) must have a USB port or a USB adapter to use this system.

Memory

A computer uses several different types of memory, **Figure 23-10.** The cache memory is physically close to the CPU, making access to it fast. The main memory the computer uses to run is called ***random-access memory*** (***RAM***). It takes longer for the CPU to access RAM than it does to access cache memory. RAM is a temporary storage place that is used to store

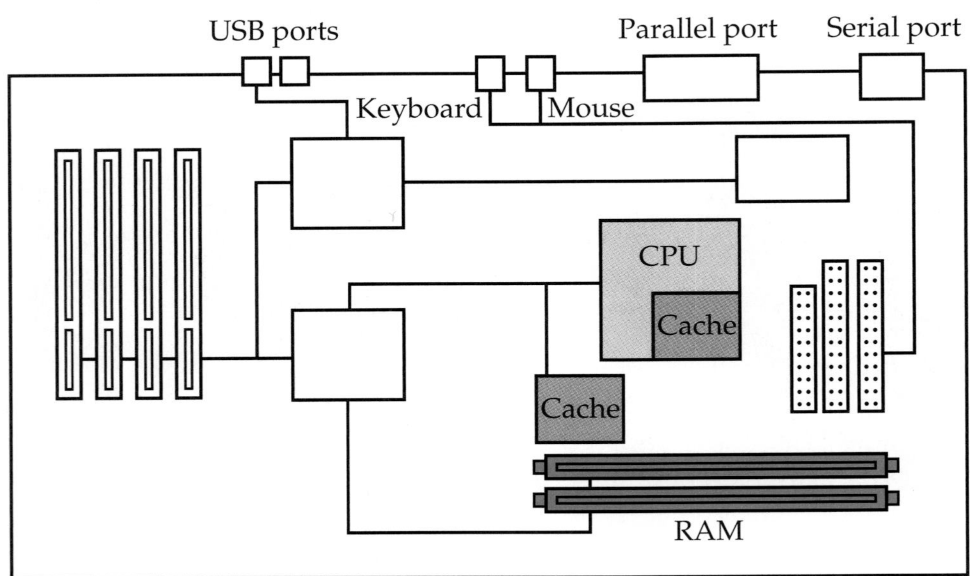

Figure 23-10. RAM and cache memory. Cache memory is typically located inside the CPU and near the CPU.

information and programs the computer is using or running at the moment.

RAM can store much more information than cache memory. Like cache, however, RAM does not store information when the computer is turned off. It loses any information in it when the computer is turned off and is empty each time the computer is turned on. This means if you are writing a paper in a word-processing program, what you have on the screen is stored in RAM. If you close that program, RAM no longer stores it, and your paper is lost. However, if you save your paper, you are moving a copy of it to a permanent storage place in the computer so the computer will remember it even if you close the word-processing program.

Hard Drive

One of the permanent storage places on a computer is called the ***hard drive*,** **Figure 23-11.** This is where all of the computer's programs and documents are stored. If you shut off your computer, your programs and documents are not lost. This is why it is referred to as *permanent storage*. Some other permanent storage devices are DVD and CD-ROM drives.

The hard drive provides the computer's slowest memory access. This is why RAM is used to store programs that are running. If the CPU had to constantly retrieve information from the hard drive to keep a program running, it would take much longer.

A hard drive is made up of several disks, called platters, stacked one above the other. There is just enough space between the platters for the read-write heads to pass between them, **Figure 23-12.** A read-write head writes the data on the platter and reads the data when the data needs to be retrieved.

The read-write heads use principles of magnetism to store and retrieve data on the hard drive. Once information is written to the disk, the read portion of the head can sense the polarity of the magnetic particles by induction.

Figure 23-11. A computer's hard drive allows it to store files even when the computer is off.

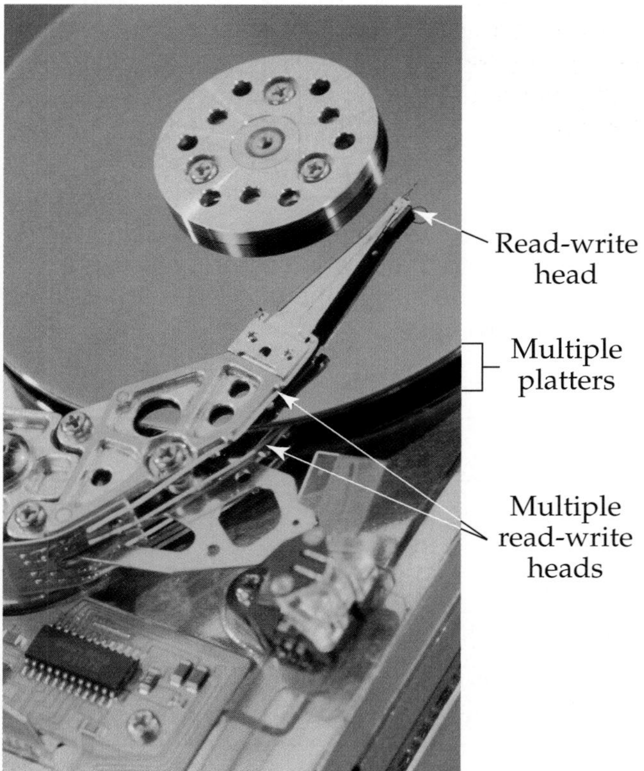

Figure 23-12. Read-write heads hover over the hard disk platters. They do not touch the platters.

The disk is divided into concentric circles called *tracks*. Each track is divided into a series of *sectors*, **Figure 23-13.** The hard drive uses the tracks and sectors to organize the information it stores. As the disk spins, the read-write head senses the information on the tracks.

Hard drive sizes have continued to increase in their ability to store data. Their sizes increased from megabytes (MB) to gigabytes (GB). Some computers are using hard drives in the size of terabytes (TB) and beyond. Look at the table in **Figure 23-14** to understand the ranges of values.

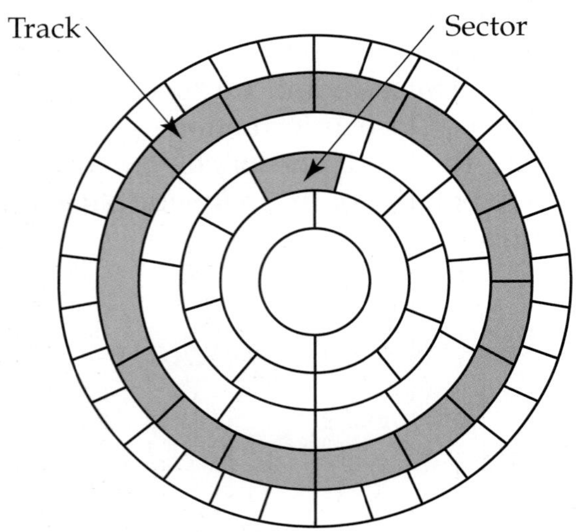

Figure 23-13. The platters of a hard disk drive are divided into tracks and sectors; although, they may not be as symmetrically divided as the platter shown here.

Value	Approximately Equal To:
Bit	Binary digit 0 or 1
Byte	8 bits
Kilobyte (kB)	1000 bytes
Megabyte (MB)	1000 kB or 1,000,000 bytes
Gigabyte (GB)	1000 MB or 1,000,000,000 bytes
Terabyte (TB)	1000 GB or 1,000,000,000,000 bytes

Figure 23-14. Typical values used in the computer industry. Hard drives are available in gigabyte (GB) and terabyte (TB) sizes.

CMOS Memory

None of the memory discussed so far can be used to store settings needed when the computer is turned on. ***CMOS memory*** is part of the motherboard and is powered by a small battery, **Figure 23-15.** It remembers the computer settings used each time the computer starts. *CMOS* stands for *complimentary metal oxide semiconductor.*

The CMOS memory stores many necessary pieces of information such as the start order of the drives in the computer, the time, and hard drive configuration settings. Since the computer must keep track of these elements even while it is turned off, the CMOS uses power from a small battery to preserve its memory. Without this battery, all stored settings would be lost when the computer is shut off.

Input Devices

Input devices are used to put information into the computer. Some important input devices include the keyboard, mouse, scanner,

Figure 23-15. Computers have a small battery and CMOS memory on the motherboard. This allows the computer to keep track of date, time, and some setup features when the computer is off. (Fry's Electronics)

DVD/CD ROM drives, digital camera compact flash card readers, and Web cameras. The keyboard, mouse, and CD-ROM drive are the most common input devices and deserve some attention.

The keyboard is one of the main input devices for the computer. When a key is pressed, a signal is sent to the CPU. Combinations of keys pressed at the same time, such as the [Shift] key and the [Y] key, are also used to send signals. In this case, the keyboard sends a signal for a capital *Y*. Each character is given a specific binary code, and the keyboard sends one of these codes when one of its keys is pressed. The code that is used is called the ***American Standard Code for Information Interchange*** (***ASCII***) and is shown in **Figure 23-16.** Notice that the ASCII code number for *Y* is different from the code number for *y*. The signal is sent to the CPU, where it is stored in the computer's temporary memory and displayed on the screen.

The mouse is another important input device for the computer. The shape of the first mouse and its connecting wire earned this device its name. It allows the cursor to move across the screen easily and allows the user to access menus and files by pressing, or "clicking," the mouse button.

Most different types of drives are both input and output devices. They give information to and receive information from the computer. There are many types of drives for computers. CD-ROM, DVD, tape drives, and USB flash drives are all common. If, for example, a CD has information on it, the information can be put into any computer with a CD-ROM drive. Also, users can save their data or documents to a CD, DVD, tape, or USB flash drive instead of saving to the hard drive.

Note

Although floppy disk drives were widely used in the past, almost all computer manufacturers have eliminated them as a data storage option.

Digit Key	Binary	ASCII Code	Digit Key	Binary	ASCII Code
0	00110000	48	V	01010110	86
1	00110001	49	W	01010111	87
2	00110010	50	X	01011000	88
3	00110011	51	Y	01011001	89
4	00110100	52	Z	01011010	90
5	00110101	53	(	01011011	91
6	00110110	54	\	01011100	92
7	00110111	55	)	01011101	93
8	00111000	56	^	01011110	94
9	00111001	57	-	01011111	95
:	00111010	58	`	01100000	96
;	00111011	59	a	01100001	97
<	00111100	60	b	01100010	98
=	00111101	61	c	01100011	99
>	00111110	62	d	01100100	100
?	00111111	63	e	01100101	101
@	01000000	64	f	01100110	102
A	01000001	65	g	01100111	103
B	01000010	66	h	01101000	104
C	01000011	67	i	01101001	105
D	01000100	68	j	01101010	106
E	01000101	69	k	01101011	107
F	01000110	70	l	01101100	108
G	01000111	71	m	01101101	109
H	01001000	72	n	01101110	110
I	01001001	73	o	01101111	111
J	01001010	74	p	01110000	112
K	01001011	75	q	01110001	113
L	01001100	76	r	01110010	114
M	01001101	77	s	01110011	115
N	01001110	78	t	01110100	116
O	01001111	79	u	01110101	117
P	01010000	80	v	01110110	118
Q	01010001	81	w	01110111	119
R	01010010	82	x	01111000	120
S	01010011	83	y	01111001	121
T	01010100	84	z	01111010	122
U	01010101	85			

Figure 23-16. Pushing a key on the keyboard sends one of these ASCII code numbers to the CPU.

The output ability of these drives is commonly used to ***backup*** all the data that is on the hard drive. This means a backup copy of everything on the hard drive is made. Sometimes hard drives can fail and lose everything stored on them. If this happens, having a backup copy of all the data files means they are not actually lost and can be copied onto the new hard drive.

A ***modem*** also works as input and output device. It is typically used to transmit digital computer data electrically over phone lines and other types of copper-core cables. These cables typically transmit analog signals. Recall that digital signals are in the form of 0s and 1s and analog signals consist of varying voltage values. Therefore, the sending modem must change the computer's digital data to an analog signal. This is known as *modulating a signal*. The receiving modem demodulates the data, or changes the signal back into digital data, **Figure 23-17.** This process of ***mo***dulation and ***dem***odulation is how the modem gets its name.

Output Devices

For a computer to be practical, a user must be able to receive output in response to input. ***Output devices*** are used to take information out of the computer. The computer ***monitor*** is the most used output device. The monitor displays the information on its screen. Many different technologies are used for monitor screens, **Figure 23-18.** The two most common are the cathode ray tube (CRT), Figure 23-18A, and liquid crystal display (LCD), Figure 23-18B.

Figure 23-18. Two common types of computer monitors are the cathode ray tube (CRT) and liquid crystal display (LCD). Notice how thin the LCD monitor is compared to the CRT monitor.

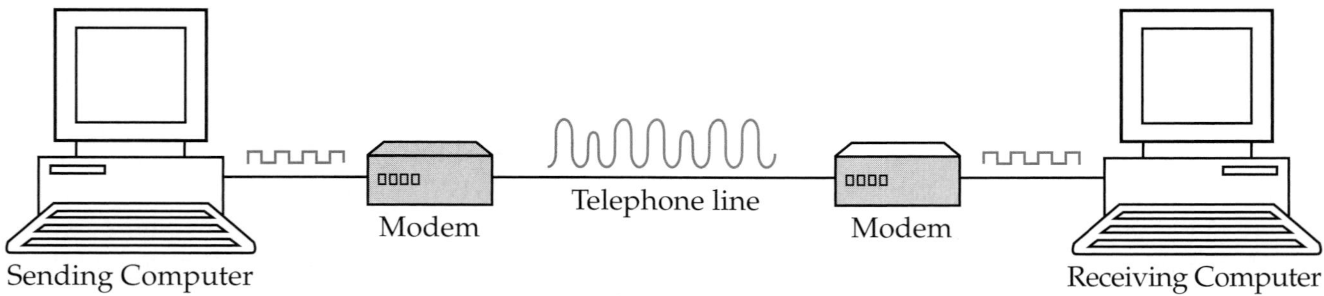

Figure 23-17. A modem converts data from digital to analog and back again.

Printers are another commonly used output device. They transfer data from the computer to paper. There are many different types of printers. Some print only in black ink, while others print in full color. Some spray the ink onto the paper, while others use magnetic principles to draw the ink to the paper. Two most common types of printers are inkjet and laser.

An ***inkjet printer*** fires (sprays) ink onto the paper when the inkjet head is electrically charged, **Figure 23-19.** The drop of ink an inkjet printer sprays is very small. Some inkjet printers can produce 4800 × 1200 dots per square inch (dpi). Some inkjet printers use a piezo crystal. The piezo crystal flexes when electrically charged, forcing ink from the reservoir. Other inkjet printers use a thermoresistor. The thermoresistor heats up when electrically charged, causing the ink around it to vaporize and a bubble to form. As the bubble grows in size, a drop of ink is ejected from the print head nozzle. When the bubble pops, more ink is drawn into the reservoir. This type of inkjet printer is commonly referred to as a *bubble jet.*

A ***laser printer*** uses magnetism to print. **Figure 23-20** shows the major parts of a laser printer. The corona wire charges the drum. This causes a negative static charge to surround the drum. Next, the laser writes onto the drum the image of what is to be printed. When the drum turns past the toner cartridge, it attracts particles of toner onto its charged regions. As the paper is taken into the printer, the transfer corona wire charges the paper more heavily than the drum, causing the paper to attract the toner particles that are on the drum. The paper then passes through the fuser, which heats the paper and toner, melting the toner onto the paper's surface.

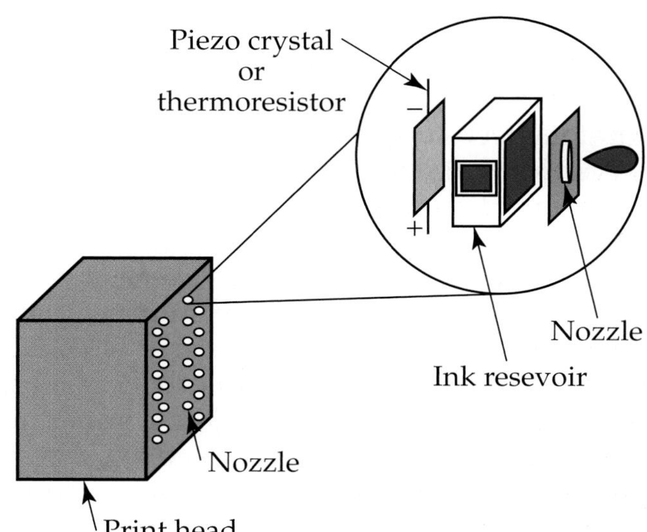

Figure 23-19. Print head operation of an inkjet printer.

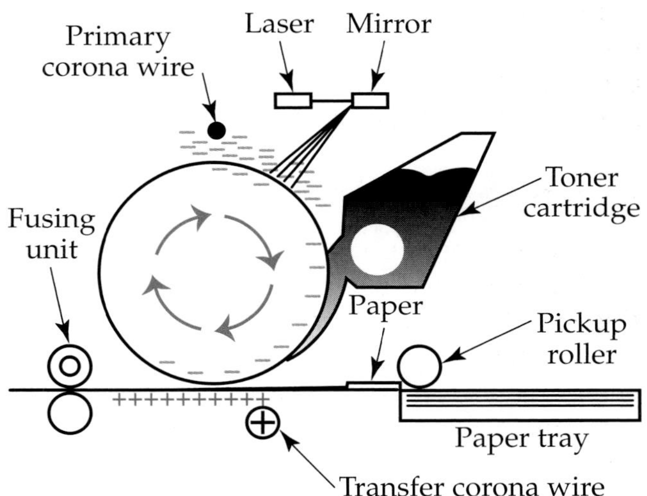

Figure 23-20. Laser printer operation.

Networks

A ***network*** is a group of computers that are connected so that information can be exchanged among them. A network in a home or office is a ***local area network (LAN).*** A LAN is used mainly for file sharing. Usually there is a main computer that stores files. This main computer is called the ***server.*** The other computers linked to the server are called ***workstations,*** **Figure 23-21.**

One advantage of a LAN is that the workstations do not need as much equipment as the server. A workstation does not need a hard drive, although it may still have one. Another advantage of a LAN is that it allows all files to be stored at a central location: the server. This allows files to be accessed easily by all users at each workstation. A LAN also means administrators need to go to only one location, the server, to restrict access to any files.

Two other classes of networks are the ***metropolitan area network (MAN)*** and ***wide area network (WAN).*** A MAN is a group of LANs connected by private communication lines. It is managed by a single party. A WAN, on the other hand, is a large group of networks connected by private and public communication lines. A WAN is managed by many different parties. The Internet is an example of a WAN.

Networks can be connected by phone lines, twisted-pair cable, fiber-optic cable, or signals broadcast by wireless cards. There are many different patterns, called *topologies,* in which a network can be connected. Efficiency and cost dictate which network topology is used. Some common network topologies are bus, ring, and star, **Figure 23-22.**

Software

Software is a set of instructions for a computer. There are thousands of kinds of software programs. These programs include games, operating systems, and other applications. All software programs must be written by a programmer. The programs are written in a programming language, or code, which the computer reads. Each line of code is a command or message for the computer. The code may tell the computer to do math functions, send sounds to a speaker, or display graphics for the user. Some languages are better suited to certain applications than others. Some common languages are C++, JAVA, and Visual Basic.

Programmable Logic Controllers

Computers are used to control the motions of robots in many factories and industries, **Figure 23-23.** Robots are used to paint car bodies, weld metal, and test ICs. They can even do simple tasks like squirting out a precise amount of glue. The computers that control these robots must be programmed to make the robot do the specified motions.

A ***programmable logic controller (PLC)*** is a type of computer that controls many of the robots just described. They are also used to monitor and control other types of

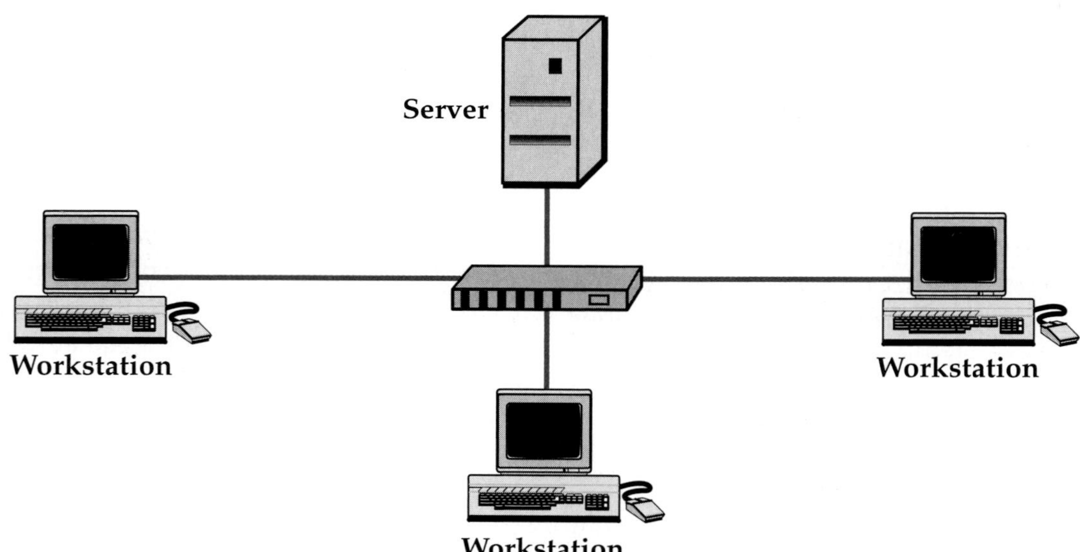

Figure 23-21. A local area network (LAN) with a server and workstations. The server provides files and other resources to the workstations.

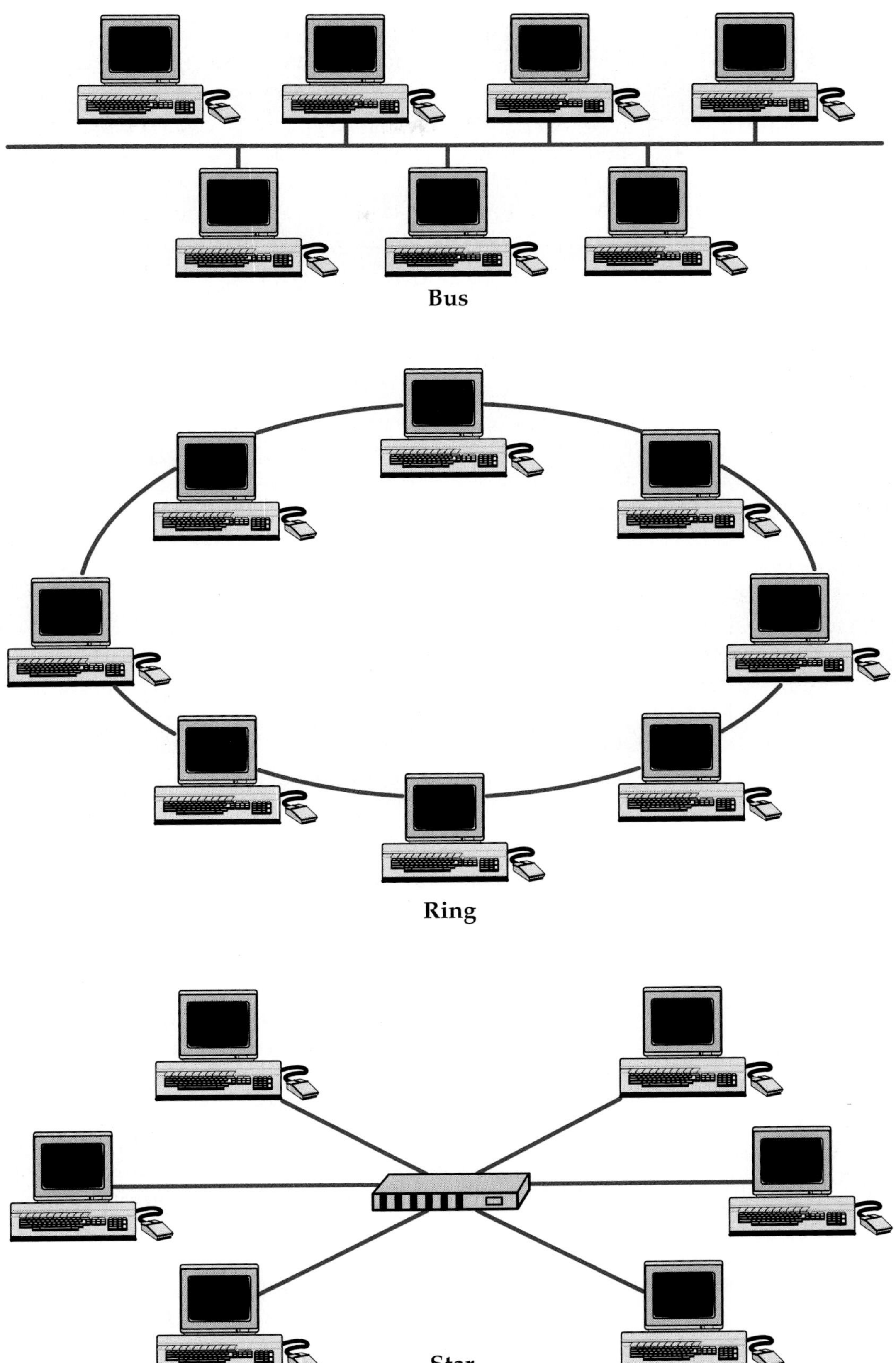

Figure 23-22. Network topologies.

Figure 23-23. These robots have been programmed to perform welding operations.

machinery. A PLC uses logic gates to monitor input and tells the rest of the machine what to do based on that input. PLCs use many of the elements of computers mentioned earlier in this and other chapters. For example, many automobiles have sensors which are tied to a PLC. The PLC uses these sensors to monitor and measure items like charging voltage, oil pressure, and revolutions per minute (rpm) of rotating devices. Based on the deviation from the measurement, the PLC will tell the computer to do any of the following:

- Turn on a warning light.
- Ring an alarm if a seatbelt is not fastened.
- Deploy an air bag during a crash situation.
- Not allow the car to start if the wrong key is used (a chip is embedded in some keys).

Many robots and other production machines can be combined into a ***computer integrated manufacturing*** (***CIM***) ***cell***. A CIM cell is a group of machines that are put together to produce a finished product. The group might include a milling machine, robots, conveyor belts, measuring devices, and sorting racks. Raw materials enter at the beginning, and a finished product emerges at the end. A CIM cell might turn a block of plastic into an artificial hip socket. Most large manufacturing companies use CIM cells to produce their products, so PLCs are very important devices. Engineering and technical schools have entire courses devoted to how to program and use PLCs.

Summary

- A computer's hardware consists of the physical components of the computer.
- Input devices put information into the computer.
- Output devices allow the user to receive information from the computer.
- The CPU receives all input and controls all of the components of a computer.
- The motherboard is the main PC board of a computer.
- The CPU, RAM, CMOS memory, and adapter cards are mounted on the motherboard.
- The hard drive provides the only permanent memory inside the computer.
- Other forms of permanent memory like CDs, floppy disks, and tapes are used to back up data.
- ASCII is a binary code for each key on a keyboard and various combinations of keys.
- A network is a group of computers that are connected.
- Software is a set of instructions for the computer.
- A PLC is a computer that uses logic gates to monitor input and, based on the input, tells the rest of the machine what to do.

Test Your Knowledge

Do not write in this book. Write your answers on a separate sheet of paper.

1. What is the difference between hardware and software?
2. The ______ processes all the input and controls all the other parts of the computer.
3. What is a bit?
4. What components would you see attached to a motherboard?
5. What is the function of CMOS memory?
6. CMOS memory get its power from a(n) ______.
7. Write the full name of each of the following parts.
   a. ALU
   b. CMOS
   c. CPU
   d. RAM
8. Describe the makeup of a hard drive.
9. The read-write heads use principles of ______ to store and retrieve data on the hard drive.
10. The information going into the computer is called ______, and the results are called ______.
11. What is the ASCII code and why is it used?
12. How is software written and what does it do?

Activities

1. Open the case of a PC with the power off. Observe the items inside. Make a sketch and label the various components.
2. Make a spreadsheet of the various components that make up a total computer system. List the components in column A. List the prices for each of the items in column B. Type TOTAL at the bottom of column A. Use a formula to calculate the total in column B. Does the total cost of the items exceed the cost of an assembled computer? Why?
3. Make a list of some of the programming languages available to you. Study the code from one of the programming

books you have in your school or library. Write a simple program and make it operate. If you cannot do this, try one of the programs you can find in a computer magazine.

4. Ask your teacher or a friend if you can have a useless computer chip to take apart. Using a great deal of care, see if you can remove the plastic cover and view its circuitry. What did you learn from doing this task?

5. Using a spreadsheet program, compare the cost and advantages of at least three word processing programs. Based on the items in your spreadsheet, which one of the programs would you buy? Why?

6. There are several other codes in use besides binary. Where would these codes be used? Create a chart using the same digits to compare these codes. What are some of the advantages of using different code systems?

7. Make a list of some of the numbers of the chips used in computers. What are their functions?

8. Using a desktop publishing program, create a one-page brochure that includes graphics. Use 8 1/2″ by 11″ sheets of paper for this task and design its finished size to be 8 1/2″ high by 5 1/2″ wide. Next, create an eight-page brochure using the same program and process. Fold up a mock brochure for the latter task. How are the pages laid out so that the page numbers follow in the correct order?

9. Describe how a laptop computer can be made so much smaller than one that sets on a desk. Make a list showing the cost of different laptop computers. Which one has the highest price and which has the lowest price? How would you account for this difference? Is the most expensive of these the best computer? Why or why not?

10. A computer is used to send a signal to a space satellite. If the message is sent to a person in another country on the opposite side of our planet, how long does it take the message to arrive? If the message were sent to a space probe in orbit around Mars, how long would it take for the message to arrive?

Energy Conservation

Learning Objectives

After studying this chapter, you will be able to do the following:

- Compare the differences between compact fluorescent and incandescent bulbs.
- Select the most energy efficient home appliances.
- List some of the different types of energy-efficient windows.
- Explain the concept behind the geothermal heating system.
- Describe two alternative energy sources.
- List some of the purposes of using an inverter for photovoltaics.

Technical Terms

albedo
alternative energy
anemometer
Annual Fuel Utilization Efficiency (AFUE)
compact fluorescent (CFL) bulb
energy conservation
geothermal system
green power
grid
net metering
photovoltaics
porous concrete
therm
thermopane glass
wind farms
wind turbine

Energy conservation involves saving the earth's natural resources for future generations. It is concerned with such things as how to reduce pollution, waste, and landfill overuse. It is also concerned with using alternative sources of energy that place less of a burden on our natural resources. There are conservation groups throughout the U.S. that are concerned about energy conservation.

This chapter highlights some of the efforts being taken to remedy the situation. It shows a wide variety of examples used to conserve energy. It also discusses alternative methods of generating energy.

Ways to Conserve Energy

If the demand for electricity were reduced, there would also be a reduction in the use of coal, oil, gas, and nuclear materials. Consider what it would mean to the environment to reduce the use of these materials. It would reduce the air pollution that comes from burning the material to produce electricity. It would also reduce the pollution that results from mining the material, transporting the material, removing waste, and cleaning the environment. All of this is affected by the choices we make concerning conservation.

There are some ways that people can help conserve energy by choices they make on a daily basis. Three areas in a home in which it is easy to waste energy, but also easy to conserve it, are lighting, hot water, and heating and cooling. This section discusses each of these areas, shows where waste is possible, and gives some simple ideas about how to conserve energy.

Lighting

Lighting is an ever present need. However, the need for light can easily turn into wastefulness by leaving unnecessary lights turned on or by using inefficient bulbs.

One simple way to save energy in the area of lighting is to turn off all lights that are not being used. If you leave a room and no one else is in it, turn off the light on your way out. Do not turn on every light in a room you enter unless you need them all. For example, if you are reading, turn on only the lights that help illuminate your book.

Another way to conserve energy is by using energy efficient bulbs. For example, a 28-W ***compact fluorescent (CFL) bulb*** produces about the same amount of light as a 100-W incandescent bulb. Since a CFL bulb has a lower wattage, it uses less energy. CFL bulbs also help save money because their lifespan is 10 to 13 times that of an incandescent bulb.

Incandescent bulbs are still useful. They should be used in locations where the light is turned on and off many times in a day or where the light is needed only for a short period of time. This may include closets or hallways.

Fluorescent lights are best used in areas where they will be turned on and left on for long periods of time. This may include a kitchen, a workshop, an office, or another area that needs a lot of light over a long period of time. Frequently turning fluorescent bulbs on and off shorten their lives.

There are other ways to make lighting more efficient, such as high-efficiency incandescent bulbs. Sometimes one bulb of a higher wattage can replace two of lower wattages, which can lower the total wattage used. For example, a 100-W bulb may be able to replace two 60-W bulbs, which lowers the overall energy use. Also, many people use motion sensors for the lights that are outside of their house, **Figure 24-1.** The lights turn on when

Figure 24-1. Using motion sensors for the lights that are outside of your house can help conserve energy.

only someone comes near the home instead of being on all night when they are not needed, which reduces energy waste.

Hot Water

Heating water is one of the biggest energy uses in a home. Hot water is used for many tasks, such as laundry and showers. Industrial uses include car washes, boilers in factories, and various uses in food production. The problem with heating water is that too much energy is wasted in the process. There are a number of ways that energy can be conserved and the cost reduced.

The ways to conserve hot water are simple. Only use hot water when needed, and try not to use it for longer than needed. The way hot water is used is not the only way it can be saved. The type of hot water heater can also make a big difference.

Normally, water is heated in a tank, making it ready for use. However, if a faucet is turned on the other side of the house, far away from the water heater, cold water goes down the drain until hot water arrives at the faucet. Some companies produce a product that is mounted under the sink that produces instant hot water when the faucet is opened. These units eliminate wasted energy.

A recirculating pump can also be used so that the cold water is not wasted. When hot water is turned on, but before the water flowing is hot, a recirculating pump returns the cold water to the hot water heater. When hot water flows from the faucet, the water flows down the drain in the normal manner. Adding extra insulation to the outside of a hot water heater can also conserve energy because it allows less of the heat to escape so the heater runs less frequently, **Figure 24-2.**

When choosing a hot water heater, select one with a high efficiency rating. All hot water heaters are rated, and the energy rating is attached to the heater. A water heater may be compared to other similar models. In this case, the rating compares it on a scale between the best and worst possible ratings for its class.

Figure 24-2. This hot water heater has been wrapped in insulation.

For example, the Energy Guide for the unit in **Figure 24-3** shows it will use 273 therms per year on a scale between 215 and 283. A ***therm*** is a unit of heat equal to 100,000 British thermal units. A British thermal unit is the amount of heat required to raise one pound of water one degree Fahrenheit. This unit is not very efficient when compared with others in its class.

Another scale that may be used is the ***Annual Fuel Utilization Efficiency (AFUE)*** rating scale. In this case, higher ratings are better. A unit that scores 94.3 out of a possible 97.6 is very efficient. You may see other rating scales for energy savings on different items found in your home.

There are energy efficient models of most appliances, such as ranges, dishwashers, and clothes dryers. Since these are large appliances

Figure 24-3. According to the Energy Guide, this unit is a high-energy user.

that are used frequently, the energy savings potential is high. Sometimes the initial cost of an energy efficient model is higher than a less efficient model. However, if the cost over time to run each of these models is calculated, it may be discovered that the energy efficient model is much less expensive. Form a habit of choosing energy efficient appliances.

Heating and Cooling

Many people use gas or electric heating and cooling systems to keep their home warm in the winter and cool in the summer. To conserve energy, people can dress for the season, wearing sweaters in the winter so the heat can be set to a cooler temperature and lightweight clothing in the summer so the air conditioning can be set to a warmer temperature.

Like hot water heaters, air conditioning units and furnaces are rated for efficiency. The Seasonal Energy Efficiency Ratio (SEER) provides a rating for air conditioning equipment, **Figure 24-4.** This rating is normally between 10 and 16 for most systems, with higher ratings signifying higher efficiency. Again, energy efficient systems may cost more, but they cost less to operate. When making a decision, compare the extra purchase cost with the annual savings on energy bills.

A ***geothermal system*** is an alternative to the traditional heating and cooling system. This system uses a series of holes between 100′ and 400′ deep near the house. Coils, called a *ground loop,* are buried in the holes. This allows the coils to access and use the temperature found deep below the earth's surface, which unlike the air temperature, is nearly constant throughout the year. See

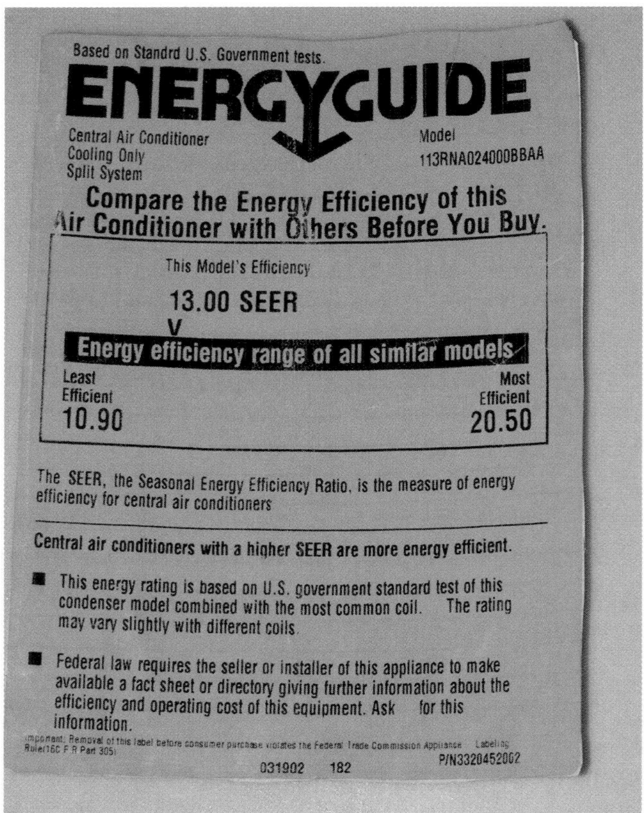

Figure 24-4. This unit uses the SEER rating scale. As you can see, it is a less-efficient model.

Figure 24-5. The fluid in the coils usually contains a mixture of antifreeze and water. Inside the house, the pipes from the coils connect to a heat exchanger.

In the summer, the fluid passes through a heat exchanger where it absorbs heat from inside the house. The temperature inside the house might be around 78° F or higher. The fluid cools in the ground loop where the temperature is around 60° F and then returns to the house. The fluid then passes through the heat exchanger where it again absorbs heat from inside the house. This process repeats itself during the hot days of summer.

In the winter, the outside temperature might range from 20° F to 30° F. However, the temperature deep below the ground is still around 60° F. The cool fluid from inside the house flows through the ground loop, picking up heat. The heated fluid flows to the heat exchanger where the heat is removed from the fluid to heat the inside of the house. This process repeats itself during the cold winter days.

Construction Techniques

Many energy saving systems involve the manner in which a house is built. While it is possible to add some energy saving systems after a building is completed, it is best to install them during construction.

There are many different potential conservation factors when building. Choosing the proper insulation, windows, and roof can all affect the conservation of energy in heating and cooling a building.

Insulation

Many of the ways to conserve energy revolve around the use of insulation. The best results are obtained if the insulation is installed as the house is being constructed.

House wrap is a barrier that goes directly underneath the exterior siding of a home, **Figure 24-6.** One of its functions is to act as a windbreaker. It prevents drafts from entering the home, which helps save energy and lowers heating and cooling bills.

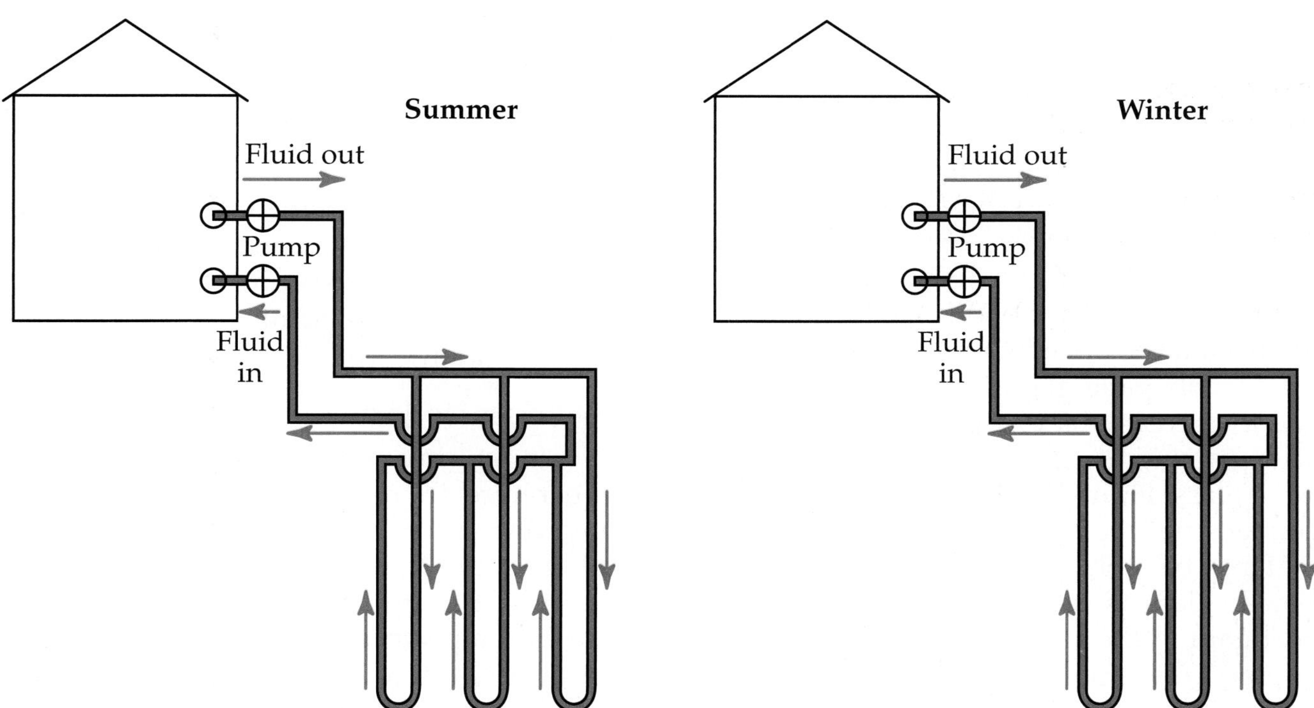

Figure 24-5. A geothermal system relies on the constant temperature found deep below the earth's surface.

Figure 24-6. This house has been wrapped and is ready for siding.

Figure 24-7. Cutaway view of a triple-pane window.

Insulation goes between the exterior wall and the interior wall. It keeps cold air out and heat in during the winter and keeps hot air out and cool air in during the summer. This conserves energy because the furnace and air conditioner run less often. It also keeps heating and cooling costs down.

Windows

Windows are one of the ways energy is easily lost. Manufacturers have designed windows to help reduce this loss. It has been estimated that poor windows can lose between 40% and 50% of the heating and cooling energy in the home. Aluminum is a good conductor. Therefore, it transfers heat out the house in the winter and allows cold air to make the windows cold inside the house. Some wooden framed windows are poorly caulked and leak air around the outside frame.

Double-pane glass is filled with an inert gas to increase its energy efficiency. This is also known as ***thermopane glass.*** Triple-pane glass, **Figure 24-7,** is used in the coldest climates, but because of their heavy weight, the seal may be broken causing a loss of efficiency. Low-E glass is used in windows to reduce heat gain in the summer and heat loss in the winter. This is possible because of the special coating that is applied to the glass. Many large office buildings use this type of glass for their exteriors, **Figure 24-8.** In addition, some have glass that is actually being used as solar cells. This concept is patented and is used for other applications, such as keeping frost off freezer windows and stopping night glare on automobile windows.

Figure 24-8. This glass building is located just outside of Minneapolis, Minnesota.

Building materials

Infrared imagery shows an increase in temperature in many areas of the country, especially near malls and large buildings. The buildings and the large blacktop parking lots absorb heat from the sun. These heat islands raise the temperature of the surrounding areas. The areas in and around these heat islands experience higher air conditioning costs.

Consequently, people have been looking for ways to reduce the heat, and thus reduce the necessary cooling of these areas. One idea is to increase the ***albedo*** (reflective ability) of the surfaces so they do not absorb as much sunlight. Increasing the albedo means a possible reduction in the peak electrical demand of the surrounding area and in cooling costs. One way to raise the albedo is by changing the roof to lighter-colored and more reflective materials.

Changing the material of parking lots could help cool the parking lot surfaces. One material with this potential is called ***porous concrete,*** **Figure 24-9.** Normal concrete and blacktop surfaces force rainwater away from the area they cover rather than into the ground. Porous concrete provides a higher void content and allows water to seep into the ground. The water can help keep the area cool. This material is designed for low-traffic areas, such as driveways and parking lots, **Figure 24-10.** Porous concrete is also lighter-colored than blacktop, which increases the albedo and reduces the heat buildup.

Figure 24-9. The holes in porous concrete allow water to run down into the ground underneath the concrete.

Figure 24-10. A porous concrete driveway in Atlanta, Georgia.

Electricity and Vehicles

Many companies have experimented with electricity-powered transportation. Some cities have made all of their public bus transportation electrical. Other cities have converted their buses to run on natural gas. Both of these plans help reduce air pollution in the city.

The most common type of automobile that is powered, at least partially, by electricity is the hybrid automobile, **Figure 24-11.** A hybrid vehicle still has a gas engine, but it also has a large capacity battery (over 200 V) that can run the vehicle. The gasoline engine, regenerative braking, and a very large generator are used to keep the battery charged.

All-electric vehicles are still very uncommon for the road; although, they are very common on golf courses, as golf carts are electric. The battery of an electric vehicle needs to be charged. The time for charging can be inconvenient for commuters, but it is

Figure 24-11. Hybrid vehicles help conserve gasoline.

easy for golf courses to recharge their golf cart batteries each night. Some vehicle charging stations are showing up at malls and at a limited number of service stations.

Some companies have been investigating electric transportation for a single person. One example of this is the Segway®, **Figure 24-12.** A Segway® uses batteries for power and gyroscopes for balance. Some police officers and some security officers in airports and on college campuses are using these to move about quickly without becoming tired. The average speed of most models is over 12 mph.

Figure 24-12. The Segway® balances itself and moves forward or backward by the rider leaning slightly in the desired direction.

Alternative Energy

Alternative energy is energy derived from a source that does not deplete the earth's natural resources or harm the environment. The use of two leading sources of alternative energy has evolved over the last few decades. One source is the sun and other is the wind. Photovoltaics are used to harness the Sun's energy, and turbines are used to harness the wind's energy. This section discusses these systems.

Photovoltaics

The total amount of energy the Earth receives from the Sun is estimated to be over 1,500,000,000,000,000,000 kWh per year. That is many thousand times more than all the energy consumed by all the countries on this planet. The Sun represents a tremendous reserve of energy, surpassing that of all the Earth's energy sources.

The use of solar energy to produce electricity is called ***photovoltaics.*** Photovoltaic (PV) cells, or solar cells, convert sunlight directly into electrical energy. This technology was initially developed in the 1950s at Bell Laboratories. Sunlight causes the electrons to move about in the solar cell. If the solar cell is connected to a circuit, an electrical current is produced.

Since there are no moving parts, these units are very reliable, operate silently, and have an expected life of 20 to 30 years. Since the sun serves as the energy source, there is no pollution.

When a number of the solar cells are connected together they form a photovoltaic module, **Figure 24-13.** Depending on the number of solar cells used, PV modules are sized by the number of watts they produce. A PV module that is roughly 20″ by 20″ might produce 1.63 A at 17.1 V and would thus be rated at 27.9 W (P = EI).

A larger size might vary the number of solar cells and be able to produce 190 W. A 1 kWp (kilowatt peak output) module

Figure 24-13. Photovoltaic modules.

would take up a space of 10 square meters. This unit would be expected to generate roughly 1000 kWh in a year. The size of the unit selected for a particular task is based on the power needed. Some PV modules are large enough to cover the entire roof of a building, **Figure 24-14.** Others are large enough to produce over one million watts of electricity.

Figure 24-14. This roof is being installed with photovoltaic roofing shingles. (Southface)

Photovoltaic systems

PV systems produce dc. Small PV systems can be used to charge batteries or run a small piece of electrical equipment. Larger systems use an inverter to change their output to ac and connect them to the grid. The ***grid*** is a wiring network that is spread across most of the United States. In the event of an electrical outage in one area, electrical power can be provided from a substation from another location. The grid provides ac electrical power, while the PV system produces dc. Therefore, an inverter is needed to convert dc power to ac. The inverter also performs the following tasks:

- Synchronizes the grid to the PV system.
- Turns on and connects to the grid after sunrise.
- Matches the frequency of the PV system and the grid.
- Monitors the voltage and current generated.
- Turns off and disconnects from the grid around sunset.
- Collects and saves energy data.
- Provides a safety disconnect during outage from storms.

When a PV system is used to supply energy to the grid, it is a two-way system. When there is a surplus of energy in the PV system, the electricity is fed into the grid. When there is a shortage of energy in the PV system, the grid feeds it electricity. If a storm causes an outage on the grid, people repairing the electrical lines could be shocked from the PV system unless it is automatically disconnected. A PV system that performs all of these functions is called a *line-tied* or *grid-connected installation*.

If a PV system is installed on a home, the owner can sell electricity to the local power company. Any excess photovoltaic energy that is not needed to run the home would be fed into the power grid. This spins the home's electric meter backward. The excess energy provides a credit against the electricity from the power company that is used in the home. This is known as ***net metering.***

However, the homeowner may choose not to sell the electricity to the power company. Stand-alone PV units provide the owner with an off-the-grid power supply that is independent from the power company.

Uses and examples

PV systems can be used in many different applications. They are ideal for areas where it is difficult or impossible to run power lines, **Figure 24-15.** For example, farmers may use photovoltaics to power electric fences in pasture areas. Some mountain homes or other isolated vacation areas use PV systems for fans, computers, lights, **Figure 24-16,** and radios. Bigger PV systems can be used to operate appliances such as furnace fans, lights, refrigerators, and security systems.

The portability of PV systems gives them another advantage and opens new possibilities for their uses. Many smaller portable devices can run using photovoltaics. There have been cell phones, calculators, watches, games, or toys that have been designed to run on solar power. PV systems can be used to run devices on boats and recreational vehicles so that those on board do not have to rely on a generator during the day.

Photovoltaic locations

Photovoltaics are used in more than a dozen countries around the world. Since photovoltaics depend on sunlight, it can be used just about anywhere on this planet. At the present time, there are more than 300 PV systems in the world. Germany has more than any other country. A useful source for the most current information can be found on the Internet at www.pvresources.com. This site provides a list of the top 300 PV systems in the world. Updates are provided as new plants start producing.

Photovoltaic devices are most cost efficient in areas where the sun shines most of each day. This includes places where the weather is cold. Locations with the most hours of sunlight provide a higher savings potential. However, in cloudy conditions, PV systems still function but at a lower rate.

Figure 24-15. Crossing gates can be powered by photovoltaic systems.

Figure 24-16. The Arizona Senora Desert Museum outside Tucson operates lights using photovoltaics.

Incentives and rebates

A majority of the states in the U.S. offer some type of incentive to residential customers and businesses that install energy saving devices, including PV systems. Power companies are offering these incentives to decrease the demand on existing power generating equipment.

To qualify for the incentive, the PV system must be installed by a contractor that is on the power company's list of approved contractors. After the system is installed, the contractor tells the power company, and the owner is issued his or her reward.

Incentives may be a certain amount of money per watt for PV systems. They could also be a certain amount per square foot for solar collectors serving as a source for heating water. The incentives and rebates for each state can be found in the Database of State Incentives for Renewables and Efficiency on the Web site www.dsireusa.org.

Wind Turbines

Wind energy has been used for centuries to move ships across lakes and oceans. During the 19th century, windmills were used in the U.S. to pump water for humans and cattle. It was during the oil shortages of the late 20th century that wind energy began to be used as a method of generating electricity. Today, wind turbines are an alternative way of generating electricity.

A ***wind turbine***, or wind generator, consists of a tower, spinning blades, and a generator. The wind turbine in **Figure 24-17** is 256′ tall, and the length of each blade is

Figure 24-17. Each blade on the wind turbine is about the height of a seven-story building.

158′. The total height of the structure is taller than a thirty-story building. The towers are tall because as altitude increases, wind speed increases. Taller towers can capture the higher wind speeds and generate more electricity.

The blades are connected to the generator by a gearbox and shaft. The generator is usually mounted behind the area where the blades meet, **Figure 24-18.** Recall that a generator turns mechanical energy into electrical energy. The blades of the wind turbine turn as the wind hits them, spinning the armature inside the generator. Generators are most efficient when the wind speeds are over 8 miles per hour (mph).

More windy days and higher wind speeds mean that more electricity can be generated. However, at very high wind speeds (over 65 mph), the wind turbine must shut itself down so it is not destroyed. Wind speed is measured with an ***anemometer,*** which is typically located above the generator. When the speed gets too high, the control system applies the brake. High wind speeds were a problem with some of the early designs. In fact, some towers were destroyed when the high wind caused the blades to deflect and actually strike the tower supporting the system.

Figure 24-18. The generator sits behind the rotors.

Wind turbine locations

The amount of wind energy that is available for a given location changes with the seasons and weather conditions. However, there are certain areas in North America where wind turbines are effectively used because the average amount of wind is enough to power them.

Many states now use wind turbines to supply part of their electricity. States where turbines are found include California, Colorado, Iowa, Minnesota, New Mexico, Oregon, Texas, Washington, and Wyoming. You may see a large number of wind turbines grouped together in one location, **Figure 24-19.** Those areas have the favorable wind conditions needed to generate electricity. There are large spaces between the wind turbines because the rotating blades change the airflow to a turbulent pattern for a short distance. If the next turbine is too close, it will be less effective in generating electricity. Proper spacing and distance between each unit is therefore necessary to allow the wind to recover its normal pattern and for the turbine to be effective. These groups of wind turbines are called ***wind farms*** or ***wind plants.***

Wind turbine costs

The cost of the wind farms includes acquiring the land, laying a foundation, building the tower, assembling the parts, and maintaining the equipment. Many times a developer will buy or lease land from a farmer or other landowner on which to build the wind turbines. The landowner may receive a one-time payment, a percentage based on the wattage produced, or a fixed amount per year.

The electricity produced by wind turbines is not usually fed directly into a home. DC is produced as the blades turn. If a home is in a remote area, its electricity may be produced by a dc wind turbine. For most wind turbines, however, dc is fed into an inverter and converted to ac. The ac is then used on-site or fed into the electrical grid. The landowners

Figure 24-19. Lake Benton, Minnesota wind farm is built along a windy ridge.

have a contract with the power company, and they may be credited for the electricity the turbines on their property put into the grid.

Advantages and disadvantages

There are some sources that say we could generate over 10 trillion kWh of electricity every year with wind turbines alone. That is more electricity than is used in the entire United States. The major problems with wind turbines, however, are their cost and the fact that the wind does not blow all the time. The wind cannot be forced to match the energy demand and will not always blow the most at the time of day when demand for electricity is highest. In addition, some consider the wind turbines eyesores and do not want them built in their community. Others are concerned because the blades on the wind turbines can kill birds that fly into them.

The major advantage of wind turbines over coal, oil, or other fuel is their ability to produce electricity without pollution. As coal and other fuels are burned, they produce carbon dioxide, nitrogen oxide, and sulfur dioxide, which contribute to air pollution and other environmental problems. Wind energy allows for the conservation of other fuels and does not pollute the air.

If you examine your electric bill, you will see that the utility companies give their customers a choice about using green power. The term ***green power*** describes electricity produced from renewable energy sources. Since green power generally costs more to produce, customers can voluntarily pay a premium of two to three cents more per kilowatt. This provides the incentive for the power companies to use renewable sources.

Another favorable feature of wind energy is that the wind turbines do not use much space. The land very near the towers can continue to be used by the landowners. Farmers can continue planting and harvesting their crops on all their land, except for a small section around each tower. This is not true of other types of power generation sources, such as power plants.

Additional Research

Another area of research for alternative energy sources being examined is tidal currents and tidal bores. Tides usually occur two times every day. That means the water moves out twice and moves back in twice each day. A tidal bore is a rapid increase in the leading edge of water returning up a river because of tidal action. Tides can be as high as 20′ or more from low to high tide. Tidal bores have been found to reach 30′ high in some rivers in China.

There are companies that have mapped the tides all around the world at different times of the year. These companies have begun installing generating equipment at specific locations where the tides can produce the most electricity. If you examine this area of research, you can determine exactly where these locations are found.

Summary

- Some ways to conserve energy in the area of lighting are turning on lights that are necessary and turning them off when they are no longer needed and using energy efficient bulbs.
- Fluorescent bulbs can provide the same amount of light at a lower wattage than incandescent bulbs.
- Energy efficient appliances can be chosen by looking at their rating and finding the model with the best rating.
- Some energy efficiency rating scales show how much energy an appliance uses compared to others in its class; others use an objective scale that rates how efficient an appliance is according to some criteria.
- Buildings are insulated to keep the outside air from affecting the temperature of the air inside the building.
- Double pane and triple pane are energy-efficient types of windows.
- A geothermal heating system extends far underground and uses the steady temperature of the earth to keep a home warm in the winter and cool in the summer.
- Solar energy and wind energy are two alternative resources to oil, coal, and other nonrenewable resources.
- The use of solar energy to produce electricity is called *photovoltaics*.
- Photovoltaic cells convert solar energy into electrical energy.
- Some of the functions of an inverter in a photovoltaic system include turning the system on after sunrise, monitoring the electricity produced, and turning the system off after sunset.
- Most states offer incentives and rebates to people and businesses that install energy-saving devices.
- Wind turbines are very tall, which allows them to catch the faster wind speeds found at higher altitudes.
- The wind turbine produces electrical energy.
- When the wind blows, it turns the blades on a wind turbine, which is attached to the armature of the generator.
- Developers who are building wind turbines build in areas with high winds and a lot of windy days.

Test Your Knowledge

Do not write in this book. Write your answers on a separate sheet of paper.

1. ______ lights are best used in areas where they will be turned on and left on for long periods of time.
2. While waiting for water to get hot, a(n) ______ can be used to return cold water to the water heater.
3. Which water heater is more efficient: one with an energy rating of 240 therms or one with a rating of 265 therms?
4. For what purpose is SEER used?
5. Which air conditioning unit is more efficient: one rated 12 or one rated 15?
6. Describe the basic concept of geothermal system operation.
7. What type of glass is used in windows to reduce heat gain in the summer and heat loss in the winter?
8. How can the albedo of a roof be increased?
9. What are two benefits of porous concrete?
10. What is the main function of a photovoltaic cell?
11. List at least four functions of an inverter used to tie a photovoltaic or wind generating system to the grid.
12. Why do taller wind turbines generate more electricity than shorter wind turbines?
13. What is done to prevent wind turbines from destroying themselves at very high speeds?
14. Why are large spaces left between wind turbines?

Activities

1. See if you can locate a demonstration home near you. Pay them a visit and list how many different energy saving devices are shown.
2. What is the latitude where you live? Are there any solar collectors mounted on buildings near you? List their locations. At what angle to the Sun are the collectors mounted? Why was that angle selected? On what side of the building are they located (North, South, East, or West)? How much water can they heat with their system?
3. Visit a house under construction near where you live. Are they using a house wrap? Which type is it? How much energy is it designed to save? How was it installed? Why?
4. Is there any location near you that is using porous concrete? What color is it? Why was that color chosen? Does porous concrete cost more or less than regular concrete? Explain why.
5. Visit a Web site that provides infrared imagery. List the Web site(s) that provided you with the information. Are they in color or black and white? What is the value of using this technology? Does your local power company provide any services to you that use infrared imagery? Is there a charge for this service? Why or why not?
6. Visit your local hardware store or home center. List the prices of their CFLs. Compare these prices with comparable incandescent bulbs. What is the cost of electricity in your town? How many years would it take to break even using the various CFLs?

7. Are there any buildings in your area that have a sky tunnel? Where are they, and why do they use them? What would they cost to purchase and install? Does weather have any effect on how they are used?
8. Visit an office building that was built with a lot of exterior glass. Compare the inside and outside temperature on the day of your visit. Is the inside temperature comfortable? What was done in the building to accomplish the comfort level? Compare the cost, matching one of the windows of that building with one used inside your house.
9. Make a list of the ways you could save energy in your house, town, or school.
10. Are there any clubs or conservation groups in your town? Have they started any projects? List them.
11. How can photovoltaics continue to produce electricity even on a cloudy day? How much electricity do you use at your house each month? What size of photovoltaic system would it take to operate your house? How much would such a system cost?
12. Search the Internet to see if there are any PV systems in your area of the country. How much electricity is being generated by this source? Are there any systems in nearby locations? How are they being used?
13. Find a Web site that will take you through the history of photovoltaics. How many years did it take to develop? When was it economical enough to actually be used as an electrical source? How much has the cost dropped over the last 10 years?
14. See if you can make a visit to a site for alternative production of electricity. Do not include a visit to a coal or nuclear facility. Where is the nearest site? What are they doing? How long has it been in operation?
15. Do you have any green power notices on your electrical bill? What do they want you to do? Is there an extra cost for using green power? How much? What does that mean in terms of added cost over a one-year period? Why is it important to participate in such a program?
16. Does your state provide an incentive program for solar power? How much is the incentive and what would it mean to your electrical bill? Would you ever recover the cost of such a program? How many years would it take to break even?

Career Opportunities

Learning Objectives

After studying this chapter, you will be able to do the following:

- List five areas of employment opportunity in the electricity and electronics field.
- Describe the training requirements for a number of electrical- and electronics-related positions.
- Explain the differences between the responsibilities of different electrical and electronics work assignments.
- Discuss what job you feel most closely matches your aptitude.

Technical Terms

assembler
chief electrical inspectors
engineers
entry-level
journeyman
maintenance worker
quality control inspector
skilled worker
supervisor
technician

The field of electricity and electronics offers many exciting possibilities. People with an education in electricity and electronics can offer repair services, engineer new and better products, and teach in a school.

Determining Your Career Path

Early in your schoolwork, you should begin the process of identifying your career path. Career counselors can help you identify the things you like to do and do well. Once you find areas that interest you and that you think you may want to pursue, choose classes in these subjects. Each year of schooling can help narrow your selection.

As you pursue these career options, do not worry if you are not sure what career is right for you. Enjoy learning more about the topics that interest you, and continue to enroll in classes that will help you should you decide to make that area of study your career.

Consider what you truly like to do and what would suit your personality. For example, a person who hates to be stuck in a building should explore careers where the majority of the work is performed outside, **Figure 25-1.** People who prefer not to be out in the sun should examine careers that keep them indoors the majority of the time.

Figure 25-1. When considering a career, examine your personality. For example, are you the kind of person who likes to be inside sitting at a workbench or outside moving about in the fresh air?

Selecting Courses

As you choose your courses in an area of interest, be sure to also enroll in classes that complement your main subject. Select courses that will move you toward your goal and provide a solid foundation for the more advanced knowledge that you will learn later.

For example, an engineer must have a solid understanding not only of engineering and design, but also of math and physics, **Figure 25-2.** If you want to be an engineer, you cannot do this without taking math and science classes. You should sign up for pre-engineering and technology classes, and you should also consider taking as much math as you can. Some careers do not require advanced classes. However, to best prepare yourself for the future, it is a good idea to take more difficult classes. In other words, when

Figure 25-2. When selecting courses, be sure to take courses that will give you a solid foundation for the more advanced knowledge you will need for your career. For example, an engineer must have a solid understanding of math and physics in order to learn the skills needed to become an engineer.

selecting classes, keep your potential career path in mind and choose those courses that build toward your chosen career.

School Performance

How you perform in school today can affect your future. Work hard and do your best so that you are provided with the best possible options when you are ready to take the next step, whether that step is entrance to advanced placement classes or entrance to college. Striving to do your best also helps turn working hard into a habit. This habit is part of what can make you a successful student and a good employee with notable potential for advancement.

As you complete your school assignments, try to explain the concepts to others who need help, **Figure 25-3.** This process will sharpen two things: your understanding and your communication skills. As you explain something to others, your understanding of what you have learned deepens. Helping others in this way also enhances your communication skills as you search for new, easily understood ways to explain something. Employers want to hire people who have solid communication skills.

Figure 25-3. Always take the opportunity to explain the concepts you have learned to any classmates who are having problems understanding them. This will help you to better learn the concept and will enhance your communication skills.

Five Areas of Opportunity

Jobs in electricity and electronics can be categorized by skill levels. Listed in order of skill, they are entry-level, skilled worker, technician, journeyman, and professional. It is a good idea to begin studying electricity and electronics early. To work in more advanced positions, one must have a solid foundation of electricity and electronics knowledge.

Entry-Level

Entry-level jobs include those that can be learned within a few hours and up to several weeks of instruction. This skill area includes assemblers who wire and solder components for industrial controls and consumer products, **Figure 25-4.** An ***assembler*** must be able to identify many electronic components and install them into circuits. Other jobs in this area include checkers, sorters, inventory clerks, and other task-oriented positions.

A person in an entry-level job starts working for a specified amount of pay. Entry-level jobs allow you to gain experience and prepare you for higher paying positions. As you gain experience, you must start your preparation for better jobs by studying and learning as much as you can. The better your education and skill, the better your chances are for promotion.

Skilled Worker

A ***skilled worker*** acquires skills over a period of several months to several years. Many of his or hers skills are learned by reading manuals, attending workshops and technical training sessions, learning from other skilled workers, and on-the-job training.

These types of workers include ***quality control inspectors*** who check the finished work of others, **Figure 25-5.** They put wired circuits and other products made by entry-level employees through a series of tests.

Figure 25-4. An assembler is an entry-level job that requires an individual be able to identify many electronic components and install them into circuits.

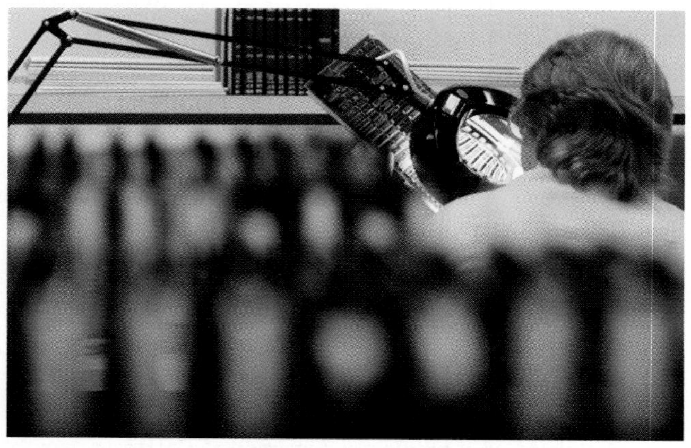

Figure 25-5. This quality control inspector is examining the solder joints on a circuit board.

Once products have been built, they must be maintained. While some products are simple enough for consumers to maintain, many products require more knowledge of electrical circuits and product operation than the average person possesses. In these cases, it is necessary to have a ***maintenance worker*** who specializes in the device or appliance. Maintenance workers are the service workers who fix many appliances such as dishwashers, ranges, washing machines, microwave ovens, and air conditioning units. Those doing maintenance work must know how to work with electrical circuits, read schematics, troubleshoot many appliances, and make the necessary repairs.

Technician

Technicians acquire their skills over a period of several years. They acquire their knowledge from two-year technical college programs, working with other technicians, attending workshops, assisting engineers and other professionals, and personal experience.

Technicians are needed to keep TV and radio stations broadcasting. They keep the transmitters working day and night. Other groups in communications fix aircraft communications systems and the telephone and microwave equipment that most of us rarely see.

A ***technician*** is trained to work with existing equipment. They may also work with engineers, helping in the design and new innovations in their field, **Figure 25-6.** To do this, technicians must be able to work with their hands and mind and continue to keep up to date. They must understand each new component discovered and built by the electronics corporations because they may have to use them in new machine designs. The new components may make it possible for a technician to design faster and more accurate machines. Millions of dollars are saved by companies whose technicians can make their production operations faster and more accurate. Technicians may be found working in the following places:

- Sound systems design departments.
- Automotive electronics design departments.
- Laboratories.

These are only a few of the hundreds of places that need the help of skilled technicians. Some companies need between three and ten technicians for every engineer they employ.

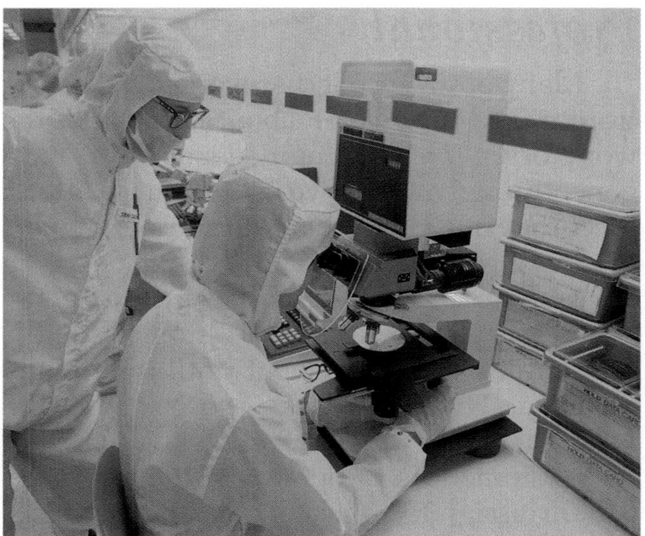

Figure 25-6. This technician (seated) is working with an engineer (standing) to develop a new product.

Figure 25-7. This journeyman is wiring a building according to the National Electrical Code.

Journeyman

A ***journeyman*** is a person who has completed an apprenticeship program. This qualified electrician wires buildings according to local safety codes and the National Electrical Code, **Figure 25-7.** He or she may work for a city as a ***chief electrical inspector*** who goes to job sites to inspect and eventually approve the wiring.

Journeyman programs usually consist of four years of experience as an apprentice. During this time, they gain on-the-job experience together with attending two year college training programs, special classes, or self-study courses. Journeyman programs have a checklist of skills and tasks the apprentice has to complete. At the end of training, the apprentice is granted journeyman electrician status and is given a journeyman card. This status gives an electrician an advantage when job hunting. Every company knows that this job hunter has had at least four years of electrical training.

Professional

Many of the recent discoveries in the electronics field have come about because of the work of ***engineers.*** Engineers usually have gone through a four-year college program. Since they must have this education, engineers get more responsibility and higher pay than the first three job opportunities mentioned.

Engineers design and supervise the building of new equipment until it is finally installed. They must make sure it will run with the least amount of problems. Once the equipment is in operation, the engineers usually assign technicians to keep it running. However, if a major problem should occur after installation, the engineers are called back. Then, they must redesign or find a better way of doing that operation, **Figure 25-8.**

Figure 25-8. These engineers are going over their design to improve its operation.

People who go into engineering must have strong math and science skills. This is necessary because so much of the work an engineer does is abstract. They have to be able to visualize things before putting their designs on paper. Engineers must have a good understanding of all the principles of physics, electronics, and English. Can you imagine what would happen if an engineer was not able to write a report on a new discovery so that people could understand it? Even the smartest engineer would be of little value to the company if no one understood what the engineer had written.

Probably the highest paying job in the field of electricity and electronics is that of a ***supervisor.*** A supervisor must have knowledge and skills related to construction, maintenance, technology, and engineering.

The supervisor has the added responsibility of making sure that others do their jobs right. Most supervisors have handled a variety of work assignments before they were supervisors. Only in this way can a supervisor gain the skill and knowledge needed to supervise efficiently.

Supervisors also gain the skill needed through study. They read technical and management reports and keep up to date on new discoveries. They attend conferences, national meetings, and seminars. Supervisors have to write hundreds of reports each year, and they must be skilled in handling people.

Once new equipment is built and installed, more job opportunities are present. People who have the needed skills are the ones who will get the best jobs. These skills and choice jobs will not be handed to you. You must work hard to earn them.

Seeking Employment

Once you have begun the process of learning the required skills for your chosen career, start looking at local employers, **Figure 25-9.** Do they hire part-time people? Do they have temporary summer positions where you could

Figure 25-9. Searching online is one of the ways to locate and find more about companies you would like to work for.

see what it is like working in the field? Some businesses hire apprentices who are still in high school. This is an excellent way to further investigate a career. Try to work in your chosen field as early as possible.

Next, start making a list of the companies in which you are interested. You can look online or in the phone book. If you do not have access to these, you can get help at your local library. Libraries offer computers you can use for your job search and books that contain information about local businesses. Get as much information about these companies as possible before you apply for a job.

Applying for the Job

Once you are ready to seek employment, prepare a résumé and apply for the job. Obtain the name of the person responsible for hiring (this is usually the human resources director) and personalize your cover letter to make a good impression. Make sure that your spelling and grammar are correct, and ask someone else to check these documents for you in case you overlooked an error. In your cover letter, be sure to tell the company what type of employment you are seeking, then stress your strengths that show why you would be a good candidate for the position.

Depending on where the company is located, either mail your résumé and cover letter or deliver it in person. If you are responding to an online advertisement, you may e-mail your résumé and cover letter or submit them on the company's Web site. Many companies tell job seekers how they would like to receive résumés. If the company you are considering has directions like this, be sure to send your documents according to the company's preferences. Many companies also accept faxed résumés and will provide a fax number.

Preparing for the Interview

Study the information that you collected about the company before you go for an interview. Learn as much as possible ahead of time and prepare a list of questions. You may want to ask about the company's scheduled working hours, medical and dental benefit programs, or educational reimbursement programs, among many other items. Make a list of these questions in the order of their importance to you and take the list with you to your interview.

When you go to the interview, dress professionally, even if you are not seeking an office job. Wear clean, neatly-pressed clothing and freshly-polished shoes. Remember that you only have one chance for a first impression, so make it a good one. Soon after the interview, send the company a short note thanking them for granting you an interview.

Accepting or Rejecting the Employment Offer

If the company offers you a job, you will usually be given some time to consider the offer. If you wish to accept the offer, let them know as soon as possible. Waiting too long to accept may give your future employer the impression that you are not interested.

If you decide not to accept it, be sure to notify the employers in a timely manner. Be courteous and give them reasons that you are not accepting their offer. If it is the pay, hours, or that you have taken other employment, let them know. As the saying goes, "Don't burn your bridges behind you." At some future date you may decide to pursue employment with them again.

Work Performance

Employers want to hire workers who have basic skills in reading, writing, and mathematics. They also expect their workers to have other traits that make them good employees. Following are some of these traits:

- Punctual.
- Dependable.
- Honest.
- Responsible.
- Productive.
- Respects authority and other employees.
- Does not drink alcohol prior to or during work.
- Does not use illegal drugs and willingly participates in drug-screening program.
- Does not engage in offensive behavior.
- Practices safe working habits in the interest of himself or herself and coworkers.

You will continue to learn new skills and knowledge as you gain more and more experience. Your willingness to develop these skills can often determine whether or not you will receive opportunities for advancement. Areas that affect advancement include the following:

- Willingness to work as a team member.
- Ability to improve skills and develop new ones.
- Problem-solving abilities.
- Ability to communicate with supervisors, coworkers, clients, and customers.
- Willingness to share knowledge and assist others when possible.

Safety

Working safely cannot be overly stressed. People take safety for granted far too often. If you have a job that includes working with electricity, you must protect yourself and others from coming into contact with live electrical circuits.

No one wants to work with or be around unsafe workers. They are a danger to themselves and to those working nearby. Companies do not want to employ unsafe workers for the sake of other employees and for liability reasons. Promotions do not usually come to unsafe workers, and those who do not follow company safety guidelines often do not keep their jobs. Protect yourself at all times in order to keep your job and your life.

Changing Employers

As you gain knowledge and experience, you may choose to change your employer if you are seeking advancement. Before leaving your place of employment, however, you should do some research. New opportunities can be found on the Internet, in newspapers, magazines, and trade journals, or at employment agencies.

It is customary to provide your employer with at least two weeks notice that you are leaving. A new employer should understand this, as they would expect the same courtesy. Prospective employers will also respect your right to privacy and will not contact a present employer unless you give them permission to do so.

You should always be courteous and respectful to your present employer. Their reference can be very important when you are changing jobs.

Continuing Education

One way to assist your career and help it advance is to continue learning. Since technology changes rapidly, someone working in the electrical or electronics field always has new things to study.

While continued learning can help your career, stopping the process of learning can end it. In rapidly changing fields, knowledge quickly becomes outdated and useless, and someone whose knowledge does not extend beyond old technology will not be hired to work with cutting-edge technology. Soon, someone with only old knowledge may not be hired at all.

Planning for Retirement

As soon as you begin working, it is a good idea to begin a long-term financial saving plan. Starting a retirement fund when you are young and investing a small portion of your income can result in a comfortable and secure retirement. There are many professional investors and bankers who can help you find and put in place the best plan for you.

The use of electricity in the world raises the need for people who know how to work with it. If you want job security, continue your study of electricity and electronics and make a job in this field work for you.

Summary

- Career planning can begin early in your life.
- Start choosing classes based on what subjects you may be interested in pursuing as a career.
- How you work in school now can later affect your life and your career.
- Use entry-level jobs to prepare yourself for advancement to a better position.
- The more training and skill you acquire, the better your chances to move into positions such as skilled worker and technician.
- To acquire a journeyman card, you have to go through a four-year apprenticeship program.
- When applying for a job, research the company first so you can write a cover letter that addresses the needs and interests of that company.
- Dress professionally for all job interviews.
- Once you have a job, continue learning about the field so your knowledge stays up-to-date.
- Start a long-term financial saving plan when you start your first job.

Test Your Knowledge

Do not write in this book. Write your answers on a separate sheet of paper.

1. Career planning should begin ______.
   a. after you receive your college degree
   b. as early as junior high
   c. when you are eighteen
   d. None of the above.
2. List the five general areas of career opportunities in the electricity and electronics field.
3. Chief electrical inspectors are sent out by the ______ to approve wiring before it is put into service.
4. ______ must be able to identify many electronic components and install them into circuits.
   a. Quality control inspectors
   b. Assemblers
   c. Construction workers
   d. Maintenance workers
5. A journeyman has to go through a(n) ______ program.

6. All building wiring is done according to the ______.
7. List four of the traits that are important for an employee to possess.
8. When leaving a place of employment, it is customary to give ______ notice.
   a. one week
   b. two weeks
   c. four weeks
   d. three days
9. *True or False?* Once you are finished with school, you will never have to take any courses.
10. Career advancement can often depend on your willingness to ______.

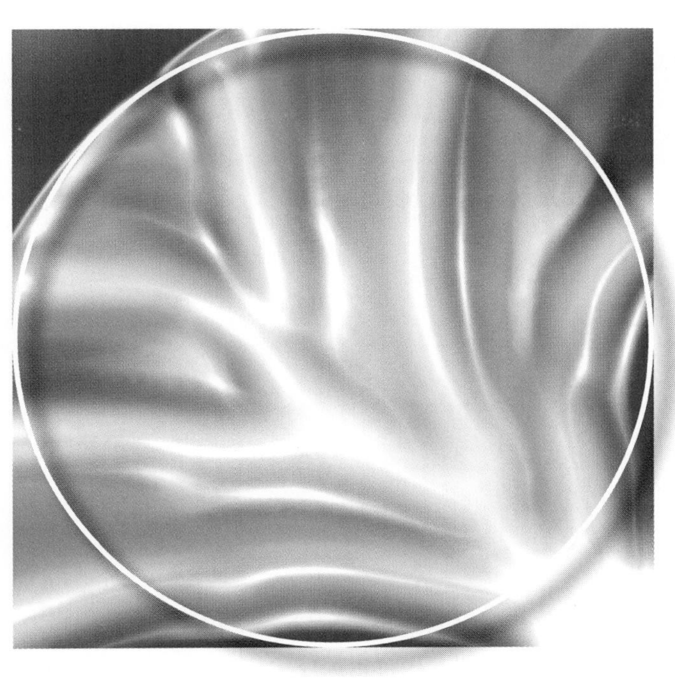

Reference Section

Formulas in Common Use

Ohm's Law for DC Circuits

$I = \frac{E}{R}$	$I = \frac{P}{E}$	$I = \sqrt{\frac{P}{R}}$
$E = IR$	$E = \frac{P}{I}$	$E = \sqrt{PR}$
$R = \frac{E}{I}$	$R = \frac{E^2}{P}$	$R = \frac{P}{I^2}$
$P = IE$	$P = \frac{I^2}{R}$	$P = \frac{E^2}{R}$

Ohm's Law for AC Circuits

$I = \frac{E}{Z}$	$I = \frac{P}{E \cos \theta}$	$I = \sqrt{\frac{P}{Z \cos \theta}}$
$E = IZ$	$E = \frac{P}{I \cos \theta}$	$E = \sqrt{\frac{PZ}{\cos \theta}}$
$Z = \frac{E}{I}$	$Z = \frac{E^2 \cos \theta}{P}$	$Z = \frac{P}{I^2 \cos \theta}$
$P = EI \cos \theta$	$P = I^2 Z \cos \theta$	$P = \frac{E^2 \cos \theta}{Z}$

Resistance

In Series $R_T = R_1 + R_2 + R_3 \ldots$

In Parallel $\frac{1}{R_T} = \frac{1}{R_1} + \frac{1}{R_2} + \frac{1}{R_3 \ldots}$

Two Resistors in Parallel $R_T = \frac{R_1 R_2}{R_1 + R_2}$

Equal Resistors in Parallel $R_T = \frac{R}{N}$

R = any one resistor

N = number of resistors in parallel

Capacitance

In Parallel $C_T = C_1 + C_2 + C_3 \ldots$

In Series $\frac{1}{C_T} = \frac{1}{C_1} + \frac{1}{C_2} + \frac{1}{C_3 \ldots}$

Two Capacitors in Series $C_T = \frac{C_1 C_2}{C_1 + C_2}$

$Q = CE$

Q = coulombs

C = farads

E = volts

Inductance

In Series $L_T = L_1 + L_2 + L_3 \ldots$

In Parallel $\frac{1}{L_T} = \frac{1}{L_1} + \frac{1}{L_2} + \frac{1}{L_3 \ldots}$

Two Resistors in Parallel $L_T = \frac{L_1 L_2}{L_1 + L_2}$

Coupled Inductance (Fields Aiding)

Series $L_T = L_1 + L_2 + 2M$

Parallel $\frac{1}{L_T} = \frac{1}{L_1 + M} + \frac{1}{L_2 + M}$

Coupled Inductance (Fields Opposing)

Series $L_T = L_1 + L_2 - 2M$

Parallel $\frac{1}{L_T} = \frac{1}{L_1 - M} + \frac{1}{L_2 - M}$

M = mutual Inductance

Mutual Inductance

$M = K = \sqrt{L_1 L_2}$

K = coefficient coupling

Reactance

$X_L = 2\pi f L$

$X_C = \frac{1}{2\pi f C}$

Resonance

$f_o = \frac{1}{2\pi\sqrt{LC}}$ or $\frac{0.159}{\sqrt{LC}}$

$L = \frac{1}{4\pi^2 f_o^2 C}$

$C = \frac{1}{4\pi^2 f_o^2 L}$

$2\pi = 6.28$

$4\pi^2 = 39.5$

Impedance

$Z = \sqrt{R^2 + X^2}$

$X = X_L - X_C$

Power Factor

$P_{true} = EI \cos\theta$

$P_{apparent} = EI$

PF = Power Factor = $\cos\theta$

Time Constants

$\tau = RC$

$\tau = \frac{L}{R}$

Peak, RMS, and Average AC Values

$E_{RMS} = E_{peak} \times 0.707$

$E_{peak} = E_{RMS} \times 1.414$

$E_{AV} = E_{peak} \times 0.637$

$E_{peak} = E_{AV} \times 1.57$

$E_{RMS} = E_{AV} \times 1.11$

Wavelength

$\lambda = \frac{3 \times 10^8}{f}$ meters

$f = \frac{3 \times 10^8}{\lambda}$ cycles

Conductance

$G = \frac{1}{R}$

$R = \frac{1}{G}$

G = Siemens

Figure of Merit Q

$Q = \frac{X}{R}$

X = capacitance or inductance

Gain

$dB = 10 \log \frac{P_{out}}{P_{in}}$

$dB = 20 \log \frac{E_{out}}{E_{in}}$

$dB = 20 \log \frac{I_{out}}{I_{in}}$

Electrical and Electronic Symbols

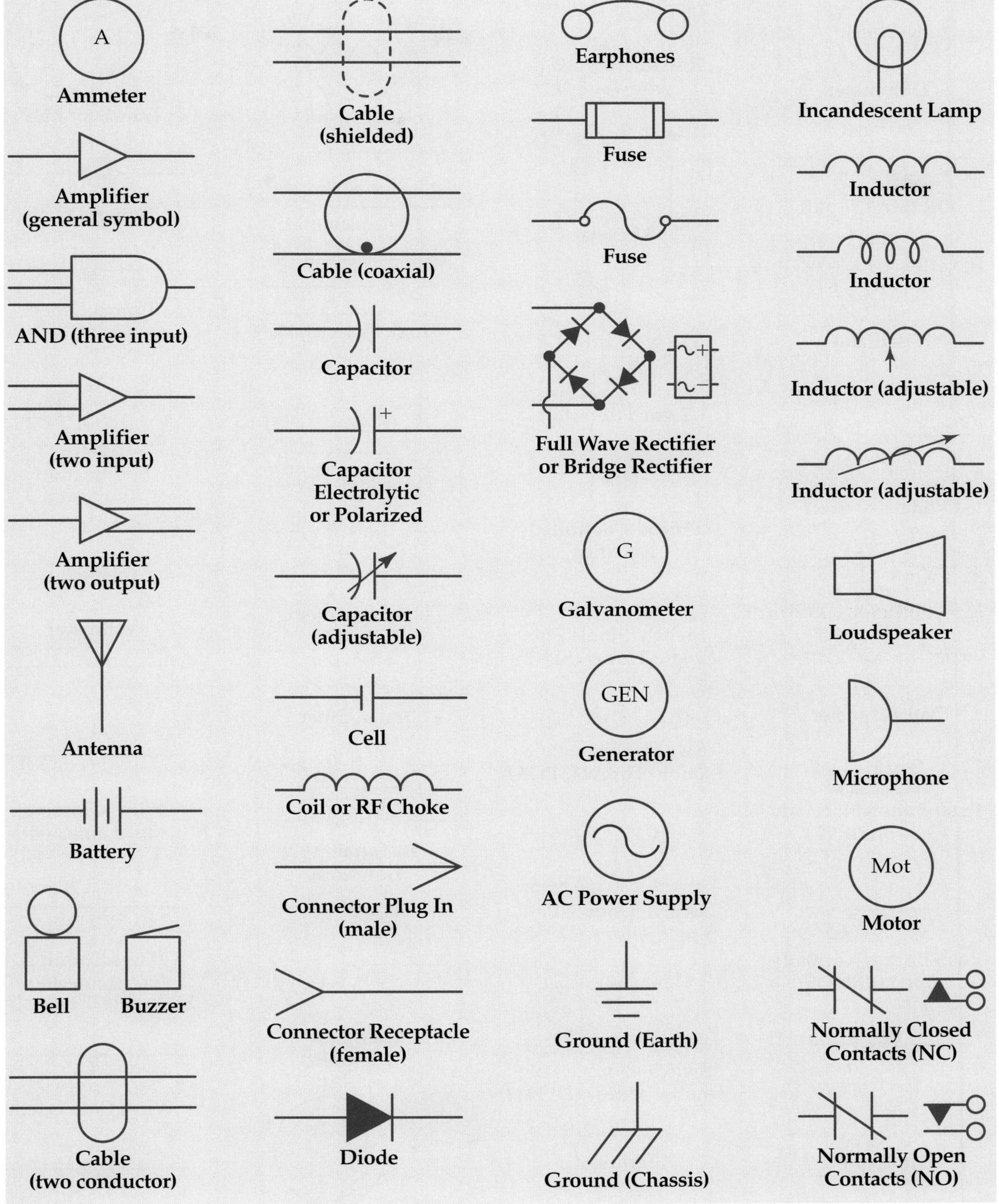

Electrical and Electronic Symbols *(Continued)*

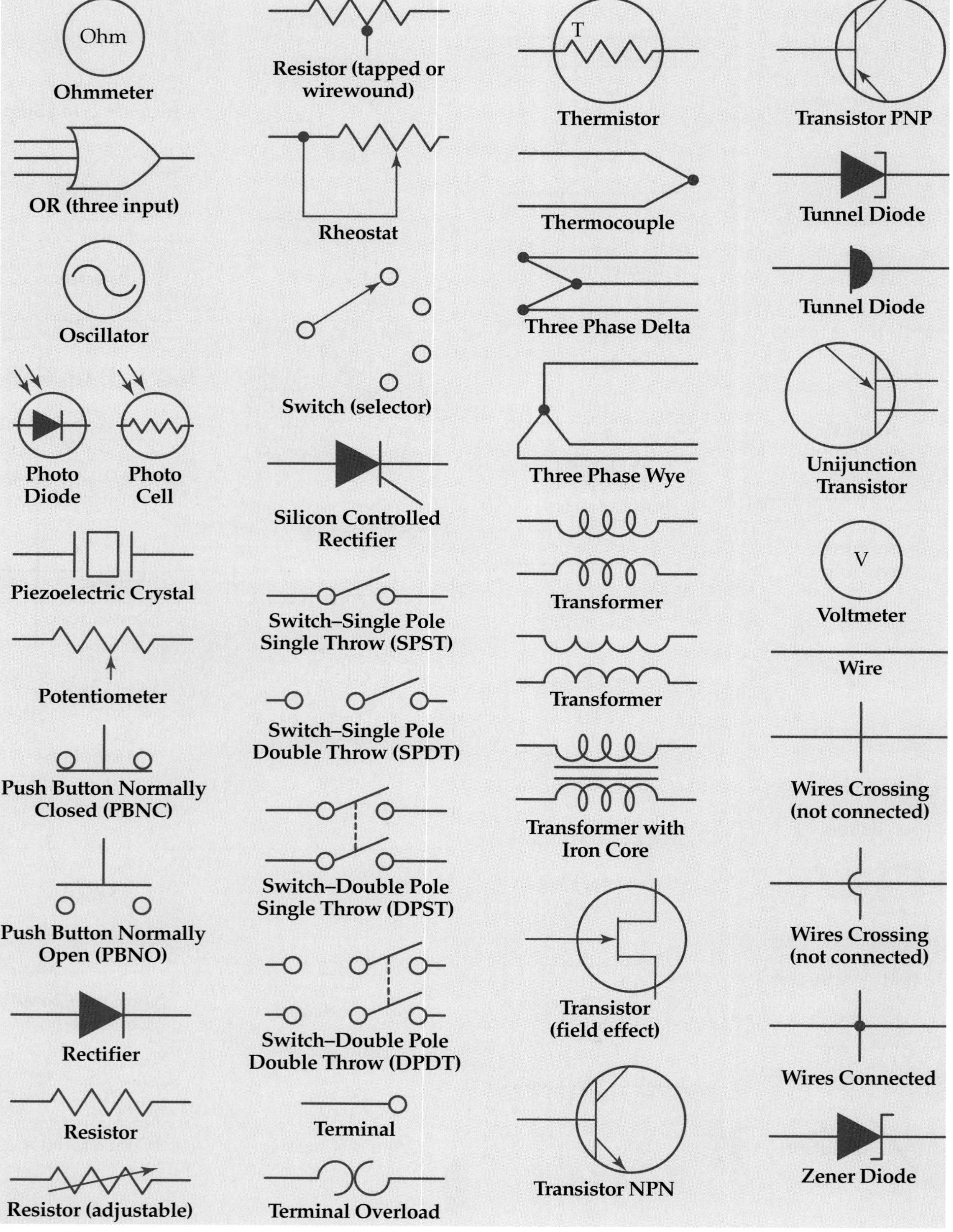

Resistor Color Code

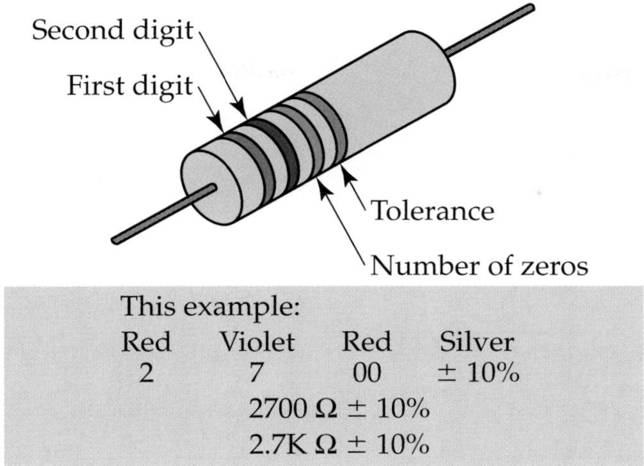

This example:

Red	Violet	Red	Silver
2	7	00	± 10%

2700 Ω ± 10%
2.7K Ω ± 10%

Color	Digit	
Black	0	
Brown	1	
Red	2	
Orange	3	
Yellow	4	
Green	5	
Blue	6	
Violet	7	
Gray	8	
White	9	
Gold	5%	Tolerance
Silver	10%	Tolerance
None	20%	Tolerance

Diode Color Code

Crystal diodes can be marked according to the EIA color code, which is the same code used with resistors. The color bands start from the cathode end with the numbers following the 1N prefix. For example, the diode in the drawing is a 1N462.

There are a number of other ways of coding diodes. Some companies use other letters and numbers. If you want to be sure of the code, use the reference catalog of the company that made the diode in question.

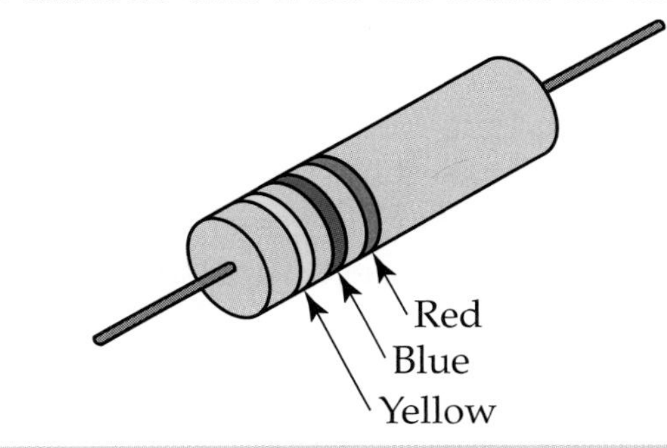

This example:

Yellow	Blue	Red
4	6	2

1N462

Color	Digit
Black	0
Brown	1
Red	2
Orange	3
Yellow	4
Green	5
Blue	6
Violet	7
Gray	8
White	9

Frequency Bands

Abbreviation	Description	Frequency
VLF	Very Low Frequency	3 kHz–30 kHz
LF	Low Frequency	30 kHz–300 kHz
MF	Medium Frequency	300 kHz–3 MHz
HF	High Frequency	3 MHz–30 MHz
VHF	Very High Frequency	30 MHz–300 MHz
UHF	Ultrahigh Frequency	300 MHz–3 GHz
SHF	Superhigh Frequency	3 GHz–30 GHz
EHF	Extremely High Frequency	30 GHz–300 GHz

Metric Prefixes

Prefix	Abbreviation	Exponential Equivalent	Decimal Equivalent
Tera	T	10^{12}	1,000,000,000,000
Giga-	G	10^{9}	1,000,000,000
Mega-	M	10^{6}	1,000,000
Kilo-	k	10^{3}	1000
Base unit	—	10^{0}	1
Milli-	m	10^{-3}	0.001
Micro-	µ	10^{-6}	0.000001
Nano-	n	10^{-9}	0.000000001
Pico-	p	10^{-12}	0.000000000001

Standard Abbreviations

A ampere, anode
ac alternating current
AF audio frequency
AFC automatic frequency control
AGC automatic gain control
AM amplitude modulation
AWG American Wire Gauge
b base
BJT bipolar junction transistor
c capacitor, collector
C Celsius
cemf counter-electromotive force
CRT cathode ray tube
CT center tap
cw continuous wave
dB decibel
dc direct current
DPDT double-pole, double-throw
DPST double-pole, single-throw
DTL diode transistor logic
e emitter
E voltage
emf electromotive force
f frequency
F Fahrenheit, farad
F_o resonant frequency
FET field effect transistor
FM frequency modulation
F_r resonant frequency
G conductance
gnd ground
H henry
HF high frequency
hp horsepower
Hz hertz
I current
IC integrated circuit
IF intermediate frequency
I_F forward biased current
IGFET insulated gate field effect transistor
JFET junction field effect transistor
K cathode
kHz kilohertz
k kilohm
kv kilovolt
kWh kilowatt-hour
L inductance
LCR inductance, capacitance, resistance
LED light emitting diode
LSI large-scale integration
LF low frequency
M mutual inductance
max maximum
MHz megahertz
mega one million
mmf magnetomotive force
µA microampere
µF microfarad
µH microhenry
µµF micromicrofarad

μV microvolt
mA milliamperes
mH millihenry
mV millivolt
mW milliwatt
min minimum
MOSFET metal oxide semiconductor field effect transistor
MSI medium-scale integration
N.C. normally closed
N.O. normally open
NTC negative temperature coefficient
OLED organic light-emitting diode
P power
pF picofarad
PF power factor
PIV peak inverse voltage
PTC positive temperature coefficient
R resistance
RC resistance capacitance
RF radio frequency
RMS root mean square
rpm revolutions per minute
SCR silicon-controlled rectifier
SPDT single-pole, double-throw
SPST single-pole, single-throw
SSI small-scale integration
τ time
UHF ultrahigh frequency
ULSI ultra large-scale integration
VOM volt-ohm-milliameter
V volts
V_{CP} clipper potential
V_p peak voltage
V_{pp} peak-to-peak voltage
V_Z zener voltage
W watt
X reactance
X_C capacitive reactance
X_L inductive reactance
Z impedance

Dimensions and Resistances of Copper Wire

Gauge No. AWG or B&S	Diameter of Bare Wire in Inches	Diameter of Bare Wire in Mils	Area in Circular Mils	Resistance in Ft./1 Ω at 68°F
0000	0.460	460.00	211,600	20,400.00
000	0.410	409.60	167,800	16,180.00
00	0.365	364.80	133,100	12,830.00
0	0.325	324.90	105,500	10,180.00
1	0.289	289.30	83,690	8070.00
2	0.258	257.60	66,370	6400.00
3	0.229	229.40	52,640	5075.00
4	0.204	204.30	41,740	4025.00
5	0.182	181.90	33,100	3192.00
6	0.162	162.00	26,250	2531.00
7	0.144	144.30	20,820	2007.00
8	0.129	128.50	16,510	1592.00
9	0.114	114.40	13,090	1262.00
10	0.102	101.90	10,380	1001.00
11	0.091	90.74	8234	794.00
12	0.081	80.81	6530	629.60
13	0.072	71.96	5178	499.30
14	0.064	64.08	4107	396.00

Dimensions and Resistances of Copper Wire *(Continued)*

Gauge No. AWG or B&S	Diameter of Bare Wire in Inches	Diameter of Bare Wire in Mils	Area in Circular Mils	Resistance in Ft./1 Ω at 68°F
15	0.057	57.07	3257	314.00
16	0.051	50.82	2583	249.00
17	0.045	45.26	2048	197.50
18	0.040	40.30	1624	156.50
19	0.036	35.89	1288	124.20
20	0.032	31.96	1022	98.50
21	0.028	28.46	810	78.11
22	0.025	25.35	642	61.95
23	0.023	22.57	510	49.13
24	0.020	20.10	404	38.96
25	0.018	17.90	320	30.90
26	0.016	15.94	254	24.50
27	0.014	14.20	202	19.43
28	0.013	12.64	160	15.41
29	0.011	11.26	127	12.22
30	0.010	10.03	101	9.69
31	0.009	8.93	79.7	7.69
32	0.008	7.95	63	6.10
33	0.007	7.08	50	4.83
34	0.006	6.31	40	3.83
35	0.006	5.62	32	3.04
36	0.005	5.00	25	2.41
37	0.004	4.45	20	1.91
38	0.004	3.97	16	1.52
39	0.004	3.53	13	1.20
40	0.003	3.15	10	0.95

Twist Drill Sizes and Decimal Equivalents

Drill	Equivalent	Drill	Equivalent	Drill	Equivalent	Drill	Equivalent
0.1 mm	0.003937	38	0.101500	B	0.238000	33/64	0.515625
0.25 mm	0.009842	37	0.104000	C	0.242000	17/32	0.531250
80	0.013500	36	0.106500	D	0.246000	13.5 mm	0.531495
79	0.014500	7/64	0.109375	E 1/4	0.250000	35/64	0.546875
78	0.016000	35	0.110000	6.5 mm	0.255905	14.0 mm	0.551180
1/64	0.015625	34	0.111000	F	0.257000	9.16	0.562500
77	0.018000	33	0.113000	G	0.261000	14.5 mm	0.570865
0.5 mm	0.019685	32	0.116000	17/64	0.265625	37/64	0.578125
76	0.020000	3.0 mm	0.118110	H	0.266000	15.0 mm	0.590550
75	0.021000	31	0.120000	I	0.272000	19/32	0.593750

Twist Drill Sizes and Decimal Equivalents *(Continued)*

Drill	Equivalent	Drill	Equivalent	Drill	Equivalent	Drill	Equivalent
74	0.022500	1/8	0.125000	7.0 mm	0.275590	39/64	0.609375
73	0.024000	30	0.128500	J	0.277000	15.5 mm	0.610235
72	0.025000	29	0.136000	K	0.281000	5/8	0.625000
71	0.026000	3.5 mm	0.137795	9/32	0.281250	16.0 mm	0.629920
70	0.028000	28	0.140500	L	0.290000	41/64	0.640625
69	0.029200	9/64	0.140625	M	0.295000	16.5 mm	0.649605
68	0.031000	27	0.144000	7.5 mm	0.295275	21/32	0.656250
67	0.032000	26	0.147000	19/64	0.296875	17.0 mm	0.669290
1/32	0.031250	25	0.149500	N	0.302000	43/64	0.671875
66	0.033000	24	0.152000	5/16	0.312500	11/16	0.687500
65	0.035000	23	0.154000	8.0 mm	0.314960	17.5 mm	0.688975
64	0.036000	5/32	0.156250	O	0.316000	45/64	0.703125
63	0.037000	22	0.157000	P	0.323000	18.0 mm	0.708660
62	0.038000	4.0 mm	0.157480	21/64	0.328125	23/32	0.718750
61	0.039000	21	0.159000	Q	0.332000	18.5 mm	0.728345
1.0 mm	0.039370	20	0.161000	8.5 mm	0.334645	47/64	0.734375
60	0.040000	19	0.166000	R	0.339000	19.0 mm	0.748030
59	0.041000	18	0.169500	11/32	0.343750	3/4	0.750000
58	0.042000	11/64	0.171875	S	0.348000	49/64	0.765625
57	0.043000	17	0.173000	9.0 mm	0.354330	19.5 mm	0.767715
56	0.046500	16	0.177000	T	0.358000	25/32	0.781250
3/64	0.046875	4.5 mm	0.177165	23/64	0.359375	20.0 mm	0.787400
55	0.052000	15	0.180000	U	0.368000	51/64	0.796875
54	0.055000	14	0.182000	9.5 mm	0.374015	20.5 mm	0.807085
1.5 mm	0.059055	13	0.185000	3/8	0.375000	13/16	0.812500
53	0.059500	3/16	0.187500	V	0.377000	21.0 mm	0.826770
1/16	0.062500	12	0.189000	W	0.386000	53/64	0.828125
52	0.063500	11	0.191000	25/64	0.390625	27/32	0.843750
51	0.067000	10	0.193500	10.0 mm	0.393700	21.5 mm	0.846455
50	0.070000	9	0.196000	X	0.397000	55/64	0.859375
49	0.073000	5.0 mm	0.196850	Y	0.404000	22.0 mm	0.866140
48	0.076000	8	0.199000	13/32	0.406250	7/8	0.875000
5/64	0.078125	7	0.201000	Z	0.413000	22.5 mm	0.885825
47	0.078500	13/64	0.203125	10 1/2 mm	0.413385	57/64	0.890625
2.0 mm	0.078740	6	0.204000	27/64	0.421875	23.0 mm	0.905510
46	0.081000	5	0.205500	11.0 mm	0.433070	29/32	0.906250
45	0.082000	4	0.209000	7/16	0.437500	59/64	0.921875
44	0.086000	3	0.213000	11.5 mm	0.452755	23.5 mm	0.925195
43	0.089000	5.5 mm	0.216535	29/64	0.453125	15/16	0.937500
42	0.093500	7/32	0.218750	15/32	0.468750	24.0 mm	0.944880
3/32	0.093750	2	0.221000	12.0 mm	0.472440	61/64	0.953125
41	0.096000	1	0.228000	31/64	0.484375	24.0 mm	0.964565
40	0.098000	A	0.234000	12.0 mm	0.492125	31/32	0.968750
2.5 mm	0.098425	15/64	0.234375	1/2	0.500000	25.0 mm	0.984250
39	0.099500	6.0 mm	0.236220	13.0 mm	0.511810	63/64	0.984375
						1	1.000000

Microelectronics and Semiconductor Web Sites

Company	Web Site
Actel	www.actel.com
Advanced Micro Devices	www.amd.com
Analog Devices	www.analog.com
AT&T	www.att.com
Cirrus Logic	www.cirrus.com
Cray	www.cray.com
Cypress Semiconductor	www.cypress.com
EXAR	www.exar.com
Hewlett Packard	www.hp.com
Honeywell International	www.honeywell.com
IBM	www.ibm.com
Integrated Device Technology	www.idt.com
Intel Corporation	www.intel.com
Kyocera Corporation	www.kyocera.com
LSI Corporation	www.lsilogic.com
Maxim Integrated Products	www.maxim-ic.com
Micron Technology	www.micron.com
MIPS Technologies	www.mips.com
Motorola	www.motorola.com
National Semiconductor Corporation	www.national.com
NEC Corporation	www.nec.com
OKI Semiconductor	www.okisemi.com
Ross	www.ross.com
Sanyo Electric Company	www.sanyo.com
Siemens	www.siemens.com
Silicon Systems	www.siliconsystems.com
Sun Microsystems	www.sun.com
Texas Instruments Incorporated	www.ti.com
Toshiba America	www.toshiba.com
Xilinx	www.xilinx.com
Zarlink Semiconductor	www.zarlink.com
Zilog	www.zilog.com

Informational Web Sites

General E&E Resources	
Name	**Web Site**
Discovery Channel	www.discovery.com
How Stuff Works	www.howstuffworks.com
Johnson Space Center	www.nasa.gov/centers/johnson/home/index.html
National Science Foundation	www.nsf.gov
Smithsonian National Air and Space Museum	www.nasm.si.edu/
Southern Company: Learning Power	www.southerncompany.com/learningpower
TechTarget	http://whatis.techtarget.com

Energy and Conservation	
Name	**Web Site**
American Wind Energy Association	www.awea.org
Database of State Incentives for Renewables and Efficiency	www.dsireusa.org
Energy Efficiency and Renewable Energy	www.eere.energy.gov
Energy Information Administration	www.eia.doe.gov
JEA	www.JEA.com
Photovoltaic Resources	www.pvresources.com
Southface Energy Institute	www.southface.org
Tennessee Valley Authority	www.tva.gov
U.S. Department of the Interior: Hoover Dam	www.usbr.gov/lc/hooverdam

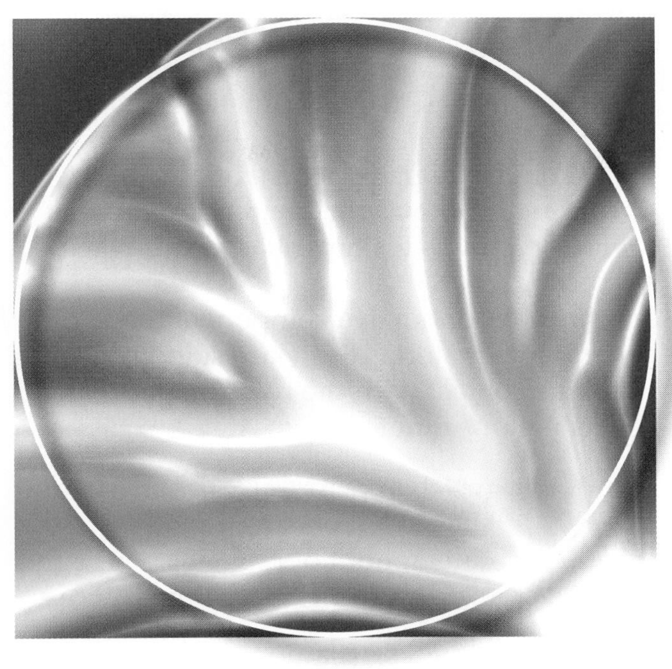

Glossary

A

ac: The abbreviation for alternating current.
adapter cards: Used to expand and enhance the motherboard. They are usually designed for one specific purpose.
AF: The abbreviation for audio frequency.
AFUE: The abbreviation for Annual Fuel Utilization Efficiency.
albedo: The reflective ability of a surface.
alternating current (ac): A flow of electrons which regularly reverses its direction of flow.
alternative energy: Energy derived from a source that does not deplete the earth's natural resources or harm the environment.
alternator: A device used to convert mechanical energy into electrical energy in the form of an alternating current.
ALU: The abbreviation for arithmetic logic unit.
American Standard Code for Information Interchange (ASCII): A number code assigned to each of the keys of the computer.
American Wire Gauge (AWG): The numbering system for wire size.
ammeter: An instrument used to measure current in ac or dc circuits.
ampere (A): The unit of measure for current based on the flow of one coulomb per second past a given point. Also, one volt across one ohm of resistance causes a flow of one ampere.
amplifying transistor: A transistor that takes a small input signal and converts it to a much larger output signal.
amplitude: The amount of change in a quantity, such as a sine wave.
analog multimeter: A meter that has multiple graduated scales (ranges of electrical values) and is capable of measuring current, voltage, and resistance.
AND gate: A logic gate that requires a signal at every input to have a signal at the output.
anemometer: An instrument for measuring wind speed.
Annual Fuel Utilization Efficiency (AFUE): A rating scale used to denote how energy efficient an appliance is.
anode (A): The general name for a positive electrode or terminal, such as that on the positive side of a diode or electrical cell.

arc: The bright spark between a live wire and another conductor.
arithmetic-logic unit (ALU): Part of the CPU that performs binary calculations.
armature coil: The coil in the rotor (the rotating part) of a generator or motor.
ASCII: The abbreviation for American Standard Code for Information Interchange.
assembler: A person who wires and solders components for industrial controls and consumer products.
attenuate: To weaken.
audio frequency (AF) waves: Frequencies between 20 Hz and 20,000 Hz.
autoranging multimeter: A multimeter that is capable of switching between ranges automatically.
AWG: The abbreviation for American Wire Gauge.

B

backup: A copy of data files created in case the original files are lost or damaged.
band-pass filter: A filter that allows only a narrow band above and below the resonant frequency to pass through the circuit.
band-stop filter: A filter that stops frequencies very close to the resonant frequency from passing through the circuit.
bandwidth: The frequency limits of a given waveband for transmitting a modulated signal.
base: The center region of a transistor. It controls the amount of current that flows through the transistor, allowing no electrons, some electrons, or many electrons to pass.
battery: Two or more cells connected to deliver dc voltage.
binary: A numbering system that uses only the numbers *0* and *1*.
binary code: System using the numbers *0* and *1* or the states "off" and "on" as a code.
bipolar junction transistor (BJT): A transistor that has either P-type material between two N-type regions or N-type material between two P-type regions. It, therefore, has two P-N junctions instead of one.
bit: A binary digit, either a *0* or a *1*.
bridge rectifier circuit: A circuit that uses four rectifiers to convert ac to full-wave rectified dc.
brushes: A conductive component made of graphite material. They are stationary and make contact with the rotating part of a generator or motor.

C

cache: A small amount of memory where the CPU stores the information it must access frequently.
capacitance: The amount of charge that can be stored by a capacitor as measured in farads (F).
capacitive reactance: Opposition offered by capacitors to alternating current.
capacitor: Two conductors separated by an insulator or dielectric. It is used to store an electrical charge.
carbon resistor: A resistor made of a carbon material held together by a binder.
cathode (K): The general name for a negative electrode or terminal, such as that on the negative side of a diode or electrical cell.
cell: A single power unit that produces electrical energy that can be used to power a circuit.
cemf: The abbreviation for counter-electromotive force.
central processing unit (CPU): An electronic component that receives all input and controls the other components of the computer. It performs all of the computer's tasks, including performing calculations and manipulating data. Also called a *microprocessor*.
CFL: The abbreviation for compact fluorescent (CFL) bulb.
channel: In reference to an oscilloscope, it is a complete pathway or circuit that allows

the frequency and amplitude of a waveform to be read. Oscilloscopes typically have two or more channels, which allow the reading of multiple waveforms. Also the part of a field effect transistor (FET) that carries current.

chassis ground: An electrical connection to the metal frame or chassis that holds the circuit.

chief electrical inspector: A person who goes to jobsites to inspect and eventually approve the wiring.

circuit analysis: The process of breaking down the full circuit network to determine its characteristics, such as the amount of current and voltage in each of its paths.

circuit breaker: A device that automatically opens a circuit when its current carrying ability has been exceeded.

circular mil: The cross-sectional area of a round wire having a diameter of 0.001 in.

clipper potential (V_{CP}): The voltage at which a clipper circuit will clip (cut off) a portion of the applied voltage.

closed circuit: A circuit with a completed path, allowing current to flow.

closed-core transformer: A transformer in which the core is a closed ring with the primary wrapped around one side of it and the secondary wrapped around the other side.

CMOS memory: Part of the motherboard that remembers the computer settings used each time the computer starts, such as the start order of the drives in the computer, the time, and the hard drive configuration settings. It is powered by a small battery.

cold solder joint: A solder joint that is dull or has lumps or small beads of solder.

collector: The part of a bipolar junction transistor (BJT) that collects current carriers from the base. The main flow of current goes out through the collector.

common-base amplifier: A transistor amplifier where the base is common to both the input and output circuits.

common-collector amplifier: A transistor amplifier where the collector is common to both the input and output circuits.

common-emitter amplifier: A transistor amplifier where the emitter is common to both the input and output circuits.

commutator: A group of bars providing connections between armature coils and brushes.

compact fluorescent (CFL) bulb: An energy-efficient bulb that uses less energy than an incandescent bulb.

computer: A device that takes in information, processes it, and outputs the results.

computer integrated manufacturing (CIM) cell: A group of machines that are put together to produce a finished product.

conductance (G): The ability for electrons to flow in a circuit.

conductor: Any item or material which is used to carry current. Also called a *wire*.

conduit: A hollow pipe used to protect wire from being cut by any sharp object along the wire's path.

coulomb: The amount of electric charge carried by a current of one ampere flowing for one second.

counter-electromotive force (cemf): The voltage that forms across the coils of an inductor due to the rising and falling of the applied alternating current. Counter-electromotive force opposes the applied voltage.

CPU: The abbreviation for central processing unit.

current: The flow of electrons through a conductor, measured in amperes.

cutoff voltage: The value of voltage at which a semiconductor will not conduct.

cycle: The change in an alternating wave from zero to a positive peak to zero to a negative peak and back to zero.

D

dc: The abbreviation for direct current.

dc generator: A device that uses rotary motion to create direct current.

dead cell: A cell that no longer produces electricity because the zinc can inside of it has been eaten away

demodulation: Converting data back to its original frequency and order.

depletion mode: A mode of operation for a metal oxide semiconductor field effect transistor (MOSFET) in which the gate voltage is used to deplete the channel and control the current in the drain. This type of mode requires the gate to be reverse biased.

depletion zone: The region in a reverse-biased semiconductor where there is a lack of electrons and holes at its junction.

DIAC: A two-lead alternating current semiconductor that conducts when its breakdown voltage is reached. It has properties of both diodes and transistors.

dielectric: The insulating material between the plates of a capacitor.

digital ICs: Integrated circuits that work with signals that are discrete.

digital multimeter: A meter that spells out the specific value of the electrical unit under test and is capable of measuring current, voltage, and resistance.

diode: A semiconductor device that has a P-N junction. Usually, it is thought of as allowing electron flow in one direction and blocking flow from the other direction. However, there are some special types which do not work that way.

direct current (dc): The flow of electrons in only one direction.

doping: The process of adding impurities to semiconductors in a controlled way.

double-pole double-throw (DPDT) switch: A switch with two terminals that can be connected to one of two other pairs of terminals, depending on how it is thrown.

double-pole single-throw (DPST) switch: A switch with two terminals that can be connected to or disconnected from another pair of terminals, depending on how it is thrown.

DPDT: The abbreviation for double-pole, double-throw.

DPST: The abbreviation for double-pole, single-throw.

drain: The part of a field effect transistor (FET) that collects electrons from the channel.

dry cell: A cell which produces electricity through chemical means. It is commonly called a *battery*.

E

eddy currents: Currents that are induced in a body. In transformers, these currents usually circulate in the iron and copper windings and produce unwanted heat.

electrolyte: The paste in a dry cell. Also, the fluid in a storage battery or material found in an electrolytic capacitor.

electrolytic capacitor: A capacitor made with an aluminum positive plate, electrolyte as a negative plate and a thin oxide layer on the aluminum as the dielectric. Its great advantage is larger capacitance from a smaller package.

electromotive force (emf): The force that causes electricity to flow because a difference in potential exists between two points.

electron: A negatively charged particle that orbits the nucleus of an atom.

emf: The abbreviation for electromotive force.

emitter: The part of a bipolar junction transistor (BJT) that emits current carriers to the base. The emitter is manufactured to have many more free carriers than the collector and to conduct easily across the collector-base junction.

energy conservation: The activity involved in saving the Earth's natural resources for future generations. It is concerned with such things as how to reduce pollution, waste, and landfill overuse. It is also concerned with using alternative sources of energy that place less of a burden on our natural resources.

engineer: A person who designs and supervises the building of new equipment until it is finally installed. They must make sure it will run with the least amount of problems. This person must have strong math and science skills.

enhancement mode: A mode of operation for a metal oxide semiconductor field effect transistor (MOSFET) in which the source and the drain are not connected by a channel, so the MOSFET is normally off. The gate voltage is used to create and enhance the channel and control the current in the drain. In the enhancement mode, the gate is forward biased.

entry-level: A type of job that can be learned within a few hours and up to several weeks of instruction.

equivalent circuit: A way of taking a complicated circuit and finding out what it is like in a simpler, more easy to understand version that is the same electrically.

F

Farad (F): The unit of capacitance that is the result of a charge of one coulomb producing a difference in potential of one volt.

ferrous metals: Metals that contain iron.

FET: The abbreviation for field effect transistor.

fiber optics: Communication using light pulses transmitted through a solid glass fiber.

field coils: Copper windings located in the stator of a generator or motor.

field effect transistor (FET): A semiconductor device that conducts from one terminal called the *source* to another terminal called the *drain* when a voltage is applied to a third terminal called a *gate*.

field magnets: Electromagnets used to make up the magnetic field of motors, generators, and other electrical devices.

filter: A circuit network made from resistors, capacitors, and inductors. It is used to block or reduce the amplitude of specific bands of frequencies.

filtering: The act of eliminating a specified group of frequencies from an electronic signal.

fixed resistor: A resistor with a single, constant value.

float switch: A switch that is activated mechanically by a rising or falling water level.

flux: A material used in soldering to clean the surface of any oxide so that a good connection will result.

forward bias: The voltage applied to a P-N junction so that it conducts.

forward biased current (I_F): The current when the diode allows electron flow to complete the circuit.

frequency: The number of complete cycles per second as measured in hertz.

fuse: An electrical device put in a circuit to protect against overloading. Current above the rating of the fuse will melt the fusible link and open the circuit.

G

gain: The increase in voltage, current, and power that an individual transistor can provide.

gate: The gate is attached to a part of the FET that is the opposite type of material from the channel. Voltage applied to the gate controls the current that passes through the channel.

generator: A rotating machine that converts mechanical energy into electrical energy.

geothermal system: A heating and cooling system with coils that extend far underground and that uses the steady temperature of the earth to keep a home warm in the winter and cool in the summer.

graticule: A grid placed at the face of an oscilloscope's display for measuring purposes.

green power: Electricity produced from renewable energy sources.

grid: A wiring network that is spread across most of the United States.

ground: The voltage reference point in a circuit.

H

hard drive: A permanent storage place that consists of platters and read-write heads. The read-write heads use principles of magnetism to store and retrieve data on the hard drive.

hardware: The physical components that make up the computer.

heat shrink: A type of tubing placed over bare wire for insulation. It will shrink in diameter when heat is applied to it.

heat sink: A metal base on which semiconductor devices are mounted to help dissipate heat. Also, a device used when soldering components to stop them from overheating. It usually is placed between the component and the joint being soldered.

henry (H): The unit of measure for inductance. A coil has one henry of inductance if an EMF of one volt is induced when the current through the inductor is changing at the rate of one ampere per second.

hertz (Hz): The unit of measure for frequency which is equal to one cycle per second.

high-pass filter: A circuit that allows only high frequencies to pass through it.

holding contacts: Contacts that are in parallel with a momentary pushbutton. When the pushbutton is pressed, a coil energizes and closes the contacts. The contacts remain closed and the coil energized until a stop button is pressed.

hole flow: The idea that a positive charge carrier in a P-type material will attract electrons which, when combined, will form new holes. It is similar to the idea that a flow of positive charges will exist in a semiconductor.

hot wire: A current carrying wire.

hybrid IC: A circuit made by combining two or more monolithic ICs. A hybrid could also be made by combining monolithic with discrete components, such as inductors and additional transistors.

I

IC: The abbreviation for integrated circuit.

IGFET: The abbreviation for insulated gate field effect transistor.

impedance: The total opposition which a circuit offers to the flow of alternating current at a given frequency. It is the combination of resistance and reactance.

in phase: The state of two or more waves that reach their maximum amplitude at the same time and zero at the same time.

induced emf: The voltage that is induced in a coil due to a changing magnetic field.

inductance: That property of a circuit or device which opposes any change in current through it. Its symbol is L.

induction motor: A motor in which the rotor turns because of its attraction to an opposite magnetic pole in the stator.

inductive reactance: The opposition of an inductor to a changing current. Its symbol is X_L.

inductor: An electrical component that has the property of inductance. The most common form of an inductor is a coil of wire.

inkjet printer: A printer that fires (sprays) ink onto the paper when the inkjet head is electrically charged.

input devices: Hardware that is used to put information into the computer. Some important input devices include the keyboard, mouse, scanner, DVD/CD ROM drives, digital camera compact flash card readers, and Web cameras.

insulated gate field effect transistor (IGFET): A field effect transistor that has very high input impedance and an insulated gate.

insulating spaghetti: A hollow insulating material similar to a plastic straw. It is useful when splicing two pieces of wire.

insulator: Any material that does not easily allow electrons to flow through it.

integrated circuit (IC): A single electronic device built on a semiconductor chip capable of holding hundreds of thousands of microscopic components. It is sometimes called a *chip.*

inverter: A circuit capable of inverting a signal.

J

JFET: The abbreviation for junction field effect transistor.

journeyman: A person who has completed an apprenticeship program. During the apprenticeship program, this person gains on-the-job experience together with attending two year college training programs, special classes, or self-study courses.

junction field effect transistor (JFET): A field effect transistor (FET) that allows current to flow without gate voltage. It is considered a "normally on" device.

K

kilowatt-hour (kWh): The amount of energy supplied by 1000 watts for one hour.

kilowatt-hour meter: A meter that records how much electrical power is used.

Kirchhoff's voltage law: In a simple circuit, when the individual voltage drops are added, the sum is equal to the applied voltage.

L

ladder diagram: A drawing which shows the logic of a circuit. The ladder diagram typically looks like a ladder. The rails of the ladder represent the main lines through which power flows. The rungs, or lines between the rails, represent circuits. A technician uses a ladder diagram when troubleshooting a system.

LAN: The abbreviation for local area network.

large-scale integration (LSI): A classification of integrated circuits that contain from 100 to 9999 components.

laser printer: A printer that uses the principles of magnetism to print.

LCR circuit: A circuit that contains an inductor, capacitor, and resistor.

LED: The abbreviation for light emitting diode.

left-hand rule: Grasp a conductor carrying current with your left thumb pointing in the direction of current flow and your fingers will show the direction of the magnetic field.

light emitting diode (LED): A P-N junction diode that emits light when it is biased in the forward direction.

limit switch: A switch that closes or opens when an outside, mechanical force contacts its lever.

linear ICs: Integrated circuits that accept a continuous range of signals.

lines of force: The flux lines given off by an electric field. Also, the magnetic field surrounding a current-carrying device.

Lissajous figures: The patterns produced on the screen of an oscilloscope by changing the amplitude and phases of sine waves and feeding those to the horizontal and vertical inputs of the scope.

load: The device that is being driven by the source of power. It absorbs the power from the supply voltage and converts it to heat, light, or such.

local area network (LAN): A single network in a home or office building.

logic family: Describes the physical makeup of the IC.

logic gate: A circuit whose output depends on the combination of digital signals applied to its inputs.

low-pass filter: A circuit that allows only low frequencies to pass through it and attenuates high frequencies.

LSI: The abbreviation for large-scale integration.

M

magnetic coupling: When the changing magnetic field in one turn of the coil produces an emf in the other turns of the coil through which the field passes.

magnetic north: The Earth's pole that attracts the north pole of a compass needle.

maintenance worker: A service worker who fixes many appliances, such as dishwashers, ranges, washing machines, microwave ovens, and air conditioning units. Those doing maintenance work must know how to work with electrical circuits, read schematics, troubleshoot many appliances, and make the necessary repairs.

MAN: The abbreviation for metropolitan area network.

mechanical energy: The ability to do work.

medium-scale integration (MSI): A classification of integrated circuits that contain 12 to 99 components.

metal oxide semiconductor field effect transistor (MOSFET): A field effect transistor that has very high input impedance and an insulated gate.

metropolitan area network (MAN): A group of LANs connected by private communication lines. It is managed by a single party.

microfarad (μF): One millionth of a farad.

modem: A device used to transmit digital computer data electrically over phone lines and other types of copper-core cables.

modulation: The process of splitting the data, changing its frequency, and adding it to a carrier signal.

monitor: Displays the information on its screen. Two common types of monitors are the cathode ray tube (CRT) and liquid crystal display (LCD).

monolithic IC: An integrated circuit that is made on a single piece of silicon.

MOSFET: The abbreviation for metal oxide semiconductor field effect transistor.

motherboard: The main printed circuit board of the computer.

motor: A device used to convert electrical energy into rotating mechanical energy.

MSI: The abbreviation for medium-scale integration.

multicasting: Sending out two or more signals at the same time.

multimeter: A meter that can measure different values such as voltage, current, and resistance.

multiplexing: Sending two or more separate signals down a single channel at a higher data rate.

mutual inductance: The condition that exists when the magnetic field of one conductor is linked to the magnetic field of a second conductor. In this way, each field has some affect on the emf of the other conductor.

N

NAND gate: A logic gate that has a NOT gate added to an AND gate. The only way to prevent an output signal in the NAND gate is to have a signal present at each input lead.

N.C.: The abbreviation for normally closed.

N.O.: The abbreviation for normally open.

negative temperature coefficient (NTC) thermistor: A thermistor that decreases its resistance value as its temperature increases.

net metering: The process of selling excess photovoltaic energy to the local power company. This spins the home's electric meter backward. The excess energy provides a credit against the electricity from the power company that is used in the home.

network: A group of computers that are connected so that information can be exchanged among them.

nonferrous metals: Metals that do not contain iron.

NOR gate: A logic gate in which a NOT gate is combined with an OR gate. The only way to get an output signal from the NOR gate is to have no signal at any input lead.

normally closed (N.C.): Contacts that are closed when the relay is in the de-energized state.

normally open (N.O.): Contacts that are open when the relay is in the de-energized state.

NOT gate: A logic gate, also called an *inverter,* outputs the opposite signal of the input signal.

O

ohm: The unit of measure for resistance. It is symbolized by the Greek letter omega (Ω). There is one ohm of resistance when an EMF or one volt causes a current of one amp.

Ohm's law: A relationship between the values for voltage (E), current (I), and resistance (R). Discovered by a scientist named Ohm, it states $E = I \times R$.

OLED: The abbreviation for organic light-emitting diode.

one-wire electrical system: An electrical system that uses its frame or chassis as a path for ground.

open circuit: A circuit with at least one break in the path for current.

open secondary: When the transformer's secondary does not have a load.

optical detector: Part of a fiber-optics receiver that converts photons into an electrical signal. Two diodes commonly used as optical detectors are the positive intrinsic negative (PIN) diode and the avalanche photo diode (APD).

OR gate: A logic gate that requires a signal to be present at one or more of its input leads.

organic light-emitting diode (OLED): A light emitting diode that uses ultraviolet light to solidify polymers in the creation of a plastic-like film of a desired color.

oscilloscope: A testing instrument designed to give a technician information about a wave, such as its waveform, frequency, period, and amplitude.

output devices: Hardware that is used to take information out of the computer.

P

parallel circuit: A circuit which contains two or more paths for current to flow.

parallel ports: A port that can transmit at least eight bits at one time. Parallel ports have a connector with 25 holes.

PC: The abbreviation for personal computer.

PDA: The abbreviation for personal digital assistant.

peak inverse voltage (PIV): The value of the voltage applied across a diode in the reverse direction.

peak voltage (V_p): The highest voltage amplitude of a wave.

peak-to-peak voltage (V_{pp}): The value of a wave from the positive peak to the negative peak.

period: The time required to complete one cycle.

peripherals: Parts that are attached to the computer.

personal computer (PC): A computer that is often used by a single person.

personal digital assistant (PDA): A small, handheld, wireless computer.

phase angle: The angle by which the voltage sine wave leads or lags the current sine wave in a circuit.

photomask: Material used in the fabrication of an integrated circuit. It has the pattern of the different parts of the IC and is used to transfer the pattern to the IC.

photons: Pulses of light.

phototransistor: A transistor that uses light applied to its base as a power source.

photovoltaics: A system that uses solar energy to produce electricity.

picofarad (pF): A way of labeling a capacitor equal to one millionth of a microfarad (one millionth of one millionth farad).

piezoelectricity: Electricity produced by the piezoelectric effect.

pinch-off voltage: The voltage at which current remains at a steady value until breakdown occurs.

PIV: The abbreviation for peak inverse voltage.

P-N junction: The line of separation between P-type and N-type semiconductor materials.

polarized: The state of having polarity.

poor solder joint: A solder joint that has any combination of the following characteristics: poor appearance, does not conduct, burned solder joint, too much solder on joint, loose wires at joint.

porous concrete: Concrete that provides a higher void content and allows water to seep into the ground. The water can help keep the area cool. Porous concrete is lighter-colored than blacktop. This increases the albedo and reduces the heat build-up.

ports: Places into which peripherals are plugged.

positive temperature coefficient (PTC) thermistor: A thermistor that increases its resistance value as its temperature increases.

potentiometer: A variable resistor used as a voltage divider. Also, an electrical device used to give an electrical output signal that is proportional to some rotary movement.

power (P): The rate of doing work.

power supply: A circuit that provides the correct voltages to an electronic device.

pressure switch: A switch that opens and closes its contacts according to changes in pressure.

primary: The winding of a transformer that is connected to the energy source.

primary cell: A cell that is not designed to be recharged.

printed circuit boards: The insulated board on which the printed circuit is formed.

programmable logic controller (PLC): A computer that is used to monitor and control various types of machinery.

PTC: The abbreviation for positive temperature coefficient.

pulsating direct current: The current that is formed by passing alternating current through the rectifier.

Q

quality control inspector: A person who checks the finished work of others.

R

radio frequency (RF) waves: Electromagnetic waves in the range of 3 Hz to 300 GHz.

RAM: The abbreviation for random-access memory.

random-access memory (RAM): A temporary storage place that is used to store information and programs the computer is using or running at the moment.

range selector switch: A rotary switch that is used to select the range of the current, voltage, or resistance being measured.

RC time constant (τ): The time period needed for the voltage across a capacitor in an RC circuit to increase to 63.2% of the source voltage or decrease to 36.7% of the voltage across the capacitor.

recovery time: The time it takes a device to turn on, turn off, and turn back on again.

relay: An electromechanical switching device that consists of a coil, an armature, and electrical contacts.

repeater station: A device that receives and resends the signal, boosting its strength to prevent the loss of any part of the message.

resistor: A device that opposes the flow of current in an electrical circuit.

resonance: When the amount of reactance offered by a capacitor equals that of an inductor. In this case, the total reactance of the circuit is zero, and the impedance of the circuit is equal to the resistance of any resistors in the circuit.

reverse bias: When voltage is applied to a P-N junction so that it does not conduct or conducts only a very small amount.

RF: The abbreviation for radio frequency.

rheostat: A variable resistor placed in series with the load to control current. It has one fixed terminal and a movable contact called a *slider.*

right angle: An angle that is 90°.

right triangle: A triangle that has a 90° angle in it.

ripple: The peaks and valleys that appear on top of pulsating direct current (dc).

RL circuit: A circuit that contains a resistor and an inductor.

RMS: The abbreviation for root mean square.

root mean square (RMS): The value of ac that would produce the same heating effect as an equal amount of dc. In ac, the value is 0.707 times the peak value, and it is the same as the effective value.

rosin core solder: A material used to join electrical components. It is a mixture of tin and lead, and it has a fluxing agent in the hollow center.

rotor: The rotating part of an electric motor or generator.

S

schematic: An electrical diagram that shows components as symbols and electrical connections of components to each other.

SCR: The abbreviation for silicon-controlled rectifier.

secondary: The winding of a transformer with the induced energy.

secondary cell: A cell which, after it has been discharged is meant to be recharged.

serial port: A port for serial communications. Bits travel through it in serial fashion, one bit at a time. A serial port typically has 9 or 25 pins. It is also called a *communications port* or an *RS-232 port.*

series circuit: A circuit with only one path for current.

series-parallel circuit: A circuit that consists of series and parallel circuits.

server: A main computer that stores files.

shell-type transformer: A transformer constructed by wrapping the primary around a cardboard core and then wrapping the secondary over the primary. Insulating paper separates each layer of windings from the other, producing a transformer with a high mutual inductance.

short circuit: A usually undesired path for current which bypasses a desired path.

signal generator: A machine that produces ac voltages with a broad range of frequencies.

silicon-controlled rectifier (SCR): A semiconductor device having an anode, cathode, and gate. It is used for current values that are usually higher than those carried by transistors.

sine wave: A wave form of a single frequency alternating current whose displacement is the sine of an angle proportional to time or distance.

single-pole double-throw (SPDT) switch: A switch used to connect one terminal to one of two other terminals.

single-pole single-throw (SPST) switch: A switch used to open or close one circuit.

skilled worker: A person who acquires skills over a period of several months to several years. Many of their skills are learned by reading manuals, attending workshops and technical training sessions, learning from other skilled workers, and on-the-job training.

slip: The difference between the actual speed of a motor and its synchronous speed.

slip rings: Devices used on an ac generator that are designed to let the rotor coil turn without twisting or breaking any wires. Each slip ring is soldered to one end of the armature coil. As the coil turns, the slip rings slide past the fixed brushes.

small-scale integration (SSI): A classification of integrated circuits that contains up to 11 components.

software: A set of instructions for a computer.

solder: An alloy of tin and lead used for providing a low resistance electrical connection that is also mechanically strong.

soldering iron: One kind of device that provides the heat to melt solder.

solenoid: A device that has a coil and a plunger and that operates on the principles of electromagnetism. The electromagnetic force causes the plunger to move. When the plunger moves, it controls something else that needs to move.

solid-state device: An electronic device that can stop, allow, and control the flow of electrons.

source: One of the leads of a field effect transistor (FET).

SPDT: The abbreviation for single-pole, double-throw.

specific gravity: A comparison of the density of a substance with the density of water. Used to measure the remaining strength of a storage battery.

SPST: The abbreviation for single-pole, single-throw.

square wave: An electrical wave which alternates abruptly between high and low values of approximately equal duration.

squirrel cage induction motor: An induction motor having a rotor that looks similar to a squirrel cage in construction.

SSI: The abbreviation for small-scale integration.

stator: The stationary part of a generator or motor.

step-down transformer: A transformer used to reduce voltage.

step-up transformer: A transformer used to increase voltage.

substrate: An insulating material on which circuits are formed.

superconductivity: A state usually reached by very cold temperatures (typically –452° Kelvin) in which a material offers no resistance to the flow of electrons through it.

supervisor: A person that makes sure others do their jobs right. They must have the knowledge and skills related to the field in which they supervise. Supervisors must be able to write reports and be skilled in handling people.

switching transistor: Transistors that are used to turn circuits on and off.

synchronous motor: A motor in which the rotor spins in exact synchronization with the rotating magnetic field of the stator. This causes the magnetic field and rotor to turn at the same speed.

T

tap: A fixed electrical connection on a transformer that is used to provide another voltage level.

technician: A person who is trained to work with existing equipment. They may also work with engineers, helping in the design and new innovations in their field.

therm: A unit of heat equal to 100,000 British thermal units.

thermistor: A thermal resistor. A device whose resistance changes with a change in temperature.

thermocouple: A device used to measure temperature by the use of two dissimilar metals which are joined at the point where the measurement is being taken. As heat is applied, the voltage generated can be read on a meter calibrated in degrees.

thermopane glass: Double pane glass that is filled with an inert gas to increase its energy efficiency.

thick-film IC: An integrated circuit that is made by depositing multiple layers of paste-like material on a substrate. A screen printing process is used to deposit the paste in layers for the various components. The material is fused together in a high-temperature oven at about 1562° F (850° C).

thin film resistor: A resistor made of nichrome or tantalum nitride, which is deposited on a substrate similar to ceramic. Also called a *chip resistor.*

thin-film IC: An integrated circuit that is made by depositing very thin layers of vapor on an insulating material known as a substrate. The process used to do this is called *sputtering.* It consists of depositing atoms two to three layers thick of silver and various oxides on a substrate, covering it with a resist material, applying a mask, and exposing it to a light source. The light hardens the material not covered by the mask. Unexposed resist material is washed off and the IC is baked to harden it.

three-phase system: A system that uses three wires to carry current and simultaneously act as the return wires for alternating current with a phase difference of 120° or one-third of a cycle.

thyristor: A broad group of devices that are made up of four or more alternating layer of P-type and N-type material. Four layers create three P-N junctions, one at each region where the P-type and N-type materials are joined together.

timed relay: A relay that provides a time delay, often as long as one minute, before the contacts change positions.

torque: A force that tends to produce rotation.

total capacitance: The combined value of all the capacitors in a circuit.

total inductance: The combined value of all the inductors in a circuit.

total resistance: The combined value of all the resistors in a circuit.

total voltage drop: The combined value of all the voltage drops in a series circuit.

transformer: A device used to step up or step down voltage by induction. Also, an electrical device used to transport energy from one circuit to another circuit at the same frequency.

transistor: A semiconductor made from germanium or silicon with leads known as emitter, collector, and base. It is made to conduct by forward biasing an emitter-base junction.

TRIAC: A three-terminal, bidirectional thyristor device that is triggered into conduction by applying a signal to its gate.

trimmer resistor: A small variable resistor that is used to fine tune the resistance in a circuit.

truth table: A table used to show the output with all the possible combinations of inputs for various logic gates.

U

ULSI: The abbreviation for ultra large-scale integration.

ultra large-scale integration (ULSI): A classification of integrated circuits that contain 100,000 to over 1,000,000 components.

undercut: When a PC board is left in etching chemicals too long so that the copper under the protective coating of the PC board is etched away.

V

variable resistor: A resistor whose resistance can be changed within a range of values.

vector: A quantity having both magnitude and direction.

voltage (E): The electromotive force which causes current to flow.

voltage divider: A circuit consisting of a series of resistors across a voltage source so that multiple voltages can be obtained.

voltage drop: The difference in voltage between two points in a circuit.

volt-ampere: The result of multiplying volts by amps.

voltmeter: A meter that measures voltage.

W

WAN: The abbreviation for wide area network.

watt (W): The unit of measure for power (rate of doing work).

Wheatstone bridge: A highly accurate circuit used to measure resistance in series-parallel circuits. It works by comparing voltages across a bridge circuit.

wide area network (WAN): A large group of networks connected by private and public communication lines. A WAN is managed by many different parties. The Internet is an example of a WAN.

wind farms: Groups of wind turbines.

wind turbine: An alternative way of generating electricity that consists of a tower, spinning blades, and a generator.

wirewound resistor: A resistor whose resistance element is a piece of resistance wire wound onto some insulating form. Many of them are then coated with an insulating material.

wiring harness: a large number of wires taped together or encased in protective tubing.

workstation: A computer that links to a server and uses its resources.

X

XNOR gate: A logic gate that is the combination of the NOT gate and the XOR gate. It has the opposite output from the XOR gate.

XOR gate: A logic gate in which there must be an input signal at exclusively the *A* input or exclusively at the *B* input for an output signal to be present. Also called an *exclusive or gate.*

Z

zener diode: A silicon diode that makes use of the breakdown properties of a P-N junction. If reverse voltage across the diode is increased, a point will be reached where the current will greatly increase.

zener knee: The sharp bend in the characteristics curve of a zener diode, which occurs at the zener voltage.

zener voltage (V_Z): The voltage at which a zener diode begins to easily conduct without much change in voltage.

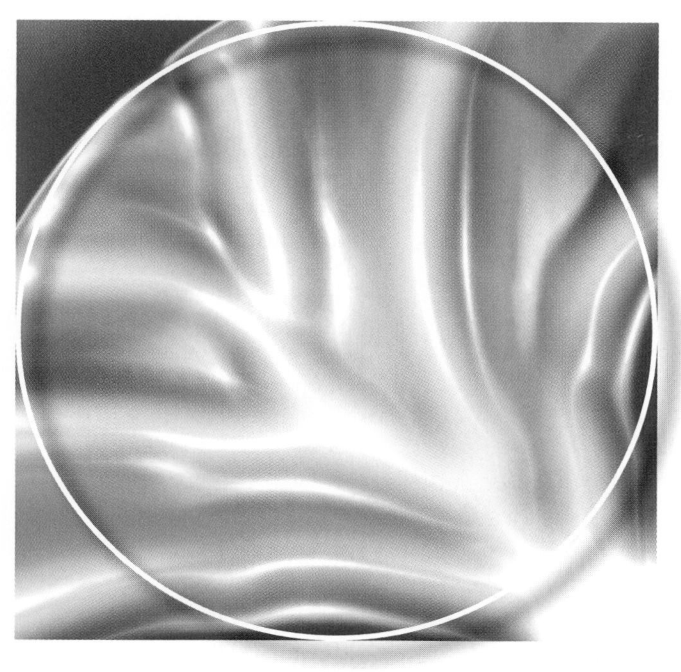

Index

C

D

E

L

M

N

O

P

Q

R

S